Roland Taugner · Eberhard Hackenthal

The Juxtaglomerular Apparatus

Structure and Function

With a contribution on the pathology
of the human JGA by R. Waldherr

With 129 Figures

Springer-Verlag Berlin Heidelberg GmbH

Prof. Dr. med. Roland Taugner
I. Physiologisches Institut der Universität Heidelberg
Im Neuenheimer Feld 326
6900 Heidelberg, FRG

Prof. Dr. med. Eberhard Hackenthal
Pharmakologisches Institut der Universität Heidelberg
Im Neuenheimer Feld 366
6900 Heidelberg, FRG

ISBN 978-3-642-88428-3

Library of Congress Cataloging-in-Publication Data
Taugner, Roland. *The juxtaglomerular apparatus : structure and function* / Roland Taugner, Eberhard Hackenthal ; with a contribution on the pathology of the human JGA / by R. Waldherr. Bibliography: Includes index.
1. Juxtaglomerular apparatus - Physiology. 2. Juxtaglomerular apparatus - Pathophysiology.
ISBN 978-3-642-88428-3 ISBN 978-3-642-88426-9 (eBook)
DOI 10.1007/978-3-642-88426-9
I. Hackenthal, Eberhard. II. Waldherr, R. (Rüdiger) III. Title. QP249.T38 1989 612'.463 - dc19

Originally published by Springer-Verlag Berlin Heidelberg New York in 1989
Softcover reprint of the hardcover 1st edition 1989

2127/3145-543210 Printed on acid-free paper

Preface

Until recently it was possible and appropriate to condense and integrate the body of data available on the juxtaglomerular apparatus (JGA) into the form of conventional reviews. Indeed, many excellent reviews covering various aspects of the JGA have monitored the progress made in this field during the last 50 years (e.g. GOORMAGHTIGH 1942, 1945a; TOBIAN 1960a; REALE et al. 1963; HARTROFT 1966; HATT 1967; VANDER 1967; FAARUP 1971; KAZIMIERCZAK 1971; ROULLER and ORCI 1971; LATTA 1973; THURAU and MASON 1974; DAVIS and FREEMAN 1976; GORGAS 1978). In particular, the reviews and updates by BARAJAS (BARAJAS 1972, 1979, 1981; BARAJAS and POWERS 1984) provided continuous state-of-the-art information on both the structure and the function of the JGA. However, as a consequence of the development of new techniques in immunocytochemistry, biochemistry, and molecular biology in the last decade, this flow of information on the renin-angiotensin system and the JGA has swelled into a flood. This large volume of data is not only overwhelming by its quantity, but also carries with it several new aspects of broader interest, such as general questions of cell differentiation and adaptation, structural and biochemical mechanisms of the secretory process, and integrated control mechanisms of secretion. Furthermore, the recognition that the renin-angiotensin system plays an important role in the maintenance of blood pressure and the development of hypertension, the widespread clinical use of inhibitors of angiotensin-converting enzyme, and the development of potent inhibitors of human renin with therapeutic potential have stimulated general interest in all aspects of the renin-angiotensin system.

Therefore we felt it timely to attempt a comprehensive overview on the present state of research on the JGA and, in order to accomodate as much of the essential original information as possible, we have chosen the form of a monograph. In keeping with the longstanding cooperation between the authors, a central aim in this enterprise has been the close integration of the description of JGA structure and function.

Nevertheless, structural and functional aspects received a somewhat different treatment with respect to the completeness of coverage of the literature. The reason for this apparent inconsistency lies in the fact that the knowledge of JGA morphology is far more advanced than that of JGA function, whilst at the same time, it is documented in a much smaller number of publications. Therefore, coverage of the literature on the morphology is more complete, whereas in the description of functional aspects a more selective approach to the literature was necessary. Obviously, some personal bias is unavoidable in this procedure, and we apologize to all those whose valuable contributions have not been cited. A subjective view is also inherent to our interpretion of new data which have not yet been accepted unequivocally as "standard" knowledge and may need further confirmation. We hope for the kind understanding of the reader if he or she does not agree with our views.

Two authors cannot give birth to such a monograph without significant help from others and, first of all, we wish to thank Dr. R. WALDHERR who contributed the chapter on the "Pathology of the human juxtaglomerular apparatus", which we feel is an excellent description of the topic and fits well into the general line of argumentation from the preceding nonclinical chapters.

We also thank Dr. KARIN GORGAS, as well as Drs. J.A. CHRISTENSEN, L. ROSIVALL, and C.P. BÜHRLE for critical reading of individual chapters and helpful suggestions for the preparation of the text. Our own work, as presented in this monograph, originated from the collaboration with many scientists whose names appear in the text and in the cited references and will therefore not be listed here. However, we wish to express our gratitude to Mrs. G. REB, Mrs. M. HARLACHER, Mr. H. WREDE, and Mr. G. ROHS for their valuable technical assistance for many years and to Mrs. M. WYBRANIEC and Mrs. S. RENFRO-KOHL for their assistance in preparing the manuscript. Our special appreciation is due Mrs. R. HACKENTHAL, who not only accompanied our experimental work by expert and unexhaustable laboratory assistance, but who also took on the burden of collecting, checking, and organizing some 1800 references, as well as typing and editing a large part of the manuscript.

Finally, we wish to thank Springer-Verlag, especially Dr. J. WIECZOREK, and K. RAU for their efficient and sensitive cooperation and the excellent layout of the book.

Heidelberg, January 1989

R. TAUGNER
E. HACKENTHAL

Table of Contents

Chapter 1

The Juxtaglomerular Apparatus - History of a Concept

Nearly a century ago, GOLGI (1889) described the essential components of the anatomical unit which is known today as the *juxtaglomerular apparatus* (JGA). He showed that the ascending limb of Henle's loop regularly returns to the original renal corpuscle and attaches itself precisely to the point where the afferent arteriole enters and the efferent arteriole exits the glomerular hilus (Fig. 1.1). Shortly thereafter, TIGERSTEDT and BERGMAN (1898) discovered the pressor-active principle of the kidney, which they named renin. Many years passed before RUYTER (1925) described a new granulated cell type in the media of the afferent glomerular arteriole and GOORMAGHTIGH (1937, 1939) established the connection between the earlier observations of GOLGI and those of TIGERSTEDT and BERGMAN by the proposal that renin may be elaborated by these cells.

The assumptions that the epithelioid cells are responsible for renin secretion and that this secretion plays an important role in both blood pressure regulation and some types of hypertension (GOORMAGHTIGH 1937, 1939, 1942; ELAUT 1934; GOLDBLATT et al. 1934; GOORMAGHTIGH and GRIMSON 1939) were supported by the demonstration of a direct correlation between the degree of epithelioid cell granulation, kidney renin content, and plasma renin concentration (see HARTROFT 1968 for review). Further support of Goormaghtigh's suggestion came from microdissection (COOK and PICKERING 1959; BING and KAZIMIERCZAK 1962; FAARUP 1967, 1968) and micropipetting (COOK 1971) of JGA elements. After the purification of renin (COHEN et al. 1972; MALLING and POULSEN 1977), specific antibodies and efficient methods for immunolabeling (STERNBERGER 1974; ROTH et al. 1978) enabled the existence of the enzyme in epithelioid cells (TAUGNER et al. 1979) and their secretory granules (TANAKA et al. 1980; CANTIN et al. 1984; TAUGNER et al. 1984) to be definitely verified.

In 1932, GOORMAGHTIGH described a new cell group placed between the afferent and efferent

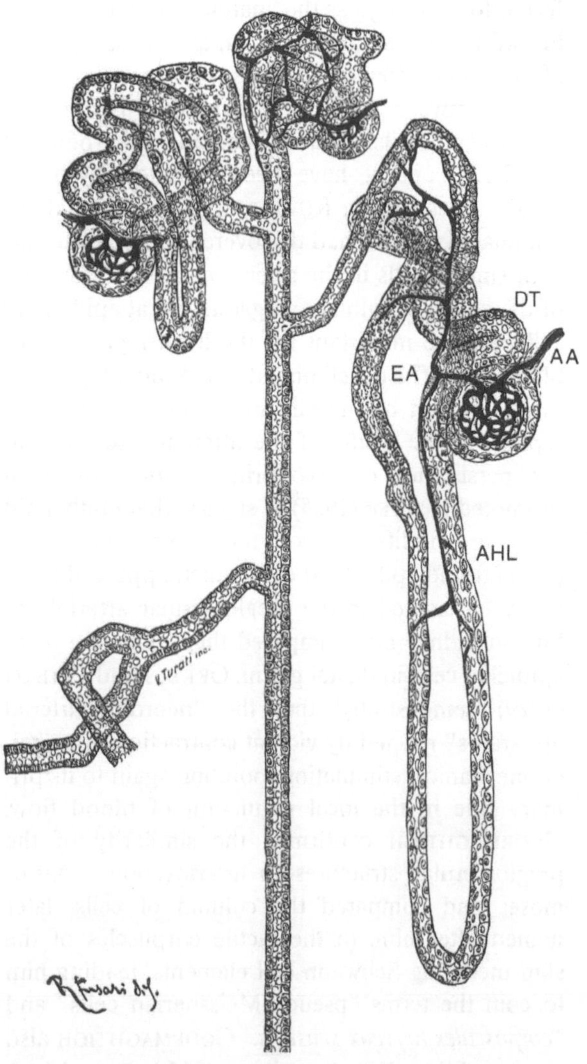

Fig. 1.1. Three nephrons, with their tubulovascular relationships, from cat kidney (GOLGI 1889). In the juxtamedullary nephron *(right)* the ascending Henle's loop *(AHL)* is most clearly shown to return to the parent renal corpuscle attaching itself to the point where the afferent arteriole *(AA)* enters and the efferent arteriole *(EA)* exits from the glomerular hilus. Somewhat upstream from the juxtaglomerular apparatus, the afferent vessel is in contact with the distal tubule *(DT)*. Downstream, the efferent arteriole accompanies the ascending limb of Henle's loop over a considerable distance. Note that the direction of blood flow is antidromic to the tubular fluid flow

arterioles of the glomerulus. GOORMAGHTIGH (1937) also coined the term *"appareil neuro-myo-artériel juxtaglomérulaire du rein"* (neuromyoarterial juxtaglomerular apparatus of the kidney). The macula densa owes its name to ZIMMERMANN (1933), applying to a body of distal tubular cells characterized by a peculiar grouping of their nuclei, known already to PETER (1907, 1909). According to these prerequisites, the JGA in a somewhat broader sense is referred to nowadays as the anatomical unit at the hilus of the renal corpuscle which consists of the afferent and efferent arterioles, the macula densa, and, in between, the GOORMAGHTIGH cells (Fig. 1.2).

Further details on the historical background of the JGA concept have been reported by HATT (1967). Accordingly, RUYTER (1925) compared the granulated cells he had discovered in mice and rats with similar cells in the arteriovenous anastomoses of the coccygian glomus, suggesting that epithelioid cells may be important for the local regulation of blood flow. The development of granulated cells via differentiation of smooth muscle cells which they replace in the media of the afferent arteriole and the persistence of myofibrils in some of them prompted RUYTER (1925) to suggest that epithelioid cells were modified smooth muscle cells. OBERLING (1927) found epithelioid cells which appeared to be richly innervated in the preglomerular arteriole of human kidney and compared them with the myoepithelial cells in digital glomi. OKKELS and PETERFI (1929) demonstrated that the "neuromyoarterial apparatus" reacted by violent contraction upon micromechanical stimulation, pointing again to its primary role in the local regulation of blood flow. GOORMAGHTIGH confirmed the similarity of the preglomerular structures to arteriovenous anastomoses and compared the column of cells, later named after him, to the tactile corpuscles of the skin including Schwann cell elements, leading him to coin the terms "pseudo-Meissnerian cells" and *"corpuscules nerveux sensitifs."* GOORMAGHTIGH also suggested that these structures could inform the afferent arteriole of modifications ocurring in the distal portion of the corresponding nephron sensed by the highly differentiated cells of the macula densa.

The first electron microscopic representation of individual JGA elements, especially of the epithelioid cells and their secretory granules, was credited to BOHLE (1959), OBERLING and HATT (1960a, b), BUCHER and REALE (1961a, b), HARTROFT and NEWMARK (1961), HATT (1961, 1963), THOENES (1961), LATTA and MAUNSBACH (1962a, b), HATT et al. (1962), BARAJAS and LATTA (1963a, b), and REALE et al. (1963). The innervation of the JGA has, among others, been described by BARAJAS (1964), the nexal junctions in the area of the JGA by BIAVA and WEST (1966a), and the development of the epithelioid cell secretory granules by BARAJAS (1966). The first reconstruction of the vascular pole with references to the tubulovascular proximities outside the JGA came from BARAJAS and LATTA (1963a). It is also due to BARAJAS (1972) that the postglomerular arteriole has since been considered one of the essential vascular JGA components. RUYTER (1925) had already assumed that epithelioid cells were modified vascular smooth muscle cells. CANTIN et al. (1977) firmly established the reversible mutual metaplastic transformation of these two cell types upon fluctuations in the stimulation niveau. The "paradoxical" effect of calcium on renin secretion was discovered by VANDONGEN and PEART (1974) (for review, see KEETON and CAMPBELL 1980; HACKENTHAL and TAUGNER 1986). The secretion of active renin by the exocytosis of mature granules was shown by TAUGNER et al. (1984, 1986b), using morphological techniques, and by SKØTT (1986b) using biochemical methods.

The favorable strategic position of the macula densa within the nephron for tubulovascular interactions had already been pointed out by GOORMAGHTIGH (1937). Based on both the structural and functional characteristics of the JGA, GUYTON et al. (1964) and THURAU (1964) developed a more precise conception of its role in the feedback regulation of renal blood flow (RBF) and glomerular filtration rate (GFR). THURAU and MASON (1974) put forward the arguments for a locally effective renin-angiotensin system (RAS); the microtopographical prerequisites for such a system were summed up later by TAUGNER et al. (1982a) on the basis of immunohistochemical data.

The above-mentioned hypotheses of GOORMAGHTIGH and coworkers and of GOLDBLATT for a role of renin in blood pressure regulation and hypertension due to renal ischemia, respectively, released the JGA from its previous modest role of a local circulatory regulator. The elucidation of the enzymatic action of renin by PAGE (1939) and BRAUN-MENENDEZ et al. (1939) leading to the release of a pressor substance - angiotensin - from a circulating plasma substrate - angiotensinogen - and the recognition of the renin-angiotensin-aldosterone interactions

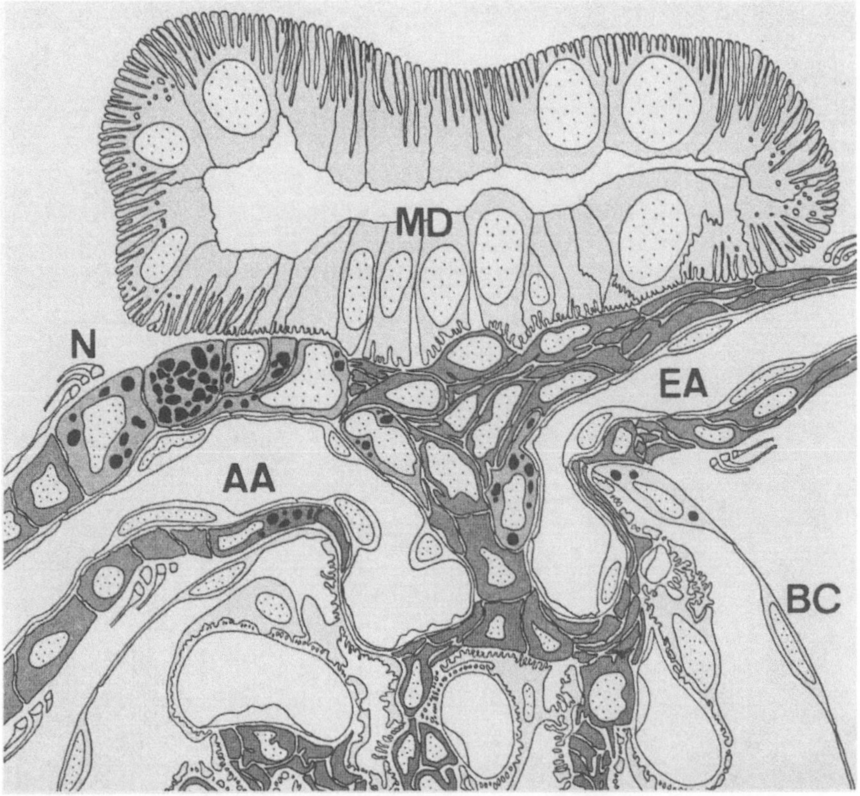

Fig. 1.2. Hilus of the renal corpuscle with the juxtaglomerular apparatus, consisting of the afferent and efferent glomerular arterioles *(AA, EA)*, the macula densa *(MD)*, and, in between, the Goormaghtigh cell field, continuous with the glomerular mesangium. The media of the afferent arteriole contains granulated cells in its juxtaglomerular segment as well as somewhat further upstream. Another granulated cell is seen in the media of the efferent arteriole, facing the Goormaghtigh cell field. At the visceroparietal junction of Bowman's capsule *(BC)*, a peripolar cell is depicted. Nerve axons *(N)* innervate both the afferent and the efferent arteriole. The figure is redrawn from GORGAS (1978a) with minor modifications including the dilatation of the intracellular spaces in the MD region and the addition of a peripolar cell. Mouse kidney; ×1300

have been reviewed in detail by LARAGH and SEALEY (1973). Angiotensin-converting enzyme (ACE) was first detected in blood plasma by SKEGGS et al. (1956a). The amino acid sequence of angiotensin was determined independently by SKEGGS et al. (1956b) and by ELLIOTT and PEART (1957). Angiotensin II was first synthesized by BUMPUS et al. (1957) and SCHWYZER et al. (1958). GROSS (1958) demonstrated the inverse relation between salt balance and kidney renin content, suggesting that renin might be involved in the regulation of adrenal cortical function. LARAGH et al. (1960) showed that infusions of angiotensin increase aldosterone secretion, thus paving the way for acceptance of the important and complex role of renin in the regulation of electrolyte balance, body fluid volume, and arterial blood pressure.

Chapter 2

Morphology of the Juxtaglomerular Apparatus

2.1 Components and Definitions

The discovery of the JGA has been discussed in detail above. This introduction also comprises the history of the terminology of both the apparatus and that of its components. We can therefore restrict ourselves to the current terms and definitions and attempt to submit these to a critical examination.

Most textbooks state that the JGA lies close to the vascular pole of the renal corpuscle, comprising the afferent and efferent arterioles, the cell group (field) of Goormaghtigh, and the macula densa (Figs. 1.2, 2.1 A). This definition, initiated by GOORMAGHTIGH (1937), appears adequate in view of our present knowledge. It should not be overlooked, however, that by the delimitation "close to the vascular pole" arbitrary bounds are drawn which may not include all functionally relevant tubulovascular connections as shown below.

For functional and linguistic reasons, it appears reasonable to distinguish between the two glomerular arterioles along with the Goormaghtigh cells as the *"vascular components"* of the JGA and the macula densa as its *"tubular component"* (BARAJAS 1979). It also appears suitable - for the sake of briefness - to adhere to the term *"lacis"* for the *Goormaghtigh cell field* (OBERLING and HATT 1960a, b). The term polkissen (ZIMMERMANN 1933), however, should be abandoned in our opinion since the details of this concept have changed in the course of time (cf. HATT 1967). Used as a synonym for the Goormaghtigh cell field it is not only superfluous but also misleading. The term "juxtaglomerular complex" (McMANUS 1942), which has among others been used as a synonym for "juxtaglomerular apparatus", is largely out of use and should be abandoned, the more so as the acronym JGA is now commonly employed.

The above definition of the JGA is based on demarcations which may appear too limiting where functional tubulovascular relations and/or renin secretion are concerned. First, the pre- and postglomerular arterioles are in contact with the distal tubule not only in the region of the macula densa, but even in areas far from the JGA. This holds for the cortical and all the more so for the juxtamedullar nephrons (BARAJAS and LATTA 1963a; BARAJAS 1970; FAARUP 1965, 1971). Second, in many species, the majority of renin-positive cells are located in the media of the afferent arteriole far upstream from the JGA defined as above. Therefore, at least as long as definite evidence for effects of the distal tubule on renal microcirculation originating exclusively from the macula densa is not provided, it might be reasonable to define a *"tubulovascular complex of the nephron,"* which, as a comprehensive term, would include the JGA as defined above and in addition cover all portions of the distal tubule and glomerular arterioles in contact with one another (Fig. 2.1 C, D).

The classical JGA concept comprises the macula densa, the glomerular arterioles, and the lacis or extraglomerular mesangium, but not the glomerular mesangium, with which the lacis is connected via the glomerular stalk. Also the peripolar cells (RYAN et al. 1979) do not belong to the JGA. With regard to possible functional interrelationships between JGA components and the glomerular mesangium or the peripolar cells (PPC), it seems appropriate to combine these elements under the term *"glomerulo-juxtaglomerular complex"* (GALL et al. 1986; Fig. 2.1 B).

The issue of a most precise and clear nomenclature for the renin-producing cells of the kidney presents intriguing difficulties, as renin-producing cells are found in various locations and - depending upon the respective level of stimulation - may vary remarkably in their morphological appearance as well as in their contractile and secretory function. In physiological and biochemical studies, the term "juxtaglomerular cells" is generally applied. As this term and its acronyms incorrectly suggest that renin-producing cells are only present in the area of

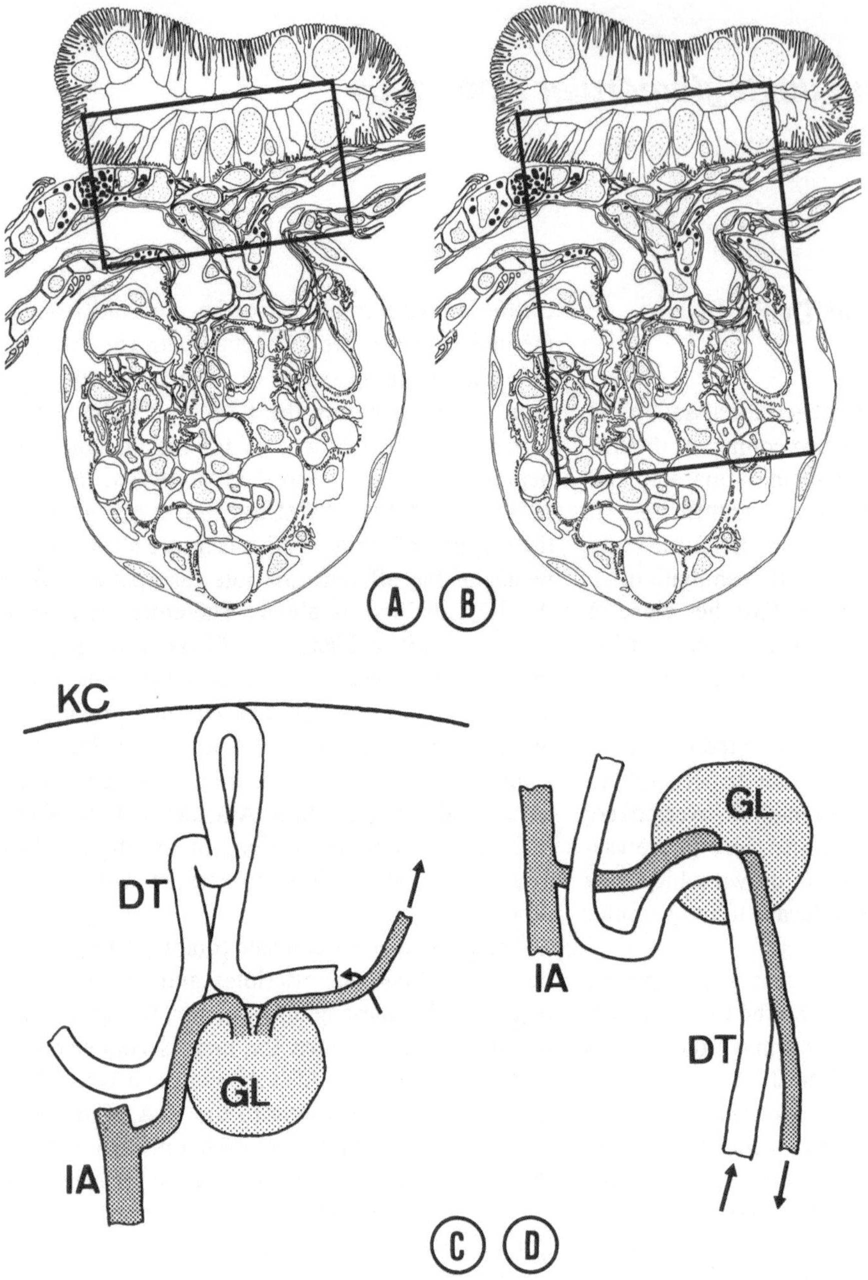

Fig. 2.1. The juxtaglomerular apparatus (**A**), the glomerulo-juxtaglomerular complex (**B**), and the tubulovascular complex of the nephron, shown separately for a superficial (**C**) and a juxtamedullar nephron (**D**). **A** and **B** are modified copies of Fig. 1.2; **C** and **D** are redrawn from reconstructions of FAARUP (1965). *KC*, kidney capsule; *DT*, distal tubule; *IA*, interlobular artery; *GL*, glomerulus

the JGA, they should be avoided unless used explicitly in this restrictive sense. Instead, the denotations *granulated (or granular) cells*, *renin-positive cells*, or *renin-producing cells* should be taken into consideration. These refer to three different techniques: electron microscopy (for secretory granules), immunohistochemistry (for renin), and hybridization (for mRNA) and are all three apposite to the respective method. Functional studies proceed on the assumption that renin release from the kidney is a regulated secretory process, i.e., a secretion by cells disposing of a granular store of the mature secretory product, renin. For this reason, the term *"granulated (or granular) cells"* appears appropriate in this connection. The designation *"epithelioid cells"* was originally coined from light-microscopic observations. It applies to cells devoid of myofibrils and fully loaded with secretory granules, carrying in their ultrastructure the metaplastic transformation of vascular smooth muscle cells to secretory cells to the extreme. *Intermediate cells* are cells which according to their shape, ultrastructure, and granular renin store display the entire sequence of transitions between both extremes: plain smooth muscle cells and epithelioid cells. Designations such as polkissen cells (of ZIMMERMANN), "granulated epithelioid cells," "granular or epithelioid cells," "afibrillar cells," "nongranular epithelioid cells," and "agranular or lacis cells" would be redundant or misleading.

2.2 Afferent Arteriole

2.2.1 Overview

By definition, only the juxtaglomerular portion of the afferent arteriole is a component of the JGA (Figs. 1.2, 2.1 A). In regard to the structure and putative functions of the preglomerular vessels, this restriction appears to be arbitrary: renin-positive, granulated cells are not only found in the juxtaglomerular portion of the afferent arteriole, but also further upstream (Figs. 2.2, 2.3), and, particularly during stimulation of the RAS, even in the interlobular artery; in addition, the tonus of the preglomerular resistance vessels is thought to influence renin secretion (cf. Sect. 7.2) and, in the opposite direction, modulations of the afferent resistances may be initiated from the JGA region (cf. Chap. 9). It therefore appears reasonable to include in this section the functional morphology of the proximal portion of the afferent arteriole and the distal portion of the interlobular artery, similar in structure and function (KÄLLSKOG et al. 1976; HEYERAAS TØNDER and AUKLAND 1979/80; BOKNAM et al. 1981; HEYERAAS and AUKLAND 1987).

Morphologically and functionally, it is possible to distinguish between a proximal - renin-negative - and a distal - renin-positive - portion of the afferent arteriole (Fig. 2.2). The media of the proximal portion consists of plain smooth muscle cells which do not differ from those of other resistance vessels. In contrast, the media of the distal afferent arteriole is composed of renin-producing, granulated cells (Fig. 2.3). At this point, it should be emphasized that the length of the renin-positive portion of the preglomerular vessel may be very different with respect to both internephron and interspecies heterogeneity (Fig. 2.2; cf. Sect. 4.1.1.2, and Chap. 12); in addition, the length of the renin-positive portion of a given afferent arteriole is extremely variable: upon stimulation of the RAS, smooth muscle cells are subject to metaplastic transformation into granulated cells and vice versa, so that the boundary between the renin-positive and the renin-negative segment of the vessel is shifted up- or downstream. As this boundary thus will comply only coincidentally with the definitional demarcation of the JGA, it would be misleading to label the renin-positive por-

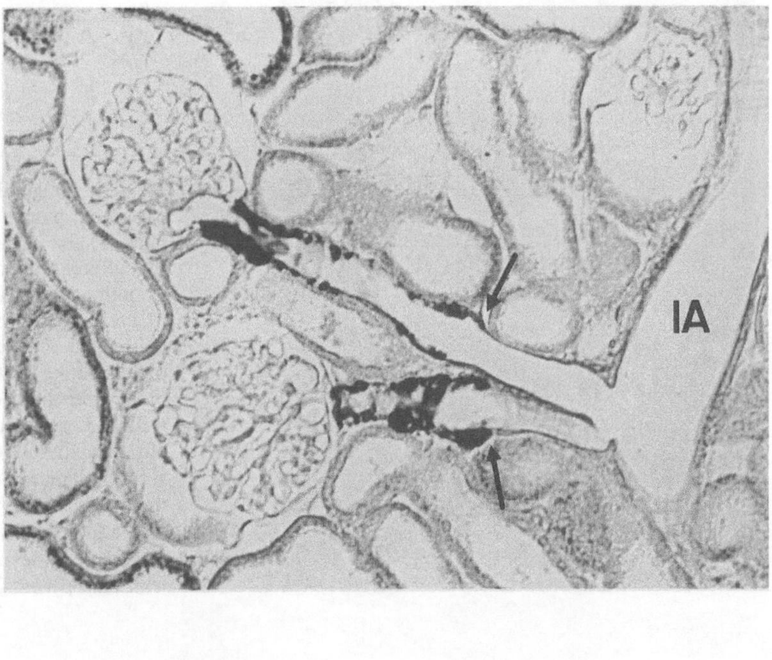

Fig. 2.2. Two afferent arterioles from mouse kidney with renin-positive media cells over a distance of 140 μm (above) and 70 μm (below) upstream from the glomerulus *(arrows)*. *IA*, interlobular artery. Antiserum dilution, 1:1000; ×300 (From TAUGNER et al. 1981)

tion of the afferent arteriole its "juxtaglomerular portion."

The general morphology of the renal arterial tree has been described in detail by FOURMAN and MOFFAT (1964, 1971), ROLLHÄUSER et al. (1964), and DIETERICH (1978). The afferent arteriole branches off the interlobular artery in juxtamedullar, midcortical, and subcapsular nephrons at different - acute, right, and finally obtuse - angles. At the origins of juxtamedullary afferent arterioles of some species, intraarterial cushions equipped with myoendothelial junctions may be found (for literature and discussion of their putative functions see MOFFAT and CREASEY 1971 and CASELLAS et al. 1982). The afferent arteriole has an unilayered media. Its lumen diameter is reported to be about 18–20 μm in rabbits (BANKIR and FARMAN 1973) and 15–16 μm in rats (GATTONE et al. 1983; WILSON 1986), its mean length being about 180 μm (STEINHAUSEN and TANNER 1976; WILSON 1986). Juxtamedullar rat afferent arterioles are significantly longer than their midcortical and superficial counterparts (WILSON 1986). Two human glomeruli out of 3000 have been found to be supplied by double afferent arterioles (MURAKAMI et al. 1985a; cf. Fig. 3.10). In the area of the JGA, the media of the afferent arteriole often indefinably fades into the Goormaghtigh cell field as well as into the mesangium of the glomerular stalk (cf. sects. 2.4, 3.2.2).

In the following, the various cell types present in the media of the afferent arteriole and interlobular artery are to be dealt with: smooth mucle cells, intermediate cells, and epithelioid cells completely transformed for the purpose of renin secretion (cf. GOORMAGHTIGH 1932, 1940; BARAJAS and LATTA 1967; CANTIN et al. 1977). Next, the contractile ability and finally a few other aspects of the functional morphology of the proximal and the distal portions of the afferent arteriole will be compared to each other.

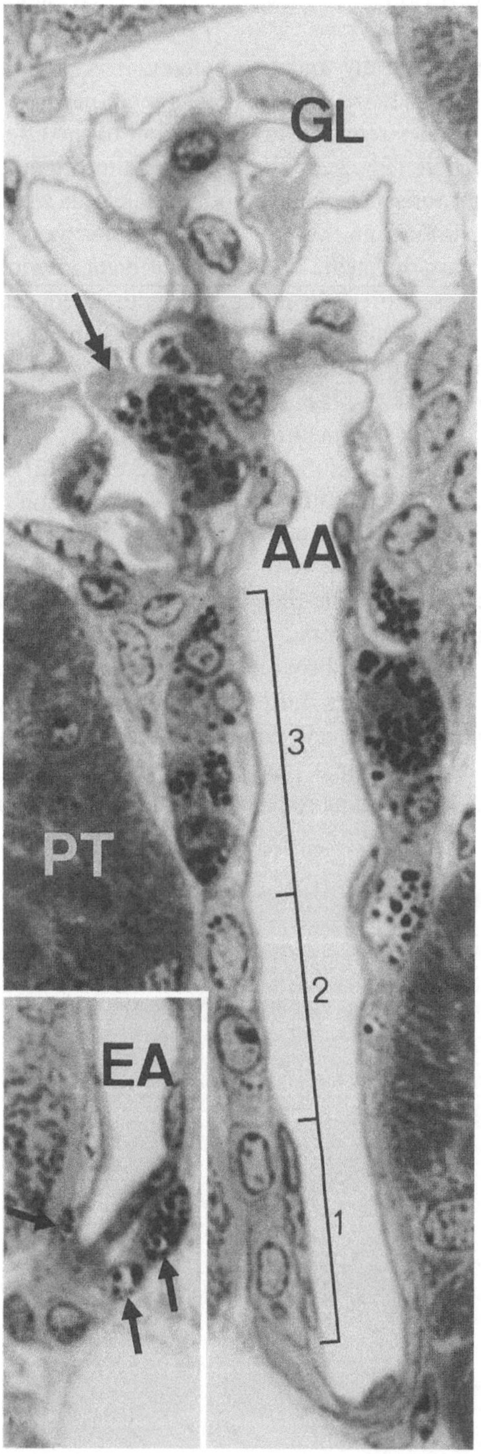

Fig. 2.3. Longitudinal section through an afferent arteriole *(AA)* from mouse kidney, exhibiting its proximal (nongranulated, *1*) and its distal (granulated, 2–3) portion. The media of the afferent arteriole consists in its proximal portion of smooth muscle cells *(1)*. In its distal portion intermediate cells *(2)* and further downstream epithelioid cells are seen *(3)*. The epithelioid cells lead to a distinct cuff-like widening of the outer contour of the vessel. The *double arrow* points to another epithelioid cell located in the area of the glomerular stalk. *GL*, glomerulus; *PT*, proximal tubule. In the inset, the efferent arteriole of the same nephron is shown in another section plane, about 13 μm apart. Note the granulated cells in the media of the vessel *(arrows)*, about 100 μm downstream of the glomerulus. Semithin (0.75-μm) sections stained with methylene blue-azure II; ×1000. (From BARGMANN 1978, with kind permission of K. GORGAS, Heidelberg)

2.2.2 Media Cells of the Afferent Arteriole

2.2.2.1 Smooth Muscle Cells

The smooth muscle cells of the proximal afferent arteriole do not differ in their ultrastructure from smooth mucle cells of resistance vessels in other organs (Fig. 2.9 A). Most characteristically, they contain a great number of leiomyofilaments and attachment sites. Their Golgi system is poorly developed and the rough endoplasmic reticulum (RER) limited to scattered small saccules studded with ribosomes. The plasmalemma of the cells is enlarged by many caveolae.

Gap junctions between smooth muscle cells of the proximal afferent arteriole and interlobular artery were found to be rare and remarkably small, consisting only of short rows of membrane particles (Fig. 2.23 C). The myoendothelial contacts of these vessel segments have the form of flat appositions or club-shaped invaginations (Figs. 2.26-2.28); the endoendothelial junctions consist of tight junction meshworks filled with gap junction elements (Fig. 2.29).

The media of the afferent arteriole and distal interlobular artery is unilayered (GORGAS 1978a). Their smooth muscle cells have the usual spindle shape and spiral arrangement, reaching, in the afferent arteriole, about 1.5 times around the circumference of the vessel (TAUGNER et al., unpublished observations). As the preglomerular arteriole approaches the parietal layer of Bowman's capsule, the media of the vessel may thin considerably, making it difficult to distinguish the wall of the afferent from that of the efferent arteriole (BARAJAS and LATTA 1963a; cf. Fig. 2.5).

2.2.2.2 Epithelioid Cells

Epithelioid cells fully loaded with renin-positive granules represent the other extreme of media cells in the wall of the afferent arteriole. Figure 2.4, as an example, shows a section through the perinuclear region of an epithelioid cell in the juxtaglomerular portion of an afferent arteriole from mouse kidney. Large round or ovoid electron-dense secretory granules are most striking, located peripherally at a distance of only about 100 nm from the cell surface. In suitable section planes, over 30 such mature granules may be found in a cell profile (BARAJAS

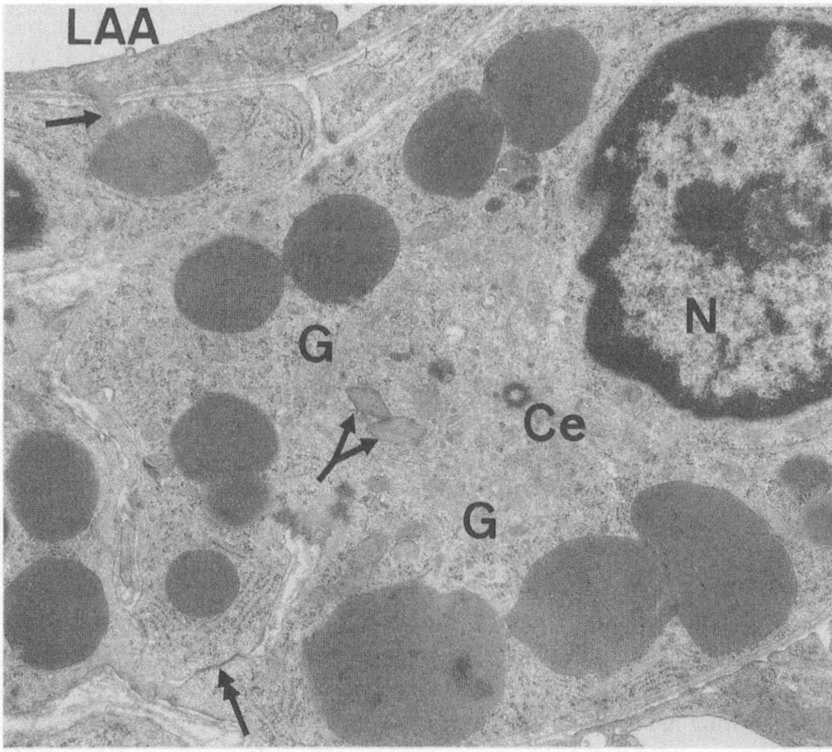

Fig. 2.4. Section through the pericaryon of an epithelioid cell in the juxtaglomerular portion of the afferent arteriole. Large mature granules of varying size are observed at the cell periphery and protogranules in the Golgi region *(split arrow)*. *N*, nucleus; *Ce*, centriole; *G*, Golgi complex; *single arrow*, myoendothelial contact; *double arrow*, reflexive gap junction; *LAA*, lumen of the afferent arteriole; ×15000

and LATTA 1963a). Besides mature granules, a few protogranules and intermediate granules are seen close to the well-developed Golgi complex. The cisterns of the RER with their pale amorphous internum are situated peripherally in the vicinity of the mature granules (BARAJAS and LATTA 1963a; BIAVA and WEST 1966a). In most mammalian species, the matrix of mature granules is homogeneously electron-dense. In *Tupaia belangeri,* as an exception, epithelioid cell granules stand out by irregular contours and characteristic, thread-like paracrystalline inclusions (FORSSMANN and TAUGNER 1977).

Besides secretory granules, the cytoplasma of the epithelioid cells contains some peroxisomes, a few small electron-dense vesicles, and many clear ones, some of which are coated (Fig. 2.4; cf. GORGAS 1978a). The scarce mitochondria of epithelioid cells are relatively small, elongated structures (BARAJAS and LATTA 1963a; BIAVA and WEST 1966a; GORGAS 1978a). The amount of glycogen is greater in granulated cells than in smooth muscle cells (CANTIN et al. 1977). Myofilaments, attachment sites, and caveolae, characteristic for the smooth muscle cells in the proximal part of the afferent arteriole, are rare or even absent in fully differentiated epithelioid cells (Fig. 2.4). Instead, sublemmal coated vesicles and coated pits are found, the number of which are increased in secretory active cells (cf. Chap. 6). In contrast to plain smooth muscle cells, relatively large macular gap junctions are regularly found between epithelioid cells (Fig. 2.23). In addition, reflexive gap junctions between processes of individual cells (Fig. 2.4) or at the neck of omega-shaped postexocytotic recesses may be seen (Fig. 6.8). Epithelioid cells of the afferent arteriole are not only coupled by gap junctions with each other, but also with adjoining Goormaghtigh cells, smooth muscle cells, and endothelial cells (BIAVA and WEST 1966a; FORSSMANN and TAUGNER 1977; TAUGNER et al. 1984c).

The roundish, fully transformed epithelioid cells show no polarity in view of the arrangement of their organelles: the Golgi complex with its dependents, including protogranules, usually lies in the vicinity of the nucleus, the mature granules in the cell periphery (Fig. 2.4). The position of the secretory granules thus does not appear to be influenced by the different structures found in the vicinity of epithelioid cells: The number of granules is neither increased close to the adjoining endothelium nor to the neighboring nerve terminals, the macula densa, or the Goormaghtigh cell field.

In contrast to the pericaryon, large epithelioid cell processes often show an asymmetrical arrangement of their organelles, with mature granules and RER cisterns along one margin, Golgi cisterns and protogranules along the other. Smaller processes of fully differentiated epithelioid cells may contain only mature granules, RER cisterns and some vesicles or Golgi elements and juvenile granules (Fig. 5.1 A).

Corresponding to this microtopography, the observed exocytoses of mature granules showed no preferential orientation, e.g., into the direction of the lumen or the abluminal interstitium of the afferent arteriole. The exocytosis of juvenile granules may be promoted by the superficial position of the Golgi complex in some cell processes. However, as with mature granules, this would not imply any preferential (luminal or abluminal) orientation of the forthcoming exocytotic event (cf. Chap. 6).

In contrast to the clear helical arrangement of the spindle-shaped intermediate and smooth muscle cells, the shape and the manifold microtopographical relations of the fully transformed epithelioid cells seem to be more complex. The plumpness of the epithelioid cell bodies, as compared with the shape of smooth muscle cells, has already been recognized under the light microscope (RUYTER 1925; OBERLING 1927; GOORMAGHTIGH 1932). Electron-microscopic findings added the information that epithelioid cells develop numerous cell processes (OBERLING and HATT 1960; LATTA and MAUNSBACH 1962a; BIAVA and WEST 1966a; CAIN and KRAUS 1969) to the point that the unilayered media of the juxtaglomerular afferent arterioles by their thickness and complexity may give the impression of being multilayered (GORGAS 1978a; cf. Sect. 2.2.3). Recently, GATTONE et al. (1984) emphasized in their extensive scanning electron microscope study the prominent soma and several long cell processes typical for juxtaglomerular epithelioid cells, thus markedly deviating from the fusiform appearance and circumferential arrangement of smooth muscle cells in the proximal portion of the afferent arteriole. This becomes especially evident in the juxtaglomerular portion of afferent arterioles which enter the renal corpuscle after a curvilinear course. Here, epithelioid cells are in an eccentric position with respect to the axis of the vessel: they are mostly localized at the distinctly longer convex bank of the vessel nearest to the macula densa and/or the Goormaghtigh cell field (ZIMMERMANN 1933; BIAVA and WEST 1966a; HARTROFT 1968), thus

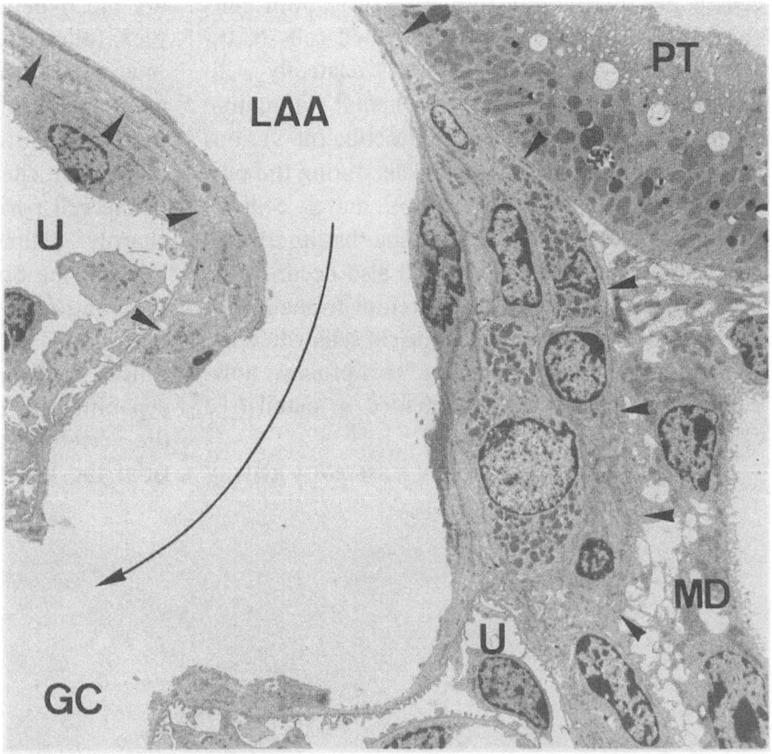

Fig. 2.5. Asymmetrical arrangement of media cells *(arrowheads)* in the wall of an afferent arteriole which enters its renal corpuscle in a curvilinear course. Note the plumpness of the epithelioid cells at the larger convex bank of the vessel disadvantageous to vasoconstriction. *LAA*, lumen of the afferent arteriole; *GC*, glomerular capillary; *U*, urinary space; *MD*, macula densa; *PT*, proximal tubule. *Tupaia* kidney; ×3500

lacking the precondition for any circumferential arrangement suitable to efficient vasoconstriction (Fig. 2.5; cf. Sect. 4.1.2).

2.2.2.3 Intermediate Cells

In the following, the term intermediate cells is introduced for cells found to be intermediate in size, ultrastructure, and granular renin store between both extremes: renin-negative, plain smooth muscle cells, and epithelioid cells fully loaded with renin granules. Hence, this cell group covers a wide spectrum of more or less completely transformed smooth muscle cells in the media of the afferent arteriole and interlobular artery.

The authors are aware that the proposed grouping of all not fully granulated preglomerular media cells under the collective term "intermediate cells" presents a simplification, the justification and usefulness of which need yet to be proven. This pertains especially to the juxtaglomerular area where, at the poorly defined border between the media of the afferent arteriole and the lacis, cells are observed which can only be classified with great difficulty (e.g., the "large, clear, agranular cells" of TURGEON and SOMMERS 1961 and the "agranular epithelioid cells" of GORGAS 1978a; cf. BARAJAS 1979). At this point, the lack of experimental evidence becomes apparent concerning the question as to whether Goormaghtigh cells can indeed be transformed into epithelioid cells and whether intermediate stages may also occur in this process (cf. Sect. 4.1.3; BARAJAS and LATTA 1963a; CHRISTENSEN et al. 1988).

In species where the media of the juxtaglomerular afferent arteriole is composed of epithelioid cells, intermediate cells are usually found further upstream close to the border between the renin-negative and the renin-positive portion of the vessel (Fig. 2.3). In other species, like the hamster or guinea pig, where epithelioid cells are very scarce under control conditions, intermediate cells are commonly found at juxtaglomerular sites (Fig. 12.3). The same holds true for mice and rats after inhibition of the RAS, e.g., by sodium loading (Fig. 12.4; cf. TAUGNER et al. 1984a). In addition, most of the scattered renin-positive cells which occur in the media of the interlobular artery or the proximal afferent arteriole upon stimulation of the RAS are of the intermediate cell type (Fig. 2.6).

One possibility for the emergence of intermediate

cells is metaplastic transformation of smooth muscle cells into renin-positive granulated cells (CANTIN et al. 1977). As this process is relatively well-known from studies with experimental stimulation of the RAS, it is convenient to describe the various manifestations of intermediate cells during the process of transformation from smooth muscle cells into epithelioid cells. We may assume that intermediate stages of granulated cells will also occur in the opposite direction, i.e., during retransformation of epithelioid cells into smooth muscle cells when the stimulation niveau is decreased. This process, however, has not yet been documented in detail (see Chap. 10).

First signs for the inception of a secretory activity are the hypertrophy of the RER and Golgi complex, followed by the occurrence of protogranules, intermediate and mature secretory granules. These components of the manufacturing machinery displace the contractile structures, myofilaments, and attachment sites, at first in the pericaryon and later in the cell periphery. At a certain transition stage, sharply delineated filament bundles may run through the body of intermediate cells, separating the contractile from the secretory portions of the pericaryon. More frequently, cells are seen exhibiting a continuous peripheral rim of myofilaments separating the secretory machinery and particularly the secretory granules of such intermediate cells from the cell membrane (Fig. 2.6). Finally, at the

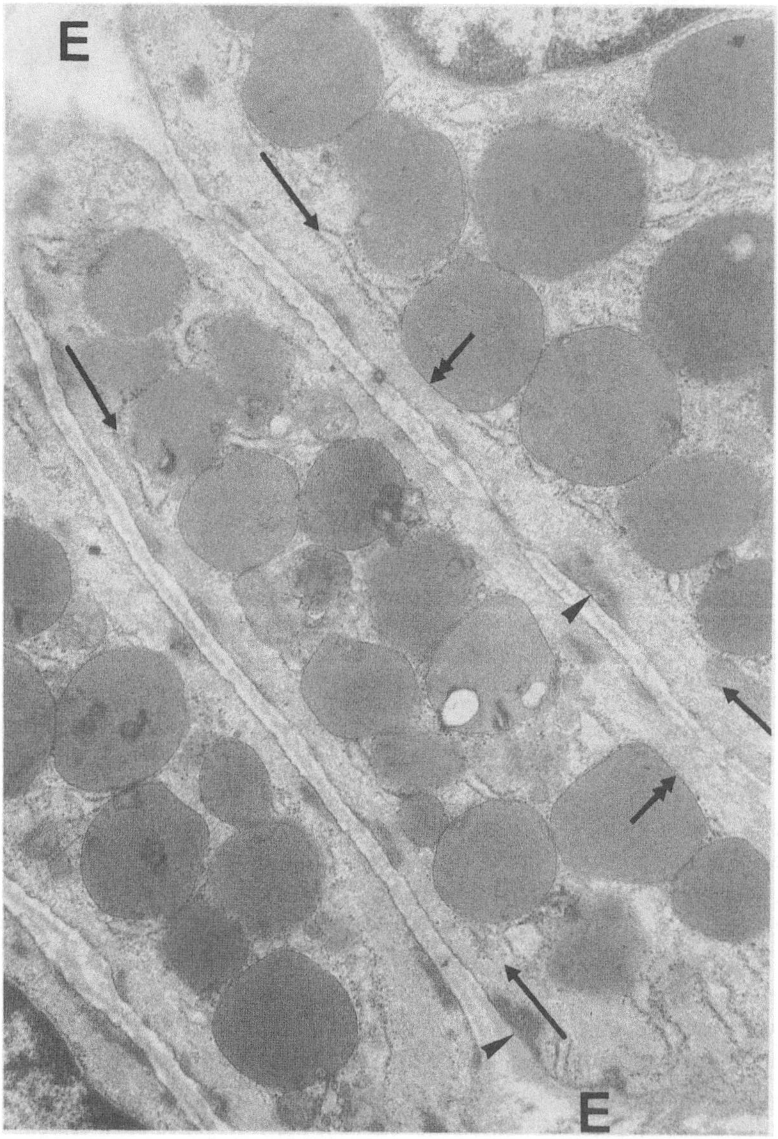

Fig. 2.6. Intermediate cells in the wall of an interlobular artery from mouse kidney exhibiting both contractile and secretory aspects. In two cells, the margins between the sublemmal rim of myofilaments and the large depot of secretory granules are indicated by *arrows*. Note that in places the renin granules are flattened by the apposed filaments *(double arrows)*. *Arrowheads*, attachment sites; *E*, extracellular space; ×26500

end of this spectrum of increasingly less contractile cells, epithelioid cells, i.e., cells with exclusive secretory activity, are found.

During transformation of smooth muscle cells into epithelioid cells, both the ultrastructure and the shape of the involved cells are subject to characteristic alterations. Intermediate cells have - just as smooth muscle cells from which they develop - at first a distinct fusiform shape to encircle the endothelial tube as these do in a helical arrangement. When the granular store of intermediate cells increases, the outer contour of the afferent arteriole or interlobular artery appears widened at their respective location (Fig. 2.3). If fully transformed - epithelioid - cells are interposed between smooth muscle cells, they may give the impression of having been placed outside upon the media.

As intermediate cells encircle the endothelial tube in a helical arrangement and are more or less well equipped with myofilaments, it may be assumed that they - in the range of possible variations in shape and ultrastructure - are provided with a variable vasoconstrictive ability. Intermediate cells, on the other hand, are also equipped with renin granules. It seems reasonable to conclude that their secretory activity increases with the size of the respective renin depots, i.e., with the number of secretory granules of the individual cells. At the end of this spectrum, mechanically passive epithelioid cells with exclusive secretory activity might be found. This hypothesis of a continuous decrease in contractile capacity connected with an increase in the secretory activity as based on morphological findings would require that the contractile system of intermediate cells functions independently from their secretory system and vice versa. This issue will be discussed in some detail in context with the paradoxical influence of calcium on renin secretion in Sect. 7.6. The morphological criteria for the evaluation of the vasoconstrictive ability of the glomerular arterioles are discussed in Sect. 2.2.3.

At this point, we would like to call to mind that the idea of a uniform reaction of all so-called juxtaglomerular cells to all secretory stimuli is no longer tenable for various reasons. Immunohistochemical experiments showed that renin-positive cells are often found 100-200 μm upstream from the JGA (Taugner et al. 1981, 1982a, c). Moreover, in the afferent arteriole - starting at the JGA and proceeding in an upstream direction - first epithelioid cells and thereafter increasingly less transformed intermediate cells are encountered (Fig. 2.3). Recent experiments with isolated glomeruli and afferent arterioles lead to the conclusion that renin release from the various portions of the afferent arteriole may be affected differently by certain stimuli (Itoh and Carretero 1985; Itoh et al. 1985b; Baumbach and Skøtt 1986). We tend to suggest that these differences can be explained by both the respective position in the afferent arteriole and differences in the ultrastructure of the involved media cells.

Unfortunately, there is only very little information on the number of intermediate cells in comparison to that of epithelioid cells in the various species. Reliable data on the respective cell numbers might only be gathered by serial thin sectioning, for the following reasons: Barajas (1981) pointed out that profiles of intermediate cells in a given section plane may not contain secretory granules, being indistinguishable from neighboring smooth muscle cells. Serial sectioning, however, would reveal the existence of characteristic granules in another portion of these cells (Gorgas 1978a). At present, it is known only from electron microscopic studies that at the border between the renin-negative and the renin-positive portion of the afferent arteriole some cells always show the ultrastructural characteristics -increasingly from proximal to distal - of a transition from smooth muscle cells into epithelioid cells (cf. Bucher and Reale 1961a; Biava and West 1966a; Barajas and Latta 1967; Barajas 1981). At 1 or 2 days after an abrupt increase of the stimulation level, e.g., by adrenalectomy, the number of intermediate cells appears to be increased as a consequence of the metaplastic transformation from smooth muscle cells to epithelioid cells. This impression still needs confirmation by quantitative data (cf. Chap. 11).

In contrast to thin and semithin sections, paraffin sections give only little information on the occurrence and number of intermediate cells, particularly when these are only poorly equipped with secretory granules. Even immunohistochemical analysis is only of limited value for the discrimination between scantily and amply granulated renin-positive cells. After immunostaining of paraffin sections, a clear borderline between the renin-negative and renin-positive portions of the afferent arteriole is usually observed (cf. Figs. 2.2, 2.8A). Obviously, granulated cells can hardly be portrayed immunohistochemically in their gradual stages, i.e., analogous to their equipment with secretory granules. This is most evident with renin-positive cells scattered sporadically among plain smooth muscle cells of the interlobular artery or the proximal afferent arteriole. These helically arranged cells are known to be scarcely granulated, intermediate cells from their ultrastructural appearance. In the immunohistochemical portrayal under the light microscope, however, they are usually seen to be circular, i.e., renin-positive throughout (cf. Figs. 2.2, 4.2). This is probably due to the fact that with routinely used immunohistochemical techniques renin leaks out of the scattered secretory granules and spreads into the entire cytoplasm of these intermediate cells. In addition, in such experiments, supraoptimal antibody concentrations are commonly used, not suited to reflect differences in the renin concentration (cf. Sect. 4.1.4).

2.2.3 Morphological Prerequisites for Vasoconstriction in the Renin-Positive and Renin-Negative Portions of the Afferent Arteriole

Reconstructions of the juxtaglomerular afferent arteriole, which would allow a definitive judgment of the microtopographical relation between the media cells and the endothelial tube of this vascular segment are not yet available. It is obvious, however, that epithelioid cells differ fundamentally from the common notion of a contractile cell population not only in their ultrastructure, but also in their shape and occasionally in their position: smooth muscle cells are fusiform and helically arranged around the endothelial tube; epithelioid cells, on the other hand, are already noticeable by their plumpness in the light microscope; they are often also arranged asymmetrically only at the longer convex bank of a winding afferent arteriole (Fig. 2.5). Schloss (1945/6) has reviewed and commented on pertinent arguments of Goormaghtigh (1932, 1936), Goormaghtigh and Handovsky (1939) on this topic. It is therefore surprising that the recent literature dealing with the morphology of the afferent arteriole does not discern between the proximal and the distal part of this vessel as far as their capability to govern the preglomerular resistance is concerned. This may, among others, be explained by the lack of comparative data relating to the equipment with leiomyofilaments of the proximal and distal afferent arteriole's media cells.

With this lack of information in mind, the prerequisites of vasoconstriction in the renin-positive (juxtaglomerular) portion of the afferent arteriole,

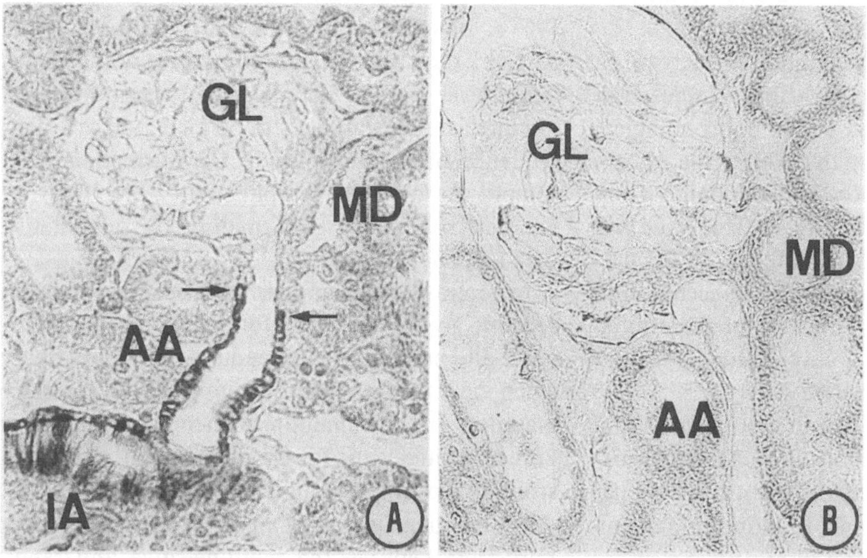

Fig. 2.7. Paraffin sections from rat kidney, reacted with native **A** and preabsorbed **B** antibodies to smooth muscle myosin; ×360. **A** After application of the native antibodies, the interlobular artery *(IA)* and the proximal afferent arteriole *(AA)* are immunostained. The *arrows* indicate the transition from the myosin-positive to the myosin-negative juxtaglomerular segment of the vessel. *MD*, macula densa; *GL*, glomerulus. Antimyosin concentration, 20 μg/ml. **B** After preabsorption with extracted myosin, the afferent glomerular arteriole *(AA)* shows only background staining. *MD*, macula densa; *GL*, glomerulus. (From Taugner et al. 1987b)

as compared with those of its renin-negative (proximal) segment, were investigated using antibodies to smooth muscle myosin as well as thin sections after fixation with osmium tetroxide (TAUGNER et al. 1987b). By contrast to the myosin-positive/renin-negative proximal portion of the preglomerular arteriole, myosin-like immunoreactivity could not be demonstrated in the distal, renin-positive part of this vessel (Figs. 2.7, 2.8A, B). As an exception, completely renin-negative afferent arterioles were found; in this case, the vessel was found to be myosin-positive throughout (Fig. 2.8C, D). In addition, thick myofilaments were found to be drastically reduced or virtually absent in partially or fully granulated media cells (Fig. 2.9; cf. BUCHER and KAISSLING 1973). This abrupt change of immunoreactivity in addition to the remarkable difference in the number of thick filaments was interpreted as a correspondingly large difference in the myosin content of both vessel segments. Since the density of myosin filaments is proportional to the tension a cell may develop upon maximal activation, the inference was drawn that the juxtaglomerular part of the preglomerular arteriole close to the macula densa, in contrast to its proximal segment, has only very limited capabilities for vasoconstriction.

As a consequence, this noncontractile segment of the vessel was proposed to represent the renal vascular receptor responsible for the increase of renin secretion during pressure reduction (Sect. 7.2). The implications of such a concept envisaging a more or less mechanically passive juxtaglomerular vessel segment for the tubuloglomerular feedback, a mechanism allegedly triggered by the so-called macula densa-signal, are discussed in Sect. 7.3.

2.2.4 Additional Aspects of the Functional Morphology of the Proximal and Distal Portions of the Afferent Arteriole

The proximal segment of the afferent arteriole differs from its distal segment not only in the ultrastructure of their media cells, i.e., in their capacity for renin secretion or vasoconstriction, but also in their equipment with musculomuscular, musculoendothelial and endoendothelial junctions (cf. Sect. 2.7). As pointed out earlier, the boundary between the renin-positive and the renin-negative por-

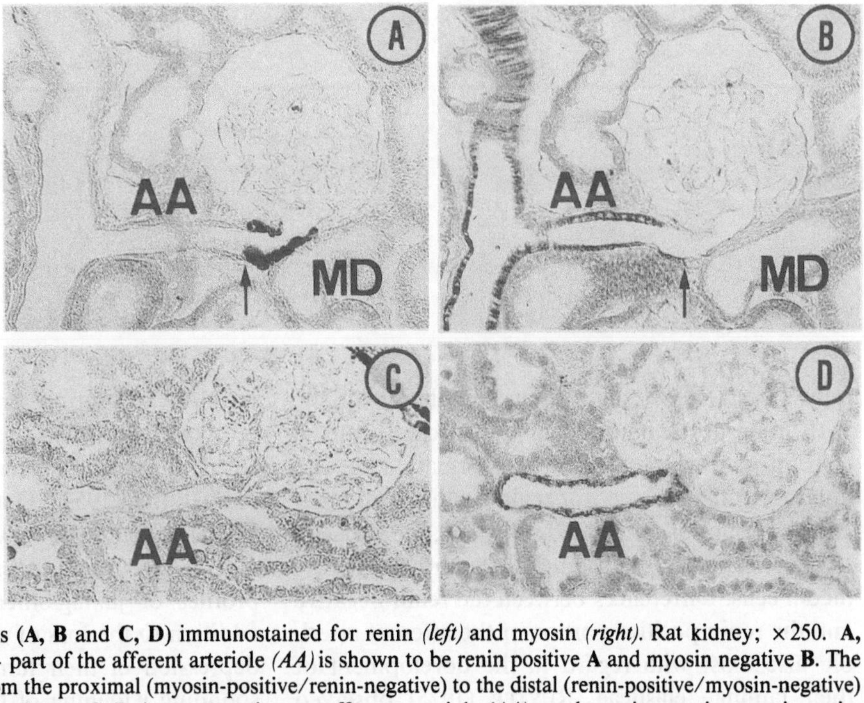

Fig. 2.8. Paired paraffin sections (**A, B** and **C, D**) immunostained for renin *(left)* and myosin *(right)*. Rat kidney; ×250. **A, B** The distal - juxtaglomerular - part of the afferent arteriole *(AA)* is shown to be renin positive **A** and myosin negative **B**. The *arrows* point to the transition from the proximal (myosin-positive/renin-negative) to the distal (renin-positive/myosin-negative) portion of the vessel. *MD*, macula densa. **C, D** As an exception, an afferent arteriole *(AA)* may be renin negative over its entire lenght **C**; in this case, the vessel is found to be myosin positive throughout **B**

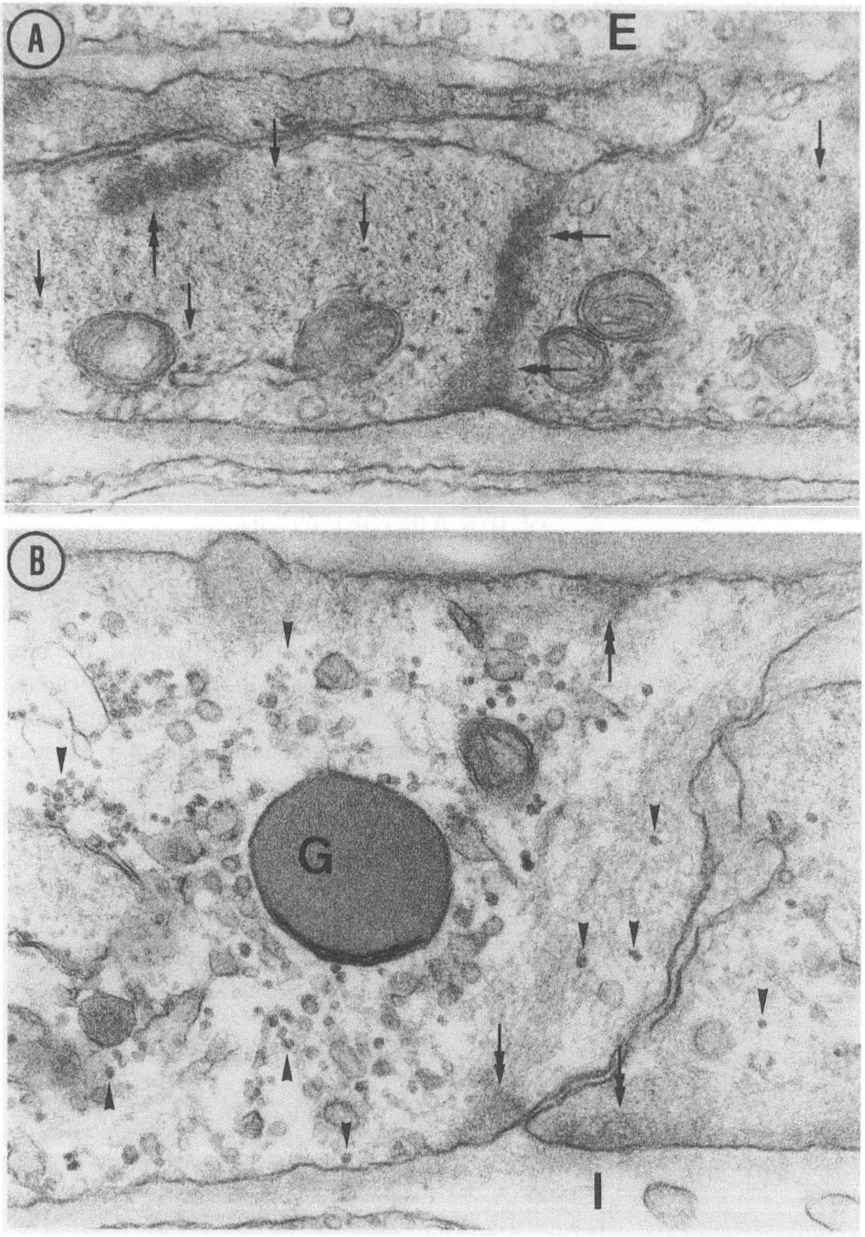

Fig. 2.9. Differences in filament equipment of media cells in the proximal, i.e., renin-negative and in the distal, renin-positive portion of the preglomerular arteriole (A and B, respectively). **A** The nongranulated plain smooth muscle cell is well equipped with thick filaments *(arrows)* and attachment sites *(double arrows)*. *E*, endothelial cell. **B** In the scarcely granulated - only partially transformed - media cell, thick filaments are virtually absent. *Double arrows*, putative remnants of attachment sites; *arrowheads*, glycogen particles; *G*, renin granule; *I*, interstitium. Rat kidney, perfusion fixation with 2.5% glutaraldehyde and 2% OsO_4, en block staining with saturated uranyl acetate; ×62200

tions of the vessel shifts upon stimulation of the RAS in an upstream direction. Significantly, media cells affected by metaplastic transformation acquire apart from the capacity for renin synthesis all other characteristics of granulated cells, e.g., the capacity to express macular gap junctions with neighboring media cells. Differences between the renin-negative and the renin-positive portion of the afferent arteriole also exist with regard to the development of their laminae elasticae. RUYTER (1925) observed that the lamina elastica interna of the juxtaglomerular afferent arteriole is missing. Later, electron microscopy confirmed the absence of internal elastic fibers in the wall of the distal preglomerular arteriole (BOHLE 1959; OBERLING and HATT 1960b; HARTROFT and NEWMARK 1961; BIAVA and WEST 1966a). According to BIAVA and WEST (1966a), the profiles of juxtaglomerular afferent arterioles, in addition to the loss of elastic fibers, present a local depression of their luminal surface, with a corresponding widening of their lumina. In the foregoing, the mechanically passive segment of the afferent arteriole was proposed to represent the renal vascular receptor. The lack of elastic fibers would be in agreement with the concept of a mechanotransductive vessel segment, sensing transmural pressure variations by the deformation of its endothelial and/or media cells (cf. Sect. 7.2).

2.3 Efferent Arteriole

As with the afferent vessel, the common definition of the JGA includes only the juxtaglomerular portion of the efferent arteriole (Figs. 1.2, 2.1 A). With respect to the structure and functions of the postglomerular vessel, this restriction is also somewhat arbitrary. Thus, renin-positive cells are not only found in the proximal (juxtaglomerular) portion of the efferent arteriole, but also further downstream (Figs. 2.3, 2.12). In addition, modulations of the efferent resistance may be initiated from the region of the JGA. Therefore, the functional morphology of the distal portion of the postglomerular arteriole will be included in this section. (For more detailed reviews see FOURMAN and MOFFAT 1964, 1971; ROLLHÄUSER et al. 1964; DIETERICH 1978; BEEUWKES 1980; KRIZ and BACHMANN 1985).

Cortical efferent arterioles are different in length: those of the superficial glomeruli ascend through the cortex corticis almost to the cortical surface, whereas those of midcortical glomeruli are shorter. The juxtamedullary efferent arterioles descend as far as into the outer stripe of the medulla to branch out into the descending vasa recta still retaining the structure of arterioles (for further details and references see BEEUWKES 1970). The cortical efferent arterioles are smaller in caliber than the corresponding afferent vessels (STEINHAUSEN and TANNER 1976). The juxtamedullary efferent arterioles at least in some species including man are larger, surpassing even their afferent counterparts (BANKIR

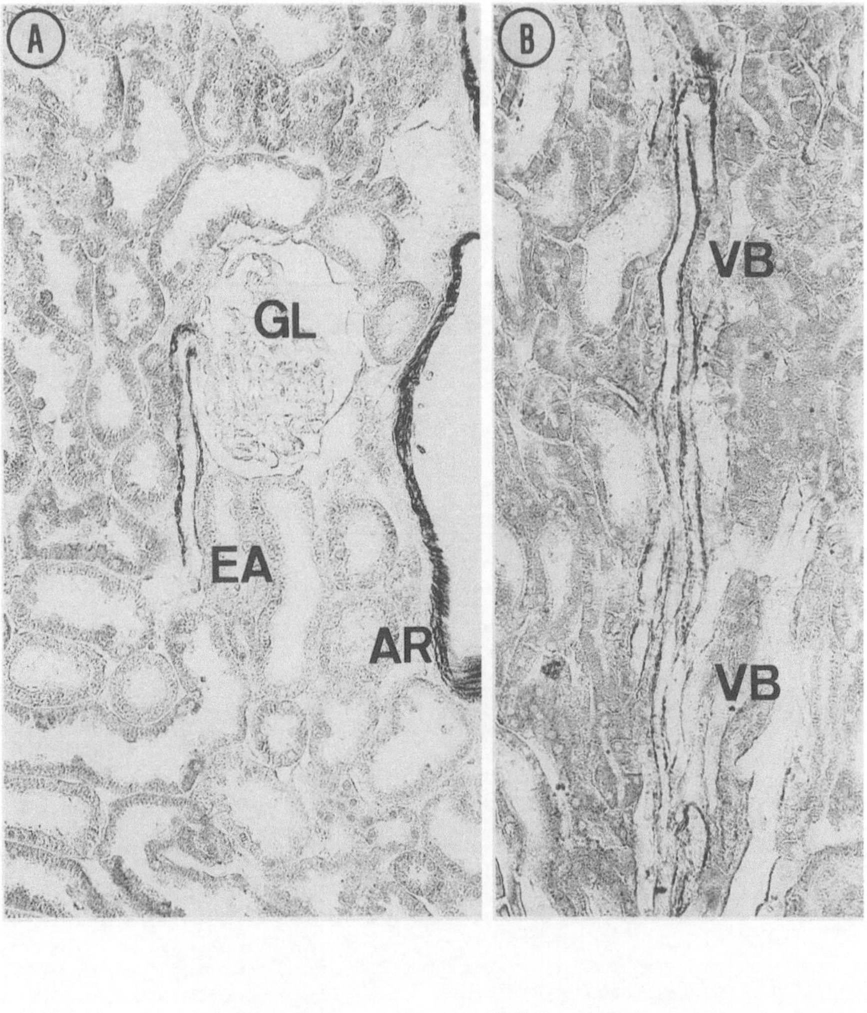

Fig. 2.10. Myosin-positive efferent arteriole *(EA)* of juxtamedullary nephron **A** and vascular bundles *(VB)* in the outer medulla of rat kidney **B**. *GL*, glomerulus; *AR*, arcuate artery. Further details in the legend of Fig. 2.7; ×230

and FARMAN 1973; for references see KRIZ and BACHMANN 1985). However, the diameters of efferent arterioles may differ largely with the distance from the parent glomerulus (STEINHAUSEN et al. 1986a, b).

Close to the glomerular hilus, the media cells of the efferent arterioles ramify into several overlapping processes, giving the impression of a multilayered sheet (GORGAS 1978a) which merges with the Goormaghtigh cell field and the mesangium of the glomerular stalk (Fig. 1.2). More distally, the media of the arterioles thins to one layer of pericyte-like cells with incospicuous filament bundles and attachment sites. In some regions, the media may even be absent, leaving only the basement membrane and endothelial cells to form the wall of the vessel (BARAJAS and LATTA 1963a). Although slender, typical media cells of the proximal efferent arteriole contain intermediate and thin filaments as well as scattered thick filaments. Correspondingly, the media of the efferent arteriole - in contrast to that of the afferent vessel - can mostly be traced with antimyosin antibodies right up to the glomerular stalk (Fig. 2.10, TAUGNER et al. 1987b).

Elastic laminae or fibers are missing in the efferent arteriole. The interstitium between media cells

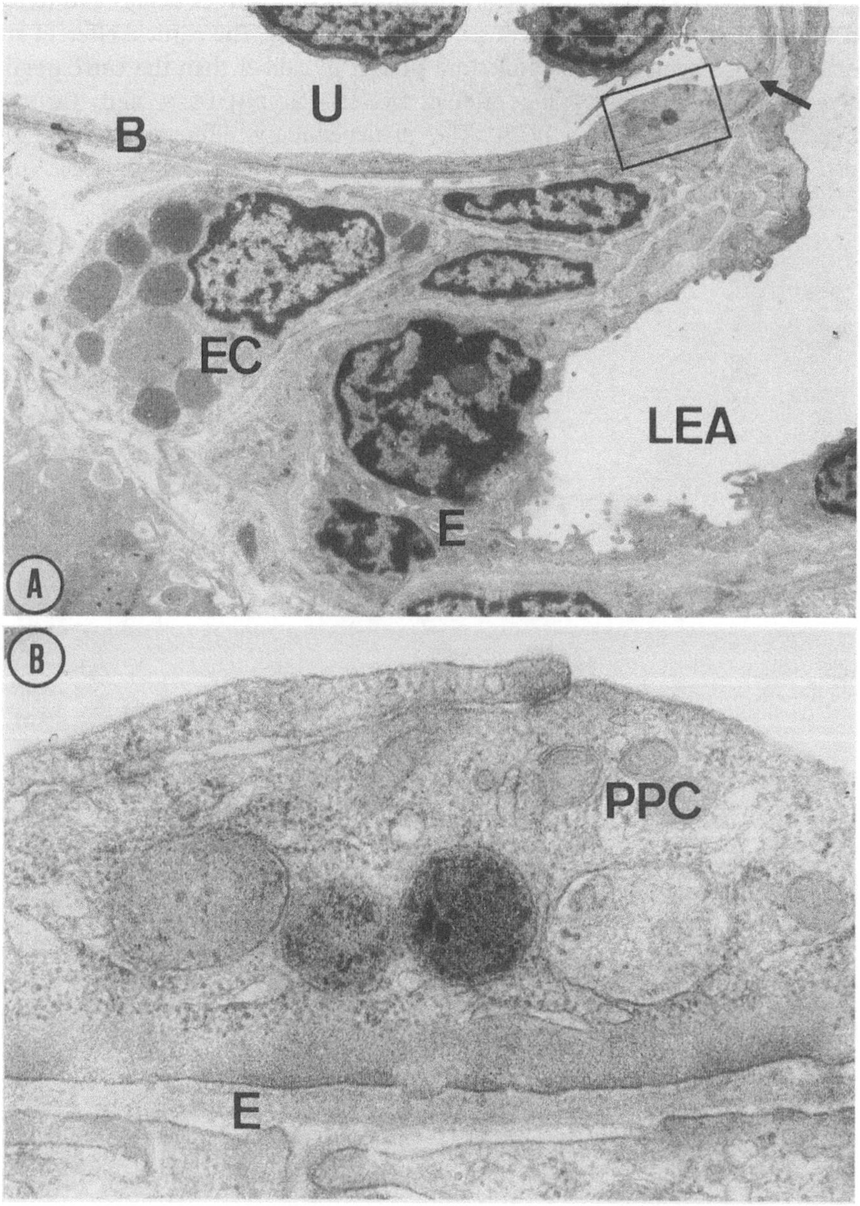

Fig. 2.11. A Epithelioid cell *(EC)* in the media of the juxtaglomerular portion of an efferent arteriole. Close to the visceroparietal junction *(arrow)* of Bowman's capsule *B*, a peripolar cell can be seen, depicted in **B** at higher magnification. *LEA*, lumen of the efferent arteriole. *U*, urinary space of the renal corpuscle. Mouse kidney; × 5700. (For overview see Fig. 2.16). **B** Peripolar cell *(PPC)* from the area marked in **A** at higher magnification. Note the differences in electron density and matrix structures of the four "granules" depicted. *E*, extracellular space; × 50500

as well as that between media and endothelial cells is filled with basal lamina material which is continuous with that of the lacis and - in part - also with that of the distal tubule (cf. Sect. 2.5). Characteristically, there are large irregular extracellular spaces filled with basal lamina-like material between the media and the endothelium of proglomerular arterioles and, less conspicuously, between media cell processes. The endothelium of the juxtamedullar efferent arterioles, reported to be rich in intermediate filaments of the vimentin type (Bachmann et al. 1983), stands out by its high endocytic activity and the great number of longitudinally arranged cells bulging into the lumen of the vessel (Gattone et al. 1984; Kriz and Bachmann 1985; cf. also Zimmermann 1932; Barajas and Latta 1963a).

Glomeruli with two to five efferent arterioles have been found in the kidneys of humans (Smith 1956; Murakami et al. 1985a) and other mammalian species (Fourman and Moffat 1964, 1971; Murakami et al. 1971, 1985b; Yang and Morrison 1980). As their number amounts only to a few percent of all glomeruli, a major functional significance of these multiple efferent arterioles may be excluded (Frank and Kriz 1982).

The numerous myoendothelial contacts of the efferent arterioles consist mainly of flat appositions, thus diverging somewhat from those of the afferent arteriole (Taugner et al. 1984c). The same holds true for the endoendothelial contacts, which are similar to those of the distal afferent arteriole only in the juxtaglomerular portion of the vessel, consisting of tight junction strands not subdivided into individual particles (Figs. 2.30c, 2.31; Mink et al. 1984). In more distant segments of the efferent arteriole, mostly combinations between solid and subdivided tight junction strands have been observed. This seemed to indicate a gradual transition to the less elaborate, "leakier" junctions characteristic for the subsequent fenestrated peritubular capillaries.

Electron microscopically, granulated cells were recognized early in the juxtaglomerular part of the efferent arteriole (Barajas and Latta 1963a; Dunihue and Boldosser 1963; Faarup et al. 1967; Simpson 1970; Barajas 1971; Christensen et al. 1975; cf. Fig. 2.11). In serial paraffin and in semithin sections, they have also been observed at considerable distances from the JGA (Adebahr 1962; see Figs. 2.3, 2.12). Since, however, granulated cells are found less frequently in efferent than in afferent arterioles, there is correspondingly little information concerning their ultrastructure, in particular with regard to the transformation of typical media cells into epithelioid cells. In addition to being renin positive (Taugner et al. 1979, 1981), another correspondence between granulated cells of the afferent and efferent arterioles is their angiotensin II reactivity (Fig. 4.13). Accidentally, granulopoiesis through protogranules as known from the afferent arteriole has been found in an epithelioid cell of the efferent arteriole (unpublished observation).

Data on the incidence of granulated cells in the postglomerular arteriole are also fragmentary. In the kidneys of mice, 20%-40% of the efferent arterioles were found to be renin positive close to the vascular pole of the renal corpuscle. In some cases,

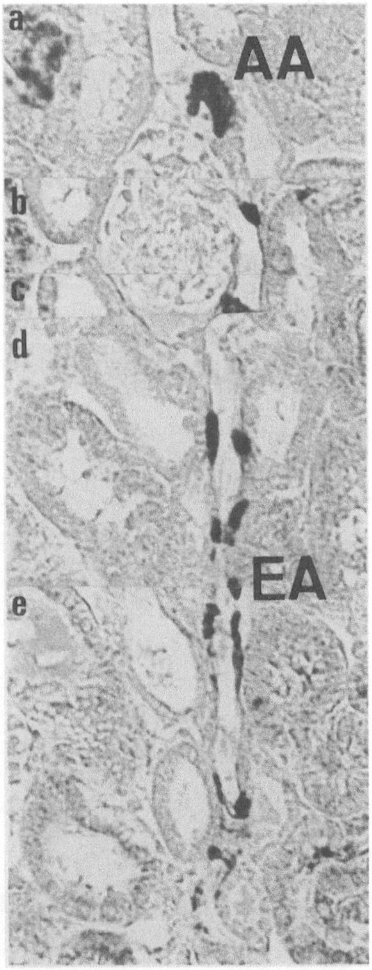

Fig. 2.12. Juxtamedullar efferent arteriole *(EA)* from mouse kidney with scattered renin-positive cells over a distance of about 200 μm from the glomerulus into the direction of the medulla. *AA*, immunoreactive afferent arteriole. The micrograph was mounted from a series of consecutive paraffin sections; ×250. (From Taugner et al. 1981)

scattered renin-positive cells were seen in efferent arterioles of juxtamedullary nephrons at a distance of more than 100 µm from their corresponding glomeruli (Fig. 2.12). Therefore, the amount of 20%-40% may represent a minimal count for renin-positive postglomerular vessels. Sodium depletion was shown to increase preferentially the number of renin-positive superficial efferent arterioles (TAUGNER et al. 1981).

In summary, we have to state that only little is known about the granulated cells in the efferent as compared with those in the afferent arteriole, reflecting their comparatively small number. This holds true in particular for the largely unknown stimuli which may lead to an increased renin synthesis in and secretion from the efferent arteriole. Owing to their small number, it is improbable that the granulated cells of the efferent arteriole substantially contribute to the plasma renin level. Instead, local intrarenal effects may be of prime importance at postglomerular sites.

2.4 Goormaghtigh Cell Field

In 1932, GOORMAGHTIGH described a group of relatively small, elongated cells in the area of the vascular pole of the renal corpuscle with scanty cytoplasm and a chromatin-rich nucleus, which he called "pseudo-Meissnerian cells" thought to be parts of a *corpuscule nerveux sensitif* (sensitive nervous body). According to their localization in the area between both glomerular arterioles and the macula densa, Goormaghtigh cells (G cells) pile up to a cone, the base of which faces the macula densa and its apex by way of the glomerular stalk merging continually into the glomerular mesangium (Figs. 1.2, 2.13; BECHER 1949). Impressed by the dense meshwork of basal lamina-like material surrounding G cells and their processes, OBERLING and HATT (1960a, b) introduced the name "lacis" (net, tissue) for section profiles of this region, otherwise called "G cell field." As G cells - by their ori-

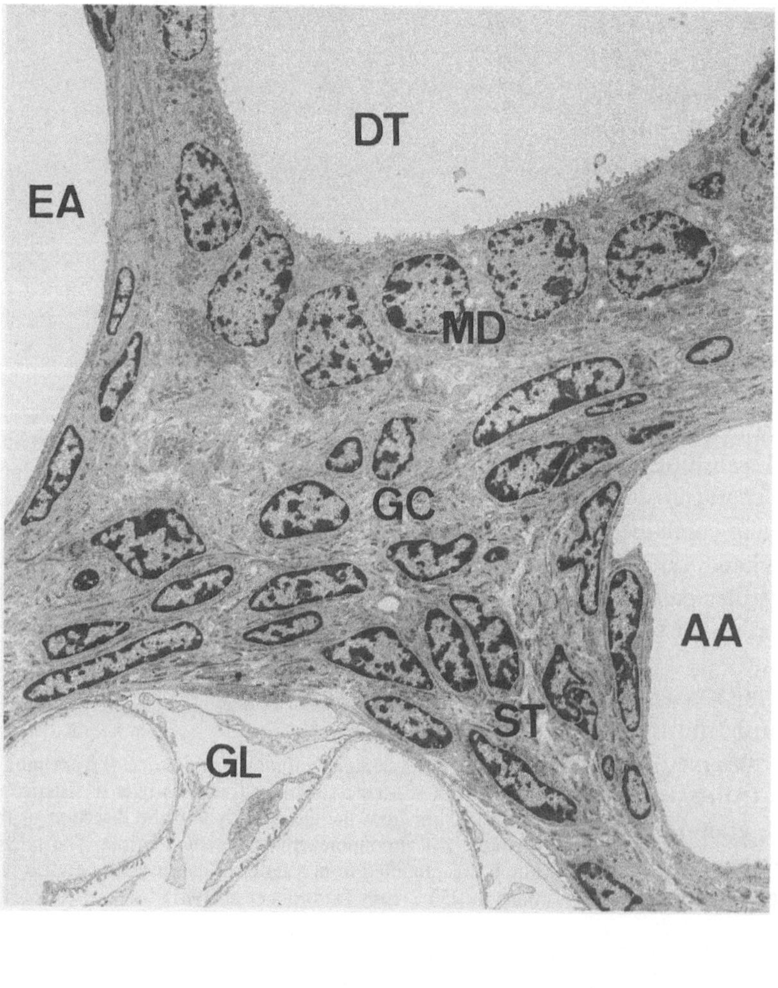

Fig. 2.13. Juxtaglomerular apparatus from rat kidney showing the extended Goormaghtigh cell field *(GC)* continuous with the glomerular stalk *(ST)*, both surrounded by the afferent and efferent glomerular arterioles *(AA, EA)*, the macula densa *(MD)*, and Bowman's capsule of the glomerulus *(GL)*. *DT*, distal tubule; ×3100

gin and morphology – are not clearly distinguishable from glomerular mesangial cells, the G-cell field may also be viewed as an extension of the glomerular mesangium and termed extraglomerular mesangium (cf. BARAJAS 1970). The affiliation of the Goormaghtigh cell field to the other vascular components of the glomerulo-juxtaglomerular complex becomes evident after unilateral ureteral ligation: in the hydronephrotic kidney, the Goormaghtigh cells still adhere to the glomerular stalk, although the macula densa, subject to atrophy, has disappeared (NOBILING et al. 1986). Before the first description of the G-cell ultrastructure by OBERLING and HATT (1960a, b), BECHER (1949) already pointed out the difficulty of distinguishing between G cells and some of the adjacent media cells of the afferent and efferent glomerular arterioles. As a consequence, in morphometrical studies the border between the G-cell field and the glomerular stalk has to be drawn arbitrarily, e.g., by the turning points of the parietal layer of Bowman's capsule into its visceral layer (CHRISTENSEN and BOHLE 1978; LATTA und JOHNSTON 1978).

The G-cell field has a differing appearance, depending mainly on the section plane and the number of G cells, which shows remarkable internephron and interspecies differences. In sections through the axis of the hilar portions of both glomerular arterioles in species with relatively large G-cell fields such as the rat (GORGAS 1978a), the rabbit (SPANIDIS et al. 1982), or man, groups of oblong G cells are often found oriented parallel to the basis of the macula densa (Fig. 2.13). Here, G cells seem to have the shape of a flatly pressed cylinder (SPANIDIS et al. 1982). Accordingly, in perpendicular sections passing the glomerular stalk and the basis of the macula densa, only small, mostly roundish G-cell profiles are encountered. Close to the glomerular stalk, where the G-cell field narrows, as well as laterally, where the lacis merges with the media of the hilar arterioles, more and more cells and cell processes oriented perpendicularily to the basis of the macula densa are found. In species where the G-cell field is smaller, e.g., in mice, the above-described arrangement of G cells parallel to the basis of the macula densa is altogether less pronounced, and, in addition, the G cells rarely have an elongated form (Figs. 1.2, 2.16).

GORGAS (1978a) pointed out that the area occupied by G cells is not strictly limited to the cone-like spandrel between the glomerular arterioles. Approaching the hilus of the renal corpuscle, the media of the preglomerular arteriole may be seen to divide, its "inner" layer merging with the mesangium of the glomerular stalk, the cells of its "outer" layer, together with and not distinguishable from G cells, forming a funnel-like extension around the hilar portion of Bowman's capsule (cf. KRIZ and SAKAI 1988). BARAJAS and LATTA (1963a) showed that the G-cell field and the glomerular mesangium may be continuous not only at the center of the hilus of the renal corpuscle, but also at its periphery next to Bowman's capsule.

SPANIDIS et al. (1982) counted 19 individual G cells in one lacis of the rabbit. FAARUP (1965) determined the number of G cells in rats, using serial paraffin sections. In subcapsular nephrons, 3–15 G cells were found (mean, 9; $n = 15$), in juxtamedullary nephrons 10–21 G cells (mean, 15; $n = 10$). BARAJAS (1971), in his series of thin sections from rat kidneys, encountered 17–34 G cells per nephron ($n = 4$). Hence, just as other parameters (e.g., extension of the macula densa, mutual contact areas between different JGA components, number of epithelioid cells), also the number of G cells and, accordingly, the size of the lacis, seem to be subject to a marked internephron heterogeneity.

The possible functional implications of these internephron and also the above-mentioned interspecies variations in the extension of the G-cell field are largely unknown as are those of the contact areas discussed in Sec. 2.6. The size of the G-cell field of course will vary with the geometry of the structures which form its boundaries. The diameters of the distal tubule and those of the glomerular arterioles are larger in humans and rabbits than, for example, in mice. The size of the respective G-cell fields seems to vary accordingly. Second, the proportion of the diameter of the distal tubule to the angle both glomerular arterioles form near the hilus will be important. Third, the way the distal tubule passes by or between the hilar arterioles may influence the size and form of the G-cell field (cf. BARAJAS and LATTA 1963a; FAARUP 1965; CHRISTENSEN et al. 1979). With the microtopographical conditions being so diverse, there are, quite obviously, large variations of the contact areas between the individual JGA elements, some of which are intercorrelated (CHRISTENSEN and BOHLE 1978). In the context of the tubuloglomerular feedback (TGF), first of all, the contact area between the macula densa and the Goormaghtigh cell field has been discussed (THURAU and MASON 1974). However, to our knowledge, there are no experiments addressing a correlation of the microtopography of an individual JGA with possible functional peculiarities of the corresponding nephron.

OBERLING and HATT (1960a, b) were the first to describe the untrastructure of G cells, noting their irregularly shaped, elongated nucleus surrounded by a relatively small rim of cytoplasm. The cytoplasmic equipment of plain G cells is inconspicuous. Their Golgi system may be large and prominent or scanty in a given section plane (LATTA and MAUNSBACH 1962a; BUCHER and REALE 1962b; BARAJAS and LATTA 1963a). Some profiles of granulated endoplasmic reticulum are regularly found. The number of relatively small mitochondria is moderate. Less frequently, centrioles, accumulations of glycogen, multivesicular bodies (MVBs), and small dense cytoplasmic or lipofuscin-like bodies are seen. Bundles of microfilaments may be found in the cytoplasm of G cells, predominantly in their processes, with areas of increased density close to the plasma membrane (LATTA and MAUNSBACH 1962a; BARAJAS and LATTA 1967; GORGAS 1978a). Typically, G cells develop numerous long processes (OBERLING and HATT 1960a, b; LATTA and MAUNSBACH 1962a; BUCHER and KAISSLING 1973; BARAJAS and LATTA 1967; GORGAS 1978a).

SPANIDIS et al. (1982) demonstrated by reconstruction of a G cell from rabbit kidney how these processes might be arranged in detail. The cell had the shape of a flatly pressed cylinder with both of its ends splitting up into a bunch of parallel processes. Interestingly, these cell processes maintained a neighboring position and did not interdigitate with those of other cells. Seven other G cells traced by SPANIDIS et al. (1982) - without performing a reconstruction - divided into similar processes, again without interdigitations. In the lacis investigated, all 19 G cells were piled up in several layers with their longitudinal axis parallel to the basis of the macula densa. Cell processes oriented perpendicularly were not found. In the reconstructed neighboring granulated cell, by contrast, bunches of processes were lacking; instead, the cell ramified into a few large processes (SPANIDIS et al. 1982; cf. GATTONE et al. 1984). GORGAS (1978a) called attention to a similar orientation of G cells in rats, the processes of these cells often reaching from the endothelium of the afferent arteriole to that of the efferent vessel. In species with less developed G-cell fields, e.g., in mice, the above-mentioned deviations from such a regular arrangement of G cells parallel to the basis of the macula densa can be clearly observed in thin sections. Here, G cells and G-cell processes perpendicular to the base of the macula densa may bridge the relatively small distance from the tubule to the mesangium in the glomerular stalk (GORGAS 1978a). Cells with this perpendicular orientation in the direction of the glomerular mesangium at the apex of the cell cone in species with large G-cell fields have been mentioned above. Apart from numerous processes, G cells show many deep invaginations of their cell membranes (GORGAS 1978a) similar to those in mesangium cells and epithelioid cells. Often, the extracellular space enclosed by these invaginations contains basal lamina material, or, seldomly, collagen fibrils; quite frequently it is bordered by reflexive gap junctions.

It should be noted at this point that the description given so far is related to G cells encountered in the lacis of individuals with no or only a moderate subchronical stimulation of the RAS. In contrast to these renin-negative G cells, heavily granulated renin-positive G cells resembling epithelioid cells have recently been found in patients with *pseudo-Bartter's syndrome* due to the chronic abuse of furosemide and/or laxatives (CHRISTENSEN et al. 1988; cf. Sect. 4.1.3). It is therefore likely that also intermediate forms of G cells exist, as anticipated by others (see below).

Characteristically, G cells are connected by gap junctions not only with other G cells (PRICAM et al. 1974), but also with media cells of both glomerular arterioles (BOLL et al. 1975; FORSSMANN and TAUGNER 1977) and mesangial cells (TAUGNER et al. 1978b). In rabbits, a broad avenue of gap junction coupled cells was found from the G-cell field through the mesangial stalk up to the glomerular mesangium, with 74 macular gap junctions in one freeze fracture plane (Fig. 2.24; TAUGNER et al. 1978b). In addition, there are numerous reflexive gap junctions bridging individual processes of the same G cell (PRICAM et al. 1974; GORGAS 1978a; SPANIDIS et al. 1982). Reflexive gap junctions were also found in glomerular mesangial and in epithelioid cells of the afferent arteriole's media (TAUGNER et al. 1984b). The finding that G cells of all species studied are consistently coupled with all other cell types of the JGA by gap junctions except macula densa cells may be considered as further evidence for the common smooth muscle type origin of G cells, mesangial cells and media cells of the glomerular arterioles. As G cells in this complex of coupled mesenchymal cells hold a central position and are widely aligned along the macula densa, it appears reasonable to assume that they may be an indispensable link in the signal transfer between the tubular and vascular components of the JGA.

G cells and their processes are separated by basal lamina-like material of irregular thickness, surrounding them like meshes of a net, the "lacis" of OBERLING and HATT (1960a, b). The basal lamina covering all G cells is only missing in areas where intercellular contacts, i.e., gap junctions, and sporadically desmosomes are found. The plait of the basal lamina continues from the G-cell field in various directions: through the wall of the glomerular arterioles to the basal lamina of their endothelia, through the hilus of the renal corpuscle to the basal lamina of mesangial cells, and, as described in detail elsewhere, to the base of the macula densa cells. According to BARAJAS (1970, 1971, 1972), the basal lamina of G cells and that of the macula densa tend to form networks, thus establishing permanent contacts between these structures. In contrast to this, BARAJAS found only simple appositions between the basal laminae of the distal tubule and the afferent glomerular arteriole, thus forming only a reversible type of contact, the area of which may be subject to changes with the volume of the distal tubule or afferent arteriole (cf. Sect. 2.6).

Processes of neighboring fibroblasts penetrate the labyrinth of basal laminae around G cells, where in some section planes they cannot be distinguished from processes of lacis or smooth muscle cells (GORGAS 1978a). Occasionally, processes of macrophages and plasma cells may also be seen at the periphery of the G-cell field. Inside the Goormaghtigh cell field and also in the areas of contact between the lacis and the other JGA components, neither free cells nor blood or lymph capillaries can be observed (GOORMAGHTIGH 1939; OBERLING and HATT 1960b; THOENES 1961; BUCHER and REALE 1962b; BARAJAS and LATTA 1963a).

LATTA and MAUNSBACH (1962a) as well as BARAJAS and LATTA (1963a) called attention to the fact that otherwise typical G cells may contain granules resembling those of epithelioid cells. Intermediate cell forms seemed to exist in the lacis, and stimulation of renin synthesis was postulated to be able to increase the granularity of these cells to the point that by their ultrastructure they may hardly be distinguished from typical epithelioid cells in the nearby media of the afferent arteriole (KROON 1960; BOHLE and SITTE 1966; cf. HATT 1967; HARTROFT 1968; ROUILLER and ORCI 1971; CHRISTENSEN et al. 1978). However, the question of whether new renin-positive cells may be recruited from the pool of renin-negative G cells could not be considered definitely resolved (OBERLING and HATT 1960b; BUCHER and REALE 1962b). In Sec. 4.1.3, difficulties involved in distinguishing G cells with regard to their position and ultrastructure from the adjacent media cells of the bordering glomerular arterioles are discussed. These difficulties may be of even greater importance if changes in the stimulation level of the RAS could alter the microtopographical relationships between the different components of the JGA. In immunohistochemical experiments, the G-cell field proved to be renin negative in control as well as in sodium-depleted and adrenalectomized animals (TAUGNER et al. 1979, 1981). This, however, did not rule out the presence of sporadic renin-positive cells in the periphery of the lacis. With immunocytochemical techniques, renin-nega-

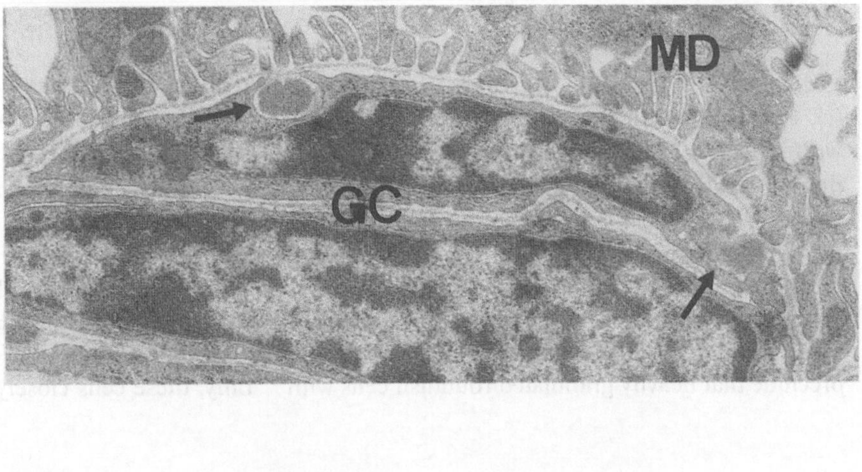

Fig. 2.14. Goormaghtigh cells *(GC)* from hamster kidney facing the macula densa *(MD)* with omega-shaped indentations containing electron-dense material *(arrows)*. Hamster kidney; ×16700

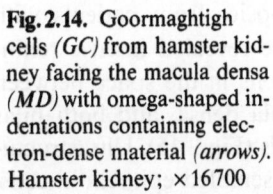

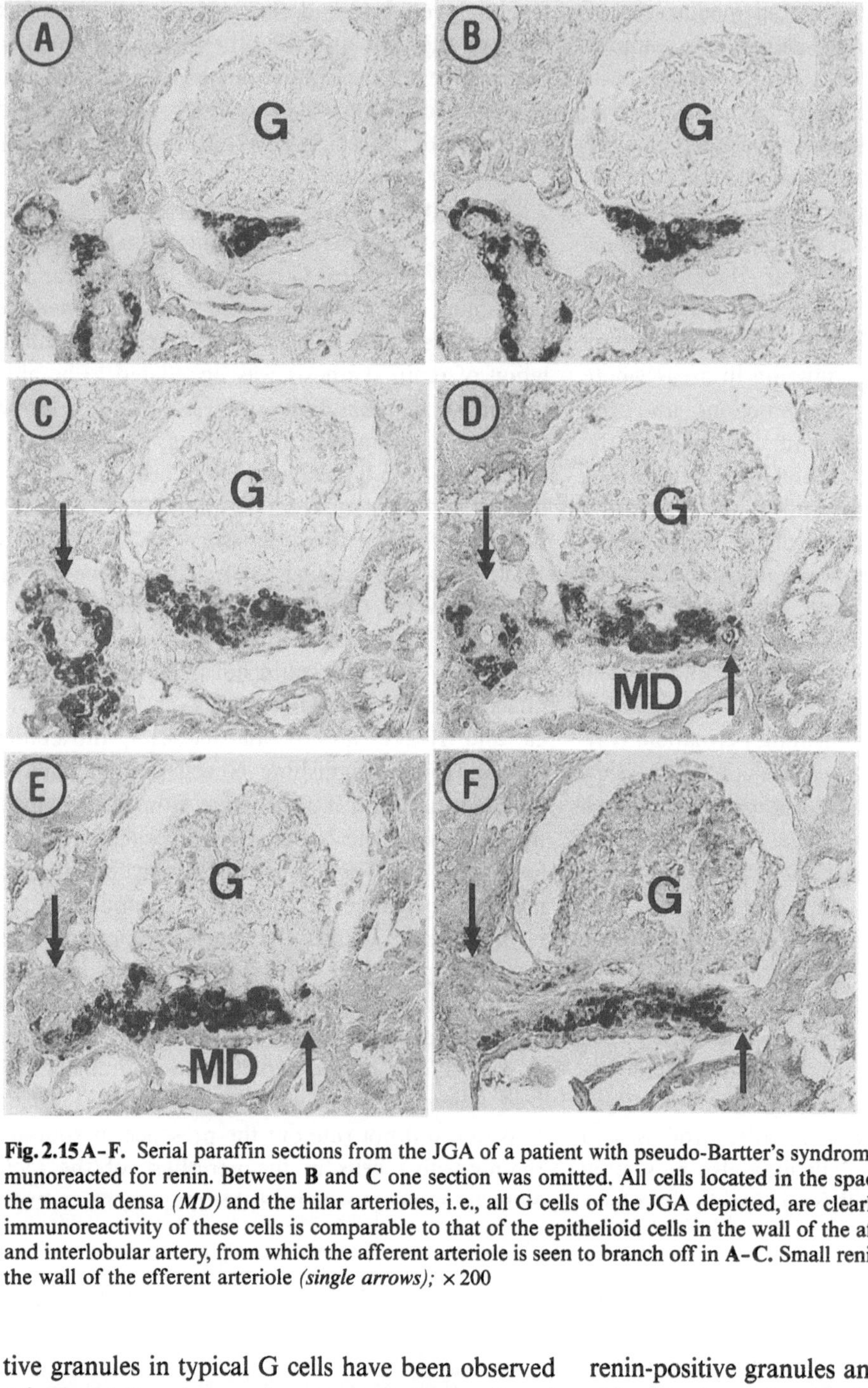

Fig. 2.15 A–F. Serial paraffin sections from the JGA of a patient with pseudo-Bartter's syndrome due to furosemide abuse, immunoreacted for renin. Between **B** and **C** one section was omitted. All cells located in the space between the glomerulus *(G)*, the macula densa *(MD)* and the hilar arterioles, i.e., all G cells of the JGA depicted, are clearly renin-positive. Note that the immunoreactivity of these cells is comparable to that of the epithelioid cells in the wall of the afferent arteriole *(double arrows)* and interlobular artery, from which the afferent arteriole is seen to branch off in **A–C**. Small renin-positive cells are also seen in the wall of the efferent arteriole *(single arrows)*; ×200

tive granules in typical G cells have been observed (cf. Fig. 4.11) and the only exocytotic images found with G cells (Fig. 2.14) were quite different from those typical for epithelioid cells (Fig. 6.1, 6.4). However, as G cells may be suspected to contain two different kinds - *specific* and *nonspecific* - granules (Biava and West 1966b), and specific markers for cells of the different vascular components of the JGA are not available, observations as these do not preclude that heavily granulated roundish cells with renin-positive granules and an ultrastructure identical with that of epithelioid cells might be transformed G cells located at the periphery of the lacis.

Recently, immunohistochemical experiments using serially sectioned biopsies from patients with pseudo-Bartter's syndrome revealed intense renin-positive reactions of all cells in the space between the glomerulus, the macula densa, and both hilar arterioles, i.e., of all G cells (Fig. 2.15). Ultrastructurally, these cells closely resembled epithelioid cells.

It was therefore concluded that renin-positive G cells might indeed be recruited from the pool of renin-negative lacis cells upon long-lasting stimulation (CHRISTENSEN et al. 1988), thus corroborating the suggestions of the authors cited above. The question remains whether an exceedingly long-lasting stimulation or some specific (G-cell directed) stimulus is necessary to induce the observed transformation of plain lacis cells into renin-producing and possibly also renin-secreting G cells. In this context it may be significant that the majority of the patients of this study had a long-lasting history of furosemide abuse.

2.5 Macula Densa

The macula densa is an oblong patch of modified epithelial cells shortly before the transition of the straight part to the convoluted part of the distal tubule facing the vascular pole of the renal corpuscle (for references see KAISSLING 1980). Because of its vicinity to both hilar arterioles and the Goormaghtigh cell field (Fig. 2.16) as well as by reason of functional aspects, the macula densa is considered to be the "tubular component" of the JGA (BARAJAS 1979).

GOLGI (1889) was the first to observe that the ascending limb of Henle's loop regularly returns to the respective renal corpuscle, coming into close contact with the vascular pole. Golgi also gave a developmental explanation for his observation, in that *"un frammento della branca ascendente ... rimane attaccato alla capsula par l'adesione contratta fin dalla prima sua origine."* PETER (1907) elaborated these findings in microdissection studies and observed that the distal tubule often shows an accumulation of nuclei and a sudden broadening where it passes through the angle between the afferent and efferent arterioles (cf. CHRISTENSEN et al. 1979). As in most cases the epithelial cells are narrower and taller and their nuclei lie closer together than in the adjacent portions of the distal tubule, the term "ma-

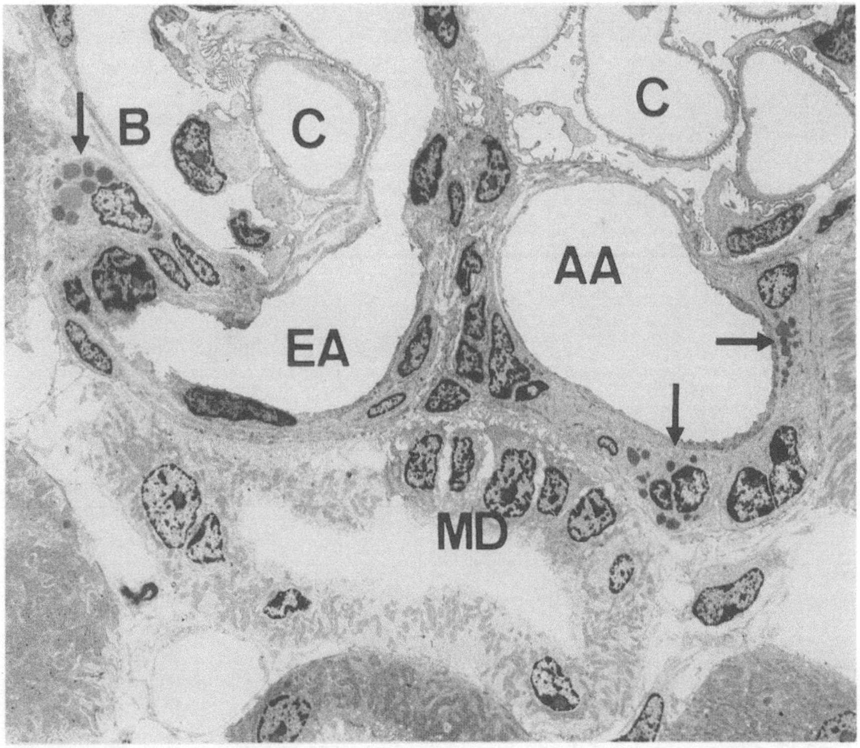

Fig. 2.16. Electron micrograph showing the vascular pole of a renal corpuscle from mouse kidney. The juxtaglomerular apparatus is shown to be composed of the afferent *(AA)* and efferent *(EA)* glomerular arterioles, the macula densa *(MD)* and the Goormaghtigh cell field or lacis situated in between. Note that the cells of the lacis, which is relatively small in mice, can hardly be distinguished from those of the hilar arterioles and the glomerular stalk. Granulated cells are seen in the media of the afferent and efferent arterioles *(arrows)*. *B*, Bowman's space; *C*, glomerular capillaries; ×5300

cula densa" was coined for the involved cell plaque by ZIMMERMANN (1933). Thin sections reveal that indeed only the part of the tubule facing the vascular pole is modified in this way, while the ultrastructure of its opposite part corresponds with that of the rest of the adjoining distal tubule (Figs. 1.2 and 2.16). Since the macula densa, as an actual "dense spot," does not cover the entire circumference of the tubule, it does not appear reasonable to speak of a "pars maculata" of the distal tubule inserted between other tubule segments.

Macula densa cells differ in several aspects from the adjoining distal tubule cells. Instead fo being cuboid, macula densa cells often have a prismatic shape, hence protruding into the tubular lumen. This is mostly underlined by their oblong palisading nuclei and their poorly developed lateral cytoplasm (Figs. 1.2, 2.16). It should be pointed out, however, that next to distinct internephron heterogeneities (FAARUP 1965) also interspecies differences in the overall morphology of the macula densa seem to exist. Thus, macula densa cells outstanding in their height are reported to occur in rabbits, whereas in the rat, mouse, and humans they often tend to have a less conspicuous nearly cuboidal form with an ovoid nucleus (Fig. 2.13) (BARAJAS and LATTA 1963a; FAARUP 1971).

In comparison with the contiguous distal tubule cells, the low expansion and order of the basal labyrinth is the most common and most conspicuous characteristic of macula densa cells: large lateral interdigitating processes are missing and the basal infoldings are fewer, shorter, and much less regularly arranged. Consequently, the smaller and less numerous mitochondria are not in columnar order as are those of the neighboring distal tubular cells, but more randomly placed in the cytoplasm between profiles of the less developed RER and free polyribosomes (OBERLING and HATT 1960a, b; THOENES 1961; BUCHER and REALE 1961b; LATTA and MAUNSBACH 1962a; BARAJAS and LATTA 1963a, b; BUCHER and RIEDEL 1965). The Golgi complex is inconspicuous and located, as in the other parts of the distal tubule, basally or laterally from the nucle-

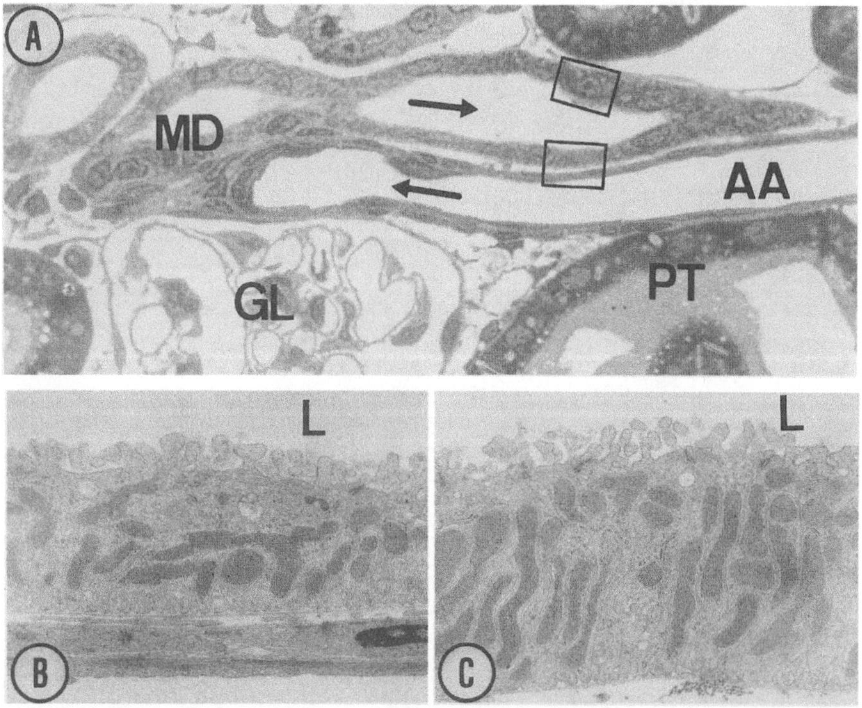

Fig. 2.17. Close proximity of the thick ascending limb of Henle's loop to the afferent glomerular arteriole **A** and deviation of the ultrastructure of the tubule by this contact **B, C**. Rat kidney; ×650 and 9700, respectively. **A** Overview. The distal tubule parallels the afferent arteriole *(AA)* over a distance of approximately 125 μm. The *arrows* indicate the (antidromic) direction of flow in both structures. *MD*, macula densa; *GL*, glomerulus; *PT*, proximal tubule. **B, C** Detailed reproduction of the areas marked in **A** at higher magnification. Note the difference in height and ultrastructure of the epithelial cells according to their microtopographical relationships. *L*, lumen of the distal tubule

us (KRIZ and KAISSLING 1985; for references from the older literature, see BUCHER and ZIMMERMANN 1960; BUCHER and KAISSLING 1973). Close to the Golgi apparatus, many smooth and coated vesicles and, sporadically, also small homogeneous electron-dense, single membran-bound organnelles are found, such as primary lysosomes, MVBs, and tubular microperoxisomes. The microvilli of the luminal plasma membrane of macula densa cells are frequently more elaborated than those of the neighboring cells of the distal tubule (BUCHER and REALE 1961b). Serial thin sections show that the transition from cells of the straight distal tubule to typical macula densa cells may not occur abruptly, but gradually by one to three intermediate cells (GORGAS 1978a). Groups of cells similar to these intermediate cells are also found in areas of close proximity between the distal tubule and the interlobular artery or a glomerular arteriole. Figures 2.17B, C and 2.18 depict the emerging deviation of the tubular ultrastructure.

Although macula densa cells proved to be capable of division, their index of autoradiographic labeling after subtotal nephrectomy was considerably lower than that of typical distal tubule cells. Thus, macula densa cells differ from the other distal tubular cells not only in appearance, but also in their pattern of proliferation (ROMEN et al. 1978).

HESS et al. (1958) were the first to report on high glucose-6-phosphate dehydrogenase activities of the macula densa (for further literature see NØRGAARD 1979). Using a quantitative fluorimetric method, NØRGAARD (1979) demonstrated the activity of this enzyme in the macula densa region to be about twice that of proximal or distal tubular cells. KROMPECHER-KISS and BUCHER (1977), comparing macula densa cells with the cells of the distal convolution, reported on specific differences in the histochemical activities of various other dehydrogenases. In contrast to adjacent distal tubule cells, no Na-K-ATPase activity could be demonstrated in cells of the macula densa (BEEUWKES et al. 1975). Interestingly, macula densa cells were also shown to be free of Tamm-Horsfall glycoprotein, although the luminal (and basolateral) surface of all other distal tubular cells showed positive immunoreactions (SIKRI et al. 1979, 1981; HOYER et al. 1979; HOYER and SEILER 1979). Likewise EGF (epidermal growth factor) and prepro-EGF mRNA, traced in the adjacent distal tubular cells, were found to be absent in macula densa cells (SALIDO et al. 1986a; BARAJAS et al. 1988).

In connection with the TGF and the so-called macula densa signal, the assumption arose that especially leaky tight junctions were present between macula densa cells as compared with those in the remaining distal tubule (GIACOMELLI and WIENER 1976). HATT (1967) had already suggested that the loose character of the intercellular bonds in the region of the macula densa may facilitate direct contact between the tubular contents and the lacis. This aspect was examined in detail by SCHILLER and TAUGNER (1979) using freeze fractures and - after intravenous injection of horseradish peroxidase (HRP) as an exogeneous tracer - thin sections from several species. The freeze fractures revealed only insignificant differences in the number of strands and the apicobasal depth of the occluding junctions when the macula densa region was compared with the rest of the distal tubule (Fig. 2.19, Table 2.1). In

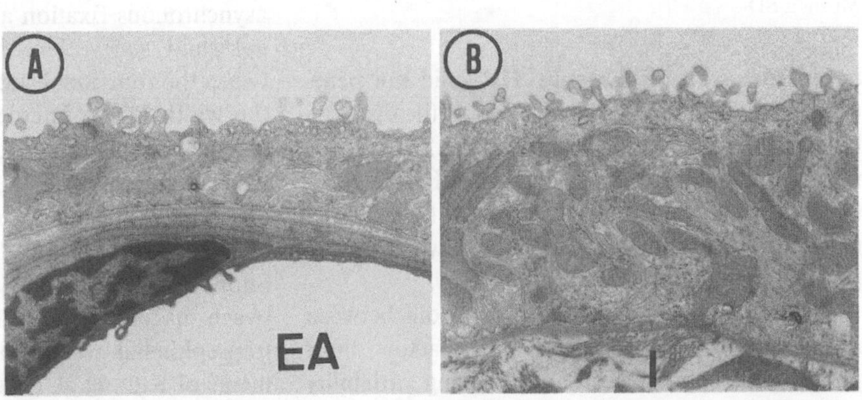

Fig. 2.18 A, B. Thick ascending limb of Henle's loop from *Tupaia* kidney, about 80 μm upstream of the JGA. **A** Indentation and ultrastructural alteration of the distal tubule as a consequence of its intimate contact with the corresponding efferent arteriole *(EA)*. **B** Unaltered appearance of the opposite epithelial cells bordering on the cortical interstitium *(I)*; ×10000

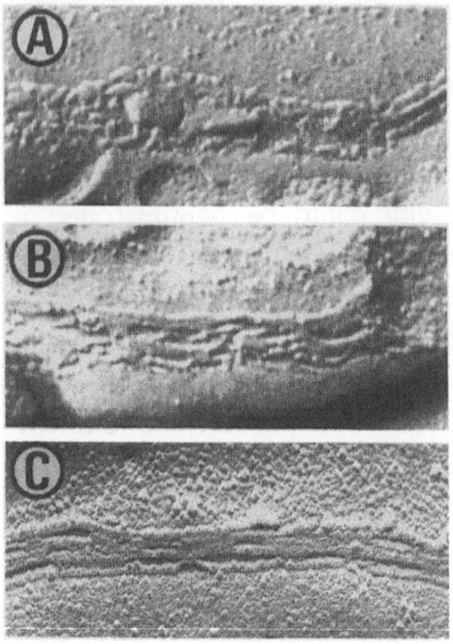

Fig. 2.19. Tight junctions between the epithelial cells of the macula densa **A**, the opposite side of the distal tubule **B**, and the thick ascending limb of Henle's loop **C** in rat kidney. Note the similarity of the junctions with respect to their apicobasal depth and the number of junctional strands. Freeze-fracture replicas; ×63000

Table 2.1. Tight junctional morphology of the different distal tubular segments including the macula densa

Species	Distal tubular segment	Junctional morphology		
		No of strands	Depth (nm)	*n*
Dog	Pars recta	6.0±1.1	122±35	15
	Macula densa	5.0±0	142±15	3
	Pars convoluta	5.7±1.6	128±42	8
Rat	Pars recta	4.8±1.4	105±38	26
	Macula densa	4.8±0.8	119±29	5
	Pars convoluta	5.8±2.1	133±51	26
Tupaia	Pars recta	4.6±1.2	114±45	9
	Macula densa	5.3±0.6	108±23	3
	Pars convoluta	4.4∓1.5	115±35	4

Means±SD

accordance with these results, HRP did not penetrate the apical junctions in any part of the distal tubule, including the macula densa (Fig. 2.20). These results were interpreted as suggesting that the TGF is more likely to be mediated by the transcellular reabsorption of solutes than by passive diffusion through the paracellular shunt pathway.

The tightness of the occluding junctions between macula densa cells is, among other factors, most probably of importance for the striking variability in the site and extension of the lateral intercellular spaces in the macula densa region. As pointed out, macula densa cells do not interdigitate with each other by large lateral processes. Thus their intercellular spaces run more or less directly from the occluding junctions toward the basis of the epithelium. However, LATTA and MAUNSBACH (1962a) and BARAJAS and LATTA (1963a) showed that the lateral plasma membranes at the base of macula densa cells - in contrast to the rest of the distal tubule - tend to separate, forming conspicuous extracellular compartments (Figs. 1.2, 2.16). Characteristically, these lateral intercellular spaces may be seen to vary in width depending on fixation conditions and - with proper perfusion fixation - according to the prevailing functional state of the kidney. In the presence of furosemide and in mannitol diuresis, i.e., during inhibition of tubular reabsorption and/or elevation of the osmolality of the luminal fluid, the lateral cell membranes of macula densa cells are narrowly apposed. In this condition, the slender lateral microplicae or microvilli are seen to lie close to corresponding processes from neighboring cells. If, in contrast, the tubular reabsorption is not impaired, more or less dilated intercellular spaces are encountered (KAISSLING and KRIZ 1982). Characteristically, these dilatations are strictly limited to the area of the macula densa in a well-preserved, i.e., adequately fixed kidney. In analyzing this finding, it is necessary to consider that even a conceivably excellent perfusion fixation does not lead to a sudden and synchronous halt to all tubular transport processes and that the base of the macula densa along with the adjoining Goormaghtigh cell field belongs to those particular regions of the kidney cortex which, lacking capillarization, are probably reached last by the fixative. It could therefore not be ruled out that the configuration of the intercellular spaces observed in thin sections is altered by asynchronous fixation allowing for some preterminal fluid shift. Nevertheless, the correlation between the functional state of the in vivo kidney and the width of the intercellular spaces after fixation lead to the suggestion that the macula densa is a water-permeable cell plaque within the otherwise water-impermeable ascending limb of Henle's loop (KAISSLING and KRIZ 1982; ALCORN et al. 1986). Support for the notion that the lateral spaces between macula densa cells may act as conduits for transepithelial water flow came from the experiments of KIRK et al. (1985). These authors perfused isolated distal tubuli with attached glomeruli under the differential interference-contrast microscope. Reduction of the luminal osmolality from 290 to

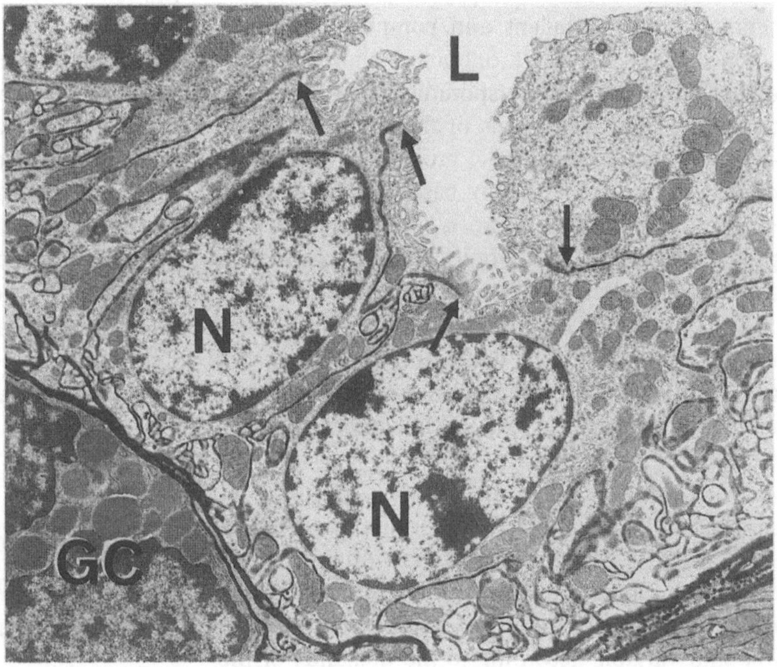

Fig. 2.20. Penetration of horseradish peroxidase *(HRP)* in the macula densa region of the mouse. The tracer enters the intercellular spaces by a retrograde route and stops at the tight junctions *(arrows)*. Immersion fixation after 5 min continuous i.v. infusion of HRP. *L*, lumen of the distal tubule; *N*, nuclei; *GC*, granulated cell; ×5400

70 mOsmol/kg by removing NaCl resulted in a reversible dilatation of the lateral intercellular spaces specific to the macula densa region. With regard to the interpretation of their findings, the authors pointed out that the degree of expansion is a function of the relative resistances to inflow and outflow to and from the lateral spaces. Thus, the observed expansion of the intercellular spaces limited to the macula densa may for one thing be the result of a relatively high hydraulic conductivity of the epithelium in this region. Alternatively, these dilatations could be due to an outstanding large resistance to outflow from the intercellular spaces. In Sect. 3.3 it will be shown that a high resistance to flow from the basal area of the macula densa into the direction of the remote peritubular capillaries is very likely to exist due to the interposition of the lacis. Fluid balance in the region of the JGA may in addition be under the influence of a competitive fluid flow from the site of the glomerular stalk. Thus further studies were required to assess the hydraulic conductivity of the macula densa cell plaque. In isolated and perfused rabbit tubules, a relatively low water permeability of the MD plaque similar to that in the cortical collecting duct has recently been observed by GONZALEZ et al. (1988).

In view of the proximity between the macula densa and the other components of the JGA (MCMANUS 1947), the existence of intercellular continuities and, hence, the possible passage of substances from one cell type to the other had been envisaged (HARTROFT and NEWMARK 1961). Subsequently, electron microscopic studies showed that the macula densa cells, although sending out basal processes, are definitely separated from the cells of the vascular components of the JGA (BARAJAS and LATTA 1963a). Later, freeze fractures revealed that macula densa cells, like the cells of the entire distal tubular system, showed no gap junction coupling whatsoever (SCHILLER and TAUGNER 1979).

Another topic of special interest in connection with the postulated functional interplay between the macula densa and the vascular components of the JGA has been the existence and morphology of the basal lamina at contact sites between these structures. MCMANUS (1947) in his light microscopic studies suspected gaps in the reticulum of the tubular basement membrane and suggested that these discontinuities might facilitate the passage of substances between macula densa cells and cells of the other JGA components. Electron microscopic studies showed that a continuous basal lamina separates these structures (BUCHER and ZIMMERMANN 1960; THOENES 1961; BARAJAS and LATTA 1963a; TISHER et al. 1968). At the same time, it was observed that the contact zone between the macula densa and the vascular components of the JGA presents a rather turbulent and complex picture in some section planes. On one hand, short branches of the basal lamina may sporadically protrude between the basal extensions of the tubule cells; on the other hand, cytoplasmic projections of macula densa cells accompanied by basal lamina material may often penetrate between Goormaghtigh cell processes, so that, as a whole, the impression of a tubulovascular labyrinth is given in this area.

BARAJAS (1970, 1971) noted that the basal laminae between the macula densa and the Goormaghtigh cell field or the efferent arteriole, respectively, tend to form networks, the so-called complex contacts, interpreted as being irreversible structures. Apart from these, there are simple appositions of the basal laminae, which, according to BARAJAS (1971, 1979, 1981), are present most often between the distal tubule and both glomerular arterioles; these simple contacts were suggested to be reversible and variable according to the demands of the TGF (cf. Sects. 2.6, 7.3). According to GORGAS (1978 a), fusions between the basal laminae of tubular and vascular structures may be found not only at the hilus of the renal corpuscle, but also further up- and downstream along the afferent and efferent arterioles, respectively.

In connection with the tubulovascular interactions suggested to exist in the region of the JGA, also the size of the macula densa in terms of cell number and/or the macula densa basal area is of interest. The pertinent data seem to be subject to remarkable internephron and interspecies heterogeneities. In rats, FAARUP (1965) found 7-39 macula densa cells in subcapsular and 5-20 macula densa cells in juxtamedullar nephrons ($n = 15$ and 10, respectively). In humans, on the other hand, the mean number of macula densa cells found in 31 randomly selected nephrons was 51 (CHRISTENSEN et al. 1975). The macula densa basal area was $834 \pm 228\ \mu m^2$ ($n = 46$) and $1271 \pm 300\ \mu m^2$ ($n = 60$) in Sprague-Dawley and BD9 rats, respectively (CHRISTENSEN and BOHLE 1978). In humans, the macula densa basal area was close to $5300\ \mu m^2$ (CHRISTENSEN et al. 1976).

In rats, narrowing of the renal artery was followed by an enlargement of the macula densa and an increase of the macula densa cell population (SCHNEIDER and THOENES 1971). In humans with primary hyperaldosteronism, the number of macula densa cells was decreased, in patients with pseudo-Bartter's syndrome increased, in parallel with the number of juxtaglomerular cells (HELBER et al. 1970). In Brattleboro rats with hereditary diabetes insipidus, the macula densa cell number and the macula densa basal area were significantly greater than those of Long Evans control rats (CHRISTENSEN and TAUGNER, unpublished observations). These observations seem to imply that the macula densa is subject to hypertrophy or atrophy depending on stimulation conditions. However, the functional implications of these changes, which, as a rule, are paralleled by similar changes of the Goormaghtigh cell field, are unknown.

2.6 Areas of Close Proximity Between Components of the Juxtaglomerular Apparatus

The transmission of the hypothetical macula densa signal as part of the TGF called for same spatial relationship between the tubular and vascular components of the JGA. As a consequence, the nature and extension of their mutual marginal surfaces or "contacts" have been thoroughly investigated.

One of the early proposals was, that a typical macula densa may only develop in regions where the distal tubule is in contact with the afferent arteriole and the Goormaghtigh cell field (BECHER 1949; cf. CHRISTENSEN et al. 1979). However, contact areas between the distal tubule and the efferent arteriole had already been described by MCMANUS (1942). Later, serial sectioning followed by reconstruction allowed for a more detailed view of the histotopographical interrelations between the distal tubule and both the afferent and efferent glomerular arterioles, although remarkable divergences still seem to exist.

BARAJAS and LATTA (1963 a), using series of semithin sections from rat kidneys, found an extended, constant and very close association between the thick ascending limb of Henle's loop and the efferent arteriole, whereas only short contacts between the distal tubule and the afferent vessel were observed. The macula densa was also found to be in more intimate contact with the efferent arteriole than with the afferent.

In the reconstructions by FAARUP (1965, 1971) using paraffin sections of rat kidneys, the distal tubule was regularly found to be close to the efferent and then, beyond the Goormaghtigh cell field, to the afferent arteriole for 15-150 and 5-165 μm, respectively (cf. Figs. 2.1 C, D; 2.17; 4.10). As the close proximity of the hilar arterioles with the macula densa measured maximally 35 μm, most of these tubulovascular contacts again were situated outside the region of the JGA. The macula densa had contact with the efferent arteriole in only 12 out of 25 cases of reconstructed nephrons, and in 16 of these 25 nephrons with the afferent arteriole. FAARUP (1971) also observed considerable internephron variations concerning the size of the macula densa, consisting, in the rat, of 5-39 cells; as an average, the macule densa of subcapsular nephrons was larger than that of juxtamedullar nephrons, amounting to 21 versus 13 cells.

Serial section electron microscopy and three-dimensional reconstruction was applied to the JGA of the rat by BARAJAS (1970). In this study, the distal tubule was found to have extensive contacts with the efferent arteriole and the Goomaghtigh cell field, but very little with the afferent arteriole. It was emphasized, however, that the absence of contact on the electron microscopic scale does not exclude close proximity as was at times the case between the afferent arteriole and the distal tubule. As mentioned earlier, two different types of contact have been described by this author: a complex, "irreversible" type with fusion of both basal laminae, and a simple "reversible" type of contact, represented by the mere apposition of the laminae. Irreversible contacts of the distal tubule were reported to be realized with the Goormaghtigh cell field and the proximal part of the efferent arteriole, reversible contacts with both glomerular arterioles. The term "distal tubule" was used instead of macula densa in this study, as the accumulation of nuclei characteristic of the macula densa was not always clear at the electron microscope level.

BARAJAS (1970, 1971) also noted that the majority of granulated cells were not in contact with the distal tubule. Out of 50 granulated cells encountered in the afferent arterioles of 4 JGAs, only 2 had such a contact; out of 25 granulated cells found in the corresponding efferent arterioles, only 9. These observations regarding the distribution of granulated cells in the JGA led BARAJAS to propose a dynamic model of JGA function based on the distinction between permanent and reversible contacts discussed in Sect. 2.4. The complex type of contact was thought to be a permanent anchoring of the macula densa to the G-cell field and to the beginning of the postglomerular arteriole (Fig. 2.21). The type of contact represented by simple appositions of the involved basal laminae, on the other hand, was interpreted as being reversible. BARAJAS (1970) therefore proposed that functional conditions may vary the extent of contact, as shown in Fig 2.21. As the gran-

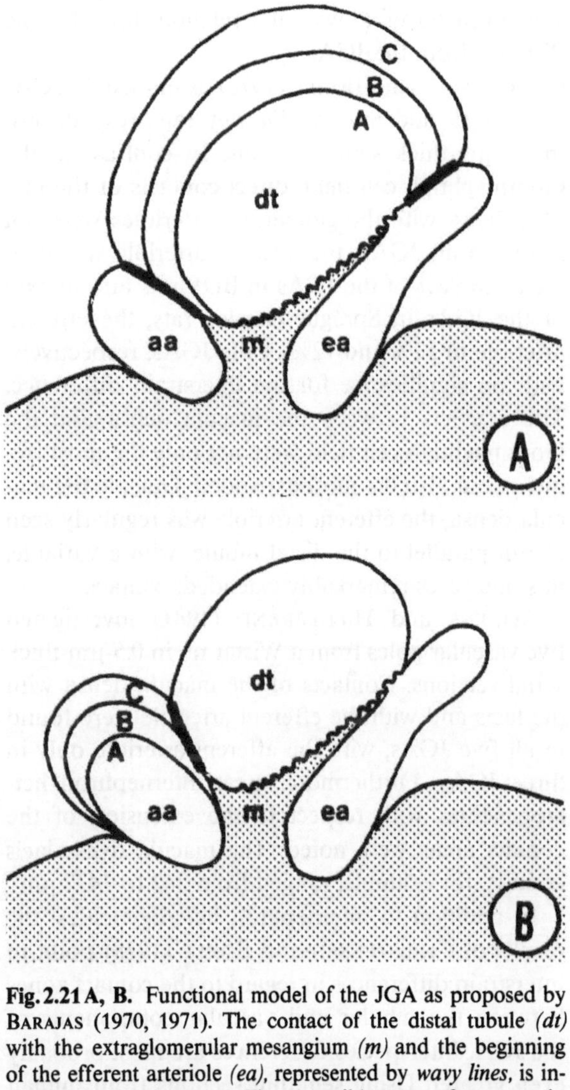

Fig. 2.21 A, B. Functional model of the JGA as proposed by BARAJAS (1970, 1971). The contact of the distal tubule *(dt)* with the extraglomerular mesangium *(m)* and the beginning of the efferent arteriole *(ea)*, represented by *wavy lines*, is interpreted as permanent. The reversible type of contact, in contrast, is marked by *heavy lines*. **A** Upon expansion of the distal tubule *(lines B, C)*, the area of "reversible" contacts with the hilar arterioles increases. **B** Changes in contact between the distal tubule and the afferent arteriole *(aa)* may also result from changes in the volume of this vessel. (With permission of the author and The American Association for the Advancement of Science)

ulated cells were found mainly at the site of reversible contacts, changes in the extent of contact were proposed to influence renin release.

The results of BOHLE et al. (1970) are at some variance to those of BARAJAS (1970) and FAARUP (1965, 1971). In 2-µm-thick serial sections of 30 vascular poles from rat kidneys, the authors found direct contacts of the macula densa with the afferent arteriole in 25 cases, yet contacts with the efferent vessel only in 4 out of 30 vascular poles; contacts of the macula densa with the Goormaghtigh cell field were found to be the most constant. No comment was given on tubulovascular relationships at some distance from the JGA.

CHRISTENSEN and BOHLE (1978) examined 25 JGAs from BD9 and Sprague-Dawley rats, respectively, in 1.5-µm-thick serial sections. In contrast to the Goormaghtigh cell field, direct contacts of the macula densa with the glomerular arterioles were not found in all JGAs: the afferent arteriole was contacted in 92% of the JGAs in BD9 rats and in 59% of the JGAs in Sprague-Dawley rats, the efferent arteriole in 52% and 72% of the JGAs, respectively, with no significance for the interstrain difference. The contacts between the macula densa and the Goormaghtigh cell field were not only the most frequent, but also the most expanded. Outside the macula densa, the efferent arteriole was regularly seen to run parallel to the distal tubule, with a variable, in some cases remarkably extended, contact.

AEIKENS and HILDEBRAND (1981) investigated five vascular poles from a Wistar rat in 0.5-µm-thick serial sections. Contacts of the macula densa with the lacis and with the efferent arteriole were found in all five JGAs, with the afferent arteriole only in three JGAs. Furthermore, great internephron heterogeneities with respect to the extension of the contact areas were noted. The macula densa-lacis contact, for example, ranged from 242 to 1427 μm^2.

In addition to internephron heterogeneities, the results of CHRISTENSEN and BOHLE (1978) point to interstrain differences in regard to the contact zones between the tubular and vascular components of the JGA. Interspecies differences are most probably even greater. Using semithin sections from human kidneys, CHRISTENSEN et al. (1975) not only found the macula densa to be in regular contact with the Goormaghtigh cell field, but also with the afferent arteriole in 47 and with the efferent arteriole in 46 of the 49 JGAs examined. In both rats and humans, the contact of the macula densa with the Goormaghtigh cell field was most constant as well as most extensive (cf. BOHLE et al. 1970).

In summary, these observations are in agreement in that a regular, especially close and mostly extensive contact exists between the macula densa and the Goormaghtigh cell field (Fig. 2.13). As far as the authors refer to this topic, they also agree that there are, in addition, extensive tubulovascular contacts outside the JGA. Their results differ when contacts between the macula densa and the glomerular arterioles are concerned. Apart from interspecies and interstrain differences, some of these divergences might be explained by the apparently considerable internephron heterogeneities noted by every author, in relation to which data from diligent morphometric evaluations are necessarily scarce. Another cause of divergences may be related to the definition of the "contacts" evaluated. BARAJAS (1978), who distinguished between "contact" and "close proximity," pointed out that, because of differences in optical resolution, areas that might appear to be in contact by light microscopy using paraffin sections would in reality be micrometers apart. The tubulovascular relationships can also be complicated by the hernia-like protrusions of the distal tubule described by PETER (1907), which in the area of the

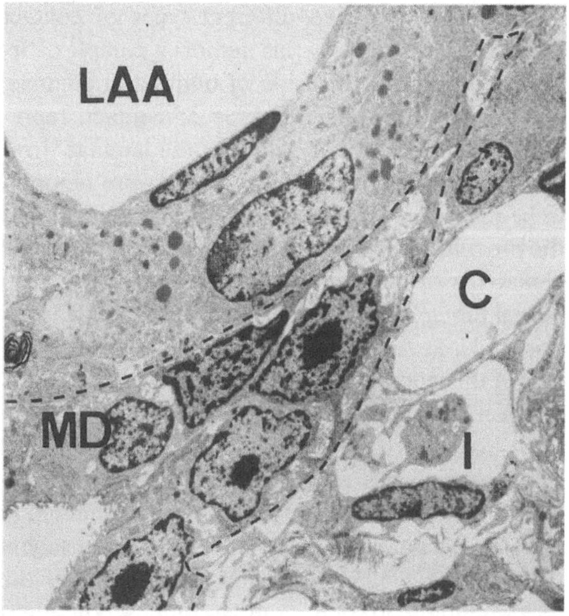

Fig. 2.22. Hernia-like protrusion of the distal tubule in the region of the macula densa *(dashed line)* in contact with the wall of the afferent arteriole. *MD*, macula densa; *LAA*, lumen of the afferent arteriole; *C*, peritubular capillary; *I*, interstitium. Mouse kidney; ×3900

JGA may be covered by macula densa epithelium (Fig. 2.22; cf. CHRISTENSEN et al. 1975, 1979).

Irrespective of differences in the mutual contact areas and in the course of the distal tubule in relation to the hilar arterioles (BARAJAS and LATTA 1963a; FAARUP 1965; CHRISTENSEN et al. 1979), the direction of flow in the distal tubule was constantly found to be antidromic to that in the glomerular arterioles (BARAJAS and LATTA 1963a; FAARUP 1965, 1971; CHRISTENSEN et al. 1975). At this point, it should be remembered that GOLGI had already depicted although not commented on many of the essentials regarding the spatial relationships between the distal tubule and the glomerular arterioles in 1889 (Fig. 1.1). Furthermore, some of the tubulovascular contacts outside the JGA discussed in detail in Sect. 3.1 may be derived from Golgi's presentation.

For the sake of completeness additional data relating to the contacts between the Goormaghtigh cell field and the glomerular arterioles should be presented here. According to CHRISTENSEN and BOHLE (1978), the contact areas - as determined in two different rat strains - are about 250-300 μm^2 for the afferent and 180-260 μm^2 for the efferent arteriole, being thus smaller than the contact areas of the lacis with the macula densa (about 500-750 μm^2). In man the macula densa as well as the Goormaghtigh cell field are larger than in the rat. Correspondingly, also the contact areas between the lacis and the afferent, viz., efferent arterioles and the macula densa, are more extensive, amounting to about 1000, 600, and 2000 μm^2 respectively (CHRISTENSEN et al. 1975, 1978).

Experiments relating to the question of whether and how these heterogeneities between the contact areas linking individual JGA elements are of any functional significance, e.g., for the gain of the feedback loop, have not been reported so far.

2.7 Gap Junction Coupling Between Cells of the Juxtaglomerular Apparatus

BIAVA and WEST (1966a) described focal contact zones between human juxtaglomerular cells with the formation of "quintuple-layered junctional complexes," at that time considered to be characteristic of "nexuses." Later, BOLENDER (1970) in his extensive analysis of the mouse JGA, showed a micrograph suggestive of gap junction coupling between granulated cells. Using freeze fractures and thin sections, PRICAM et al. (1974) and BOLL et al. (1975) unequivocally demonstrated the presence of nexal contacts in the JGA of rats for Goormaghtigh cells and epithelioid cells, respectively. Figures 2.23 A, B are examples of gap junction coupling between epithelioid cells of the juxtaglomerular afferent arteriole in mouse kidney. Later, FORSSMANN and TAUGNER (1977) found evidence for mutual gap junction coupling between all cell types of the vascular JGA components, i.e., granulated cells, smooth muscle cells of both glomerular arterioles, and Goormaghtigh cells. PRICAM et al. (1974) had already described gap junctions between mesangial cells, while TAUGNER et al. (1978b), studying the vascular pole of the renal corpuscle in rabbits, found a wide avenue of gap junctions connecting the Goormaghtigh cell field with the glomerular stalk and the intercapillary mesangium. In one freeze fracture plane, 74 macular gap junctions were encountered (Fig. 2.24). Gap junctions at these different locations were found not only to bridge two different cells, but to occur also between processes of the same cell (so-called reflexive gap junctions), known among others, for vascular smooth muscle cells of the aorta (IWAYAMA 1971). These results inferred the existence of a metabolic and/or electrotonic coupling between all the above-mentioned cellular components of the glomerulo-juxtaglomerular complex, allowing them to act as a functional syncytium relevant for renin release and/or glomerular hydrodynamics.

Initially, it was assumed that in analogy to the juxtaglomeruler portion of the afferent arteriole, gap junction coupling may also be present farther upstream, i.e., between the smooth muscle cells of the proximal afferent arteriole and the interlobular artery. However, in a comprehensive study of thin sections and freeze fractures from control rat and mouse kidneys, this assumption could not be confirmed (TAUGNER et al., unpublished observations). Equivocal results were obtained in thin sections (Fig. 2.25). In freeze fractures, only relatively short rows of gap junction particles were found on the membranes of smooth muscle cells of the proximal afferent arteriole and interlobular artery. Figure 2.23 C shows the most elaborate of these gap junctions encountered in mice while inspecting a

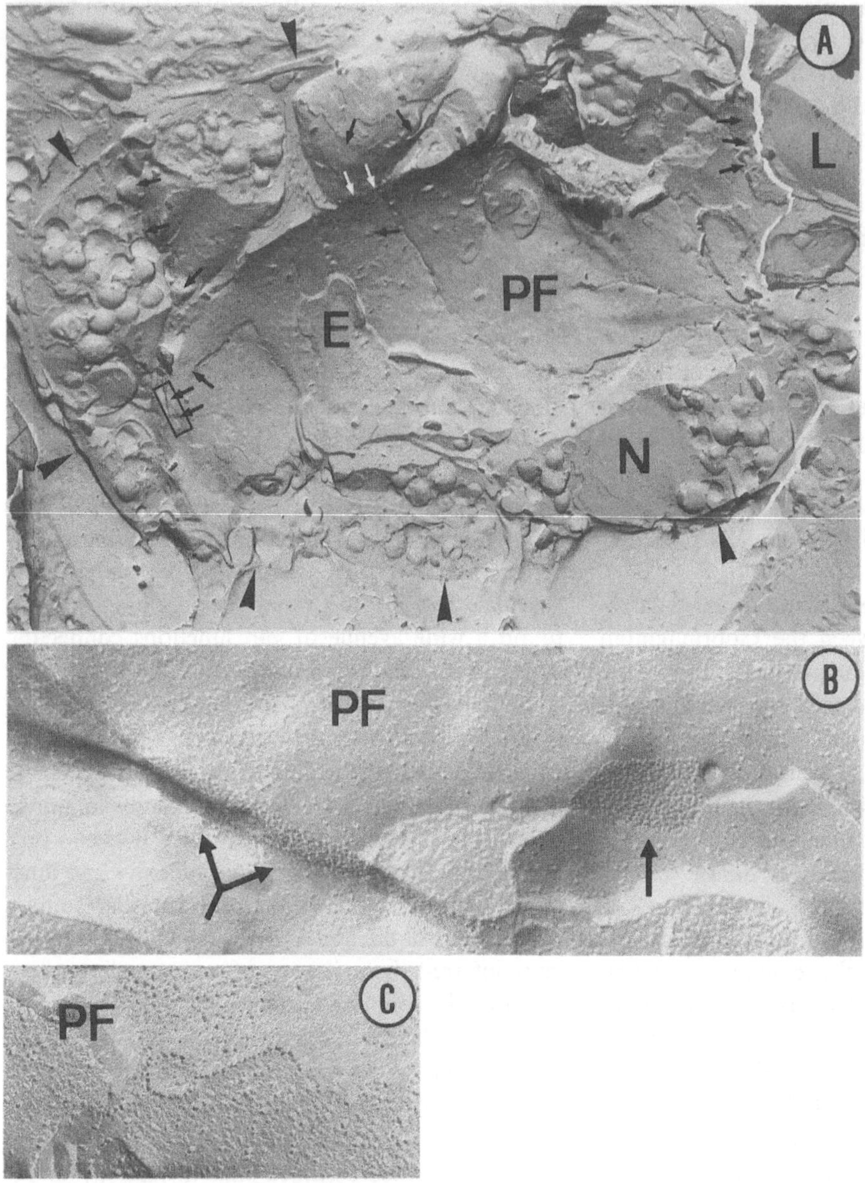

Fig. 2.23. **A, B** Gap junction coupling *(arrows)* between media cells of the juxtaglomerular afferent arteriole in mouse kidney as revealed by freeze fracturing. **A** overview, **B** area encircled in **A** at higher magnification. In **A** the outer contours of the vessel are delineated by *arrowheads*. *L,* lumen of the arteriole close to its entry into the glomerulus. *PF,* P-face of the luminal membrane of media cells, by their content of secretory granules identified as epithelioid cells. *E,* small remnant of endothelial cell, bypassed by the fracture process. *N,* nucleus of epithelioid cell; ×3300. In **B,** two of the gap junctions shown in **A** at higher magnification. *PF,* P-face of epithelioid cell; ×72000. **C** Smooth muscle cell in the media of an interlobular artery from mouse kidney. Gap junction-like linear arrangement of intramembrane particles. *PF,* P-face of the cell membrane. Freeze fracture; ×72000

membrane area of about 750 μm^2. In rats, similar results were obtained.

After stimulation of the RAS, e.g., by adrenalectomy, groups of granulated cells may be found far upstream of the JGA in the proximal afferent arteriole and the interlobular artery. Characteristically, these granulated cells emerging from poorly coupled smooth muscle cells after metaplastic transformation are found to be equipped with macular gap junctions. Thus, metaplastic transformation seems to include definite changes in the quality of gap junction coupling of the involved media cells.

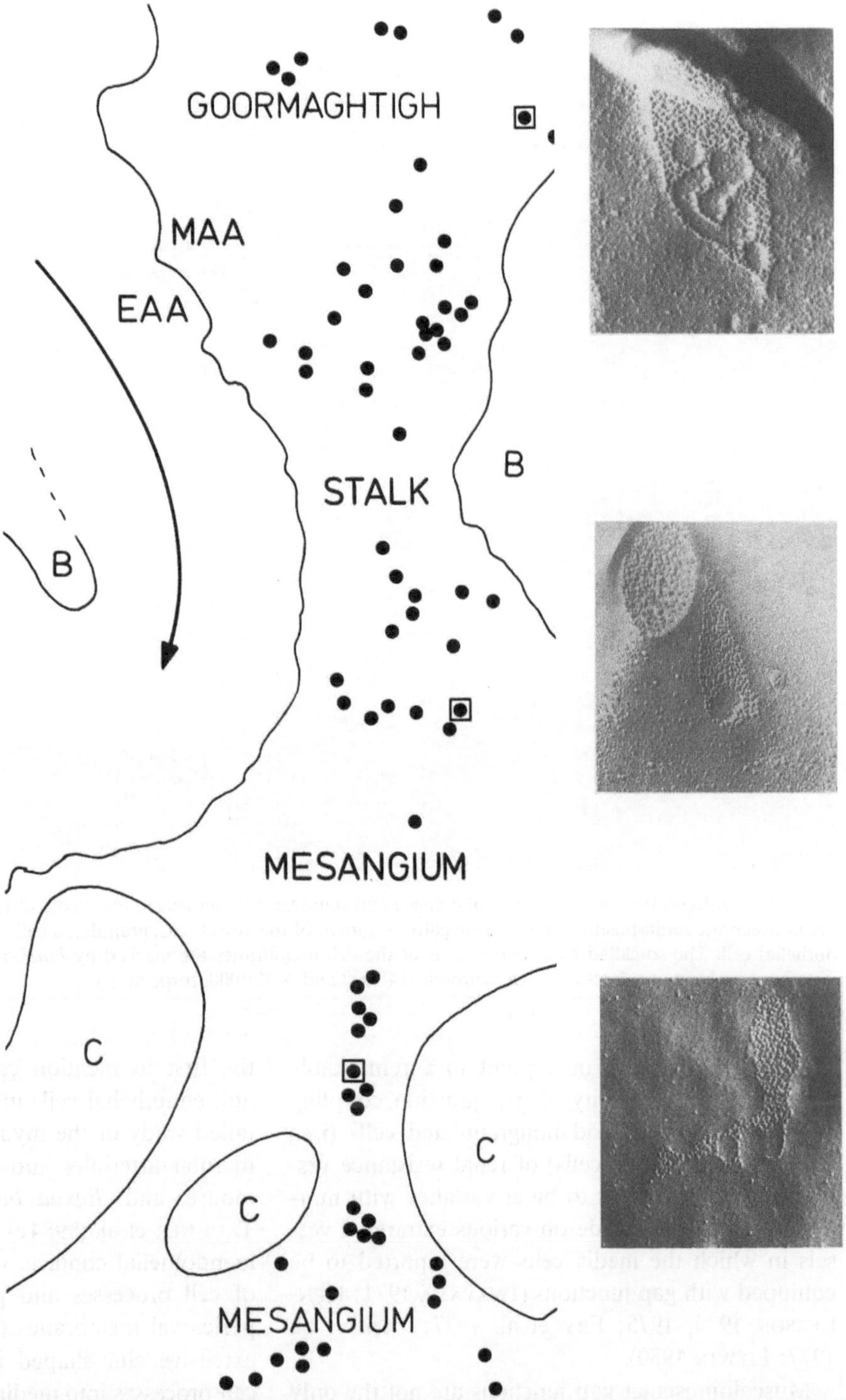

Fig. 2.24. Sketch of the glomerular stalk as found in a freeze-fracture replica from rabbit kidney *(left)* with examples of gap junctions encountered in the fracture plane *(right)*. The fracture exposed the Goormaghtigh cell field *(above)*, the glomerular stalk, and the glomerular mesangium *(below)*. The positions of the 74 gap junctions identified in the fracture plane are marked by *dots*. The three surrounded areas correspond to the examples of gap junctions shown on the right. *MAA*, media of the afferent arteriole; *EAA*, endothelium of the afferent arteriole; *arrow*, entrance of the afferent arteriole into the renal corpuscle; *B*, Bowman's capsule; *C*, glomerular capillaries. Sketch, ×2700; gap junctions from above to below; ×51000, 50000, and 45000, respectively

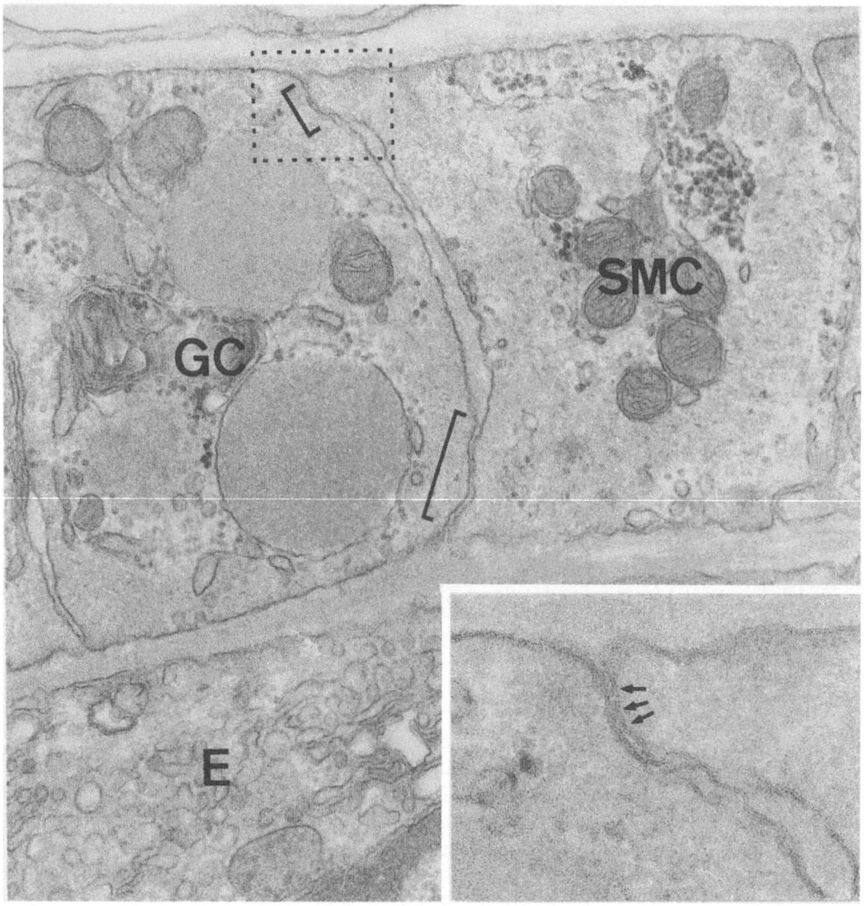

Fig. 2.25. Configuration of the extracellular space and putative cell contacts in the media of rat afferent arteriole at the boundary between the renin-positive and renin-negative segment of the vessel. *GC*, granulated cell; *SMC*, smooth muscle cell; *E*, endothelial cell. The so-called close appositions of the cell membranes are marked by *brackets*. At higher magnification *(inset)* putative membrane contacts are seen *(arrows)*; ×40000 and ×120000, respectively

These findings not only point to a remarkable difference in the quality of gap junction coupling between granulated and nongranulated cells (i.e., plain smooth muscle cells) of renal resistance vessels. They also appear to be at variance with morphological studies made on various extrarenal vessels in which the media cells were reported to be equipped with gap junctions (IWAYAMA 1971; HENDERSON 1974, 1975; FRY et al. 1977; TANI et al. 1977; LITWIN 1980).

Musculomuscular gap junctions are not the only possibility for gap junction coupling in the wall of blood vessels. Myoendothelial and endoendothelial junctions may also be involved (RHODIN 1967; SIMIONESCU et al. 1975, 1976).

BIAVA and WEST (1966a), JACOBSEN et al. (1966), NEWSTEAD and MUNKACSI (1969), MOFFAT and CREASEY (1971), ROJO-ORTEGA et al. (1973a) were the first to mention cell contacts between media and endothelial cells in kidney vessels. A more detailed study of the myoendothelial contacts in glomerular arterioles and interlobular arteries of rat, mouse, and *Tupaia belangeri* was performed by TAUGNER et al. (1984c). The shape of these musculoendothelial contacts varied from flat appositions of cell processes and pericarya (Fig. 2.26) or two pericaryal membranes (Fig. 2.27A) to more or less extensive club-shaped indentations of endothelial cell processes into media cells (Figs. 2.27B, 2.28). As an exception, media cell processes indenting an endothelial cell were seen (Fig. 2.27C). In the interlobular artery, club-shaped contacts predominated. Flat appositions, in contrast, were somewhat more numerous in the afferent arteriole and almost exclusively found in the efferent vessel. In the juxtaglomerular portion of the preglomerular arteriole, the

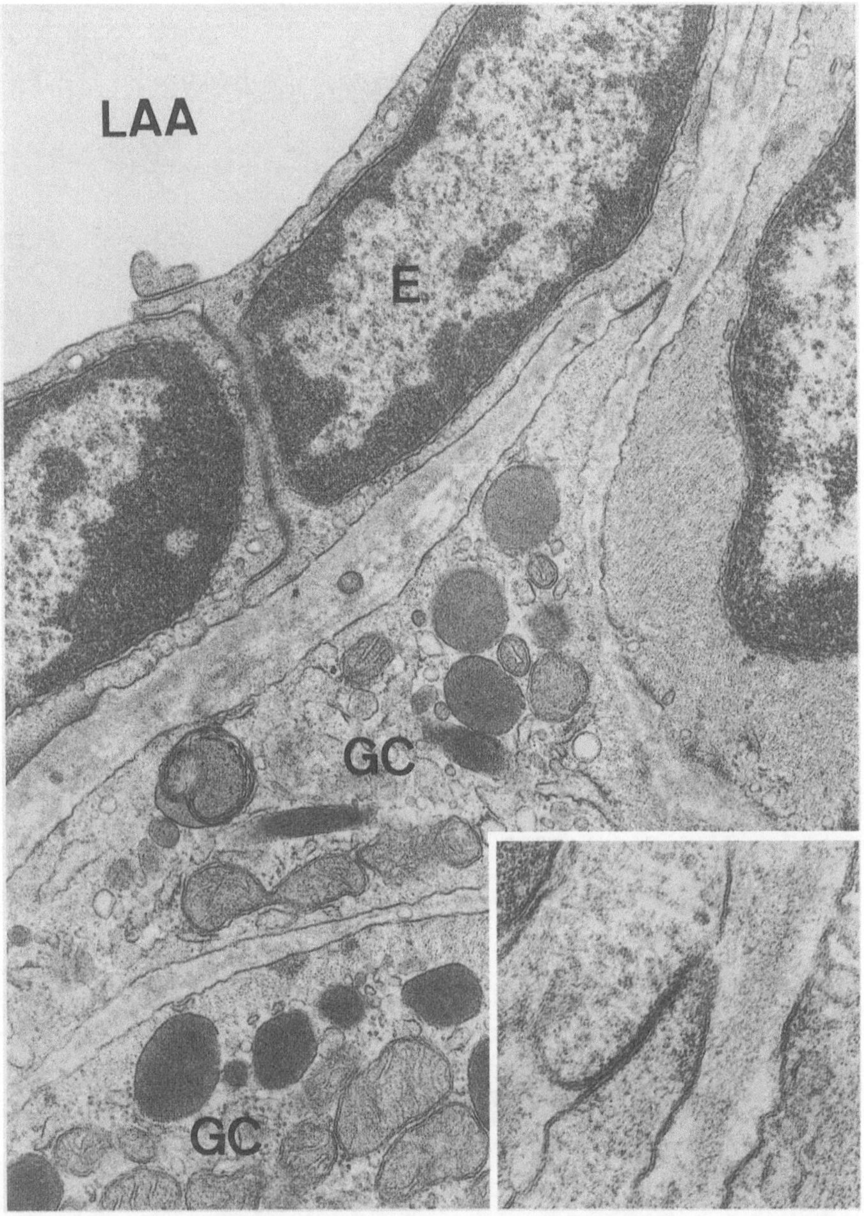

Fig. 2.26. Myoendothelial contact in the juxtaglomerular portion of the afferent arteriole in *Tupaia* kidney. The "flat apposition," established between the intimal endothelium and an epithelioid cell of the media, is equipped with a gap junction. *LAA*, lumen of the afferent arteriole; *E*, endothelial cell; *GC*, granulated cells; × 40000. *Inset:* Higher magnification of the cell contact; × 120000

contacts were established between endothelial and granulated cells. As gap junctions seemed to be a standard attribute of the membrane contacts between intima and media cells in kidney resistance vessels, it was proposed that the myoendothelial contacts, besides allowing the detection of mechanical (e.g., autoregulatory) stimuli (BOUSKELA and WIEDERHIELM 1979; JOHNSON 1980), might also be important for stimulus propagation.

As far as endothelial contacts are concerned, the pattern of junctions described by SIMIONESCU et al. (1975, 1976) might be considered as being characteristic for the whole arterial tree from the aorta to the arterioles of different organs. These endoendothelial junctions consist of a peculiar tight junction network, most meshes of which are filled with gap junction particles. Similar interendothelial contacts have been found in the renal arterial tree of eight

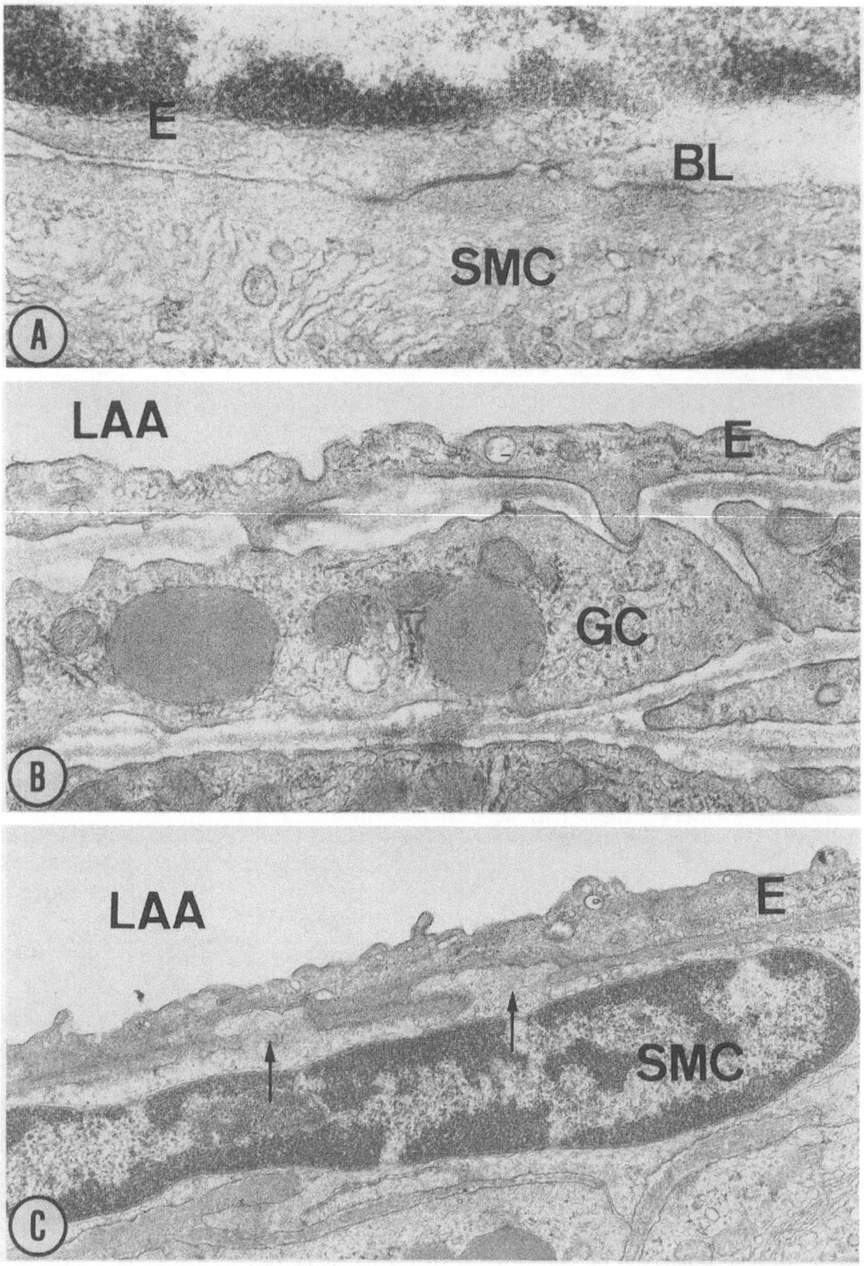

Fig. 2.27. Different forms of myoendothelial contacts in the afferent arteriole of mouse (**A, B**) and rat kidney (**C**). **A** Flat apposition equipped with gap junction between endothelial *(E)* and smooth muscle cell *(SMC)*. *BL*, basal lamina; ×48600. **B** Fingerlike processes of endothelial cell *(E)*, one of them projecting into a smooth indentation of the granulated media cell *(GC)*. *LAA*, lumen of the afferent arteriole; ×35000. **C** Rare form of myoendothelial contact with media cell processes indenting an endothelial cell *(arrows)*. *E*, endothelium; *SMC*, smooth muscle cell; *LAA*, lumen of the afferent arteriole; ×26000

species studied by MINK et al. (1984), extending from the interlobular arteries downstream into the proximal part of the afferent arterioles (Fig. 2.29). In the juxtaglomerular regions of the afferent and efferent arterioles, in contrast, quite different "classical" interendothelial tight junctions lacking gap junction equivalents have been observed (Fig. 2.30).

The presence of gap junctions arranged as "zonulae communicantes," i.e., an almost continuous band of connections, in the intima of the interlobular arteries and proximal afferent arterioles suggests that the endothelial cells of these vessel segments are functionally coupled. Although the musculo-muscular contacts of these vessel segments are

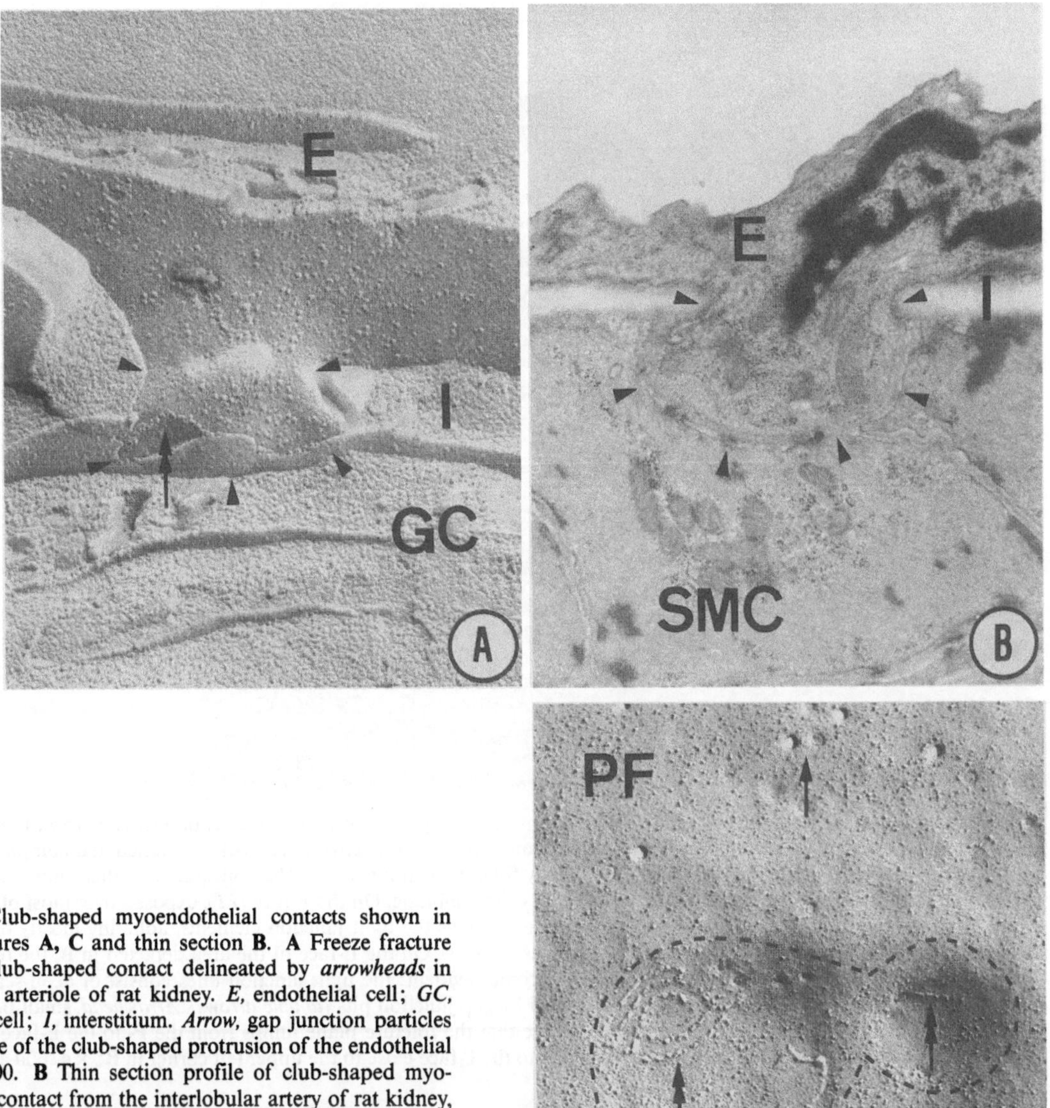

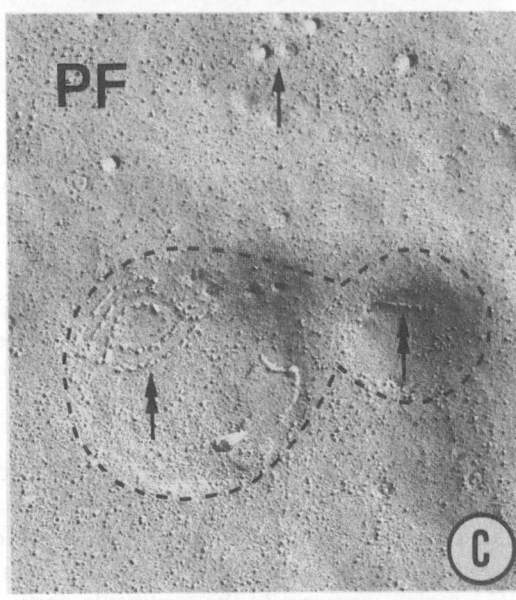

Fig. 2.28. Club-shaped myoendothelial contacts shown in freeze fractures **A, C** and thin section **B**. **A** Freeze fracture profile of club-shaped contact delineated by *arrowheads* in the afferent arteriole of rat kidney. *E*, endothelial cell; *GC*, granulated cell; *I*, interstitium. *Arrow*, gap junction particles on the P-face of the club-shaped protrusion of the endothelial cell; ×42500. **B** Thin section profile of club-shaped myoendothelial contact from the interlobular artery of rat kidney, delineated by *arrowheads*. *E*, endothelial cell; *SMC*, smooth muscle cell; *I*, interstitium with interna elastica; ×15700. **C** En face view of double-headed myoendothelial contact from interlobular artery of *Tupaia* kidney, encircled by a dashed line. On the P-face of the involved smooth muscle cell *(PF)*, putative gap junction particles *(double arrows)* and caveolae *(arrow)* are seen; ×42500

barely developed, it seemed possible that the smooth muscle cells of the interlobular arteries and proximal (renin-negative) afferent arterioles are involved in a widespread electronic and/or metabolic cooperation between cells of the renal microcirculation by means of the gap junction-equipped myoendothelial and interendothelial contacts of the renal resistance vessels (MINK et al. 1984).

In the distal (renin-positive) portion of the afferent arteriole, the situation, as judged from the ultrastructure of cell contacts, is different. Although interendothelial gap junctions are missing in the area of the JGA, epithelioid cells, Goormaghtigh cells, and mesangial cells are connected by macular gap junctions, so that the ultrastructural prerequisites for a second - distal - communication system would be at hand. As epithelioid cells and Goormaghtigh cells seemed to be adjoined both by gap

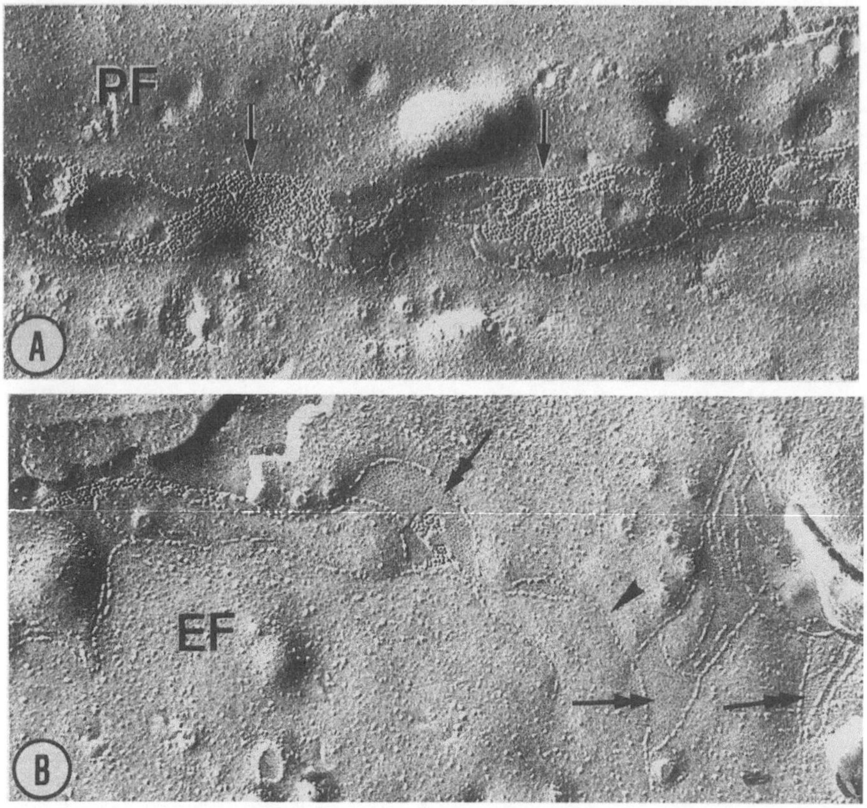

Fig. 2.29. Combined interendothelial occluding junctions consisting of tight and gap junction elements from the interlobular artery **A** and proximal afferent arteriole **B** of rat and mouse kidney, respectively. The belt-like junctional complex consists of tight junctions, enclosing irregularly shaped areas mostly filled by gap junctions. The appearance of this combined junction is different on the P-face and the E-face of the endothelial cell membrane. On the P-face *(PF)*, exposed over most of the area depicted in **A**, the gap junction particles are clearly visible, whereas the tight junction elements, adhering mostly to the E-face, tend to be obscured by the adjacent gap junctional areas *(arrows)*. On the E-face of the area depicted in **B**, the texture of the tight junction component of the combined junction becomes evident: the tight junction lines consist of deep-seated rows of particles enclosing areas of irregular shape, mostly filled by gap junction pits *(double arrows)*. *Arrowhead*, small portion of the junction consisting of only one tight junction line. (Note that the fracture plane jumps from the P- to the E-face close to the lower border of the junctional belt in **A** and from the E- to the P-face close to the upper left corner in **B**; ×47600 and ×56000, respectively

junctions in an upstream direction to the smooth muscle cells of the afferent arteriole (FORSSMANN and TAUGNER 1977), the morphological basis for a signal transfer between both systems, i.e., from the JGA region upstream to the contractile portion of the afferent arteriole, could be supposed to exist. However, the hypothesis of such an extensive functional syncytium over relatively large distances of the afferent arteriole has not yet been verified by electrophysiological techniques.

Using simultaneous intracellular recordings from two different cells of the afferent arteriole to assess the degree of electrotonic current spread, BÜHRLE, NOBILING and TAUGNER (unpublished observations) did not find any sign of coupling within the resolution limits of their measurements (coupling coefficent, 0.01). This was true for the juxtaglomerular as well as for the proximal part of the vessel and held for distances between the current injection and the membrane potential recording electrodes down to 25 μm. On the other hand, control experiments performed with the same technique on the arterial tree of the guinea pig ileum submucosus plexus lead to positive results, thus confirming data on the electrotonic coupling of media cells in this system (HIRST and NEILD 1978). Experiments to verify the electrotonic coupling between Goormaghtigh and glomerular mesangial cells have not been performed, nor has the metabolic coupling in the region of the JGA or between the cells of kidney resistance vessels been tested by intracellular tracer injection.

2.8 Tight Junction Barriers in the Region of the Juxtaglomerular Apparatus

The properties of the tight junction barriers in the region of the JGA may be important in several regards: the route renin takes after secretion from epithelioid cells, the propagation of the hypothetical macula densa signal, and the route substances reaching the kidney via the bloodstream are bound to take until they arrive at the individual components of the JGA.

Three groups of tight junctions have to be considered in this context. First, the tight junctions between the epithelial cells of the macula densa; second, the tight junctions between the cells of the parietal layer of Bowman's capsule near the vascular pole of the renal corpuscle; and, third, most important, the interendothelial junctions of the glomerular arterioles.

The interepithelial junctions of the macula densa have been discussed in Sect. 2.5. Briefly, these junctions, well deserving of their name, by all known odds reliably separate the lumen of the distal tubule from the interstitium of the JGA. Therefore, changes in the flow or composition of the distal tubular fluid will have to be translated by the macula densa cells into still another signal if the vascular components of the JGA are to be affected.

At the vascular pole, the visceral layer of Bowman's capsule, composed of podocytes, folds over in a collar-like manner in its parietal layer. As the glomerular arterioles usually curve around this visceroparietal junction when entering the renal corpuscle, a close proximity results between the parietal layer and the concave outer surfaces of the afferent and efferent arterioles. It has been mentioned earlier that in this region the media of the preglomerular arteriole is often much thinner than that of the opposite site facing the Goormaghtigh cell field and/or the distal tubule. Nevertheless, epithelioid cells and cell processes densely packed with renin granules can often be found close to Bowman's capsule (Figs. 1.2, 2.11, 3.6). In addition, peripolar cells may occur at the visceroparietal junction of Bowman's capsule which are thought to extrude their secretory product into the urinary space of the renal corpuscle (RYAN et al. 1979). However, it appears appropriate to call to mind that the epithelial cells of the parietal layer of Bowman's capsule are equipped with typical zonulae occludentes consisting of two to five anastomizing tight junction strands, thus reliably separating the urinary space from the extraglomerular interstitium at least for larger molecules (TAUGNER et al. 1976). An exception of this rule is represented by those parts of the parietal layer that, as described in Sect. 3.3, are lined by podocyte foot processes, which are especially frequent in *Tupaia belangeri* (ROSIVALL and TAUGNER 1986; cf. Fig. 3.6). In the region of the glomerular stalk, epithelioid cells facing podocyte processes and fenestrated endothelial cells are common in *Tupaia* as well as in other species (Fig. 3.8). The question of whether secreted renin may enter the urinary space at these sites will be discussed in connection with the pressure gradients existing in the region of the vascular pole (Sect. 3.3).

In Sect. 2.7 the interendothelial junctions of the afferent and efferent arterioles have been discussed with regard to gap junction coupling. Here, the question is whether the interendothelial junctions of the glomerular arterioles present barriers which can be assumed reliably to separate the lumen of the glomerular arterioles from the surrounding interstitium.

In the interlobular artery and the proximal, renin-negative portion of the afferent arteriole, the interendothelial junctions are made up of a continuous network of two to five junction lines with rows of particles adhering predominantly to the E-face of the endothelial cell plasma membranes. Typically, most of the meshes of this tight junction network are occupied by gap junctions (Fig. 2.29). Although the junction lines of these interendothelial contacts were composed of individual particles, the relatively large number of lines, the remarkable apicobasal depth of the junctional belt, and the densely arranged intercalated gap junction elements lead MINK et al. (1984) to the conclusion that the endoendothelial junctions of the proximal part of the afferent arterioles provide good tightness and intercellular adhesion.

In the distal - renin-positive - portion of the afferent arteriole, the interendothelial junctions consist of smooth tight junction strands not subdivided into individual particles on the P-face, with corresponding particle-free grooves on the E-face of the endothelial cell membranes (Fig. 2.30 A, B). This type of occluding junction, otherwise typical of transporting epithelia, was suggested to provide

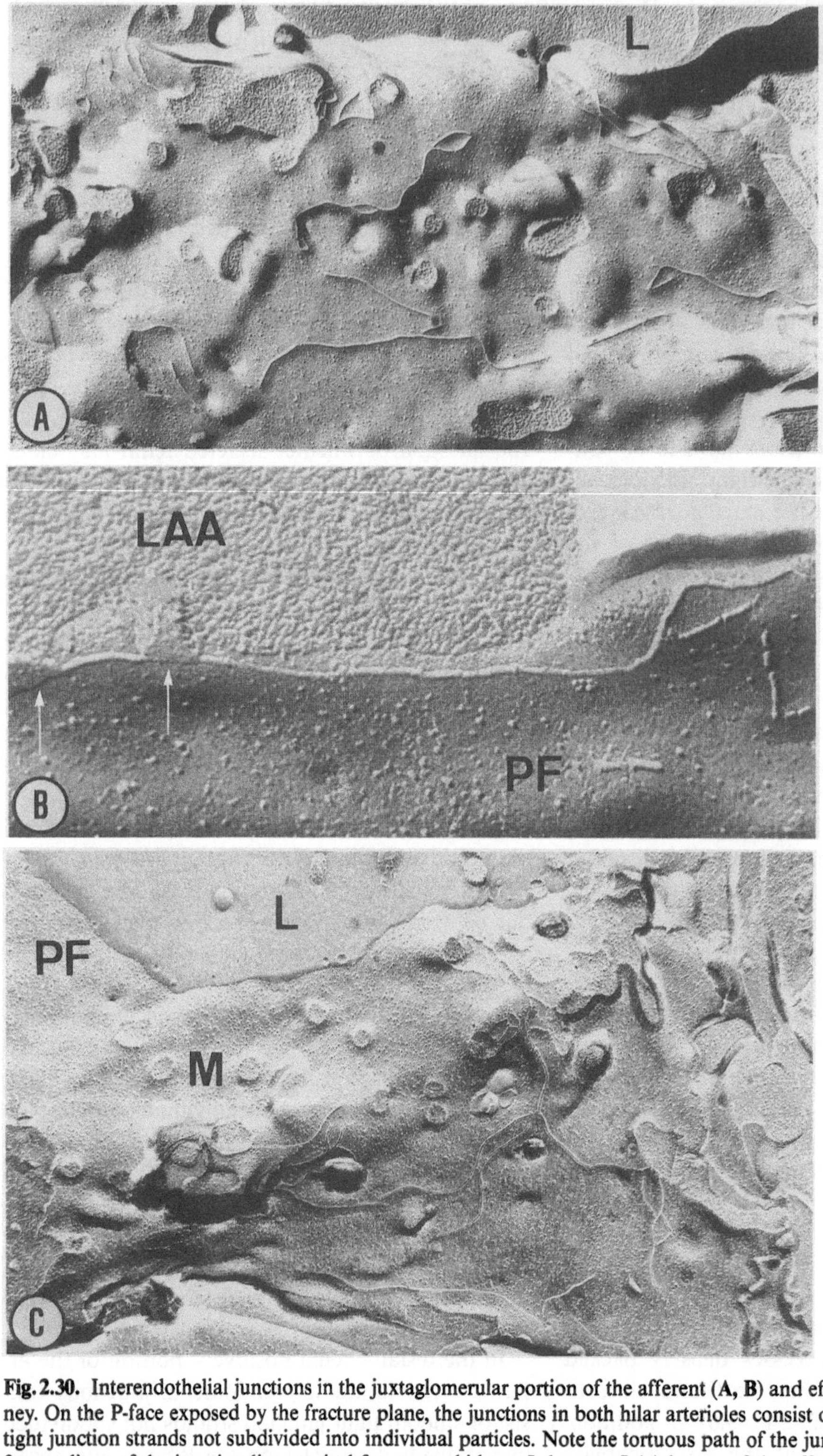

Fig. 2.30. Interendothelial junctions in the juxtaglomerular portion of the afferent (**A**, **B**) and efferent arteriole **C** of mouse kidney. On the P-face exposed by the fracture plane, the junctions in both hilar arterioles consist of one or two smooth "fibrillar" tight junction strands not subdivided into individual particles. Note the tortuous path of the junctions with discontinuities and free endings of the junction lines typical for mouse kidney. *L*, lumen; *LAA*, lumen of the afferent arteriole; *PF*, P-face of the endothelial cell plasma membrane; *M*, fractured microvilli; ×31000, ×114000, and ×28000, respectively

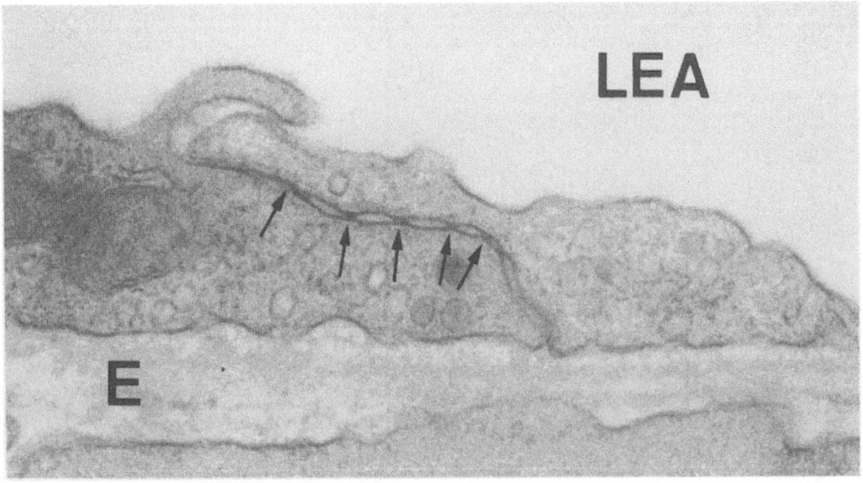

Fig. 2.31. Highly developed interendothelial tight junctions in the juxtaglomerular portion of the efferent arteriole from *Tupaia* kidney. The *arrows* point to several loci of fusion between the involved endothelial cell membranes. *LEA*, lumen of the efferent arteriole; *E*, extracellular space with basal lamina material; ×32000

good mechanical strength with minimal leakiness for macromolecules (MINK et al. 1984). For example, renin, which is secreted by adjacent epithelioid cells, probably does not enter the circulation at this - preglomerular - site (FORSSMANN and TAUGNER 1977; cf. also Chap. 9).

At its origin, i.e., in the juxtaglomerular part of the efferent arteriole, interendothelial junctions very similar to those of the distal afferent arteriole have been found with prominent strands not subdivided into individual particles on the P-face (Figs. 2.30C, 2.31). However, in more distal segments of the postglomerular arteriole, various patterns of interendothelial junctions have been observed, the most common being a combination between solid and subdivided tight junction strands. These patterns suggested a gradual transition to the more leaky junctions of the subsequent peritubular capillaries, probably allowing for the passage of substances from the lumen of the vessel to the media cells of the efferent arteriole at this site.

Concerning the functional morphology of the occluding junctions in the renal arterial tree, MINK et al. (1984) suggested that substances entering the kidney via the arterial bloodstream probably do not reach the media cells of the interlobular artery and afferent arteriole at the outset, i.e., from the luminal side. Instead, these substances were thought to act first and in highest concentration on the media cells of the efferent arteriole, after having passed through the interendothelial partially leaky junctions of this vessel, or by way of the initial peritubular capillaries. Concerning the glomerular structures cf. Chap. 3.2, 3.3.

Chapter 3

Elements with Structural and/or Functional Relationships to the Juxtaglomerular Apparatus

3.1 Tubulovascular Proximities Outside the JGA as Parts of the Tubulovascular Complex of the Nephron

As already mentioned in Sect. 2.6, the close spatial relationship between the glomerular arterioles and the distal tubule is not confined to the macula densa. The thick ascending limb of Henle's loop frequently contacts the efferent arteriole for a considerable distance before reaching the JGA area and the distal convolution or rather, the connecting tubule, is often in close proximity with the afferent vessel, with the direction of the tubular flow always being antidromic to the blood flow in the glomerular arterioles (Golgi 1889; Barajas and Latta 1963a; Faarup 1965; Barajas 1970; Christensen et al. 1975, 1979). Although it cannot be ruled out that these tubular segments located further up- or downstream from the macula densa participate in the tubuloglomerular feedback (TGF) or in changes of renin secretion, except for two reports, numerical data on the extension, or at least on the incidence, of such tubulovascular "contacts" or "proximities" outside the JGA are missing.

Faarup (1965) reported on the incidence and the minimum and maximum lengths of the tubulovascular contacts in rat for both the afferent and efferent arterioles in superficial as well as in juxtamedullary nephrons. Table 3.1 shows that contacts of the distal tubule outside the macula densa with the efferent as well as with the afferent arteriole are always present. In addition, these tubulovascular contacts are distinctly more extensive than the facultative contacts in the JGA area. However, the figures give information neither on the respective surface areas of the tubulovascular proximities nor on the "quality," i.e., the characteristics of the observed contacts. In any case, Barajas and Latta (1963a) mentioned that the contact with the efferent arteriole may be so intimate that the tubule is indented by the vessel.

Barajas et al. (1986) reinvestigated the anatomical relationships between the renin-containing vas-

Table 3.1. Incidence and extension of the contacts between both glomerular arterioles and the different parts of the distal tubule

	Efferent arteriole				Afferent arteriole			
	Contacts with the TAL		Contacts with the MD		Contacts with the MD		Contacts with the distal tubule[b]	
	Incidence (%)	Length[a] (µm)	Incidence (%)	Length (µm)	Incidence (%)	Length (µm)	Incidence (%)	Length (µm)
Superficial nephrons ($n=15$)	100	15- 80	53	0-30	73	0-35	100	10- 90
Juxta-medullary nephrons ($n=10$)	100	15-150	40	0-25	50	0-30	100	5-165

MD, Macula densa; TAL, Thick ascending limb

[a] Maximal and minimal lengths of the contacts encountered

[b] From the original description it is not clear whether the connecting tubule was included in the evaluation

(Modified from Faarup 1965, 1971)

cular and the kallikrein-containing tubular structures of the rat kidney, using double immunohistochemical labelling. They found that the kallikrein-reactive late distal tubular cells belong to the connecting tubule. These kallikrein-containing cells were adjacent to the afferent arteriole in 67%-80% of the arterioles surveyed. On the other hand, depending on their position in the cortex of the kidney, 73.3%-90% of the afferent arterioles came within 3 μm of kallikrein-positive late tubular segments. Close proximity between renin-positive cells of the afferent arteriole and kallikrein-containing connecting tubule cells was observed in 26.7%, 46.6% and 66.6% of juxtamedullary, midcortical, and superficial nephrons, respectively. According to BARAJAS et al. (1986), this spatial relationship suggests an anatomical basis for a possible interaction between the involved renin-positive vascular and the kallikrein-positive late distal tubular elements.

The possible implications of such findings with respect to the role of the distal tubule in TGF and renin secretion will be dealt with in Sect. 3.3 and 7.3. At this point, only the anatomical prerequisities will be compared, considering the hypothesis that not only the macula densa, but also other portions of the distal tubule, are involved in tubulovascular interactions.

The distal tubule including the macula densa is equipped with very tight interepithelial junctions (SCHILLER et al. 1980). Thus, changes in the distal tubular fluid flow or composition probably cannot directly affect the surroundings of the tubule; rather they have to be translated into another macula densa or distal tubular signal by the respective epithelial cells so that they may influence the vascular tone or renin secretion by mediation of the interstitial fluid.

Apart from these similarities, the remaining prerequisites of tubulovascular interactions are more or less different for the individual segments of the distal tubule. This includes (a) the spatial relationship of the individual components of the renal microcirculation to the different segments of the distal tubule, (b) the nature of the tubular signal, and (c) the characteristics of the "proximities" or "contacts" and in this, the susceptibility of the assumed tubulovascular interactions to interference from neighboring structures.

(a) Regular contacts seem to exist between the thick ascending limb of Henle's loop and the efferent arteriole, as well as between the distal convolution and/or the connecting tubule and the afferent arteriole. The macula densa, on the other hand, has contact with both glomerular arterioles, although with variable frequency (BARAJAS and LATTA 1963a; FAARUP 1965; BARAJAS et al. 1986). Here, the regular and extensive contact with the Goormaghtigh cell field might be of special relevance (Sect. 2.4-2.6).

(b) Concerning the nature of the "tubular signal" as a response to changes of the distal tubular fluid flow or composition, two concepts may be considered. First, it may consist in corresponding changes of the fluid reabsorbed by the respective distal tubular segments. Second, this signal might consist of substances produced by the involved tubule cells, e.g., adenosine (OSSWALD 1984) or kallikrein (BARAJAS et al. 1986). One may assume that the various segments of the distal tubule differ significantly from one to another in both these respects.

(c) The contacts of the macula densa with the glomerular arterioles are subject to considerable internephron heterogeneities and comprise only relatively small areas. However, they might be very intimate, especially as far as the efferent arteriole is concerned (BARAJAS 1970). Particularly regular and very close is the relationship between the macula densa and the Goormaghtigh cell field. The Goormaghtigh cell field is considered to function as a transducer or amplifier of the macula densa signal. On the other hand, its presence also means that the macula densa is bordering on a spacious cell mass not supplied with capillaries. It can therefore be assumed that the signal transfer from the macula densa via the interstitial fluid to the vascular components of the JGA may be susceptible to interference by competitive fluid flow from the glomerular stalk (cf. Sect. 3.3), but hardly so from peritubular capillaries or neighboring tubules. The tubulovascular proximities upstream and downstream from the macula densa apparently cover relatively large distances (cf. Table 3.1; Figs. 2.1, 2.17C, D). These contacts, however, are usually not as intimate as those of the macula densa. Any signal transfer thus might here be comparatively more susceptible to interference by competing fluid flow from neighboring structures. Nevertheless, it should be recalled that some of the tubulovascular contacts outside the JGA can also be very close (BARAJAS and LATTA 1963a; GORGAS 1978a).

3.2 Glomerulus

3.2.1 Overview

The renal corpuscle consists of the glomerular capillaries attached to the mesangium, both enclosed by Bowman's capsule, a pouch-like extension of the proximal tubule. In the context of the JGA, the vascular pole of the renal corpuscle is of outstanding interest. Here, at the glomerular hilus, the afferent and efferent arterioles are separated by mesangium, which constitutes the axis of the glomerular tuft and is continuous with both the Goormaghtigh cell field (the so-called extraglomerular mesangium) and the centrolobular (true glomerular) mesangium (cf. Fig. 1.2). Functional interactions in the context of a glomerulo-juxtaglomerular complex may result from these intimate microtopographical interrelations. They may also pertain to the peripolar cells at the visceroparietal junction of Bowman's capsule. For further details of the glomerular anatomy, the reader is referred to the comprehensive review of KAISSLING and KRIZ (1985). Some phylo- and ontogenetic aspects of the glomerulo-juxtaglomerular complex are outlined in Chap. 13.

3.2.2 Mesangial Cells in the Glomerular Stalk and the Centrolobular Region of the Glomerulus

The Goormaghtigh cell field and the media of both glomerular arterioles merge into the glomerular stalk and the stalk into the centrolobular mesangium without clear demarcations (Figs. 1.2, 2.13; cf. Sect. 2.4). Hence, there are three routes by which the vascular components of the JGA might be functionally coupled with the glomerular mesangium: by way of the bloodstream, by the transcellar route through gap junctions, and by the pericellular way from the mesangial interstitium via the glomerular stalk to the interstitium of the JGA (cf. Sect. 3.3). In the following, the anatomical prerequisites for such a functional interplay within the scope of the glomerulo-juxtaglomerular complex will be outlined, a cooperation which, though possible and even probable, has not yet been fully verified by experimental means.

ZIMMERMAN (1929) used the term *"interlobuläres Zwischengewebe"* to denote connective tissue extending from the glomerular stalk to the space between the glomerular capillaries. The long-lasting controversy about the mesangial or centrolobular region of the glomerulus has been reviewed by MICHIELSEN and CREEMERS (1967) and by ZAMBONI and DEMARTINO (1968). Clarification was achieved by serial sectioning (BOHLE and HERFARTH 1958) and electron microscopic studies (BENCOSME et al. 1959; LATTA et al. 1960; LATTA and MAUNSBACH 1962b; FARQUHAR and PALADE 1962; MICHIELSEN 1962; SUZUKI et al. 1963).

Mesangial cells are irregularly shaped cells which, with respect to the glomerular capillaries, occupy a centrolobular position (Fig. 3.1). Besides branching spinous processes, deep pocketlike invaginations of the plasmalemma similar to or even more pronouced than those of Goormaghtigh cells are easily found in mesangial cells. The indented nucleus of mesangial cells may be round or oval, having a light chromatin pattern with a moderate accumulation of dense chromatin near the nuclear envelope. The cytoplasm contains inconspicuous organelles: short mitochondria, a Golgi complex, some endoplasmic reticulum, and a certain number of ribosomes. Besides, one or more membrane-bound granular structures are seen resembling the granules of Goormaghtigh cells (cf. Sect. 2.4). Mesangial cells show bundles of filaments and associated cytoplasmic densities near the cell membrane, suggesting that they are modified smooth muscle cells (FARQUHAR and PALADE 1962; LATTA and MAUNSBACH 1962b).

Mesangial cells are coupled with each other and with lacis cells by a dense array of gap junctions (Fig. 3.2; cf. Sect. 2.7). They are surrounded by the mesangial matrix, continuous with the intercellular substance of the glomerular stalk and the lacis. As to the question of whether there is more than one type of mesangial cell, one non-phagocytic and one phagocytic, see the reviews of MICHAEL et al. (1980), KREISBERG (1983), and KREISBERG and KARNOVSKY (1983). The total cell number of the capillary network in a single rat glomerulus was found to be about 550, 33% of them mesangial cells, 51% endothelial, and 16% visceral epithelial cells (HELMCHEN et al. 1977).

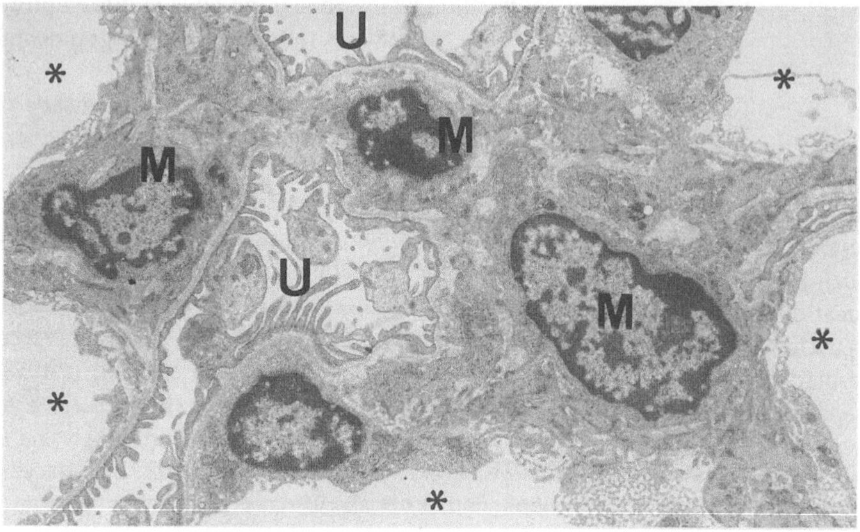

Fig. 3.1. Centrolobular (true glomerular) mesangium from mouse kidney. *M*, mesangial cell nuclei; *U*, urinary space; *asterisks*, glomerular capillaries; ×6000

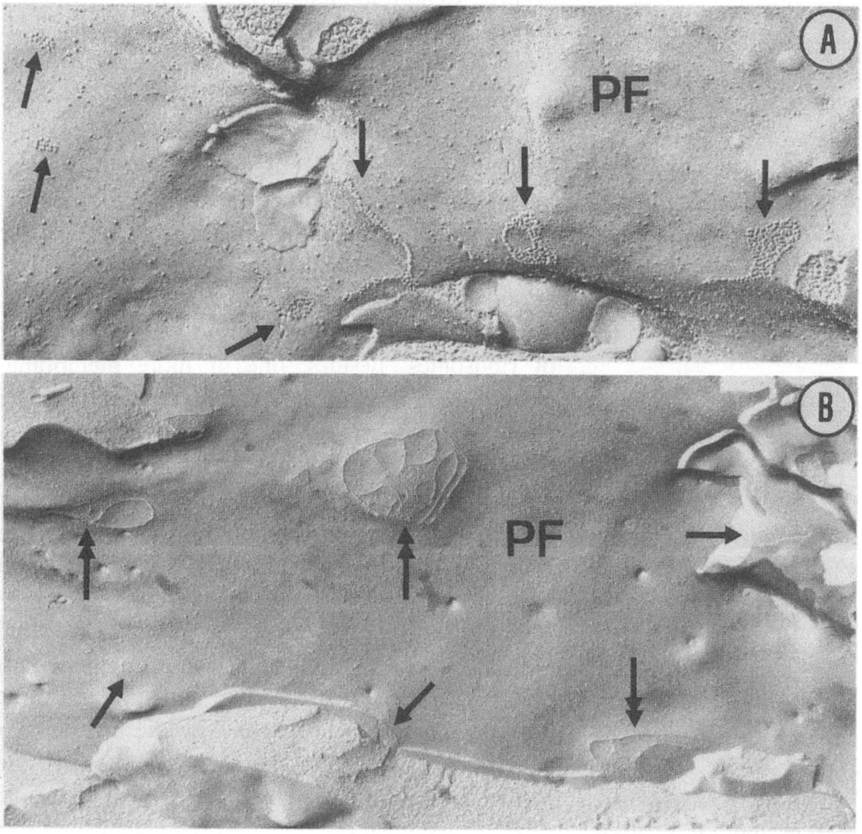

Fig. 3.2. Gap junctions *(arrows)* between glomerular mesangial cells in cat **A** and *Tupaia* kidney **B**. In the tree shrew, some of the contacts are gap junction/tight junction combinations *(double arrows)*. *PF*, P-face of the exposed plasma membrane. Freeze fracture replicas; ×58000 and ×27400, respectively

The centrolobular position of the mesangium is important with regard to both the possible effects of a mesangial contraction and to the fluid compartments involved in the exchange with the mesangial interstitium. On the one hand, the mesangium borders on the attached glomerular capillaries, on the other, on the urinary space of the glomerulus (Figs. 3.1, 3.3 A). From the urinary space, the mesangium is separated by the glomerular basement membrane, largely impermeable for macromolecules, covered by podocyte processes. In its farther course, the filtration barrier then surrounds the urinary portion of the capillary wall, thus constituting a common surface cover, wrapping together the capillaries and the mesangium. Accordingly, the mesangium is separated from the capillary lumen only by the endothelial layer (Figs. 3.1, 3.3 A). The interposed capillary endothelium shows fenestrations, thus permitting a plasmic flow, carrying macromolecules into the mesangial interstitium (LATTA et al. 1960; LATTA and MAUNSBACH 1962b; FARQUHAR and PALADE 1962; LATTA and FLIGIEL 1985). It has repeatedly been shown that due to these preconditions and the prevailing pressure gradients, a convective fluid flow results through the mesangial interstitium (the so-called mesangial channels). Although neither the quantity of this flow nor the factors regulating it are fully understood, there is agreement that it is directed not only to the urinary space but also toward the interstitium of the glo-

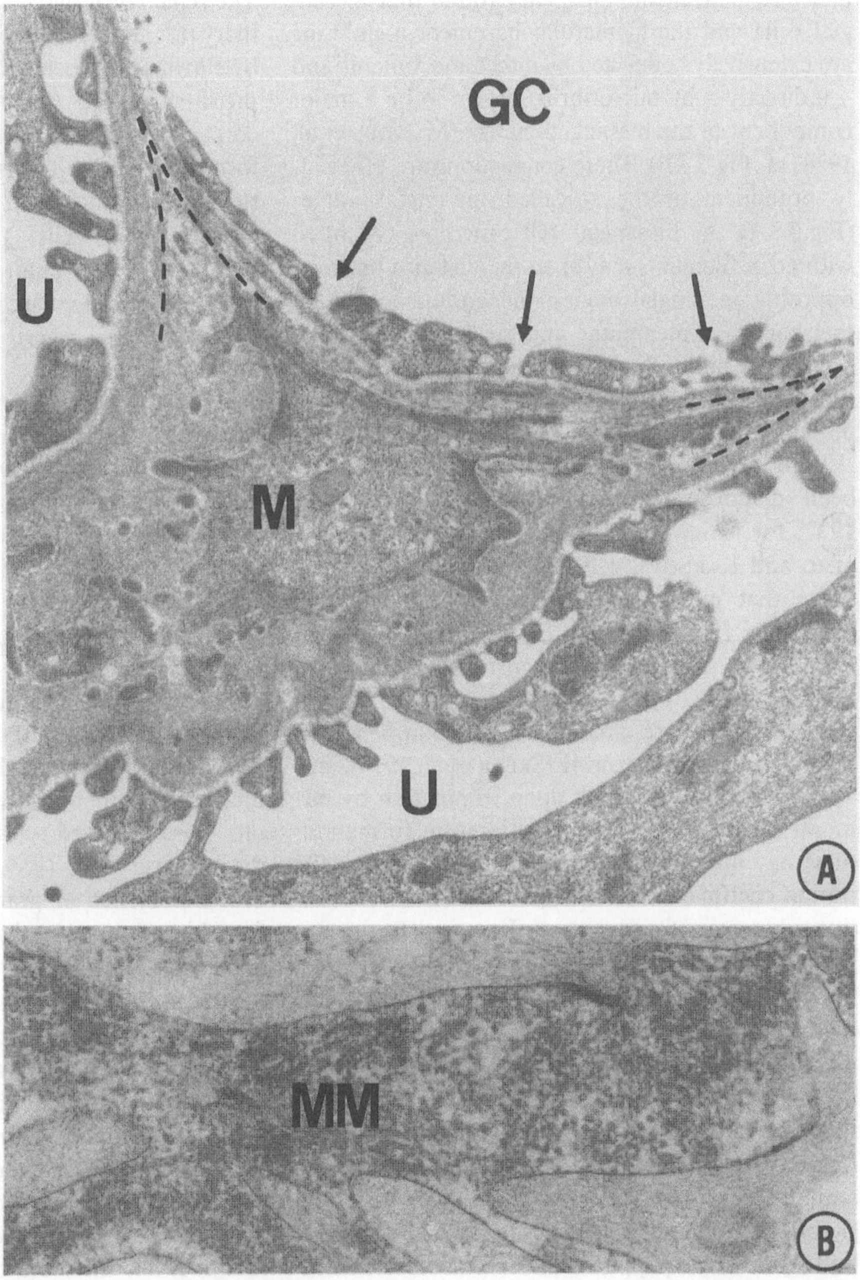

Fig. 3.3. **A** Connections between a mesangial cell process *(M)* and the glomerular basement membrane by direct attachments and by microfibrils, especially prominent in the "mesangial angles" marked by *dashed lines*. *GC,* glomerular capillary; *U,* urinary space. The *arrows* point to endothelial fenestrations; × 19 800.
B Mesangial matrix *(MM)* with microfibrils as their major component at higher magnification; × 71 500. (Courtesy of W. KRIZ, Heidelberg)

merular stalk and from there to that of the JGA (LATTA and MAUNSBACH 1962b; MICHIELSEN and CREEMERS 1967; LEIPER et al. 1977; LEE and VERNIER 1980; LATTA and FLIGIEL 1985; ROSIVALL and TAUGNER 1987; for reviews on the fate of macromolecular substances entering the mesangial interstitium see MICHAEL et al. 1980; RAIJ and KEANE 1985).

In regard to possible effects of mesangial contractions on the glomerular filtration dynamics, SAKAI and KRIZ (1987) reinvestigated the geometry of mesangial cells and their relationship to the glomerular basement membrane. They found that mesangial cells and the glomerular basement membrane are extensively connected by direct attachments and - indirectly - by microfibrils, shown to be a major component of the mesangial matrix (MUNDEL et al. 1988; cf. Fig. 3.3 B). These connections are especially prominent at the so-called mesangial angles (Fig. 3.3 A). As mesangial cell processes equipped with actin filaments may span the distance between opposing mesangial angles, the authors conclude that basement membrane and mesangial cells may establish a biomechanical unit capable of changing the geometry of the glomerular capillaries upon mesangial contraction.

With immunofluorescent microscopy, myosin has been demonstrated in mesangial cells (BECKER 1972; for review see MICHAEL et al. 1980; KREISBERG and KARNOVSKY 1983). The pertinent proposition that mesangial cells may have contractile properties (BERNIK 1969; LATTA 1973) has been verified by several groups (for references see SINGHAL et al. 1986; SAKAI and KRIZ 1987). As mesangial cells are, among others, equipped with receptors for angiotensin II (SREAR et al. 1974), and suggested to release renin upon stimulation by humoral mediators with local angiotensin formation, they may mediate alterations of the glomerular filtration coefficient induced by a number of vasoactive substances (for review see KON and ICHIKAWA 1985; BRENNER et al. 1986). The contractility of mesangial cells has been proposed to be controlled by the balance between angiotensin and prostaglandins (cf. SCHLONDORF et al. 1985; SCHARSCHMIDT et al. 1986; FOIDART and MAHIEU 1986) as well as by that between angiotensin and atrial natiuretic factor (ANF; BIANCHI et al. 1986) or angiotensin and dopamine (BARNETT et al. 1986; cf. Sect. 9.3.2). It should be recalled that besides mesangial cells also podocytes are equipped with contractile elements (actin filaments, heavy meromyosin) and suggested to participate in the regulation of the filtration coefficient, e.g., by changing the shape of their foot processes (ANDREWS and COFFEY 1983). Distinct localization of ANF-binding sites mainly on podocyte foot processes has recently been demonstrated by BIANCHI et al. (1986).

3.2.3 Peripolar Cells

The renal corpuscle consists of the glomerular capillary tuft attached to the mesangium, both enclosed by Bowman's capsule, a pouchlike extension of the proximal tubule. At the vascular pole of the renal corpuscle, the visceral layer of Bowman's capsule, formed by podocytes, is reflected to become its parietal layer, which then consists of polygonal squamous epithelial cells.

In 1979, RYAN et al. described a distinctive type of epithelial cell encircling the orgin of the glomerular tuft at the visceroparietal transition of Bowman's capsule in the sheep kidney. These peripolar cells (PPC) are well integrated within Bowman's capsule by forming junctional complexes with podocytes as well as with parietal epithelial cells, but stand out in sheep by the presence of numerous cytoplasmic granules staining with the periodic acid-Schiff technique, Bowie's stain, brillant crystal scarlet, and methylene blue. Electron microscopically, the membrane-bound roundish PPC granules, having diameters of 100–1500 nm, contain electron-dense homogeneously fine fibrillogranular material (RYAN et al. 1979, 1982).

In sodium-depleted sheep, PPC showed signs of increased synthetic activity. In addition, PPC granules were observed to discharge their contents into the urinary space (RYAN et al. 1982; HILL et al. 1983). Evidence of granule release has also been found in newborn lambs subjected to acute volume expansion and diuresis by intravenous infusion of dextrose solution (ALCORN et al. 1984). In summary, the ultrastructure of PPC including the morphology of granule discharge was judged to be similar to that of zymogen-type epithelial cells, e.g., in the acini of salivary glands or in the exocrine pancreas (RYAN et al. 1982; HILL et al. 1983). No evidence of granule discharge toward the basal lamina of Bowman's capsule, i.e., in the direction of the JGA, was found. It has therefore been suggested that factors

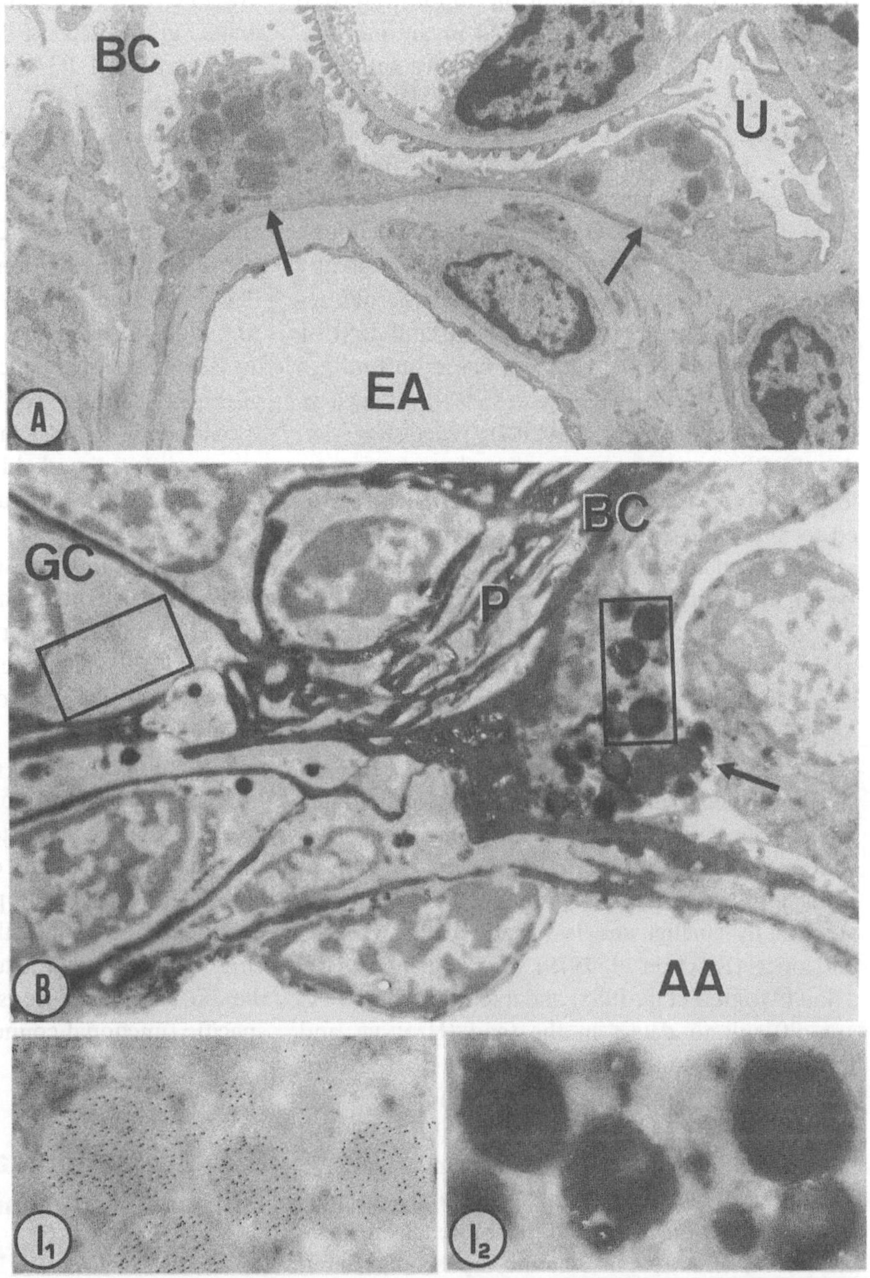

Fig. 3.4 A, B. Peripolar cells from mouse kidney. **A** Conventional electron micrograph with peripolar cell processes *(arrows)* facing the efferent arteriole *(EA)* shortly after its entry into the renal corpuscle. *U*, urinary space; *BC*, Bowman's capsule; ×7400. **B** Section through the hilus of a renal corpuscle, showing a peripolar cell *(arrow)*, Goormaghtigh cell processes *(P)*, and a granulated cell *(GC)* 40 min after the intravenous injection of 20 mg horeseradish peroxidase. After the histochemical reaction, the tissue was postfixed, embedded in Lowicryl, and immunoreacted for renin. *BC*, Bowman's capsule; *AA*, afferent arteriole. I_1 and I_2 are higher magnifications of the areas marked in **B**. Note that the organelles contained in the peripolar cell (I_2) are distinctly HRP positive and renin negative, while the epithelioid cell secretory granules (I_1) are HRP negative and renin positive. Uncontrasted section; ×7800 and ×24500, respectively

released from PPC pass with the ultrafiltrate into the tubular fluid and are involved in the modulation of tubular functions, e.g., in glomerulotubular balance and/or electrolyte homeostasis (RYAN et al. 1982; HILL et al., 1983).

Later, immunohistochemical studies indicated that sheep PPCs may contain kallikrein or a kallikrein-like polypeptide (GALL et al. 1984). Therefore, according to the fact that PPCs are frequently found close to epithelioid cells of the glomerular arterioles (RYAN et al. 1982; GARDINER and LINDOP 1985; GARDINER et al. 1986) and based on findings by CARRETERO and BEIERWALTES (1984) as well as by OKAMURA and INAGAMI (1984), it was suggested that PPC secretion may stimulate the selective release of active renin. As kallikrein has been reported to inactivate atrial natriuretic factor (BRIGGS et al. 1984), related effects upon PPC secretion into Bowman's capsular space have also been proposed in sodium-depleted sheep (GALL et al. 1986). Immunostainable renin (GARDINER and LINDOP 1985) or traces of other components of the RAS have not been detected in PPC granules (unpublished observations, cf. Fig. 3.4 B and inserts).

The incidence of PPC is subject to considerable interspecies variations. According to the first observations, PPCs were reported to be outstanding in ovine kidney; sometimes prominent in man, PPC are usually smaller and less readily found in rats and mice (RYAN et al. 1979). Conspicuous in the avian (MORILD et al. 1988) and axolotl kidney, PPC are difficult to detect in the toad (HANNER and RYAN 1980). A quantitative evaluation of PPC incidence using SB-trichrome-stained sections was performed by GALL et al. (1986). Although PPC could be detected in 16 out of 17 mammalian species, the PPC index (i.e., the percentage of PPC-positive, randomly sectioned glomeruli) differed from 0.15, 0.22, 0.44, and 0.99 in echidna, opossum, rabbit, and man, respectively, to 11.86 in sheep. The question of whether PPC can be found in all renal corpuscles of humans was examined by GARDINER and LINDOP (1985) in series of 2-μm-thick sections from six kidneys. At least 3% of the renal corpuscles were PPC positive, at most 28%. Only occasionally were two PPCs present in the same renal corpuscle. Consequently, in humans, only a fraction of glomeruli are equipped with these cells and even in positive renal corpuscles, the glomerular hilus is not encircled by PPC. One may assume that conditions are similar in other species with a correspondingly low PPC index. In our studies, only about 10 distinctly granulated PPC were observed in nearly 1000 profiles of JGAs from mice and rats, respectively (Fig. 2.11, 3.4).

In summary, the sheep seems to be outstanding, displaying more PPC-positive renal corpuscles, more PPC in positive glomeruli (HILL et al. 1984), and more granules per cell, similar in size and electron density. In addition, ultrastructural changes of PPC have been verified in sheep upon stimulation. Although the secretory product is not yet known, these findings are suggestive of a specific function of the PPC, waiting for its final definition. In other species, e.g., in rats and mice, PPC granules are scarce, differ remarkably in size and matrix structure (Fig. 2.11 B), and can be shown to avidly take up exogeneous tracers (Fig. 3.4 B). Consequently, PPC granules in these species may be suspected to be lysosomes, similar to those in the proximal tubule. In this case, their peculiar frequency in the epithelial cells of the peripolar region might be explained by their large distance from the filtration membrane, resulting in a relatively slow exchange and, therefore, atypical composition of the ultrafiltrate in this remote compartment of the urinary space. It may be of historical interest that MATSUHASHI et al. (1977 a) had observed heterogeneous organelles in cells of the parietal layer of Bowman's capsule in mice, which they classified as lysosomes. On the other hand, the localization of the PPC near the JGA is indeed conspicuous and suggestive of specific functional connections between both structures.

3.3 Morphological Basis of Fluid Balance in the Interstitium of the Juxtaglomerular Apparatus

The intimate connection between tubular and vascular elements at the vascular pole of the renal corpuscle gave rise to suggestions that the JGA plays an important role in the regulation of renal blood flow and glomerular filtration rate (OKKELS and PETERFI 1929; GOORMAGHTIGH 1937). Later, functional investigations supported this idea, indicating the existence of a TGF mechanism for the regulation of the glomerular filtration rate (GFR; for reference see SCHNERMANN and BRIGGS 1985). Some uncertainties, however, remained about the correlation between structure and function in the region of the JGA and the role of renin secretion in TGF.

Experimental evidence suggests that, during TGF, Goormaghtigh cells, situated in the center of the JGA, receive signals from the distal tubular lumen via reabsorptive processes in the macula densa region. Since one of the most striking features of the JGA is the lack of capillaries (GOORMAGHTIGH 1937; OBERLING and HATT 1960; BUCHER and REALE 1961a; THOENES 1961; BARAJAS and LATTA 1963a), a rather stable interstitial milieu was inferred to exist in the area of the Goormaghtigh cell field governed by the reabsorptive function of the macula densa which, in turn, is influenced by changes in the flow or composition of the distal tubular fluid (for references see WRIGHT 1984). As most epithelioid cells are regularly present in the zone where the lacis and the wall of the afferent arteriole merge, the interstitial milieu in the region of the JGA may, apart from TGF, also be important for the secretion of renin. With regard to this outstanding role of the extracellular fluid, it appeared appropriate to reevaluate the morphological basis of fluid balance in the interstitium of the JGA. According to ROSIVALL and TAUGNER (1986), in addition to the close proximity of the distal tubule, three ultrastructural features in the region of the renal corpuscle could be important for the fluid balance in this region: (a) the mesangial-type lining of the glomerular stalk; (b) podocyte foot processes in the parietal layer of Bowman's capsule, and (c) endothelial fenestrations in the wall of the incoming afferent arteriole facing Goormaghtigh and epithelioid cells.

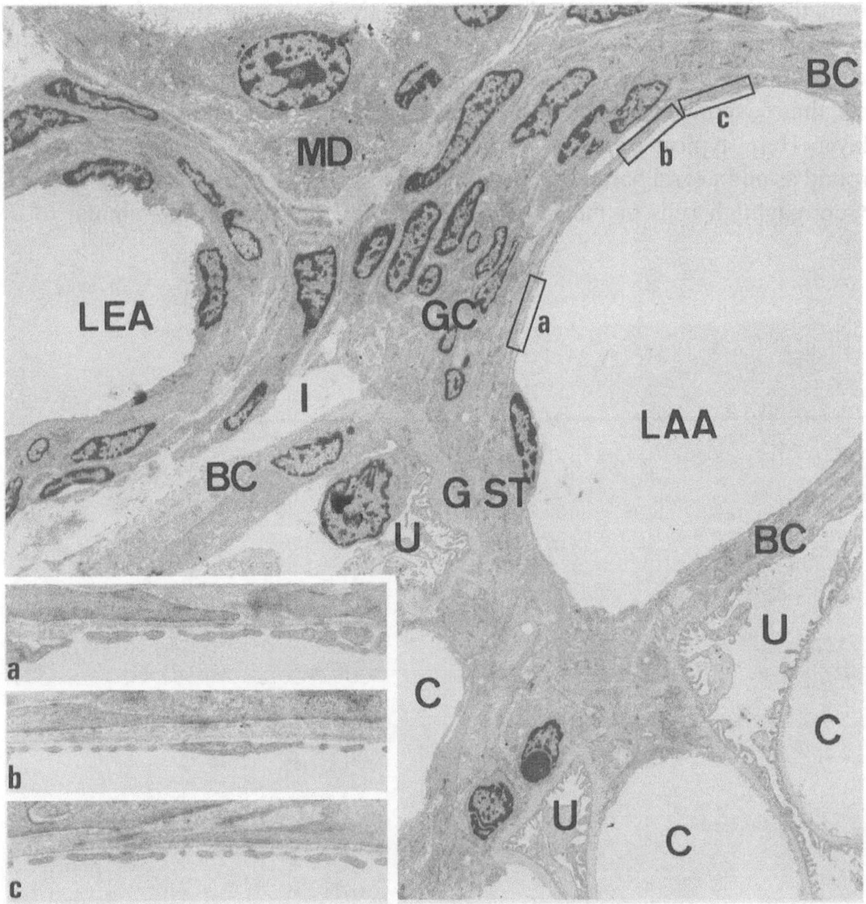

Fig. 3.5. Rat juxtaglomerular apparatus and hilus of the renal corpuscle showing endothelial fenestrations in the afferent arteriole facing Goormaghtigh cells *(GC)*. In addition, podocyte foot processes and a thin-layered endothelium bound the glomerular stalk *(G ST)*. *LAA*, lumen of the afferent arteriole; *LEA*, lumen of the efferent arteriole; *MD*, macula densa; *U*, urinary space rimmed by podocyte foot processes; *C*, glomerular capillary; *BC*, Bowman's capsule; ×2500. *Insets a, b, c:* Areas marked in the wall of the afferent arteriole at higher magnification; ×18000

The *glomerular stalk* in all species studied is bounded by podocytes and endothelial cells separating the region occupied by mesangial cells from the urinary space and from the lumen of the glomerular capillaries (Figs. 3.5, 2.13). Closer inspection reveals that the limiting structures of the stalk are very similar to those of the centrolobular region of the glomerulus, i.e., the true glomerular mesangium (LATTA et al. 1960), consisting mostly of podocyte processes and a thin layered endothelium (Figs. 3.1, 3.3). At higher magnification, it can be seen that the foot processes facing the glomerular stalk are identical to those of the filtration membrane as to size, shape, spacing, and the interposed slit membrane, and that the limiting endothelial layer in some places shows gaps or holes similar to those of the fenestrated endothelium lining all peripheral glomerular capillary loops.

Close to the turning point of Bowman's capsule from the visceral to the parietal layer, part of the lacis, or a segment of the afferent arteriole's media may continue beyond the neck of the renal corpuscle, thus covering a segment of the adjacent parietal layer. Here, typical *podocyte foot processes* can be found as an integral part of the perietal layer, facing Goormaghtigh cells or media cells of the afferent arteriole, including epithelioid cells with renin-positive secretory granules (Figs. 3.6, 3.7, 3.8). Although also found in rats and mice, podocyte foot processes in the parietal layer seemed to be common in *Tupaia*. Here, protrusions of Bowman's space rimmed by podocyte processes may bulge into the region of fusion between the media of the afferent arteriole and the lacis, thereby increasing the contact area between the urinary space and JGA elements (Fig. 3.7).

At the hilus of the renal corpuscle, the broadening glomerular stalk is continuous with the lacis region and with segments of both glomerular arterioles, the media of which splits into a sheet of loosely arranged cells that cannot be distinguished from intra- and extraglomerular mesangial (i.e., Goormaghtigh) cells. In this vessel segment, oriented along the axis of the glomerular stalk, the endothelium of the incoming preglomerular arteriole may show fenestrations, facing mesangial cells of the stalk, Goormaghtigh cells, and, occasionally, also epithelioid cells (Figs. 3.5, 3.8). Higher magnification reveals that these *endothelial fenestrations,* the existence of which has been confirmed by CASELLAS (1986a) in scanning electron microscope studies, are similar to those of the filtration mem-

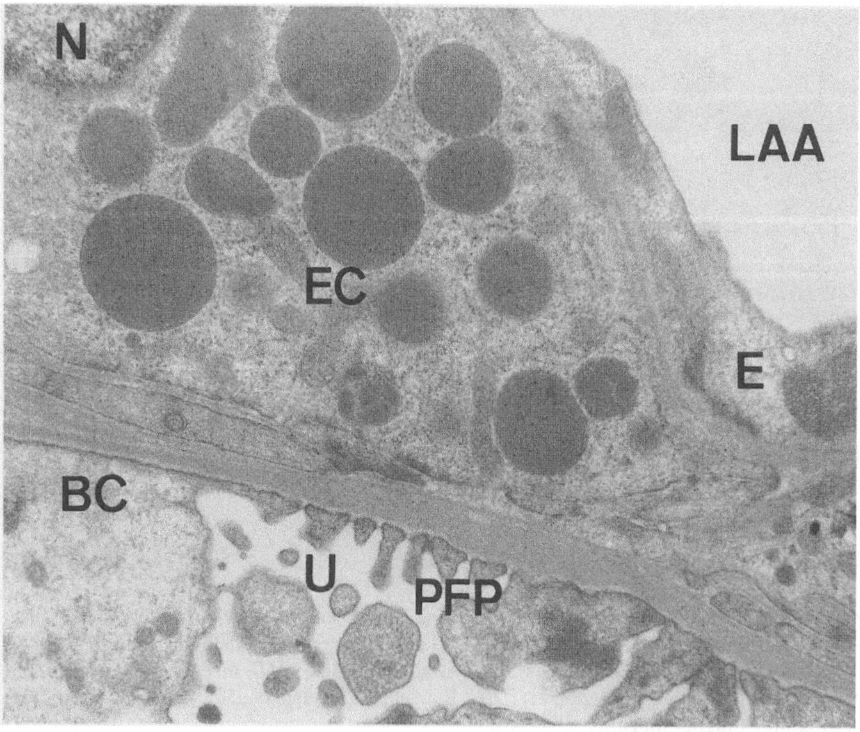

Fig. 3.6. Rat podocyte foot processes *(PFP)* in the parietal layer of Bowman's capsule *(BC)* facing an epithelioid cell *(EC)* replete with renin-positive secretory granules. *LAA*, lumen of the afferent arteriole; *E*, endothelium; *N*, nucleus; *U*, urinary space; ×22000

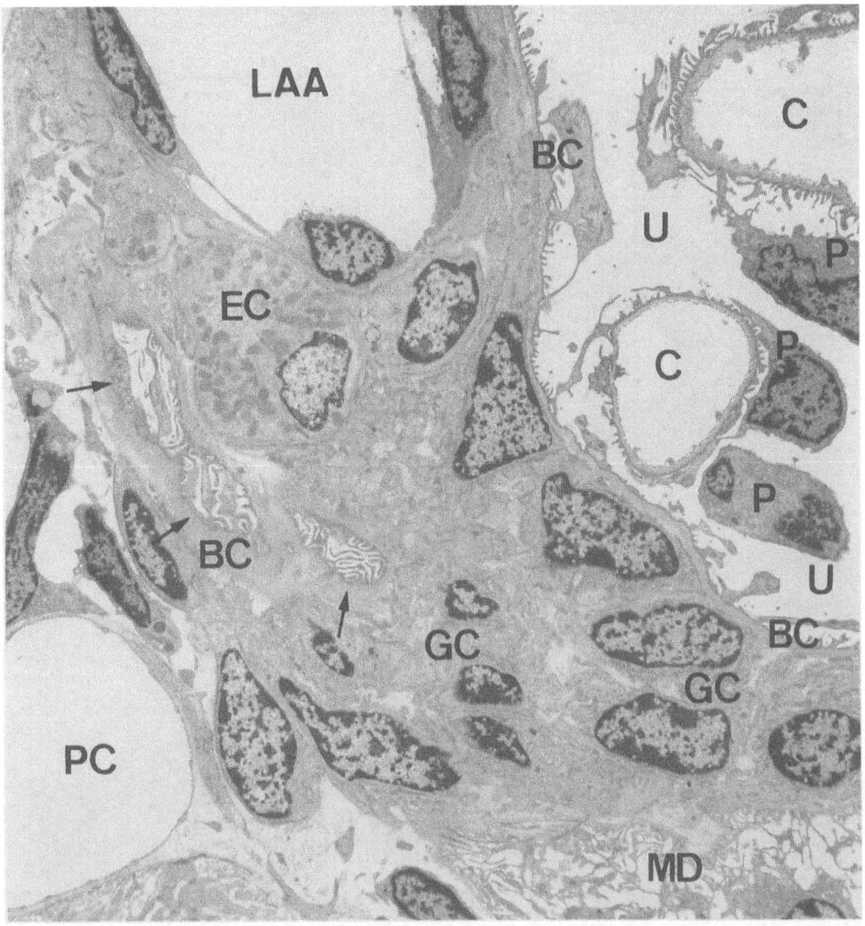

Fig. 3.7. Part of a juxtaglomerular apparatus from *Tupaia* kidney sectioned close to the entrance of the afferent arteriole in the renal corpuscle. Note the extended area of the parietal layer of Bowman's capsule *(BC)* formed by podocyte foot processes and finger-like protrusions of the urinary space *(arrows)* into the region of fusion between the wall of the afferent arteriole and the lacis. *LAA*, lumen of the afferent arteriole; *EC*, epithelioid cells; *GC*, Goormaghtigh cells; *MD*, macula densa; *BC*, Bowman's capsule; *U*, urinary space; *C*, glomerular capillary; *PC*, peritubular capillary; *P*, podocytes; ×4300

brane with regard to their varying size and the lack of a diaphragm (inset, Fig. 3.5). However, internephron heterogeneities as well as interspecies variations were found concerning the occurrence of endothelial fenestrae in the wall of the incoming afferent arteriole. Fenestrations facing the lacis seem to be rare in the efferent vessel. Based on these findings, ROSIVALL and TAUGNER (1986) reconsidered the correlation between structure and function as far as the interstitial fluid balance, TGF, and renin secretion are concerned.

The most widely accepted suggestion, originally based on the histology of the JGA, is that the macula densa is able to sense changes in the incoming distal tubular fluid and to transfer them in some unknown way to the interstitium of the JGA. Here, the Goormaghtigh cells are thought to be receptor cells that convert the response of the macula densa into a signal, which finally is propagated to the glomerular vasculature. With these assumptions in mind, most morphological studies devoted to the TGF have emphasized the mutual contact areas between the macula densa and the vascular components of the JGA, the glomerular arterioles, and the so-called lacis or Goormaghtigh cell field (BARAJAS and LATTA 1963a; BARAJAS 1970; CHRISTENSEN and BOHLE 1978; AEIKENS and HILDEBRAND 1981; cf. Sect. 2.6).

Whatever the mediator(s) of TGF, the first step in signal transfer from the distal tubule can only be of a humoral nature because of the lack of membrane contacts between the cells of the tubular and vascu-

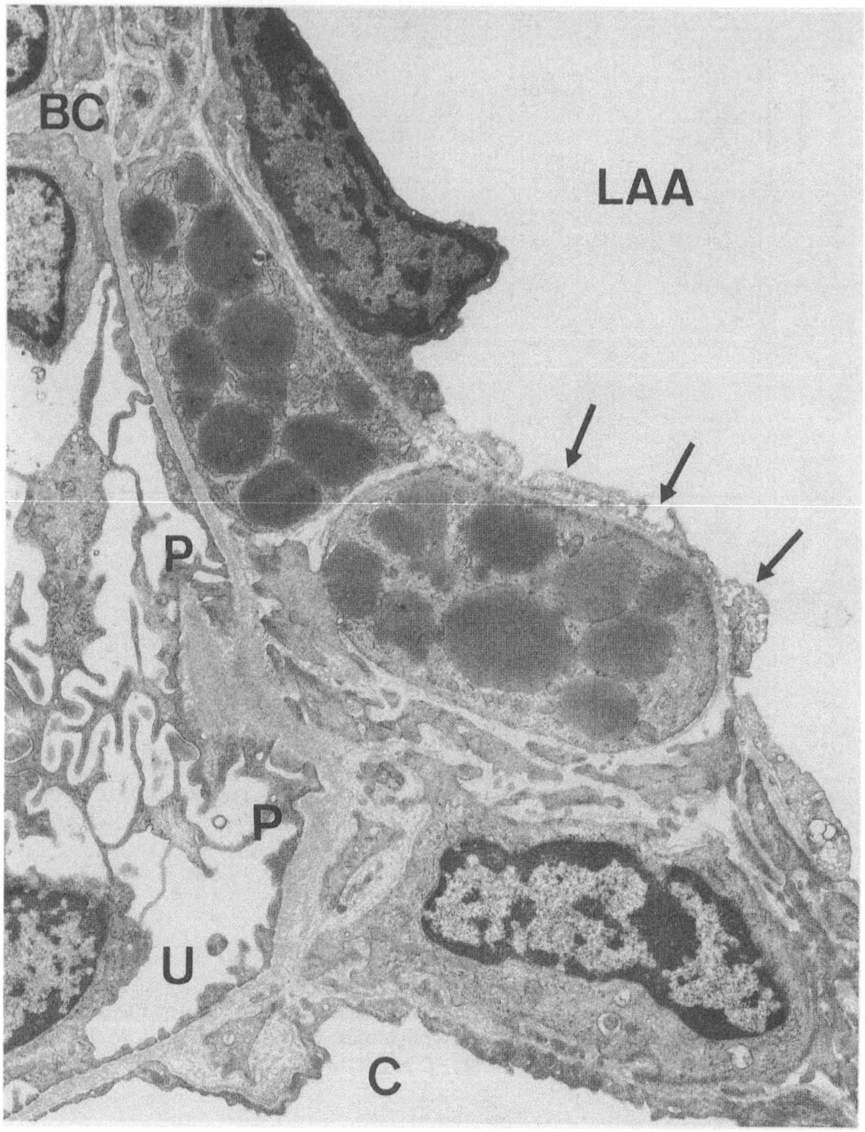

Fig. 3.8. Epithelioid cell processes in the juxtaglomerular portion of the afferent arteriole interposed between the fenestrated endothelium of the vessel *(arrows)* and podocyte foot processes *(P)* close to the visceroparietal junction of Bowman's capsule *(BC)*. *LAA*, lumen of the afferent arteriole; *C*, glomerular capillary; *U*, urinary space; ×9000

lar components of the JGA. Therefore, factors influencing fluid flow or fluid balance in the interstitium of the JGA may play an important role in TGF, e.g., by altering mediator transport to, or mediator concentration at, the involved target cells. Because neither capillaries nor lymph vessels exist inside the JGA, it has been assumed that fluid flow as well as fluid and solute balance in the interstitium of the JGA depend primarily upon reabsorptive transport processes driven by the macula densa as a part of the distal tubule (for references see SCHNERMANN and BRIGGS 1985). A mathematical model has been used by MOORE et al. (1988) to explore the transport-coupling hypothesis of TGF signal transmission from the macula densa to the "lacis." If the macula densa reabsorbate creates a variable pressure gradient for fluid flow through the mostly narrow interstitial spaces of the lacis region, it finally would probably be directed to peritubular or lymph capillaries outside the JGA (THOENES 1961).

Based on the findings of (a) fenestrated endothelium in the hilar portion of the afferent arteriole, (b) interpodocyte filtration slits close to the lacis region, and (c) the mesangial-type lining of the glomerular stalk, ROSIVALL and TAUGNER (1986) suggested that fluid flow and fluid balance in the interstitium of the JGA are determined in a more complex way than assumed previously. Since these fenestrations and filtration slits are morphologically similar to those in the filtration membrane of glomerular capillaries, they were suggested to allow bulk fluid flow into the interstitium of the adjacent JGA components, i.e., the Goormaghtigh cell field and the media of the hilar arterioles, provided that the pertinent pressure gradients are present. In addition, there should be some fluid flow from the mesangium of the glomerular tuft draining via the glomerular stalk into the interstitium of the adjacent lacis.

According to micropressure measurements in rat kidney, the hydrostatic pressure in the afferent glomerular arteriole is approximately 50-60 mm Hg, in the efferent arteriole 45-50 mm Hg, in peritubular capillaries 9-13 mm Hg, and in Bowman's space 13-15 mm Hg (cf. NAVAR et al. 1983). The pressure drop in glomerular capillaries is considered to be minimal. The colloid osmotic pressure in the afferent arteriole is approximately 20 mm Hg, in glomerular and peritubular capillaries about 25 mm Hg, and in the urinary space close to zero. Interstitial pressures inside the JGA cannot be measured. If they were similar to those found in the adjacent cortical interstitium, the hydrostatic pressure in the interstitium of the JGA would be 1-3 mm Hg, and the colloid osmotic pressure 2-3 mm Hg. Based on these data, one could further assume an effective filtration pressure of 30-35 mm Hg from the hilar portion of the afferent arteriole, 20-25 mm Hg from glomerular capillaries, and 14-15 mm Hg from Bowman's space, all directed to the interstitium of the JGA. From the JGA area a negative (reabsorptive) effective pressure of 12-15 mm Hg directed to peritubular capillaries might exist to allow drainage of the fluid reabsorbed by the macula densa and/or filtered from the above-mentioned areas into the interstitium of the JGA (THOENES 1961).

These assumptions have recently been tested by ROSIVALL and TAUGNER (1988), using ferritin particles injected into the renal artery. Characteristically, particle density profiles developed with gradients directed from the glomerular mesangium and the hilar portion of the afferent arteriole to the base of the macula densa. As soon as 25 s after the injection, the tracer appeared in the interstitium of the lacis; during the next 2 min, an increase of the particle density close to the base of the macula densa was oserved. While a great number of ferritin particles could be seen in the subendothelial interstitium of the afferent arteriole and glomerular capillaries bordering on the mesangium, there were no particles close to the wall of the efferent arteriole, indicating a marked difference in penetration of the tracer through the wall of the hilar segments of the pre- and postglomerular arterioles. Macromolecular transport through the glomerular mesangium has already been shown by several authors using different tracers (for review see MICHAEL 1980). Although most of these studies concentrated on the uptake of the tracer by mesangial cells, particles were observed in the interstitium of the JGA hours after the injection (LATTA and MAUNSBACH 1962b; LEIPER et al. 1977; LEE and VERNIER 1980).

Taken together, not only the morphological data and the calculated pressure gradients in the hilar region of the renal corpuscle, but also the distribution of ferritin particles shortly after injection into the renal artery, indicate the existence of a variable bulk fluid flow from the hilar portion of the afferent arteriole, from the lumen of some glomerular capillaries, and also from the urinary space into the interstitium of the JGA (Fig. 3.9). These flows were suggested to be important codeterminants of the interstitial fluid balance and fluid composition in the lacis area, thus modulating the influence of the macula densa and may, themselves, act as signals to Goormaghtigh or epithelioid cells. The important feature of the suggested signals arising from the vascular side would be that they respond immediately to hemodynamic changes, whereas signals of tubular origin would have a greater delay. This suggestion seemed to be supported by the fact that the area of the fenestrated endothelium of the afferent arteriole in some JGAs, and that of the glomerular stalk and lacis covered by podocyte foot processes, is of the same order of magnitude as the contact area between the macula densa and the other JGA elements (CHRISTENSEN and BOHLE 1978; AEIKENS and HILDEBRAND 1981).

Several recent physiological studies support the assumption that the reabsorptive function of the macula densa may not be solely responsible for the efficiency of TGF. After NaCl loading, for instance, the apparent feedback sensitivity is reduced (DEV et al. 1974; MÜLLER-SUUR et al. 1975), despite continuous or even enhanced NaCl reabsorption, suggesting that the macula densa signal has been diminished or counteracted. Observations that fluctuations of the interstitial pressure in kidney cortex and changes in the osmolality of distal tubular fluid

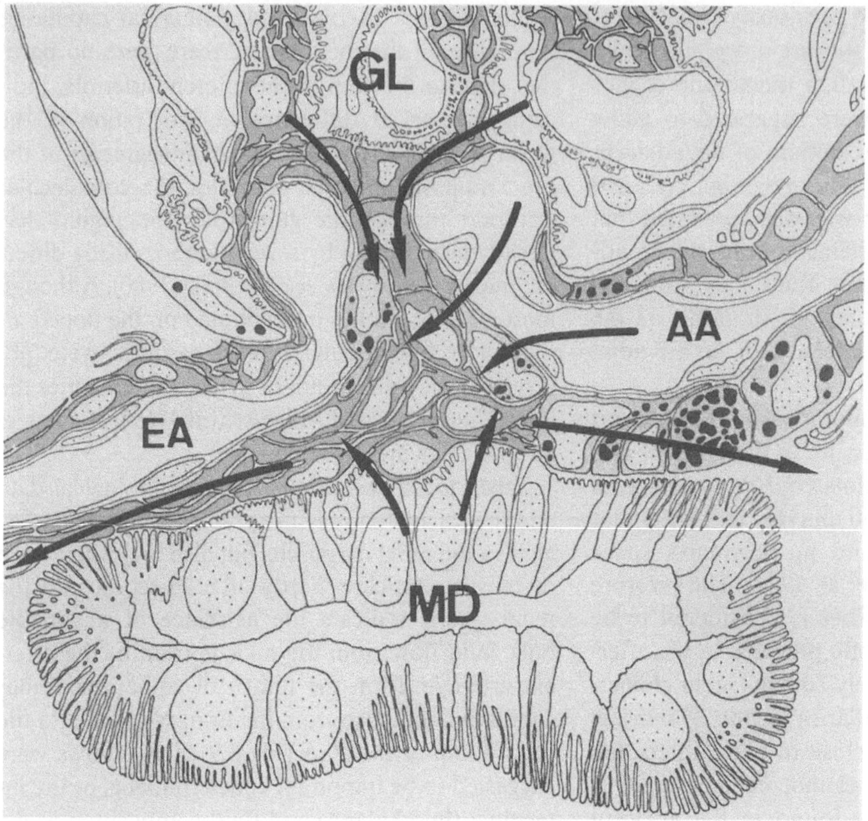

Fig. 3.9. Suggested direction of fluid flow into and out of the interstitium of the JGA. *GL*, glomerulus; *AA*, afferent arteriole; *EA*, efferent arteriole; *MD*, macula densa

can alter or trigger TGF responses point in the same direction (PERSSON et al. 1982; NAVAR et al. 1978).

Irrespective of their significance for TGF, the results of ROSIVALL and TAUGNER (1986) could also be relevant for renin secretion from juxtaglomerular epithelioid cells. Fluid flow from the afferent arteriole, glomerular capillaries, and the urinary space directed into the interstitium of the JGA would also affect the extracellular milieu of these cells known as an important determinant of renin release (see KEETON and CAMPBELL 1980). Therefore, changes in pressure gradients directed to this interstitium, i.e., predominantly changes in the afferent arterioles and glomerular filtration pressure, may alter renin secretion by changing the composition and movement of fluid in the interstitium of the JGA. As the fluid competing with the macula densa reabsorbate is derived from blood plasma, any changes in plasma composition relating to mediators, hormones, electrolytes, and osmotically active solutes may influence renin secretion from juxtaglomerular epithelioid cells also by this short way and not only by mediation of the peritubular capillaries and the interstitium of the cortical labyrinth.

According to LEYSSAC (1986), a decrease of distal tubular flow may be followed by an increase of renin concentration in the proximal tubular fluid of the same nephron. This observation was interpreted to result from renin secretion by epithelioid cells into the lumen of the corresponding afferent arteriole followed by glomerular filtration of the enzyme. As pointed out previously, the pressure gradients existing would not permit any convective fluid flow carrying renin from the interstitium of the JGA, where it is secreted, into the lumen of the preglomerular arteriole. One of the few microtopographical arrangements compatible with the assumption of some leakage of renin into the urinary space of the glomerulus would be that shown in Fig. 3.8, where endothelial fenestrations in the wall of the incoming afferent arteriole are apposed over a relatively short distance to the process of an epithelioid cell followed by podocyte foot processes lining the glomerular stalk. Another possibility is shown in Fig. 3.10, with a granulated cell process located in

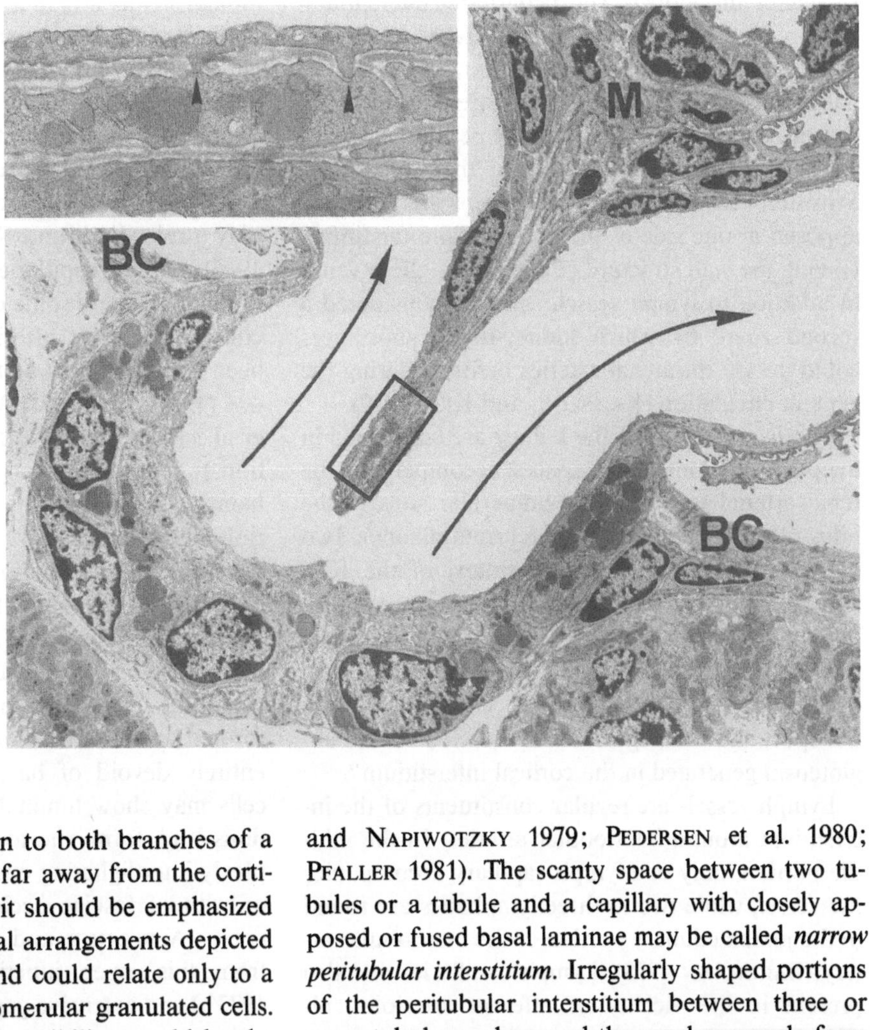

Fig. 3.10. Afferent arteriole from mouse kidney, branching out shortly before the renal corpuscle. *Inset:* process of granulated cell located in the narrow media common to both branches of the vessel, far away from the cortical interstitium and peritubular capillaries but close to the glomerular stalk. *Arrowhead,* myoendothelial contact; *BC,* Bowman's capsule; *M,* mesangial cells; ×3000 and ×14000, respectively

the narrow media common to both branches of a dividing afferent arteriole far away from the cortical interstitium. However, it should be emphasized that the microtopographical arrangements depicted in these figures are rare and could relate only to a minute fraction of juxtaglomerular granulated cells. Still another hypothetical possibility would be the escape into the urinary space of renin secreted by mesangial cells (cf. Sect. 3.2.2 and Chap. 9).

3.4 Cortical Interstitium and Lymph Vessels

The cortical interstitium has been studied with the electron microscope by ROMEN and THOENES (1970) and DIETERICH (1973) in rats, by KRIZ and DIETERICH (1970) in eight different species including dog, cat, and guinea pig, and by BULGER and NAGLE (1973) in rabbits (for reviews see PEDERSEN et al. 1980; KRIZ and KAISSLING 1985; KRIZ 1987). Distinction can be made between a peritubular and a periarterial interstitium.

The *peritubular interstitium* amounts to 7%-9% of the cortical volume, including interstitial cells (KRIZ and NAPIWOTZKY 1979; PEDERSEN et al. 1980; PFALLER 1981). The scanty space between two tubules or a tubule and a capillary with closely apposed or fused basal laminae may be called *narrow peritubular interstitium.* Irregularly shaped portions of the peritubular interstitium between three or more tubules and around the renal corpuscle form the so-called *wide peritubular interstitium* (BULGER and NAGLE 1973; PEDERSEN et al. 1980). Morphometric data suggest that a major part of the tubular reabsorbate has to traverse the wide interstitial spaces before entering the fenestrated peritubular capillaries. Besides microfibrils, collagenous fibrils are found as single units, in small groups, and in larger bundles in the narrow and wide peritubular interstitium, respectively (BULGER and NAGLE 1973). Most interstitial cells of the renal cortex are of the fibrocyte type, some of them exhibiting phagocytotic activity (ROMEN and THOENES 1970).

The *periarterial interstitium,* best visible in quick-frozen kidneys (KRIZ and DIETERICH 1970; SWANN and NORMAN 1970) has been estimated to occupy 15% of the cortical volume. Continuous with the peritubular interstitium, it represents a loose layer of connective tissue which surrounds the branches of the renal arterial tree and contains the lymph vessels of the kidney. The periarterial interstitium, large in the region of the arcuate artery, attenuates peripherally and ends at the vascular pole of the renal corpuscle. Constituents of the periarterial interstitium, besides those found in the peritubular interstitium, are nerve processes and their associated Schwann cells. The interlobular veins are generally apposed at one side of the pariarterial interstitium. Having the wall structure of capillaries, these veins, in addition to lymph vessels, may be considered a second route by which kidney-borne substances could act on intrarenal arteries before entering the general circulation (KAISSLING and KRIZ 1979).

The *lymph vessels* of the kidney are embedded in the periarterial interstitial spaces, accompanying the renal arterial tree from preglomerular sites to the hilus of the organ in a countercurrent manner. Two questions are relevant in the context of the JGA and of renin secretion: what are the topographical relationships between the JGA or, rather, between the granulated cells and the lymph vessels and, secondly, do lymphatics represent channels for the transport and drainage of renin secreted and of angiotensin generated in the cortical interstitium?

Lymph vessels are regular constituents of the interstitium around interlobular arteries. In the middle portion, only one lymph capillary accompanies the vessel, while two to three lymphatics are found in the juxtamedullary portion of the interlobular artery (BARGMANN 1978). Lymph capillaries are also present in the vicinity of afferent arterioles, although it was a matter of dispute whether they already begin at the vascular pole of the renal corpuscle (for a review of the older literature see ROJO-ORTEGA et al. 1973b). Also, the classification of lymph vessels into "interlobular" and "intralobular" lymphatics is viewed differently (cf. KRIZ and KAISSLING 1985). Here, species differences may play a role (KRIZ and KAISSLING 1985; NIIRO et al. 1986). Lymph capillaries beginning in the periglomerular area and some of them in the immediate vicinity of afferent arteriolar epithelioid cells have been found by ROJA-ORTEGA et al. (1973b), GORGAS (1978a, cf. Fig. 3.11), O'MORCHOE (1978), NIIRO et al. (1986), and TAUGNER (unpublished observations); these findings pertain to rat, dog, golden hamster, and mouse. In the area of the efferent arteriole, lymphatics have never been encountered. Likewise, a continuation of the periarterial fluid-rich tissue sheath along the postglomerular arteriole has not been observed (KRIZ 1987).

Characteristically, and in contrast to peritubular capillaries, the wall of lymphatic capillaries is formed by a nonfenestrated endothelium, largely or entirely devoid of basal lamina. The endothelial cells may show luminal and/or abluminal projections and form end-to-end, overlapping or complexly interdigitating endoendothelial contacts with or without fasciae occludentes or fasciae adhaerentes (ALBERTINE and O'MORCHOE 1980). Apart from this, open junctions (ROJO-ORTEGA et al. 1973b) and extensive gaps in the wall of lymph cap-

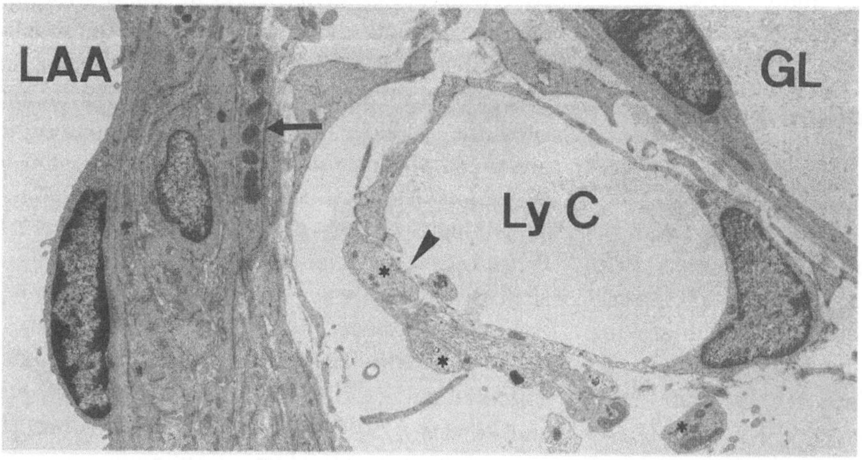

Fig. 3.11. Lymph capillary *(Ly C)* close to the juxtaglomerular portion of the afferent arteriole in rat kidney. *GL*, glomerulus; *LAA*, lumen of the afferent arteriole; *arrow*, granulated cell in the media of the vessel; *arrowhead*, gap in the wall of the lymph capillary, covered by one of several adrenergic axon terminals or varicosities seen in the section plane *(asterisks)*; ×4600. (From GORGAS 1978a)

illaries can be observed with a certain regularity, although at somewhat larger distances (GORGAS 1978a; KRIZ 1987; TAUGNER unpublished observations; cf. Fig. 3.11).

Although it is uncertain which mechanisms are essential for the exchange of water and solutes between interstitium and lymphatics, lymph vessels in the kidney certainly represent, as they do in other organs, an important drainage system of the interstitium (KRIZ and DIETERICH 1970). It is therefore possible, with due reservation, to infer the composition of the cortical interstitial fluid from the constitution of kidney lymph, considering that lymph fluid, as far as it is derived from the area of the renal corpuscles, is "diluted" by the tubular reabsorbate, perhaps to an extent varying among the species (NIIRO et al. 1986).

Renin is contained at higher concentrations in kidney lymph than in the blood plasma (LEVER and PEART 1962; HIGGINS et al. 1964; HOSIE et al. 1970; HORKY et al. 1971; BAILIE et al. 1971; 1972; KHURI et al. 1978; O'MORCHOE et al. 1978, 1981; PROUD et al. 1984; INAGAMI et al. 1986). From this, it was concluded that renin is at least partially secreted into the interstitium of the renal cortex. In addition, kidney lymph contains angiotensin I and II (BAILIE et al. 1971, 1972; PROUD et al. 1984; INAGAMI et al. 1986). However, since kidney lymph also contains angiotensinogen (HORKY et al. 1971; TAUGNER et al. 1982) and converting enzyme (HORKY et al. 1971; ROJO-ORTEGA et al. 1973b; PROUD et al. 1984), it is possible that the angiotensins, aside from their generation in the interstitium, are also released later during lymph passage. The difficulties involved in the assessement of angiotensin II in the interstitium of the kidney will be discussed in connection with the intrarenal RAS in Chap. 9.

Nevertheless, it appears certain that the concentrations of angiotensin II in the interstitium of the organ are larger than those of angiotensin I (MENDELSOHN 1979, 1982). INAGAMI et al. (1986) on the other hand found angiotensin I values in renal lymph more than one order of magnitude larger than those of angiotensin II. This discrepancy is likely to be due to the release of the decapeptide during the drainage process. Immunoreactive kallikrein has been found by PROUD et al. (1984) in the renal lymph of rats. However, kallikrein levels in renal lymph, thoracic lymph, and plasma did not show the observed differences observed for renin.

Lymph vessels accompany the renal arterial tree in a countercurrent manner. The distribution of tracers injected beneath the renal capsule indicates that there is a remarkable exchange between the lymphatics and the periarterial interstitium (KRIZ 1969; KRIZ and DIETERICH 1970; PEDERSEN et al. 1980), considered as a functional entity by KRIZ (1987). Therefore, it is very probable that the resistance vessels of the kidney come into contact with substances contained in the cortical interstitium, and are affected by them, e.g., by angiotensin II, the concentration of which probably increases during lymph passage. While pursuing such considerations, it should be kept in mind that lymph flow is rather slow and that even brisk changes in the concentration of angiotensin II as they may conceivably occur in the juxtaglomerular area can only become effective with a noticeable delay in the existing countercurrent arrangement.

3.5 Innervation of the Juxtaglomerular Apparatus

GOORMAGHTIGH (1942), calling attention to the analogy between what he termed the "juxtaglomerular neuro-myo-arterial segment" and the subungual neuro-myo-arterial glomi of MASSON (1924), originally thought that the cells of the lacis were Schwann cells, enclosing an intricate network of fine nerve fibers. The existence of a dense innervation of the JGA has subsequently been confirmed by several light microscopists (for references see DE MUYLDER 1952; MAILLET 1959; cf. ROUILLER and ORCI 1971). As expecially the classical silver impregnation methods may artifactually bring out basement-like material, which is abundant in the lacis region, it required the advent of fluorescence histochemistry for biogenic amines and the electron microscope to firmly establish the presence of nerve fibers in association with the JGA, called the "neural component" of the JGA by BARAJAS (1979; for references see GORGAS 1978a). More recently, immunohistochemical investigations added important details to this body of knowledge, especially as far as neuropeptides are concerned. In the following, first the microtopography of the sympathetic innervation of the JGA and the related structures will be dealt with. Subsequently, the ultrastructure of the involved nerve terminals and, finally, the nature of the transmitters and putative cotransmitters will be discussed.

There have been several comprehensive fluorescence microscopic investigations on the distribution of the postganglionic catecholaminergic axons innervating the renal resistance vessels (NILSSON 1965; MCKENNA and ANGELAKOS 1968; DOLEŽEL 1966; DOLEŽEL et al. 1976; BARAJAS and WANG 1975; BARAJAS et al. 1976; BARAJAS 1978; GORGAS 1978a, b). The interlobular artery which passes radially from the arcuate artery to the surface of the kidney is accompanied by a thick anastomosing plexus of adrenergic fibers (Fig. 3.12). Axonal varicosities are found mainly at the boundary between the media and the adventitia of the vessel, occasionally in a niche-like excavation between adjacent smooth muscle cells. The afferent arteriole shows an even denser network of fluorescent axons (Figs. 3.13, 3.14), especially in its distal segment, devoid of internal elastic fibers (GORGAS 1978a, b). From preglomerular sites, some axon bundles are reported reaching the postglomerular arteriole (NILSSON 1965; WÅGEMARK et al. 1968; MUNKACSI 1969; DOLEŽEL et al. 1976; BARAJAS 1978), the fibers of which may be traced along the efferent vessel a distance through the cortical labyrinth or - in juxtamedullar nephrons - with the vasa recta until into the outer medulla (MCKENNA and ANGELAKOS 1968; DOLEŽEL et al. 1976; GORGAS 1978a). For the rest, the innervation of the vascular poles of cortical and juxtamedullar renal corpuscles does not differ (GORGAS 1978a).

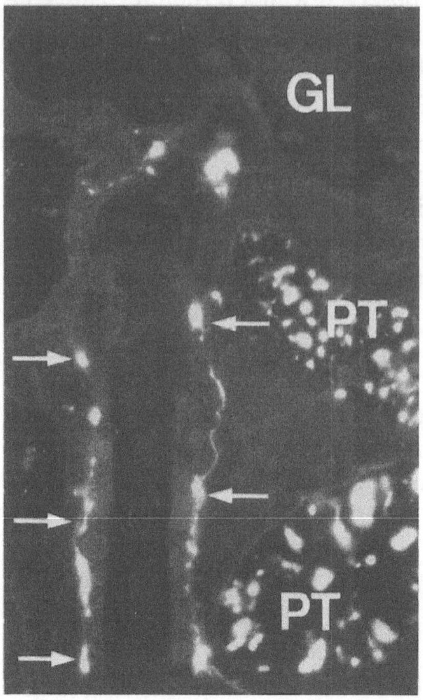

Fig. 3.13. Monoaminergic innervation of an afferent arteriole from rat kidney *(arrows)*. *GL*, glomerulus; *PT*, proximal tubules, showing autofluorescence. Formalin-induced fluorescence; ×800. (From GORGAS 1978)

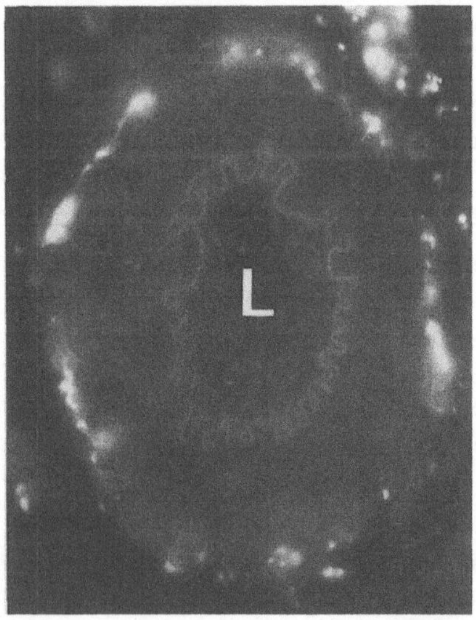

Fig. 3.12. Adrenergic innervation of cross-sectioned interlobular artery from rat kidney. *L*, lumen of the vessel. Formalin-induced fluorescence; ×800. (With kind permission of K. GORGAS, Heidelberg)

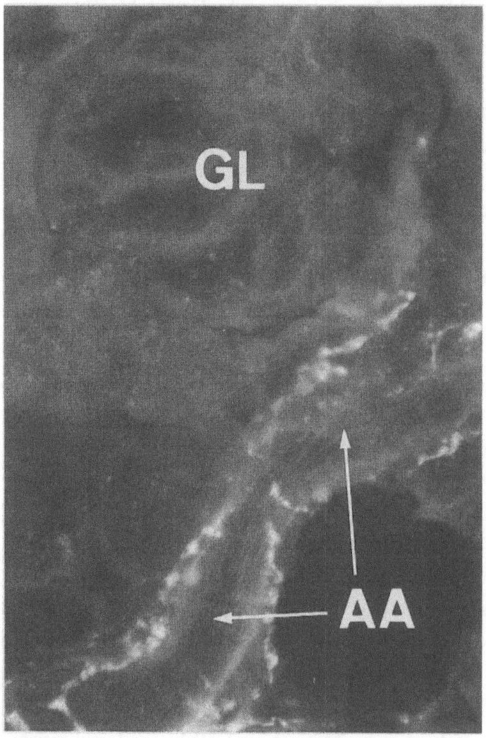

Fig. 3.14. Extensive monoaminergic plexus innervating the afferent arteriole *(AA)* of *Tupaia* kidney. *GL*, glomerulus. Frozen section from a glyoxylic acid-paraformaldehyde perfused kidney; ×450

Details on the innervation of the different JGA components were disclosed by the electron-microscopic findings of BARAJAS (1964), BIAVA and WEST (1966a), SIMPSON and DEVINE (1966), ROJO-ORTEGA et al. (1968), and especially through the examination of serial thin sections by BARAJAS and MÜLLER (1973). According to this study, in the area of the JGA, about one-third of the media cells of the glomerular arterioles and half of the granulated cells were innervated (Figs. 3.15, 3.16). Most of the innervated cells had contact with more than one axon; besides, axons were seen to contact not only the media of the afferent arteriole, but also cells of the efferent vessel. According to GORGAS (1978a), multiaxonal synaptic contacts of individual media cells are most numerous in the preglomerular arteriole. It is not known if these multiple contacts are of multineuronal origin. Apart from the media cells of the glomerular arterioles, BARAJAS and MÜLLER (1973) and GORGAS (1978a) also found cells of the lacis to be innervated; the fact that only about one-tenth of the Goormaghtigh cells show synaptic contacts was attributed to the relatively small amount of free external surface of the lacis (BARAJAS and MÜLLER 1973).

From the interlobular artery downstream, the paravasal connective tissue diminishes continually. Consequently, the tubular structures come into closer contact with the afferent arteriole and the accompanying catecholaminergic nerve plexus (GORGAS 1978a). Only in some species, such as the monkey, a separate tubular innervation arising from the periarteriolar axon bundles has been found (MÜLLER and BARAJAS 1972; BARAJAS 1978). In the others,

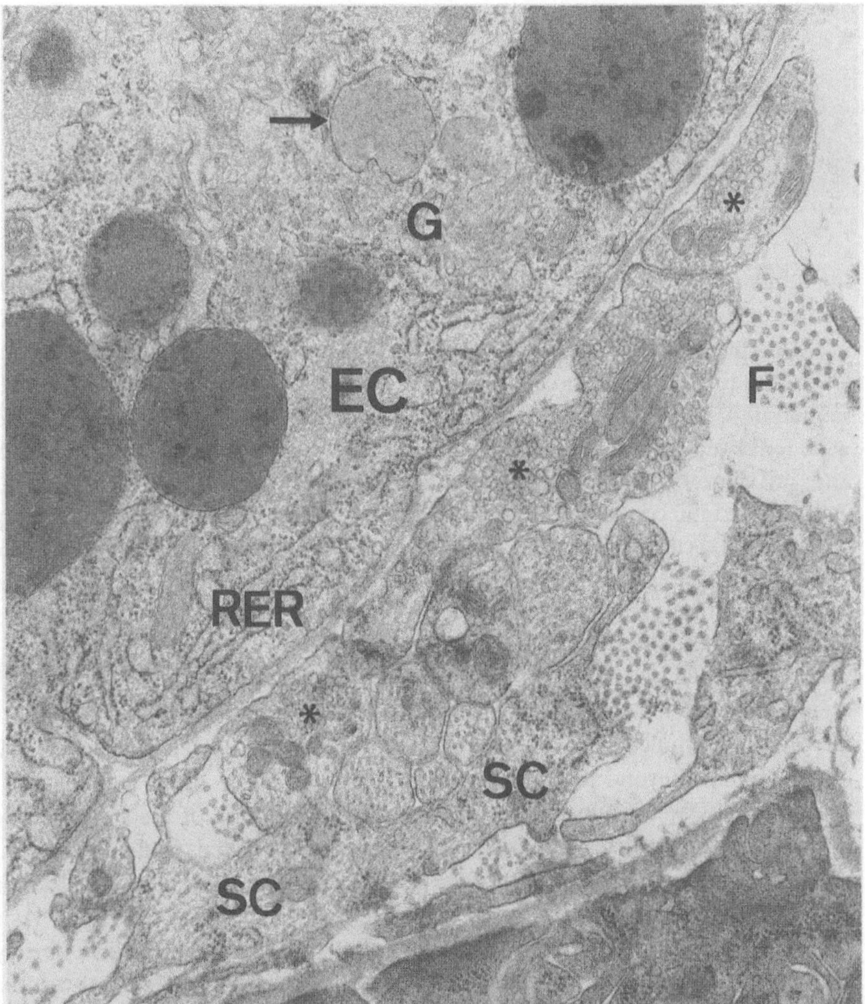

Fig. 3.15. Axon-Schwann cell complex, associated with an epithelioid cell *(EC)* in the juxtaglomerular portion of the afferent arteriole from mouse kidney. Between the vesicle-filled terminals or varicosities *(asterisks)* and the Schwann cell process *(SC)*, preterminal axon segments are seen. *RER*, rough endoplasmic reticulum; *G*, Golgi complex; *arrow*, coalescing protogranules; *F*, collagen fibrils. It should be noted that the fixation procedures used were not favorable for the conservation of the dense core otherwise characteristic for small vesicles of catecholaminergic axon terminals; ×27200

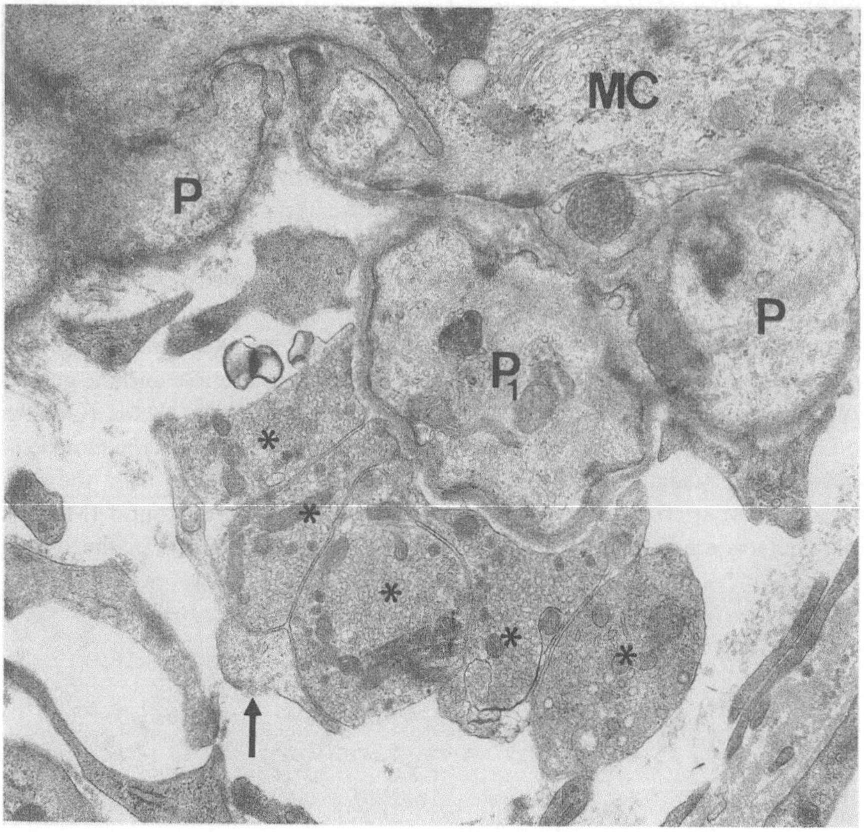

Fig. 3.16. Axon terminals *(marked by asterisks)*, associated with one *(P_1)* of several media cell processes in the proximal portion of the afferent arteriole in *Tupaia* kidney. The synaptic cleft has a constant width of about 130 nm. *MC*, perikaryon of media cell; *arrow*, Schwann cell process; ×21000

the tubular innervation seems to be confined mostly to the course of the glomerular arterioles, especially in the juxtaglomerular region (Barajas 1978). Thus, according to Reinecke and Forssmann (1988), extremely few nerve fibers are exclusively associated with renal tubules in *Tupaia*, dog, pig, guinea pig and rat. Nevertheless, distal tubular segments as well as proximal convolutions and the parietal layer of Bowman's capsule are contacted by axon varicosities, suggesting that all portions of the cortical tubular nephron are under some degree of nerval influence. The highest relative frequency of innervation occurred in the thick ascending limb of Henle's loop, followed by the distal convoluted and the proximal tubule (Barajas et al. 1984). From the distal nephron beyond the thick ascending portion of Henle's loop, an outstanding dense monoaminergic innervation of the late distal convoluted tubule-connecting tubule was observed (Barajas et al. 1985). In comparison with the innervation of the vascular components, however, tubular contacts were found much less frequently (Barajas and Müller 1973; Gorgas 1978a). Catecholaminergic axons were never seen to penetrate the lacis or the contact area between the tubular and vascular components of the JGA, thus innervating the distal tubule only outside the macula densa, beginning with the intermediate epithelial cells. Fibers extending to Bowman's capsule are reported to fade before reaching the urinary pole of the renal corpuscle (Gorgas 1978a; Barajas 1979).

In the course of the proximal segment of the afferent arteriole, the accompanying Schwann cells lose their basal laminae and branch off into several finger- or plaque-like processes which thereafter only partially cover the axon bundles or their varicosities. Thus at the hilus of the renal corpuscle, mostly free or only partially covered axon bundles are seen (Figs. 3.15, 3.16). Between varicosities or true nerve endings and the effector cells in the region of the JGA, neuroeffector junctions are established with a synaptic cleft of about 100-200 nm traversed by basal lamina material (Barajas 1964; cf. Figs. 3.15, 3.16). Aside from Schwann cell processes, often also fibrocyte processes difficult to distinguish from these make contact with the axon bundles. In the area of the JGA, flat processes of Schwann cells may demarcate the axon bundles from the interstitium in a roof-like manner (Fig. 3.15). Individual axons extending through the interstitium to the neighboring tubular segments are mostly devoid of any enveloping cell processes.

According to BARAJAS (1979), the nerve endings and varicosities of the juxtaglomerular region contain many small dense-cored and clear vesicles, with large dense-cored and clear vesicles in the minority (GORGAS 1978a, b; cf. Figs. 3.15, 3.16). The small dense-cored vesicles are considered to contain norepinephrine and thus to be characteristic of adrenergic nerve fibers.

A number of results indicate that glomerular arterioles and the JGA are in fact, by far, if not exclusively, innervated by catecholaminergic axon bundles. These are: the uptake of tritiated norepinephrine by virtually all axons encountered in the region of the JGA (DOLEŽEL et al. 1976; BARAJAS and WANG 1978, 1979); the disappearance of small dense-cored vesicles after reserpine injection or denervation of the kidney, correlating with the abolition of the catecholamine fluorescence and the depletion of norepinephrine from the cortical tissue (McKENNA and ANGELAKOS 1968; SILVERMAN and BARAJAS 1974; DOLEŽEL et al. 1976); and, finally, the disappearance of all axon bundles in the region of the JGA after 6-hydroxydopamine treatment (BARAJAS and WANG 1975; BARAJAS 1978; TAUGNER, unpublished observations; cf. Sect. 4.1.2). In addition, serial section electron microscopy revealed that, at one point or another, all axons contained small dense-cored vesicles (BARAJAS and MÜLLER 1973; GORGAS 1978a, b). From these results it was suggested that the innervation of the juxtaglomerular region, at least in the species studied in detail, the rat, is exclusively adrenergic. In accordance with this conclusion, the existence of a cholinergic innervation of the JGA as suggested by acetylcholinesterase-positive histochemical reactions, has been ruled out (BARAJAS and WANG 1975).

Most of the referred results have been obtained in rats, some also in the mouse, the dog, and the monkey. The only interspecies comparison with a wider scope was performed by DOLEŽEL et al. (1976). In their fluorescence histochemical study, the authors found a quite dense afferent nerve plexus in the guinea pig, the rat, and the dog, but only one or two fluorescent fibers in mice, cats, pigs, monkeys, and humans. Also, fluorescent axons were constantly observed along the efferent arterioles of all species studied except guinea pig and cat, where the fluorescent fibers could not always be followed beyond the vascular pole of the renal corpuscle. As compared with rats, a strikingly denser sympathetic innervation of the kidney and especially of the JGA was found by BARAJAS (1964) in monkeys *(Macaca nemistrina leonina)* and by TAUGNER et al. (1980) in tree shrews *(Tupaia belangeri)*, concurrent with a significantly higher catecholamine content of the kidney and other organs (Figs. 3.13, 3.14). As the JGA of *Tupaia* contains a great number of epithelioid cells abounding in renin granules and as the renin content of *Tupaia* kidney is considerably higher than in the rat, it was speculated that intrarenal neuroendocrine interactions might favor the development of acute renal failure known to occur in this species under certain conditions of social stress (von HOLST 1972).

Recently, nerve growth factor (NGF) immunoreactivity has been observed in the connecting tubule of the mouse (SALIDO et al. 1986b). In view of the tubulovascular proximities described in Sect. 3.1, NGF from the connecting tubule, possibly activated by kallikrein (BARAJAS et al. 1986), was suggested to be involved in the regulation of the density of the adrenergic innervation in the region of the afferent arteriole (BARAJAS et al. 1988).

As postsynaptic counterparts of adrenergic nerve endings specific receptors are expected to exist in kidney tissue. The role of α- as well as of β-receptors in relation to the function of the JGA, especially as far as renin secretion is concerned, will be dealt with in Chap. 7. Briefly, renal vasoconstriction upon sympathetic stimulation has been shown to be mediated by α_1-receptors, renin secretion by β-receptors. In this section, we will only refer to results relating to the distribution of adrenoreceptors in kidney vessels and the JGA. According to autoradiographic studies using in vitro labeling of rat kidney slices, high concentrations of β-receptors are associated with glomeruli (SUMMERS and KUHAR 1983; MÜNZEL et al. 1984). These highly localized adrenoceptors were found to be of the β_1 subtype in rats as well as in guinea pigs (SUMMERS et al. 1985; LEW and SUMMERS 1985). Obvious high silver grain densities over cortical blood vessels have not been observed except at the vascular pole of the renal corpuscle (SUMMERS and KUHAR 1983) and the afferent arteriole as it enters the glomerulus (MÜNZEL et al. 1984). It is tempting to assume that this high autoradiographic labeling is related to β-adrenoceptors associated with epithelioid cells.

Earlier autoradiographic studies performed after in vivo administration of adrenergic antagonists like propranolol should be judged critically, as these are weak bases, i.e., acidotropic substances, which may accumulate in lysosomes and probably also in renin granules independently from receptors (CRAMB 1986; cf. Sect. 5.3).

In contrast to β-receptors, α_1-receptors were observed also associated with kidney arteries (MUNTZ et al. 1985). However, pertinent information for the juxtaglomerular portion of the afferent arteriole is not available.

Pharmacological evidence for dopamine (DA_1) receptors in the canine renal vasculature (GOLDBERG 1972; GOLDBERG et al. 1978) prompted the search for a counterpart dopaminergic innervation. To delineate a renal dopaminergic system, it would be necessary to localize dopamine to neuronal elements that synthesize, store, and release dopamine not merely as a precursor for noradrenaline (DINERSTEIN et al. 1983). The canine kidney contains dopamine in a relatively high proportion to norepinephrine. Whereas increasing rates of renal nerve stimulation resulted in an approximately linear increase in norepinephrine efflux from the kidney, abolished by chronic denervation, the effects of these procedures were less definitive for dopamine (DINERSTEIN et al. 1983). Using pharmacological methods in combination with histofluorescence, BELL et al. (1978) suggested that the kidney contains a distinct dopamine pool consistent with the existence of dopamine-containing nerves. On the basis of microspectrofluorimetric measurements of individual axon bundles in dog kidney, DINERSTEIN et al. (1979) concluded that those of the vascular pole of the renal corpuscle contain predominantly dopamine as the neuronal catecholamine, those of the periadventitial layer of the arcuate arteries mainly norepinephrine, while those accompanying the interlobular arteries seemend to be composed of a mixture of both kinds of fibers. The question is if the present findings are sufficient for the verification of a specific, physiologically significant dopaminergic innervation of the kidney or if there is only a subpopulation of noradrenergic neurons with dopamine as a precursor and possibly cotransmitter of norepinephrine (BELL 1982; cf. Chap. 7).

In addition to catecholamines, several so-called neuropeptides or hormonal polypeptides have been localized in renal nerves by immunohistochemical techniques, namely vasoactive intestinal polypeptide (VIP), neuropeptide Y (NPY), neurotensin (NT), somatostatin (SOM), calcitonin-gene related peptide (CGRP) and substance P (SP).

Besides many other organs and organ systems, VIP-immunoreactive nerves have also been found in the genitourinary system. HÖKFELT et al. (1978) reported the presence of a sparse plexus of VIP-positive nerve fibers in association with renal blood vessels of the guinea pig kidney. Several other groups, in contrast, reported negative results in different species (for references, see BARAJAS et al. 1983). BARAJAS et al. (1983) found VIP-immunoreactive fibers in dog and rat kidney, associated with the renal artery and its branches. Only in the dog were VIP-positive fibers occasionally seen close to small blood vessels suggestive of arterioles. The existence of neuroeffector junctions of VIP-immunoreactive fibers with juxtaglomerular epithelioid cells

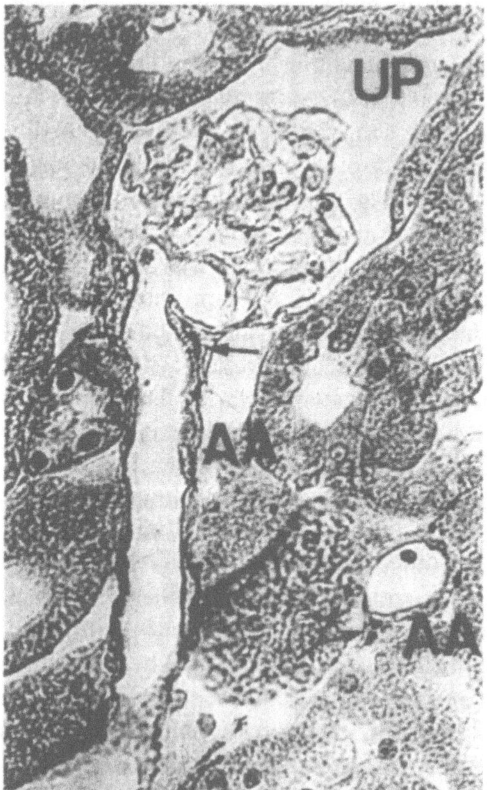

Fig. 3.17. Neuropeptide Y-positive nerve plexus *(arrows)* innervating two afferent arterioles *(AA)* of mouse kidney. *UP*, urinary space of the renal corpuscle. Note that some immunoreactive nerve endings are also found close to neighboring tubules. Seven-micrometer paraffin section from mouse kidney, PAP-method after perfusion fixation with Bouin's fluid containing 0.1% glutaraldehyde; ×450

could not be ascertained. In the guinea pig, VIP- and neurotensin-positive varicosities close to all cortical arteries with an especially dense peptidergic innervation of the juxtaglomerular afferent arterioles were found by HOCK and FORSSMANN (1984) and FORSSMANN and REINECKE (1984). Recently, REINECKE and FORSSMANN (1988) observed VIP-immunoreactive fibers in moderate densities at the adventitia-media border of the entire renal arterial tree including the hilar arterioles and the JGA in five additional species. PORTER et al. (1983) were able to extract radioimmunoassayable VIP from dog and rad kidney, and VIP has been shown to increase renin secretion in vivo as well as from superfused glomeruli, i.e. probably by a direct action of the peptide on granulated cells (PORTER et al. 1984). However, experiments with renal nerve stimulation in anesthetized dogs yielded no evidence suggesting that the effects of VIP on renin secretion are mediated by release of the peptide from renal nerves; instead, the plasma level of VIP was shown to increase under certain conditions sufficiently to stimulate renin secretion by circulating VIP (PORTER et al. 1985, 1988).

NPY, a 36 amino acid peptide originally isolated from the brain (TAKEMOTO et al. 1982), is present in peripheral noradrenergic neurons and exhibits vasoconstrictor properties (LUNDBERG et al. 1982). As could be expected, high concentrations of NPY have also been found in the renal artery and those areas of the kidney which are known to have a sympathetic supply, i.e., the cortex and the juxtamedullary region (RAINE et al. 1984). In addition, NPY-reactive nerve fibers and nerve terminals have been localized by immunohistochemical techniques within the kidney in correspondence with the above-mentioned course of the sympathetic plexus from the renal artery to the region of the vascular pole of the renal corpuscle and the vascular components of the JGA (Fig. 3.17; BALLESTA et al. 1983, 1984; REINECKE and FORSSMANN 1988; own unpublished observations). As the NPY immunoreactivity is markedly depleted after chemical sympathectomy by 6-hydroxydopamine, there can be little doubt that NPY is also associated with the catecholaminergic neuroeffector junctions established with epithelioid cells (BALLESTA et al. 1984). Although data on the sucellular distribution of NPY in the varicosities and terminals of the renal sympathetic fibers are lacking, it may be inferred by analogy that the peptide is contained in large dense-cored vesicles. The role of NPY in the control of renin secretion investigated by HACKENTHAL et al. (1987) will be discussed in Chap. 7.

NT-immunoreactive nerve fibers were found in the kidneys of dogs (REINECKE 1985) and five other species (REINECKE and FORSSMANN 1988), supplying the entire arterial tree and, most abundantly, the JGA. Here, according to these authors, they may play a role in renin secretion, thus constituting the morphological correlate of the suggested effect of NT on the renin-angiotensin system (RIOUX et al. 1982). NT-receptors have been localized in kidney cortex by QUIRION et al. (1982).

SOM immunoreactivity was only observed in single varicosities at the adventitia-media border of large arteries in the region of the renal hilus (REINECKE and FORSSMANN 1988).

CGRP, a polypeptide known from nucleotide sequence studies (AMARA et al. 1982) has reportedly been localized in sensory nerves (ROSENFELD et al. 1983; WIESENFELD-HALLIN et al. 1984); its potency as a vasodilator substance (BRAIN et al. 1985) may, on the other hand, favor the hypothesis of a neuroeffector peptide acting on vascular smooth muscle. In the kidney, CGRP-immunoreactive profiles which, according to REINECKE and FORSSMANN (1988), may represent both afferent and efferent nerves, were observed mainly around large hilus vessels and in the region of the preglomerular arteriole including the JGA.

SP-immunoreactive nerves were found around large renal blood vessels, occasional fibers also at the margins of the glomeruli in rats (FERGUSON and BELL 1985) and, in six species, also around the hilar arterioles (REINECKE and FORSSMANN 1988). As postulated for a variety of other organs (cf. PERNOW 1983), the renal SP-immunoreactive fibers may represent an afferent system, carrying sensory information of several types from the kidney (see FERGUSON and BELL 1985 for references, and REINECKE and FORSSMANN 1988 for discussion of the putative importance of this neuropeptide for the efferent and afferent innervation of the kidney).

The intrarenal distribution of the different peptides showed close resemblances among the six species studied by REINECKE and FORSSMANN (1988). Differences were quantitative rather than qualitative; e.g., in the pig, the general peptidergic innervation seemed to be relatively low, while in the rat, the incidence of NT-immunoreactive fibers was only moderate.

Chapter 4

Intrarenal Distribution of the Different Components of the Renin-Angiotensin System

4.1 Renin

4.1.1 Intrarenal Distribution of Renin

4.1.1.1 Introduction

GOORMAGHTIGH (1939) was the first to suggest the production and secretion of renin by the epithelioid cells of the afferent arteriole, originally described by RUYTER (1925). This suggestion has since been supported by cell culture studies (ROBERTSON et al. 1965), ultracentrifugation experiments (CHANDRA et al. 1964; SCHMIDT et al. 1971a, b, 1972), and, more convincingly, from microdissection (COOK and PICKERING 1958, 1959; BING and KAZIMIERCZAK 1962; VIKHERT and SEREBROVSKAYA 1964; FAARUP 1967, 1968) and micropipetting (COOK 1971) of elements of the JGA. Direct evidence, however, by the demonstration of renin in these cells without disrupting their microtopographical environment, was still lacking at this time.

There have been early attempts to localize renin by immunofluorescence methods (NAIRN et al. 1959; EDELMAN and HARTROFT 1961; HARTROFT 1963; HARTROFT et al. 1964; SUTHERLAND 1970). Since the renin preparations and, consequently, also the antibodies used in these studies were of questionable purity, there was some uncertainty left as to the fluorescent staining reported being due to renin-antirenin complexes (MENZIE et al. 1978).[1]

[1] Although until 1978/1979, it was common practice to view these early immunohistochemical experiments with reservation because of the doubtful specifity of the antigens used, today they are occasionally cited without critical comment. One explanation for this is apparently the partial concurrence between the earlier localizations and those found later using specific antirenin sera. However, as the secretory granules of epithelioid cells not only contain renin, but also lysosomal enzymes which might react with multivalent antisera (cf. Sect. 5.3), doubts pertaining to the validity of the earlier findings with antisera of uncertain specifity cannot be eliminated by this coincidence.

With the purification of mouse submaxillary renin to homogeneity (COHEN et al. 1972; MALLING and POULSEN 1977), monospecific antibodies against this enzyme became available which cross-reacted completely with mouse kidney renin (MICHELAKIS et al. 1974; MALLING and POULSEN 1977). MENZIE et al. (1978), using the indirect immunofluorescence technique, determined the distribution of renin in the mouse submaxillary gland and confirmed the reactivity of their antibody with kidney renin. The authors noticed a prominent granular fluorescence at the vascular pole of the renal corpuscle, corresponding to the juxtaglomerular apparatus. The juxtaglomerular fluorescence, however, could not be localized with absolute certainty to the walls of afferent arterioles nor to the macula densa. TAUGNER et al. (1979), using an antibody to purified submaxillary renin in combination with the peroxidase-antiperoxidase (PAP) technique of STERNBERGER (1974), were able to show that, in mice, renin occurred in high concentrations in the granulated cells of afferent arterioles and, less frequently, of efferent arterioles and interlobular arteries (Fig. 4.1); at high antiserum concentrations, also proximal tubules exhibited a specific reaction, while mesangial and Goormaghtigh cells as well as the macula densa were renin negative (HACKENTHAL et al. 1980a).

MÉNARD et al. (1979), using fluoresceinated antibodies against renin from a juxtaglomerular cell tumor, reported the staining of some interlobular arteries and all afferent arterioles of obsolescent glomeruli in a patient with a partially infarcted kidney.

TANAKA et al. (1980), TAUGNER et al. (1982c, 1984d), CANTIN et al. (1984) and FARAGGIANA et al. (1982) showed that the immunostaining demonstrated in these early experiments is essentially confined to the secretory granules of epithelioid cells in mice as well as in rats and humans, the cytoplasm being renin negative under very different experimental conditions. In animals with stimulated renin

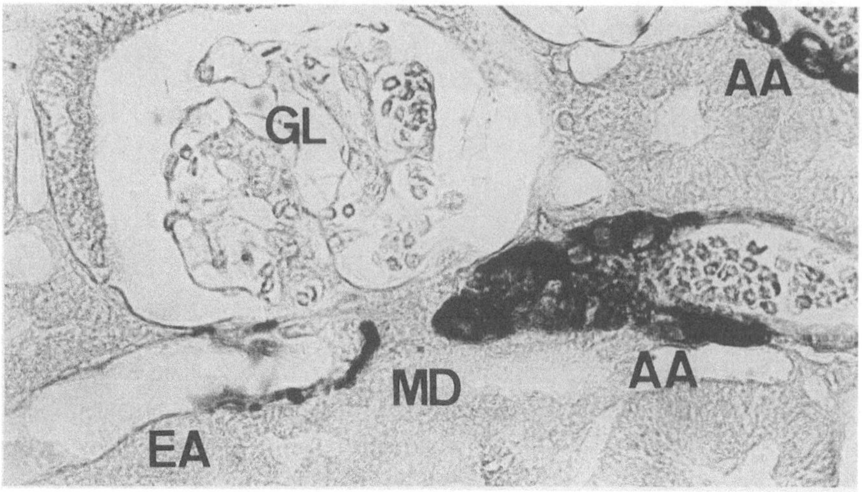

Fig. 4.1. Renin-positive cells in the media of the afferent *(AA)* and efferent glomerular arteriole *(EA)* of mouse kidney. The Goormaghtigh cell field located between the hilar arterioles, the glomerulus *(GL)*, and the macula densa *(MD)* is not immunoreactive. Antiserum dilution, 1:10000; ×520 (From TAUGNER et al. 1979)

synthesis, renin may, in addition, be traced in the cisterns of the RER and the Golgi complex (LACASSE et al. 1985; own unpublished observations; cf. Fig. 4.11). Using antibodies against the prosegment of renin, the question whether the secretory granules of epithelioid cells store renin or prorenin could also be answered. As the prosegment is already cleaved during granulopoiesis in juvenile and intermediate granules, the mature epithelioid cell granules of humans were shown to harbor mature renin (TAUGNER et al. 1986b, 1987a).

The basic knowledge in which of the different vascular components of the nephron granulated cells are localized is only a first step toward a discussion of functional aspects. For the assessment of stimulus-synthesis and stimulus-secretion relationships, quantitative data on the intrarenal distribution of renin-positive cells are required. As to the intrarenal RAS, it is essential to compare the distribution of renin with that of angiotensinogen, converting enzyme, and the so-called angiotensinases, respectively. Before this aspect is discussed in greater detail, two groups of experiments should be remembered, the results of which might be quite valuable in this context when considered with the appropriate reservation:

(1) Experience shows that in the immunocytochemically well examined species - rat and mouse - practically all of the larger granular structures in the epithelioid cells of the glomerular arterioles are renin positive, i.e., secretory granules. It seems therefore justified to integrate the results of the numerous earlier experiments relating to the ultrastructure of granulated cells in the media of the pre- and postglomerular arterioles in these species into the following discussion on the intrarenal distribution of renin. As for Goormaghtigh and in particular mesangial cells, reservations are appropriate in this regard (cf. Sect. 4.1.3). In man, atypical shapes of epithelioid cell granules are apparently much more frequent than in laboratory animals (BIAVA and WEST 1966b, 1967; cf. also Chap. 12). As the problem of "specific" versus "nonspecific" granules has not been sufficiently investigated by immunocytochemical techniques, ultrastructural results relating to the human kidney may also be considered with some reservation when the intrarenal distribution of renin is discussed.

(2) Valuable findings are also doubtlessly contained in the many early attempts to assess the renin status of the kidney with histochemical methods (cf. HARTROFT 1968). Although paired histochemical and immunohistochemical experiments are not yet at hand for final evaluation of the specifity of the histochemical reactions, it would be wrong to put aside such earlier data. This especially applies to histochemical studies on the distribution of epithelioid cells for which parallel determinations of kidney renin are available (cf. Sect. 4.1.4).

4.1.1.2 Renin in the Preglomerular Arteriole

Synopsis. The intrarenal distribution of immunostainable renin has been investigated most extensively in mice and rats (TAUGNER et al. 1981, 1982a, b). As already expected from the histochemical and

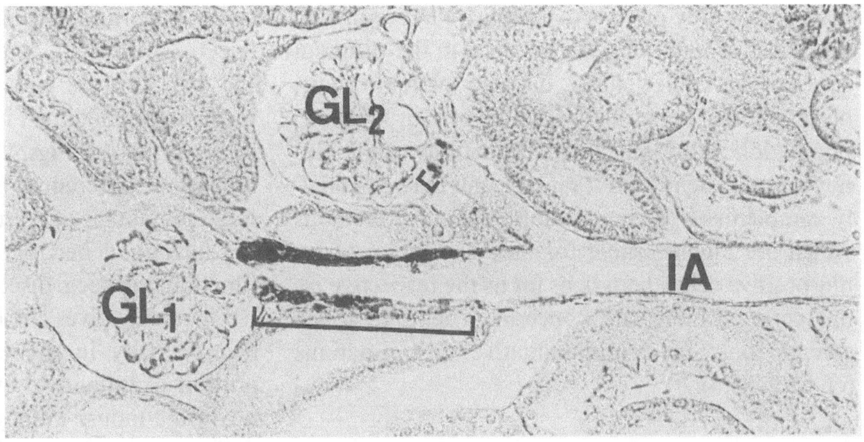

Fig. 4.2. Internephron heterogeneity in the length of the renin-positive portion - marked by brackets - of two afferent arterioles from mouse kidney. *GL*$_1$, *GL*$_2$, the pertinent glomeruli; *IA*, interlobular artery; ×340

ultrastructural studies, the bulk of renin-positive cells was found in the afferent arteriole (Figs. 4.1, 4.2). Usually well integrated in the media of the vessel, fully transformed epithelioid cells - in accordance with their plump appearance in the electron microscope - tend to form a distinct local thickening of the preglomerular arteriole, first of all an enlarged cuff around the intima of the vessel shortly before its entry into the renal corpuscle (RUYTER 1925; BOHLE and SITTE 1966; BÜHRLE et al. 1984, 1986a; cf. Fig. 2.3). With afferent arterioles curved in their juxtaglomerular portion, an asymmetrical arrangement of the renin-positive cells is generally found (Fig. 2.5), thus representing, together with the adjacent Goormaghtigh cells, the polkissen of ZIMMERMANN (1933; cf. Sect. 2.2).

According to Figs. 2.2, 2.8 A and 4.2, the renin-positive portion of the afferent arteriole contrasts clearly with the renin-negative segment. This indicates a dinstinct jump in the concentration of renin between these two portions of the vessel. In opposition to the immunohistochemical appearance, an abrupt transition between the proximal and the distal portion of the afferent arteriole may frequently be missing in the electron microscope. This could probably be explained by the fact that the intermediate cells placed at the transition between both vessel segments may not be differentiated from epithelioid cells with postembedding immunostaining using supraoptimal antibody concentrations (cf. Sect. 4.1.4). Differentiation may be improved by preembedding staining in conjunction with the PAP method followed by thin sectioning. With this procedure, we found cells scattered in the media of the interlobular artery which did not differ ultrastructurally from plain smooth muscle cells, their Golgi cistern, however, being renin positive (unpublished observations). First reports on the distribution of mRNA coding for renin appear to support the assumption that cells involved in renin synthesis without having a noticeable renin store may be found far upstream from the JGA (COGHLAN et al. 1984; DARBY et al. 1985). However, DESCHEPPER et al. (1986) report only on renin mRNA-positive cells of juxtaglomerular location.

The first immunocytochemical evaluation confirmed earlier observations that stimulation of renin synthesis leads to the recruitment of new renin-positive cells in the media of the afferent arteriole. In addition, the "granularity," i.e., the extent of the renin store, and thus the renin concentration in preexisting immunoreactive cells may increase. The question is which of the different procedures would be suited best to assess such fundamental changes in the renin status of the afferent arterioles by quantitative evaluation, reflecting, with only minor errors, the renin status of the respective nephrons. Although the results obtained with different methods seem to correlate well with each other and also with the kidney renin content (cf. Sect. 4.1.4), there are reasons to prefer the measurement of the renin-positive portion of the afferent arterioles. (a) The length of the renin-positive portion most closely reflects the recruitment of new granulated cells upon stimulation of the RAS. (b) The determination of the renin-positive portion of the afferent arteriole permits conclusions as to the stimulus-synthesis or stimulus-secretion relationship of renin-containing cells, especially when tubulovascular proximities are discussed. (c) Unlike with other procedures, this method not only permits to establish the average renin status of nephrons, for instance in the different zones of the kidney cortex, but also the renin status of individual nephrons within a certain nephron population. Biochemical determinations of the kidney renin content cannot replace, but may essentially complement, the immunohistochemical methods which are indispensable for the localization of renin-positive cells, though, as far as the respective renin concentrations are concerned, believed to be suitable only for semiquantitative determinations (cf. Sect. 4.1.4).

Internephron Heterogeneities. The measurement of the length of the renin-positive portion of the afferent arteriole (TAUGNER et al. 1981, 1982c, d) revealed differences between individual nephrons, which would not fit the notion commonly associated so far with the term "internephron heterogeneity" on the basis of morphological, biochemical, and functional data.

There have been several reports on significant heterogeneities in glomerular function, e.g., the single-nephron glomerular filtration rate (SNGFR) and the tubuloglomerular feedback (TGF), between superficial and juxtamedullary nephrons (for review see LAMEIRE et al. 1977; MIMRAN and CASELLAS 1987). By comparison, functional differences between individual glomeruli appear to be quite small within one and the same nephron population (OKEN et al. 1985; HUGHES and ICHIKAWA 1986). However, when the renin status of the respective afferent arterioles is examined, the picture is entirely different: The quantitative evaluation of the renin-positive portion reveals surprising differences between individual preglomerular arterioles (cf. Fig. 4.2), whereas, in contrast, the variations of the mean values obtained in different cortex regions are rather small (cf. Figs. 4.5, 4.6). In the following, the intrazonal differences in the renin status of individual preglomerular arterioles are compared with the interzonal heterogeneities, and some pertinent functional aspects discussed.

Figure 4.3 gives an example of the remarkable heterogeneity in the length of the renin-positive portion of individual afferent arterioles in the cortex of the mouse kidney, reflecting a corresponding difference in the number of granulated cells. In control animals, a certain fraction of afferent arterioles is renin-negative throughout. In about one-third of the vessels, the location of renin is restricted to a distance of only 10-20 μm, in another third to 20-30 μm upstream from their parent glomerulus. In a remarkable number of vessels, renin is even detected over distances of more than 50 or 100 μm. In 5% of the afferent arterioles of mice, an additional group of renin-positive cells may be found at the origin of these vessels from the interlobular artery (TAUGNER et al. 1981; cf. HATT 1967). Besides, about 10% of the profiles of interlobular arteries contain renin-positive cells in control mice as well as in control rats (cf. ROSENBAUER 1965).

Figure 4.4 shows that very similar differences are found in the length of the renin-positive portion of the afferent arteriole when - instead of a mixed nephron population as in Fig. 4.3 - superficial, intermediate, and juxtamedullary nephrons are evaluated separately.

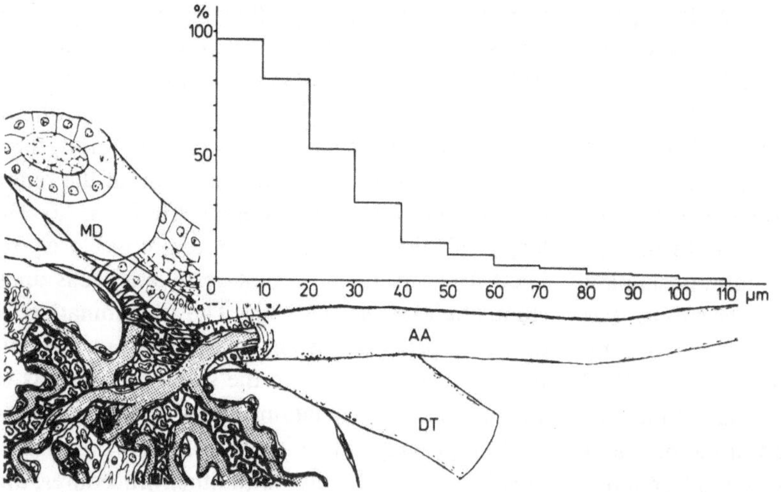

Fig. 4.3. Schematic juxtaglomerular apparatus with histograms showing the fraction (%) of renin-positive afferent arterioles *(AA)* up to the respective length in mouse kidney. *MD*, macula densa; *DT*, distal tubule

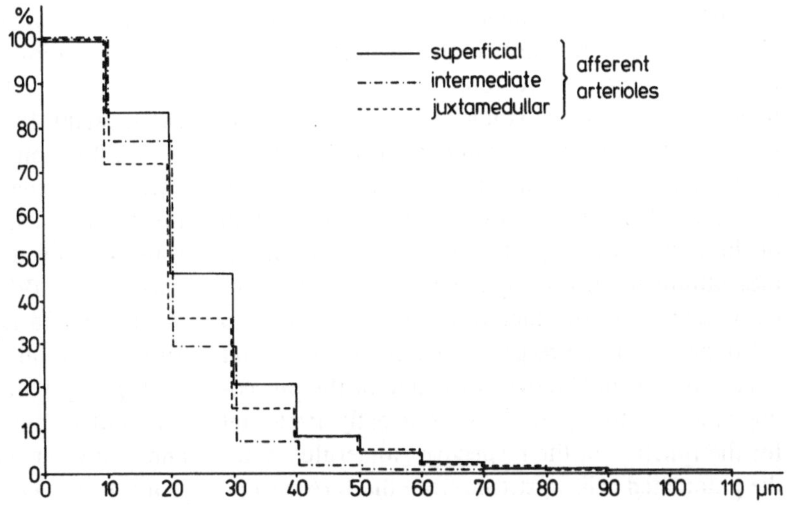

Fig. 4.4. Histograms showing the fraction (%) of renin-positive afferent arterioles up to the respective lengths for superficial, intermediate, and juxtamedullar nephrons from the kidneys of male NMRI mice. PAP method, dilution of the renin antiserum, 1:1000; 249 superficial, 110 intermediate, and 151 juxtamedullar afferent arterioles from six animals were evaluated. (WURFER et al. 1988)

Large internephron heterogeneities with respect to the renin-positive portion of the afferent arteriole within one and the same nephron population apparently exist not only in mice and rats but also in other species, including cat, dog, hog, rhesus monkey, man (WURFER et al. 1988; unpublished observations), and several other species (KON et al. 1986b; cf. Sect. 4.1.4 and Chap. 12).

Referring to some deceptively simple diagrams of the JGA, SIMPSON (1970) suggested that theories of the functional relationship between granulated cells and the macula densa should take into account that these cells are not always confined to a circumscribed collar around the entry of the afferent arteriole into the renal corpuscle (cf. ROSENBAUER 1965). The possible functional implications of the heterogeneities in the length of the renin-positive portion of the preglomerular arteriole shown in Figs. 4.3 and 4.4 become quite obvious when comparing the microtopographical relationships between the granulated cells and the sites of the action of the different stimuli influencing renin secretion. Most probably, humoral factors, the sympathetic outflow, and a decrease of pressure in the renal arterial tree may affect all the granulated cells of the afferent arteriole, irrespective of their distance from the parent glomerulus. The situation is completely different with the so-called macula densa signal. It has been mentioned above that, in mice and rats, more than one- or two-thirds of the afferent arterioles contain renin over a distance of more than 20 or 30 μm, respectively. Twenty to thirty micrometers may be assumed to be the maximal distance over which, under favorable conditions, a humoral signal from the macula densa could be propagated in order to have a direct effect on target cells in the media of the afferent arteriole. Thus, a major fraction of renin-positive cells is located outside the conventional boundaries of the JGA at considerable distances upstream from the macula densa, and can therefore only be influenced by neurohumoral or mechanotransductive stimuli for renin secretion, or, at most, by other portions of the distal tubule (cf. Sect. 3.1). TAUGNER et al. (1982a) have speculated that these granulated cells outside the JGA may, by way of their secretion, predominantly serve the systemic RAS, the contribution of the individual nephrons to the plasma renin being extraordinarily different depending on the length of the renin-positive segment of the respective afferent arterioles. Granulated cells of juxtaglomerular location were, on the other hand, thought additionally to participate in the functions of the intrarenal RAS, such as the TGF mechanism (cf. Chap. 9). Subchronic stimulation of the RAS, e.g., by sodium restriction, by the administration of diuretics, or by adrenalectomy, was found to shift the front of the granulated cells even further away from the JGA (Figs. 4.6, 4.12; cf. TAUGNER et al. 1982a, b, 1983b; HACKENTHAL et al. 1987), thus probably stressing the importance of the mechano- and β-adrenoreceptor-mediated, systemically relevant renin release. GOORMAGHTIGH (1939, 1945) and GOORMAGHTIGH and GRIMSON (1939) already pointed out that upon constriction of the renal artery the granular transformation of media cells extends upstream until into the interlobular arteries (cf. FRIEDBERG 1965b; BOHLE and SITTE 1966).

It has already been mentioned that the remarkable intrazonal differences in the renin status of individual nephrons have found little, if any, resonance in functional studies. If the functional heterogeneities between glomeruli of the same region were, in fact, insignificant, this could mean that the well-documented differences in the length of the renin-positive portion of the afferent arterioles, although functionally relevant, are compensated by alterations of other variables with respect to the dynamics of the respective glomeruli. A second assumption would be that the length of the preceding renin-positive portion is principally irrelevant for the function of the respective glomerulus, since the granulated cells upstream from the JGA, as discussed in detail above, might predominantly serve the systemic and not the intrarenal RAS.

However, differences in the renin status of individual afferent arterioles may principally be based not only on differences in the number, but also in the renin content, of the adjoining granulated cells. Large variations of the so-called granularity of individual JGAs have already been noted by SIMPSON (1970). Differences in the immunoreactivity of individual afferent arterioles at suboptimal antiserum concentrations have been observed in mice (TAUGNER et al. 1981), rats, and rabbits (unpublished observations), suggesting that intrazonal heterogeneities may indeed exist with regard to the renin store of the respective granulated cells. However, these evaluations only refer to the juxtaglomerular portion of the afferent arteriole, immediately before its entry into the renal corpuscle.

ITOH and CARRETERO (1985), ITOH et al. (1985a), and BAUMBACH and SKØTT (1986) observed differences in the reactions of granulated cells close to, as opposed to those far from, the glomerulus to secretory stimuli in experiments on isolated preparations. One is tempted to trace back these differences to the above-mentioned variations in the microtopographical relationships between the proximal and distal afferent arteriole. Nevertheless, it has to be born in mind that the extent of metaplastic transformation of the media cells increases with the distal (peripheral) direction up to the glomerulus, and that the reaction of the intermediate cells to secretory stimuli may diverge from that of epithelioid cells (cf. Sects. 2.2.2, 7.6).

In contrast to intrazonal heterogeneities, several papers have been published dealing with differences in the function and morphology of superficial, intermediate, and juxtamedullary renal corpuscles, including interzonal heterogeneities in the renin status of the respective nephrons (for review see LAMEIRE 1977).

Stratification of the renin content in the kidney cortex, with the associated amount being largest in the outer and smallest in the juxtamedullary nephrons, has been reported by VIKHERT and SEREBROVSKAYA (1964), BROWN et al. (1966), GAVRAS et al. (1970), and DRUKKER et al. (1983) in rabbits; by HORIUCHI et al. (1971) and SCHRYVER et al. (1984) in dogs; by VIKHERT and SEREBROVSKAYA (1964) in cats and in humans; by FRIEDBERG (1964, 1965a) and TAUGNER et al. (1981) in mice; and by DE ROUFFIGNAC et al. (1974), FLAMENBAUM and HAMBURGER (1974), GILLIES and MORGAN (1978), and GILLIES et al. (1982) in rats. In some contrast to these findings, VIKHERT and SEREBROVSKAYA (1964) and GRANGER et al. (1972) observed no significant difference between the renin activity of superficial and juxtamedullary glomeruli with adherent afferent arterioles in rat kidney.

These results have been obtained partly by biochemical determinations in dissected glomeruli, partly by histochemical estimation of the juxtaglomerular index (JGI; cf. Sect. 4.1.4), and, finally, also by the assessment of the renin concentration in immunohistochemical experiments. All these techniques may be subject to critical objections. With biochemical determination, for example, renin activity may be altered by the preparation procedures (GRANGER et al. 1972; FLAMENBAUM and HAMBURGER 1974); besides, depending on the length of the renin-positive portion of the afferent arteriole, a variable number of granulated cells may be lost during microdissection, thus escaping detection. With the histochemical procedures, unspecific reactions cannot be excluded. Using immunohistochemical experiments (TAUGNER et al. 1981), the renin content is not recorded as reliably as the localization of the granulated cells. In spite of such objections, the correspondence between the results of the experiments mentioned above is so close that there can be little doubt about the existence of interzonal heterogeneities, i.e., of differences in the renin status of nephrons located in different zones of kidney cortex. Nonetheless, the evaluations of immunohistochemical experiments to be described in the following yield an unexpectedly complex picture.

It has already been mentioned that differences in the renin activity associated with individual nephrons may principally be based on differences in the number and/or in the renin content of the adjoining granulated cells. There are findings which suggest that differences in renin activity associated with superficial and juxtamedullar nephrons at least in some species may be due to differences in the renin content of individual granulated cells. Thus, a gradient of "granularity" in the kidney cortex of several species had already been noted in histochemical experiments (GOORMAGHTIGH 1945; VIKHERT and SEREBROVSKAYA 1964; FRIEDBERG 1964, 1965a, b; FAARUP 1965; SATO et al. 1977). In the semiquantitative immunohistochemical evaluation of TAUGNER et al. (1981) at high antiserum concentration (10^{-3}), over 95% of both superficial and juxtamedullary afferent arterioles of mice were immunoreactive at their entry into the renal corpuscle (cf. DAUDA et al. 1976). With low antiserum concentration (10^{-5}), however, only about 50% of the juxtamedullary as compared with 75% of the superficial afferent arterioles were immunoreactive, suggesting that the renin content of the individual cells in the superficial vessels is significantly higher than that in their juxtamedullary counterparts. Similar observations were made in rats. The difference between both cortex regions was even more obvious in the rabbit: e.g., at an antibody concentration of 10^{-4}, 50% of the superficial, but only 25% of the juxtamedullar nephrons showed immunoreactivity (WURFER et al. 1988). The fact that only a certain population of juxtaglomerular afferent arterioles was immunoreactive suggests that in addition to interzonal differences there are also intrazonal heterogeneities in the renin content of the corresponding epithelioid cells.

The number of immunoreactive cells, as judged by the length of the renin-positive portion of the afferent arteriole, can be determined more reliably with immunohistochemical methods than the renin content of the granulated cells. However, with this technique only slight and in most cases insignificant interzonal heterogeneities could be detected. Figure 4.5 shows that the length of the renin-positive portion of the afferent arteriole was not significantly different in superficial, intermediate, and juxtamedullary nephrons in mice, rats, monkeys, and humans. In contrast to control and sodium-loaded mice, indications of an interzonal heterogeneity in the length of the renin-positive segment of the vessel were found in sodium-deprived animals, although only the difference between the superficial and the intermediate nephrons was significant (Fig. 4.6). It may be noteworthy that the length of the renin-positive portion of the afferent arteriole generally tends to be shortest in intermediate nephrons (WURFER et al. 1988).

Taken together, the available knowledge about both the intrazonal and the interzonal heterogeneities in the renin status of the afferent arteriole has to be considered rather incomplete. This holds in particular for the renin content of the individual cells adjoining the respective nephrons, of which it is uncertain whether it varies proportionally with the respective number of the granulated cells. Upon

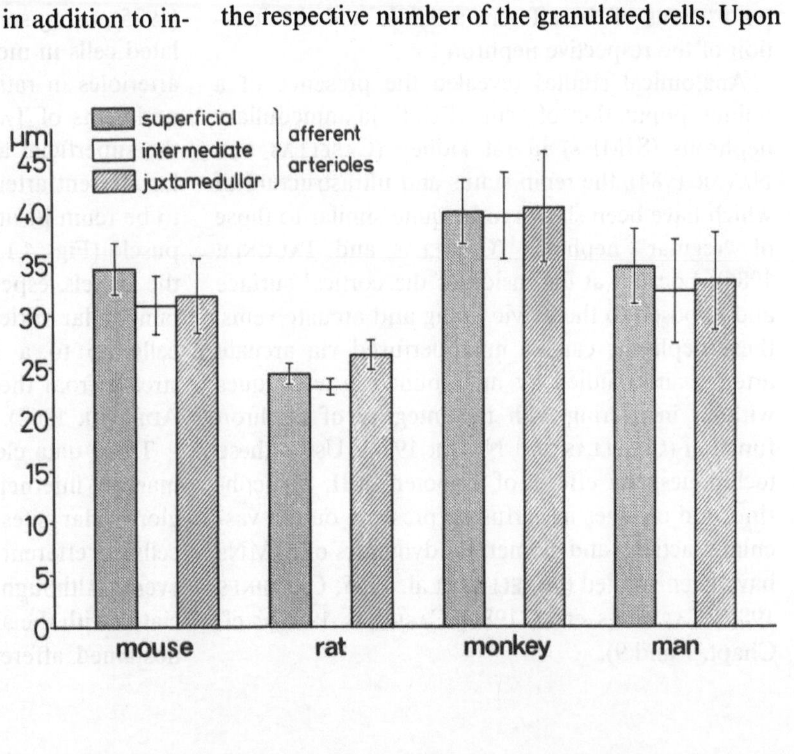

Fig. 4.5. Lengths of the renin-positive portion of superficial, intermediate, and juxtamedullar afferent arterioles in the kidneys of six mice, six rats, two rhesus monkeys, and four humans; Evaluated were: 228, 36, 10, and 56 7-µm paraffin sections, respectively

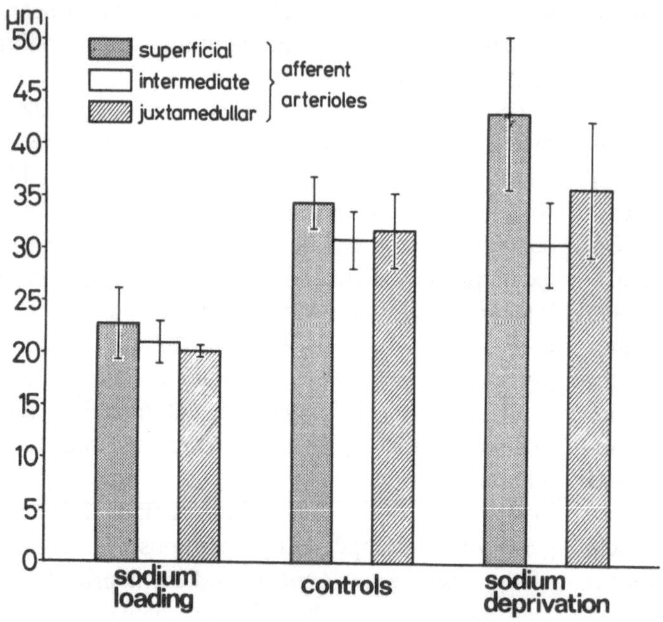

Fig. 4.6. Lengths of the renin-positive portion of the superficial, intermediate, and juxtamedullar afferent arterioles in three sodium-loaded, six control, and three sodium-deprived NMRI mice. Evaluated were: 60, 228, and 140 7-µm paraffin sections, respectively. The synthetic diet contained 2300, 1500, and 50 mg Na/100 g, respectively

stimulation of the RAS, an increase in the number of immunoreactive cells would be equivalent to the recruitment of new intermediate cells, whereas an increase in the renin content of individual cells has to be attributed to the transformation of additional intermediate cells into epithelioid cells. Hence, possibly both processes are not functionally equivalent (cf. Sect. 2.2.2). However, fundamental is the lack of knowledge concerning the relationship between the renin status of the afferent arteriole and the function of the respective nephron.

Anatomical studies revealed the presence of a unique population of "superficial" juxtamedullary nephrons (SJMNs) in rat kidney (CASELLAS and NAVAR 1984), the renin status and ultrastructure of which have been shown to be quite similar to those of "regular" nephrons (CASELLAS and TAUGNER 1986). Located at the inside of the cortical surface and apposed to the pelvic lining and arcuate veins, these nephrons can be microperfused via arcuate arteries and studied by micropuncture techniques without interfering with the integrity of nephron function (CASELLAS and NAVAR 1984). Using these techniques, the effects of angiotensin II, epinephrine, and changes in perfusion pressure on the vascular reactivity and glomerular dynamics of SJMNs have been studied (CASELLAS et al. 1985; CARMINES 1986; CARMINES et al. 1986; CASELLAS 1986b; cf. Chapt. 7 and 9).

4.1.1.3 Renin in the Postglomerular Arteriole in Mesangial and Goormaghtigh Cells

Casual reports on the occurrence of epithelioid cells in the media of the efferent arteriole have been dealt with in Sect. 2.3. CHRISTENSEN et al. (1975), using serial semithin sections, found granulated cells in more than 50% of the postglomerular arterioles in rats. In the immunohistochemical experiments of TAUGNER et al. (1981), about 25% of the superficial and nearly 40% of the juxtamedullar efferent arterioles in mouse kidney were found to be renin-positive at their exit from the renal corpuscle (Figs. 4.1, 4.7B, 4.13). In addition, some of the vessels, especially efferent arterioles of the juxtamedullar variety, showed scattered renin-positive cells up to a few hundred micrometers downstream from their parent glomerulus (Fig. 2.12; cf. ADEBAHR 1962).

These data clearly demonstrate that in spite of a marked internephron heterogeneity also at postglomerular sites, the occurrence of renin-positive cells in efferent arterioles is by no means a rare event: Although the bulk of kidney renin is associated with the afferent arterioles, so that the immunostained afferent vessel renin and the biochemically determined kidney cortex renin parallel each other in control as well as in high- or low-sodium animals, the efferent arterioles are nevertheless equipped with their own albeit much smaller renin store which could be conceived to react differentially (TAUGNER et al. 1981). In this context, it merits attention that, in contrast to smaller effects on kidney renin and the immunoreactivity of the afferent vessels, sodium depletion was followed by a highly significant increase in the number of renin-positive superficial efferent arterioles (TAUGNER et al. 1981). Therefore, it was speculated that this effect may be involved in the preferential efferent vasoconstrictor tone sustained by angiotensin II especially under the condition of sodium restriction (cf. also Sect. 9.3).

For efferent vessels, the same arguments concerning stimulus-secretion relationships which have been discussed for afferent vessels should be considered. Besides JGA renin, extra JGA renin was found especially along some juxtamedullar efferent arterioles. The generalization and thus the importance of this finding for, e.g., the blood supply to the inner medulla and the countercurrent system, however, can only be evaluated when quantitative studies using serial sections are available. It is of interest to note in this connection that sympathetic fibers likewise accompany the juxtamedullar efferent vessels over a considerable distance at least until their branching into vasa recta (NEWSTEAD and MUNKACSI 1969; DOLEŽEL et al. 1976).

Immunohistochemical evaluations with different antiserum concentrations suggest that in control mice the renin status of the superficial and juxtamedullary efferent arterioles at their entry into the renal corpuscle is similar (TAUGNER et al. 1981). However, a zonal heterogeneous response of the postglomerular arterioles seems to result from the above-mentioned preferential increase in the immunoreactivity of the superficial efferent vessels upon sodium restriction.

Inspecting about 5000 glomeruli in control and sodium-depleted mice, only five clearly renin-positive mesangial cells were seen, all of them close to the glomerular stalk (Fig. 4.18 B). The Goormaghtigh cell field, although smaller in mice than in rats, could be identified when the section plane was parallel to the axis of both glomerular arterioles; it was never found to be unequivocally immunoreactive in these species (cf. Sects. 2.4, 4.1.3). However, renin-positive Goormaghtigh cells have been observed in kidney biopsies from patients with pseudo-Bartter's syndrome due to the long-term abuse of furosemide and/or laxatives (CHRISTENSEN et al. 1988; cf. Sect. 2.4, 4.1.3).

4.1.1.4 Tubular Renin

Using monospecific antibodies against mouse renin and the PAP method of STERNBERGER (1974), TAUGNER et al. (1979) observed reaction product not only in juxtaglomerular epithelioid cells, but also in the apical portion of proximal tubule cells (cf. Fig. 4.7). With a more refined technique, in addition some cells of the connecting and cortical collecting tubules were seen to be renin positive (Fig. 4.8; TAUGNER et al. 1982b). Only occasionally was staining observed in short collecting duct segments of the inner stripe of the outer medulla or in scattered collecting duct cells of the inner medullary zone. The cells of the thick ascending limb of Henle's loop including the macula densa and the cells of the distal convolution were consistently found to be renin negative. Similar observations were made in other species, e.g., in the rat and in humans (MÉNARD et al. 1979; GALEN et al. 1980; TAUGNER et al. 1987). By stepwise dilution of the primary antiserum, the immunohistochemical staining of the proximal tubules disappeared first, followed by that of the connecting and cortical collecting tubule. The reaction of the juxtaglomerular epithelioid cells disappeared only after an additional 10- to 100-fold dilution of the antiserum (Fig. 4.8 A, B).

These experiments clearly demonstrate that renin or immunoreactive components of renin are also present in certain tubular epithelial cells, although in considerably lower concentrations than in the juxtaglomerular epithelioid cells. There are several indications that this tubular renin - in contrast to vascular renin - is not synthesized by the immunoreactive epithelial cells themselves, but represents glomerularly filtered renin, which was reabsorbed by the respective tubule cells from the tubular fluid by pinocytosis.

That the kidney is not only the site of synthesis and secretion, but also that of the breakdown of renin was already known from the fact that the clearance rate of renin is delayed by nephrectomy (SCHAECHTELIN et al. 1964; PETERS-HAEFELI 1971;

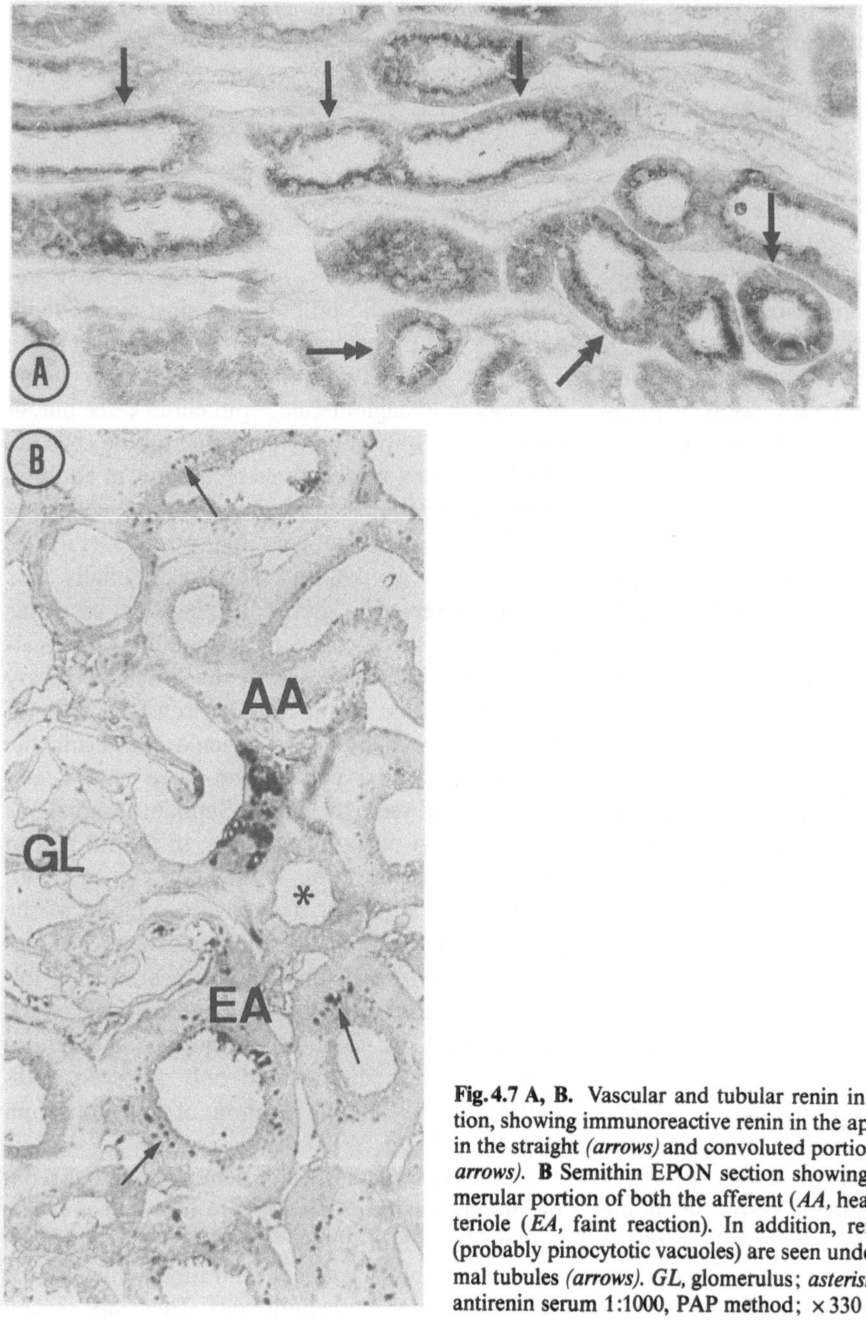

Fig. 4.7 A, B. Vascular and tubular renin in mouse kidney. **A** Paraffin section, showing immunoreactive renin in the apical region of the epithelial cells in the straight *(arrows)* and convoluted portion of the proximal tubule *(double arrows)*. **B** Semithin EPON section showing renin granules in the juxtaglomerular portion of both the afferent (*AA*, heavy reaction) and the efferent arteriole (*EA*, faint reaction). In addition, renin-positive granular structures (probably pinocytotic vacuoles) are seen under the brush border of the proximal tubules *(arrows)*. *GL*, glomerulus; *asterisk*, macula densa. Dilution of the antirenin serum 1:1000, PAP method; ×330 and ×580, respectively

BING and NIELSEN 1973; YOSHIDA et al. 1975; BOYD 1979) in correspondence with the observation that exogeneously administered renin accumulates in the kidney in large amounts (YOSHIDA et al. 1975; IWAO et al. 1983a). Another major organ involved in the metabolism of circulating renin is the liver (HEACOX et al. 1967; SCHNEIDER et al. 1970; KIM et al. 1987a).

As other proteins, also renin is preferentially pinocytosed and degraded by the cells of the proximal tubules. Autoradiographic studies revealed that labeled renin after intraaortal or intravenous injection accumulates in kidney cortex, with the silver grains located preferentially over the apical portion of proximal tubular cells (Fig. 4.9; TAUGNER et al. 1982b; IWAO et al. 1982b, 1983b). This localization corresponds to that of other endogeneous or exogeneous proteins known to be reabsorbed by pinocytosis and subsequently metabolized in the lysosomal system of these cells (cf. also Sect. 4.4 for

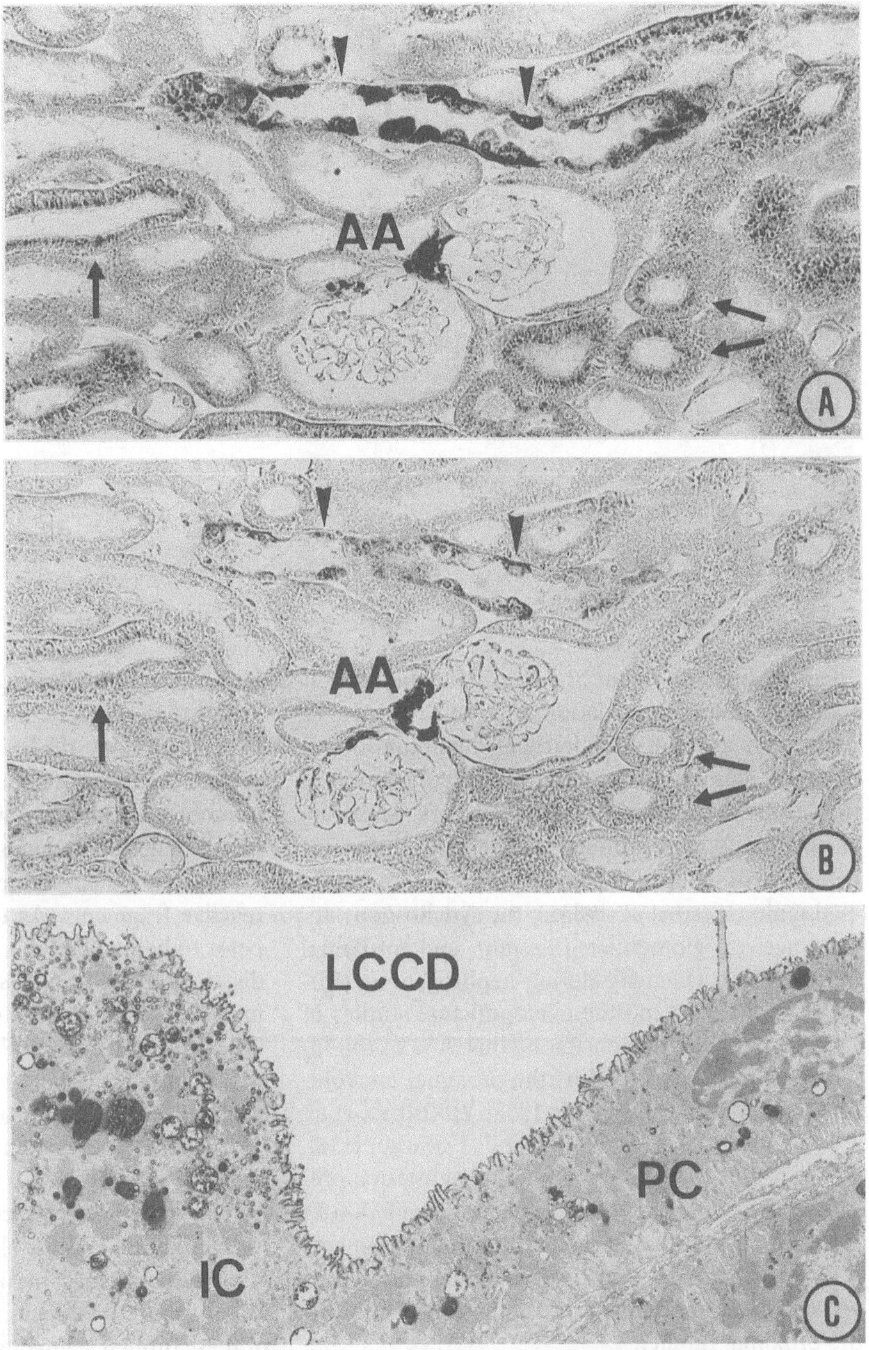

Fig. 4.8. **A, B** Paired 7-μm paraffin sections from mouse kidney immunostained for renin with different antiserum dilutions. In **A**, with an antiserum dilution of 1:1000, renin-positive cells are seen in the afferent arteriole *(AA)*, the proximal tubules *(arrows)*, and a connecting or cortical collecting tubule *(arrowheads)*. After dilution of the antiserum to 1:10000 in **B**, the immunostaining of the proximal tubule disappears, while the staining of the connecting or cortical collecting tubule is markedly decreased as compared with that in **A**; ×230. **C** Pinocytotic uptake of filtered horseradish peroxidase by intercalated *(IC)* and principal cells *(PC)* of a cortical collecting tubule in mouse kidney. *LCCD*, lumen of the cortical collecting tubule. Electron micrograph; ×5400. (From TAUGNER et al. 1982b)

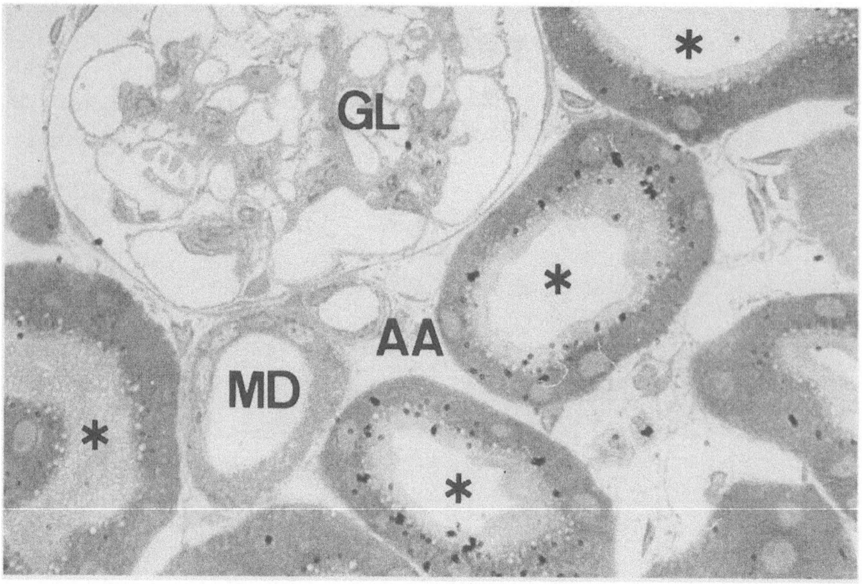

Fig. 4.9. Semithin section autoradiograph of mouse kidney cortex 2 min after bolus injection of ^{125}I-labeled renin. The radioactivity is almost exlusively localized in the apical region of the epithelial cells of the proximal tubule *(asterisks)* close to the brush border. *GL*, glomerulus; *MD*, macula densa; *AA*, afferent arteriole. Perfusion fixation with glutaraldehyde-formaldehyde-picric acid, EPON embedding; K2 emulsion (Ilford), dipping method; ×600

angiotensinogen). Other arguments for the pinocytotic uptake and degradation of renin by the proximal tubule are: the granularity of the immunocytochemical staining observed in the apical portion of the involved epithelial cells (Fig. 4.7 B; cf. TAUGNER et al. 1982c), corresponding to the subcellular distribution of exogeneously administered radiolabeled renin (IWAO et al. 1983a); the synchronous appearance of glomerular filtration and proximal tubular renin reactivity during nephrogenesis (MINUTH et al. 1981) and the micropuncture studies of LEYSSAC (1978a), demonstrating that 90% of the filtered renin is reabsorbed by the proximal convolution (cf. RAPPELLI and PEART 1968; YUKIMURA et al. 1984). IKEMOTO et al. (1982) and TAKAORI et al. (1982) found that the renin-binding substance present in the cytosol fraction of renal cortex (TAKAORI et al. 1981) is located in tubules but not in glomeruli. It was suggested that this substance may be involved in the mechanism of renin reabsorption in the proximal tubule.

In comparison to the proximal tubule, the renin-like immunoreactivity of the connecting and cortical collecting tubule is much less understood. Although the pinocytotic capacity of the distal tubular segments is relatively small (STRAUS 1964, 1967, 1971, 1979; MAACK et al. 1979; CHRISTENSEN et al. 1981; TAUGNER et al. 1982b; cf. Fig. 4.8c), and renin is predominantly reabsorbed in the area of the proximal tubule (LEYSSAC 1978a), the intensity of the renin-like immunostaining of the connecting and cortical collecting tubules was repeatedly found to be higher than that of the proximal tubule (TAUGNER et al. 1982b, 1988c). To explain this apparent discrepancy, one is reminded that the immunohistochemical staining intensity of pinocytosed material depends on the balance between uptake and internal degradation to nonimmunoreactive fragments. As the peptidolytic activity appears to be much higher in the proximal than in the distal tubule (SUDO 1981), it was suggested that a higher degradation rate of renin in the proximal tubule leads to a lower immunoreactivity despite a higher uptake.

However, as studies on the handling of renin by the distal tubular system are not available, it cannot be ruled out beyond all doubt that the high immunoreactivity of the connecting and cortical collecting tubules is due to a tubular synthesis of renin. Similar arguments may pertain to the atrial natriuretic factor found to be localized in the intercalated cells of these tubular segments (MCKENZIE et al. 1985).

4.1.2 Factors Influencing the Localization of Granulated Cells in the Renal Arterial Tree

Most renin-positive cells are located in the wall of the distal afferent arteriole close to the glomerulus. However, this localization is not invariable. Upon stimulation of the RAS, additional granulated cells are recruited by metaplastic transformation from the pool of plain smooth muscle cells, thus lengthening the renin-positive portion of the vessel in the upstream direction. As a consequence, it may be assumed that principally all media cells of the preglomerular vessels are endowed with the ability of renin synthesis but this ability is expressed only when the sum of all stimuli converging to an individual media cell exceeds a certain limit. Questions addressing the cause of the typical preglomerular arrangement of renin-positive cells along the renal arterial tree are therefore closely related to, if not identical with, those aiming at the stimuli for renin synthesis. This is of heuritistic importance inasmuch as the stimuli for the induction of renin synthesis are, due to methodological difficulties, less known of than stimuli for renin secretion, and it cannot be assumed automatically that both are identical (cf. Chap. 11).

Basically, three causes may be assumed for the selective localization of granulated cells in the media of the afferent arteriole close to the glomerulus which, with the pertinent modifications, are also discussed in connection with renin secretion: (1) the proximity of the macula densa, (2) pecularities of the innervation, and (3) hemodynamic factors.

(1) In some JGAs, the microtopographical relation between the macula densa and the epithelioid cells is highly suggestive (cf. Fig. 2.5). In addition, renin secretion can be altered by variations of the NaCl concentration perfusing the juxtaglomerular portion of the distal tubule (Skøtt and Briggs 1987). As renin secretion and synthesis generally change in parallel (cf. Chap. 11), it may be assumed that not only renin secretion, but also renin synthesis - and therefore also the localization of renin-positive cells - can be influenced via the macula densa. A number of observations, however, clearly speak against any crucial influence of the distal tubule and the macula densa on the localization of epithelioid cells. Several authors have pointed out that granulated cells can also be found in the afferent arteriole far upstream from the macula densa (cf. Sect. 4.1.1; for numerical data see Taugner et al. 1981, 1982c). Barajas (1971) showed that by far the largest number of granulated cells has no contact with the macula densa, not even in the limited area of the JGA. According to Hatt (1967), in major renal ischemia accompanied by tubular degeneration, hyperactivity of the epithelioid cells can be observed despite atrophy of the macula densa. Also, in mice and rats, unilateral ligation of the ureter leads to tubular atrophy including the macula densa, yet the renin status of the hydronephrotic kidney does not change significantly and the typical localization of the renin-positive cells in the distal portion of the afferent arteriole is unaltered (Bührle et al. 1986a; Nobiling et al. 1986). In addition, renin synthesis in the hydronephrotic kidney appears to react to stimulation of the renin-angiotensin system, e.g., by adrenalectomy or salt depletion, in a similar fashion as the contralateral, untouched kidney (unpublished observations).

It is true that the asymmetrical arrangement of epithelioid cells in the area of the convex bank of a curved afferent arteriole, remainding the polkissen of Zimmermann (1933), could be considered a compelling argument for the influences of the macula densa and/or the Goormaghtigh cell field on the localization of epithelioid cells (Oberling 1944; Biava and West 1966a; Hartroft 1968, cf. Fig. 2.5). It should be borne in mind, however, that the media of curved afferent arterioles on the contrary is often especially thin in the area of the concave bank, i.e., in the vicinity of Bowman's capsule (Barajas and Latta 1963a). Therefore, among other reasons, the asymmetrical arrangement of granulated cells may be due to the different spatial conditions prevailing in both shores of the vessel. The hydronephrotic kidney of mice and rats does not permit a clear decision between these alternatives, as the number of winding afferent arterioles decreases with the progressive atrophy of the distal tubule and the macula densa.

(2) Three observations speak against an essential importance of the innervation for the typical preglomerular localization of granulated cells. (a) Although the juxtaglomerular portion of the afferent arteriole may be most densely innervated, the varicosities or nerve terminals encountered here seem to differ from those found more upstream neither in their ultrastructure nor in regard to the transmitters or neuropeptides identified so far (cf. Sect. 3.5). In addition, the characteristics of juxtaglomerular epithelioid cells verified by electrophysiological methods seem to be similar, if not identical, to those of granulated cells or plain smooth muscle cells further upstream (Bührle et al. 1985, 1986a, b). (b) Scattered granulated cells in the media of the interlobular artery are not clearly correlated microtopographically with the arrangement of varicosi-

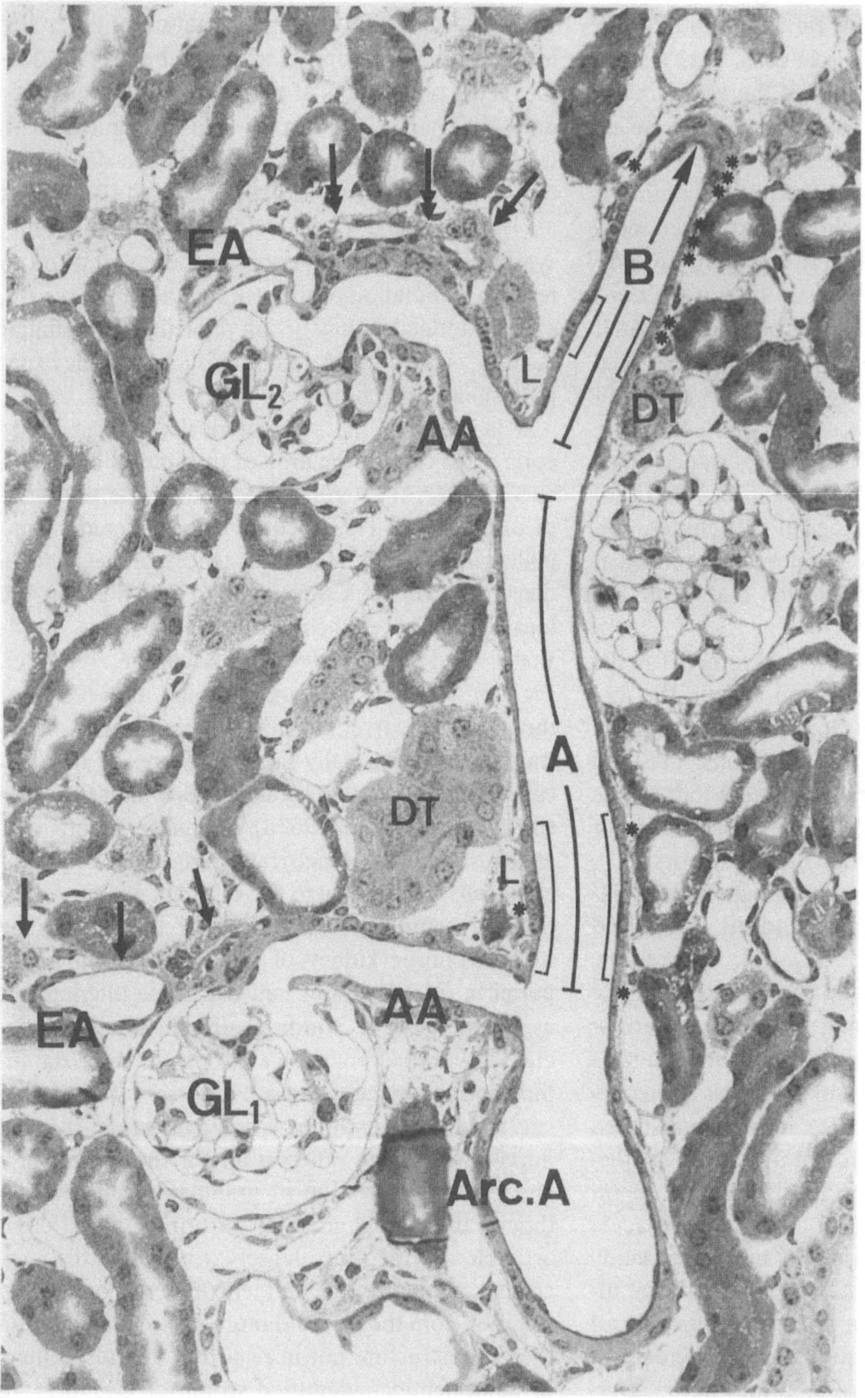

Fig. 4.10. Semithin section from mouse kidney 7 days after adrenalectomy showing an interlobular artery, with two portions *(A, B)* studied in detail by serial thin sectioning. Granulated cells have been found in the vessel segments marked by brackets; in addition, the media of both afferent arterioles *(AA)* encountered in the section plane was composed exclusively of granulated - i.e., mostly epithelioid - cells. In both efferent arterioles *(EA)* no granulated cells were found. As verified by serial sections, the efferent arteriole of one glomerulus *(Gl_1)* was in contact with the thick ascending limb of Henle's loop *(arrows)* over a remarkable distance outside the JGA. The same was valid for the afferent arteriole of the other glomerulus *(GL_2)* and the distal convoluted tubule *(double arrows)* of the pertinent nephron. *Arc.A*, arcuate artery; *asterisks*, axon terminals or varicosities at the adventitia - media border of the interlobular artery; *DT*, distal tubules; *L*, lymph vessels. For further details on the microtopographical relationships in the vessel segments marked by *A* and *B*, see Table 4.1 (M. Müllensiefen and R. Taugner, unpublished observations); ×370

ties and/or nerve terminals in the adventitia of this vessel (Fig. 4.10, Table 4.1). (c) Postnatal sympathectomy with 6-hydroxydopamine, the success of which had been controlled ultrastructurally and by formalin-induced fluorescence, was followed by a significant decrease in blood pressure, kidney renin, and plasma angiotensin II (Table 4.2); however, in the electron microscope, intermediate and epithelioid cells were found in the sympathectomized animals which differed neither in position nor in their ultrastructure from those of control animals.

Table 4.1. Microtopographical relationships of the granulated cells and the smooth muscle cells in the media of the interlobular artery depicted in Fig. 4.10.

Periarterial structure \ Type of media cells	Granulated cells ($n=31$)	Smooth muscle cells ($n=33$)
Interstitium	100%	100%
Peritubular capillaries	74%	64%
Lymph vessels	19%	–
Nerve terminals	90%	36%
Proximal tubules	52%	58%
Distal tubules	16%	30%

The incidence (%) of the different periarterial structures found in close proximity to the individual media cells of the vessel are given. The data are from mouse kidney, 7 days after adrenalectomy. For further details see legend to Fig. 4.10 and text.

(3) Coincidental observations in immunoreacted paraffin sections are in favor of the assumption that hemodynamic factors may be of prime importance for the localization of renin-positive cells in the distal (juxtaglomerular) portion of the afferent arteriole: Outstandingly narrow preglomerular arterioles were repeatedly equipped with an especially long renin-positive portion; in neighboring afferent arterioles, similar in caliber but very different in length, the renin-positive portion began at the same distance from their branching off from the interlobular artery. These observations seem to be in agreement with the assumption that the beginning of the renin-positive portion in individual afferent arterioles depends upon the respective preceding vascular resistances, renin-containing cells appearing downstream from a point where the transmural pressure falls below a certain threshold. Such a notion, resembling the stretch receptor hypothesis of Tobian (1962), would be in accordance with various other observations. Thus, the degree of metaplastic transformation of media cells appears to increase in a downstream direction in parallel with the pressure decrease within the afferent arteriole, until, finally, fully transformed epithelioid cells appear close to the glomerulus. Constriction of the renal artery shifts the front of granulated cells in an upstream direction in parallel with the critical pressure sup-

Table 4.2. Blood pressure and renin status of rats after postnatal chemical sympathectomy

	Blood pressure (mm HG)				Kidney renin[a]	Plasma renin activity[b]	Plasma angiotensin II[c]
Days after birth	50	64	78	92	45	45	45
Controls ($n=10$)	112±10.9	110.2± 9.9	114.8± 9.1	120.5± 9.1	119.3±15.4	3.74±2.1	54.5±31
	$p<0.01$	$p<0.05$	$p<0.02$	$p<0.005$	$p<0.01$	NS	$p<0.025$
Sympathectomized animals ($n=10$)	95±10.1	99.2±11.7	99.5±13.4	102.5±12.1	95.9±22.8	3.66±0.8	29.9±18

Twenty-four Wistar rats of both sexes from four litters were treated by daily subcutaneous injections of 100 µg 6 OH-dopamine/g for the first 10 days of their life; 24 animals from 4 other randomly selected litters were injected with a corresponding volume of 0.9% NaCl. In ten rats of both groups, blood pressure was measured using a pulse detector and an occluding tail-cuff at days 50, 64, 78, and 92 after birth. In ten other rats of both groups, kidney renin, plasma renin activity, and plasma angiotensin II were determined at 45 days after birth. The success of the chemical sympathectomy was ascertained by comparing four 6 OH-dopamine-treated animals with four control rats with respect to the formalin-induced fluoresecence within their kidneys, and in other four rats from each group to the presence or absence of axon terminals in series of thin sections between 45 and 92 days after birth.

[a] ng angiotensin I/g per hour

[b] ng angiotensin I/ml per hour

[c] pg/ml

(D. Ganten and R. Taugner, unpublished observations)

posed to induce transformation. TOBIAN (1962) has already pointed out that also in various other experimental models with hypergranulation and hyperreninemia - e.g., by NaCl depletion or adrenalectomy - the pressure in the renal arterial tree is comparatively low, meaning that the threshold pressure decisive for the stimulation of renin synthesis, according to this hypothesis, is shifted in an upstream direction. If such a hypothesis were to be verified by determination of the vascular resistance preceding the renin-positive portion of the afferent arteriole, the question would arise whether relaxation of the vascular wall might stimulate renin synthesis directly or via endothelial mediators. Further details pertaining to the preglomerular resistance, e.g., the modifying role of autoregulation and the musculoendothelial cooperativity, will be dealt with in Chap. 7 in connection with renin secretion.

Summing up, the observations reported suggest that among the three factors mentioned above, hemodynamic influences, i.e., the decrease in pressure along the renal arterial tree, are of prime importance for the localization of renin-positive cells in the distal afferent arteriole. However, the fact that granulated cells are not only present close to the glomerulus, but sporadically or in small groups also further upstream in the proximal afferent arteriole and in the interlobular artery, indicates that other influences must also be present.

In order to establish possible influences of neighboring structures on the tendency of vascular smooth muscle cells for metaplastic transformation, the microtopography of an interlobular artery, the media of which contained both renin-negative and renin-positive cell groups, was investigated in serial sections (Fig. 4.10). Table 4.1 shows that the microtopographical relations of the granulated cells were not clearly different from those of smooth muscle cells, i.e., the various structures observed surrounding the interlobular artery, e.g., proximal and distal tubules, nerve endings, and blood and lymph capillaries were nearly just as frequent in the vicinity of both cell types. The question whether and exactly which neurohumoral factors - aside from transmural pressure, i.e., the mechanical strain of media cells in kidney vessels - are responsible for the degree of their transformation and retransformation, may perhaps only be safely answered with cultured cells (cf. Chap. 10).

4.1.3 Renin-Positive Cells of Different Origin?

Renin-positive cells in the media of the preglomerular vessels are by far the most frequent and therefore best known. In Sects. 2.2 and 10 their development from smooth muscle cells by metaplastic transformation is described. Thus, upon adequate stimulation of the RAS, first intermediate cells and, with increasing granularity and decreasing myosin content, epithelioid cells emerge.

Renin-positive cells are also encountered in the media of the efferent arteriole (cf. Sect. 4.1.1). As the number of these renin-positive cells appears to increase and decrease with the niveau of stimulation, it can be assumed that here again we are dealing with metaplastic transformation, although in this case from pericyte-like media cells into granulated cells and vice versa. Because of their rarity, the ultrastructural characteristics of the intermediate cell types in the media of the efferent arteriole are not well known. However, at the end of the transformation process, cells very similar to the epithelioid cells of the afferent arteriole may be encountered in the juxtaglomerular portion of the postglomerular arteriole (cf. Fig. 2.11).

In contrast to the glomerular arterioles, it has been very difficult to assign immunohistochemically renin-positive cells to the Goormaghtigh cell field or to the mesangium of the glomerular stalk (cf. Sect. 2.4). In the afferent and efferent arteriole, doubts as to the correct classification of individual cells could, at most, arise in the area of the glomerular hilus, where the media of the glomerular arterioles merges with the lacis and the glomerular stalk in a fashion difficult to define. However, as the afferent arteriole does contain renin-positive cells at more upstream - and the efferent arteriole at more downstream - locations, there are principally no difficulties involved in ascribing renin-positive cells to the media of the glomerular arterioles. The situation is different with the Goormaghtigh cell field, with the glomerular stalk, and, to a certain extent, also with the peripheral centrolobular mesangium.

It has already been pointed out that in animal experiments individual renin-positive cells surrounded by renin-negative Goormaghtigh cells were just as little observed as a lacis doubtlessly consisting only of renin-positive cells. On the other hand, immunohistochemically renin-positive cells seen at the periphery of the Goormaghtigh cell field could also be media cells of the glomerular arterioles, particularly when it is considered that changes in the stimulation level of the RAS may alter the microtopography of the lacis (Barajas 1981). The situation is similar for renin-positive cells in the area of the glomerular stalk. Under the electron microscope, cells by their ultrastructural appearance are easier to classify in spite of the suboptimal fixation permitted for immunolabeling. However, in immunocytochemical experiments, clearly identifiable Goormaghtigh or mesangial cells exhibiting renin-positive granules have not yet been observed. When clearly identified Goormaghtigh cells were found to contain granules, these granules proved to be renin negative (Fig. 4.11). On the other hand, cells loaded with renin-positive granules could still be counted as media cells of the glomerular arterioles. In view of such difficulties, the controversial problem of Goormaghtigh cells and mesangial cells of the glomerular stalk having the capacity for transformation into epithelioid was still awaiting its final proof (for references see Sect. 2.4).

Recently, immunohistochemical studies using serially sectioned biopsy specimens from patients with pseudo-Bartter's syndrome due to the abuse of

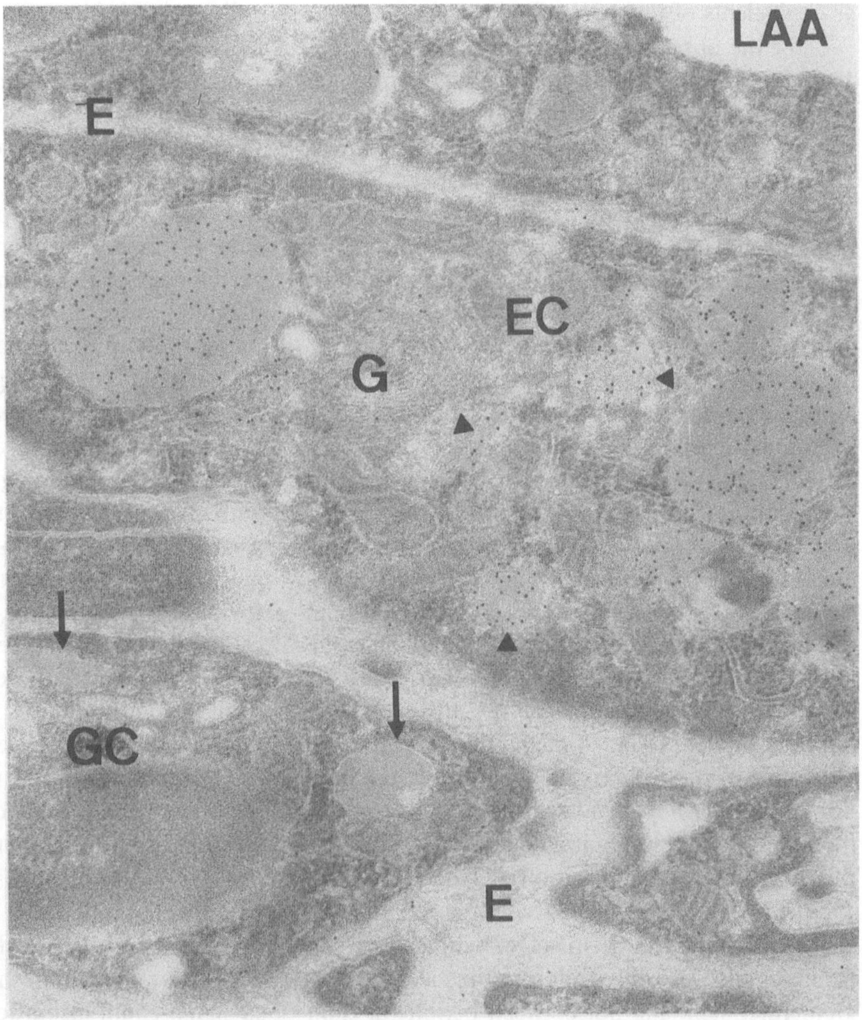

Fig. 4.11. Mouse epithelioid cell *(EC)* and Goormaghtigh cell *(GC)* close to the entrance of the afferent arteriole into the renal corpuscle, immunoreacted for renin. Note the jump in the intensity of immunolabeling from stacked Golgi cisterns *(G)* to juvenile epithelioid cell secretory granules *(arrowheads)*. From juvenile to mature granules, there seems to be only a slightly furter increase of the immunoreactivity. The granule-like organelles of the Goormaghtigh cell *(GC)* are not labeled *(arrows)*. *LAA*, lumen of the afferent arteriole; *E*, extracellular space. Perfusion fixation with 1% glutaraldehyde, embedding in London white resin, protein A-gold method. Dilution of the renin antiserum 1:100; ×42500

furosemide and/or laxatives revealed intense renin-positive reactions of all Goormaghtigh cells without exception (cf. Fig. 2.15). Under the electron microscope, the hyperplastic lacis was found to consist of polygonal or globular, densely granulated cells (Christensen et al. 1988), similar to those encountered by Schürholz et al. (1969) in a patient with pseudo-Bartter's syndrome and by Christensen et al. (1978) in a kidney with polar artery stenosis. These findings strongly suggest that long-lasting stimulation might indeed be followed by the recruitement of renin-producing cells from the pool of renin-negative Goormaghtigh cells (cf. Sect. 2.4). The question remains whether exocytotic events such as those shown in Fig. 2.14 may be classified as equivalents of renin secretion.

Since the observations of Goormaghtigh (1939, 1945) several reports indicated that glomerular mesangial cells may also have the capacity for transformation into typical epithelioid cells. The well-known paper by Dunihue and Boldosser (1963) on cats allows hardly any doubt that some of the heavily granulated cells depicted are "true," i.e., peripheral mesangial cells. Barajas et al. (1976) and Taugner and Ganten (unpublished observations) found granulated cells in rats at a considerable distance from the hilus, though not between peripheral glomerular loops. Interestingly, all these observations relate to chronic experiments during drastic stimulation of the RAS: mineralocorticoid deficiency for 6–24 months (Dunihue and Boldosser), 4 weeks of unilateral renal artery constriction (Barajas et al.), and 2 weeks of water restriction in diabetes insipidus rats (Taugner and Ganten). This may be an indication that an essentially stronger and/or longer lasting stimulation of the RAS is required for the metaplastic transformation of mesangial cells as opposed to media cells of the afferent arteriole.

In spite of these observations, however, the question of whether mesangial cells have the capacity to transform into typical epithelioid cells deserves further attention in our opinion (cf. Kon et al. 1984, 1986b; Morild et al. 1985b, 1987; and Chap. 13). In the immunohistochemical experiments, renin-positive cells were only found in the vicinity of the glomerular stalk and not between peripheral capillary loops (cf. Fig. 4.18). Immunocytochemical observations of mesangial cells with renin-positive granules are not available. On the other hand, similar to Goormaghtigh cells, mesangial cells with renin-negative "granules" have been observed (own unpublished observations). In *Tupaia belangeri*, renin granules in the epithelioid cells of the afferent arteriole stand out by irregular contours and their extraordinary characteristic thread-like paracrystalline structures (Forssmann and Taugner 1977). Apart from this, "granulated" mesangial cells are often also present in this species. However, the mesangial cell "granules" are round and homogeneous, thus differing essentially from the epithelioid cell granules. The fact that in cell cultures mesangial cells synthesize and release renin into the medium contributes very little to the question of whether they are capable of transformation into epithelioid cells in vivo, as the ability of renin synthesis pertains to an astounding and still-growing variety of cell types (cf. Chap. 7).

The cell types discussed here are all of mesenchymal origin and have a number of structural and supposedly also functional characteristics in common. Although the capacity for transformation into epithelioid cells as a further similarity is not certain in every case, it does appear justified - and for the sake of brevity, also practical - to summarize the media cells of the glomerular arterioles, the Goormaghtigh cells, and the glomerular mesangial cells under the term "vascular components" of the JGA or the glomerulo-juxtaglomerular complex. There are particularly good reasons for bringing the morphologically similar extraglomerular mesangial cells, i.e., the Goormaghtigh cells, into connection with the glomerular, true mesangial cells, thus speaking of an intra- and extraglomerular mesangium (the precise Goormaghtigh cell field or lacis). It may be imperative, however, to differentiate clearly between cell types, when dealing with stimuli affecting a certain cell or with effects resulting from it. In this context, it is important to remember that renin release experiments conventionally do not differentiate in this respect. As most of the epithelioid cells are located in the media of the afferent arteriole, the known release reactions accordingly would refer in essence only to the preglomerular cell pool. If - and how - a decrease in perfusion pressure, an increase in the sympathetic activity, or changes in the hypothetical macula densa signal could also affect renin release from other cell types is largely unknown. The same holds for the stimuli of renin synthesis. Basic differences according to the particular location of the respective cells would, in any case, exist in regard to the local intrarenal effects of renin secretion.

Examples of how enticing nomenclature can be, are recent publications where the content of the term "mesangium" was incorrectly conceived, envisaging a dense contractile syncytium which extends from the base of the macula densa to invest both hilar arterioles as well as individual glomerular capillaries, thus being able to produce widespread contraction waves controlled by the sympathetic outflow. However, the only known morphological prerequisite for such conclusions is the gap junction coupling between media cells of the glomerular arterioles, Goormaghtigh cells and mesangial cells, the functional importance of which is not yet understood.

4.1.4 Quantitative Evaluation of Histochemical and Immunohistochemical Experiments in Relation to the Renin Status of the Kidney

In Chap. 1 we outlined the role morphological methods played in the first experiments to establish the secretory function of granulated cells. GOORMAGHTIGH (1939), GOORMAGHTIGH and GRIMSON (1939) and subsequently DUNIHUE and CANDON (1940) interpreted the increased number of granular cells in the preglomerular arterioles of rabbits and dogs after partial ligation of the renal artery as an indication that these cells may secrete the renal pressor principle, renin (cf. GOORMAGHTIGH 1940; DUNIHUE 1941). Somewhat later, Dunihue's studies (1946, 1947) revealed a relationship between granulated cells and the adrenal glands. DUNIHUE (1949) also found that the increase of granularity caused by adrenalectomy could be prevented or reversed by desoxycorticosterone acetate (DOCA). MCMANUS (1950) extended these observations to kidneys from patients with Addison's disease.

DUNIHUE (1941) counted a certain number of glomeruli with identifiable JGAs; the percentage of JGAs containing cells with few, moderate, or many granules served as an index reflecting the amount of granulated cells in the respective kidney. Subsequently, HARTROFT and HARTROFT (1953), studying the effects of sodium restriction, sodium loading, and DOCA treatment, introduced the frequently used "index of granulation of the juxtaglomerular cells," later called "juxtaglomerular granulation index," "juxtaglomerular index," or JGI. In representative areas of Bowie-stained sections, all glomeruli and, separately, all juxtaglomerular granulated cell units were counted regardless of whether the respective glomeruli were included in the section plane; upon classification of the juxtaglomerular cell units according to their degree of granulation from one-plus to three-plus, the totals thereby recorded were multiplied by the factors 1, 2, and 4 respectively, and the sum of the weighted totals expressed as an index per 100 glomeruli. TURGEON and SOMMERS (1961) introduced the "juxtaglomerular cell count" or JGCC, giving the total number of cells in 25 juxtaglomerular bodies and, in addition, separate subtotals for each of four cell types as defined by the authors, including "large agranular cells with clear, watery cytoplasm," often encountered in abnormal human kidneys. The comprehensive review of HARTROFT (1968) also includes the descriptions of the "juxtaglomerular cell rating" of SCHMID and GRAHAM (1962) and the "juxtaglomerular activity" of ITSKOVITS et al. (1963), both, like the JGI, giving preference to the granularity of the cells in the region of the vascular poles - or glomeruli - inspected.

A positive correlation between the JGI of HARTROFT and HARTROFT (1953) and the bioassayable renin was found by PITCOCK et al. (1959), TOBIAN et al. (1959), TOBIAN (1960b), DEMOPOULOS et al. (1960), FISHER (1961), and CHANDRA et al. (1965) in rat kidneys with stimulated as well as with suppressed renin synthesis, by EDELMAN and HARTROFT (1961) in the rabbit kidney, and by HARTROFT (1966) in the cat kidney. A satisfactory correlation between the JGI and the bioassayable renin was not found by FAARUP (1967) in cats and by ROJO-ORTEGA et al. (1973a) in dogs. To avoid misinterpretations by similar staining affinities of specific and nonspecific granules (BIAVA and WEST 1966b), ENDES et al. (1969) introduced a new combined trichrome staining method used by DÉVÉNYI et al. (1971). BOHLE and SITTE (1966) preferred Movat's silver impregnation to assess the granularity of epithelioid cells. ROSENBAUER and KRÖNIG (1967) evaluated the square area of the polkissen upon vital staining of the granulated cells. MEYER (1972) used PAS-stained sections and a planimetric method for comparative investigations of the juxtaglomerular cell complex including lacis cells (SKAANE et al. 1975; HARA and MEYER 1975; see BOHLE et al. 1982 for review). Various other modifications of the histochemical staining and evaluation procedures have been recommended for the assessment of the renin status of the kidney.

In spite of some criticism, one may consider the mentioned histochemical procedures to be suitable for the semiquantitative evaluation of the renin status of the kidney at least in the commonly used laboratory animals, rats and mice. Taking into account the so-called granularity, it appeared possible to estimate not only the number of renin-producing cells, but also the size of the renin store of the individual cells. Uncertainties existed with regard to the JGAs and afferent arterioles of humans, reported to contain considerably more so-called nonspecific granules (BIAVA and WEST 1966b). Due to the lack of paired experiments, it also remained open as to how sensitive histochemical methods react in comparison with immunohistochemical procedures. It is nevertheless remarkable that the earlier procedures allowed for a basically correct interpretation of changes in the renin status of the kidney under a number of conditions. This, for example, applies to renovascular hypertension (TOBIAN et al. 1958, 1959; TOBIAN 1960b; DEMOPOULOS et al. 1960; FISHER 1961; TURGEON and SOMMERS 1961; DÉVÉNYI et al. 1971; MEYER 1972), to changes in sodium balance including adrenalectomy and mineralocorticoid effects (DUNIHUE 1947; HARTROFT and HARTROFT 1953; DUNIHUE and ROBERTSON 1957; TOBIAN et al. 1958; PITCOCK et al. 1959; CHANDRA et al. 1960; FISHER 1961; FRIEDBERG 1965b; BOHLE and SITTE 1966; JOHNSTON et al. 1967; MOLTENI et al. 1976; ROSENBAUER and KRÖNIG 1976), and, in addition, to a series of diseases associated with an increased or decreased renin synthesis (for review see BOHLE et al. 1982).

The immunohistochemical procedures have the advantage of both a high sensitivity and an unquestionable specifity (cf. Sect. 4.1.1). In addition, they are not significantly more laborious than the histochemical methods (STERNBERGER 1986). One disadvantage of the immunohistochemical methods, however, is the reduction of immunoreactivity of renin upon optimal fixation. Mild fixation on the other hand tends to allow the stored renin to distribute within the entire cytoplasm so that secretory granules in the paraffin section may hardly be recognized (cf. Sect. 4.1.1). Accordingly, the immunohistochemical methods are very well suited for the localization and also for the estimation of the number of renin-positive cells. However, in order to estimate the renin content of granulated cells, it is necessary to perform comparative experiments with different antiserum concentrations. The suitability of semithin sections for the immunohistochemical staining of renin has not yet been examined systematically (cf. TAUGNER et al. 1982b, c).

Figure 4.12 shows the result of three different procedures for the evaluation of immunohistochemical experiments in the mouse kidney where renin was demonstrated in paraffin sections with the PAP method of STERNBERGER (1974) using supraoptimal antiserum concentrations. Kidneys of control animals were examined in comparison with kidneys of sodium-loaded and sodium-deprived mice. The following methods were used for the evaluation: (a) determination of the immunohistochemical JGI, i.e., the number as a percentage of renal corpuscles with renin-positive vascular poles; (b) determination of the renin-positive portion of the afferent arteriole; and (c) the computer-assisted determination of the volume density of granulated cells in kidney cortex (cf. GERSTHEIMER et al. 1987). As the majority of granulated cells are located in the distal portion of the afferent arteriole, it was reasonable to assume that the results of the three evaluation procedures correlated positively with one another, even if the volume density of renin-positive cells is determined indirectly with methods (a) and (b) (cf. Sect. 4.1.1.2). Figure 4.12 shows that the effects of sodium deprivation and sodium loading are reflected very similarly by all three procedures. A rough calculation shows that a similar amount of cortical tissue would be required in order to obtain reliable results using these different methods. Figure 4.12 also shows that the results of the immunohistochemical procedures concur with the determination of kidney renin at various levels of sodium loading. Internephron heterogeneities in the equipment with renin-positive cells, however, can only be shown with method (b), i.e., by the determination of the renin-positive portion of the afferent arteriole. In order to demonstrate local differences in renin concentration, additional immunohistochemical experiments with different antibody concentrations would have to be performed (cf. Sect. 4.1.1).

Immunofluorescence may also be used in the determination of the renin status of the kidney. For semiquantitative evaluations, the following methods have been used: the ratio between the number of immunofluorescent JGAs and the number of glomeruli encountered in the section, the ratio between the number of JGAs containing six (or ten) immunoreactive cells to the total number of immunoreac-

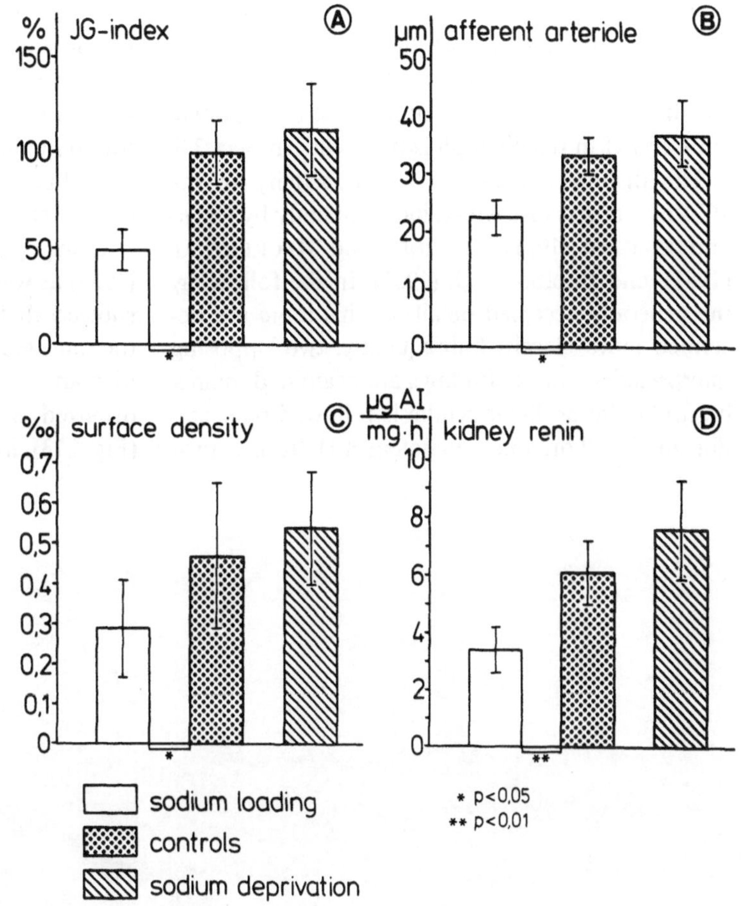

Fig. 4.12 A–D. Kidney renin status of three sodium-loaded, six control, and three sodium-deprived mice as reflected by four different methods. **A** Immunohistochemical JG-index ($n=30$, 114, and 70 PAP-stained sections, respectively); the values of the sodium-loaded and sodium-deprived mice are normalized to the JG-index of controls (33.6%). **B** Length of the renin-positive portion of the afferent arterioles ($n=60$, 228, and 140 sections, respectively). **C** Surface resp. volume density of renin-positive cells in kidney cortex ($n=6$, 12, and 10 sections, respectively). **D** Renin concentration in kidney cortex ($n=6$ tissue specimens). From WURFER et al. (1988)

tive JGAs, and also the incidence of renin-containing cells in the wall of the interlobular arteries as compared with the incidence of immunofluorescent JGAs (NOCHY et al. 1983; KIMURA and SASANO 1986). CORMAN and MICHEL (1986) determined the "immunofluorescent renal renin index" by comparing the number of immunoreactive JGAs plus afferent and interlobular arterioles with the number of glomeruli present in the studied section. MICHEL et al. (1986) found a positive correlation between this index and the renin content in the kidneys of one-clip two-kidney hypertensive rats.

Some findings indicate that stimulation of renin synthesis can affect granulated cells of different locations to a different extent, e.g., with sodium-loaded or sodium-depleted mice in the afferent as compared with the efferent glomerular arteriole (TAUGNER et al. 1981). The above-mentioned methods that basically reflect only the renin status of the preglomerular arterioles would not be suited to record such differences.

4.2 Angiotensins

Only few biochemical data are available on the existence and concentration of the angiotensins in kidney tissue. FINKIELMAN and NAHMOD (1969) found considerable angiotensin-like activity in rat glomeruli following incubation and extraction. Later, MENDELSOHN (1976) developed specific and sensitive extraction and assay procedures for angiotensin II in the kidney. In the blood-free perfused organ, he found an average content of 0.5–1.0 pmol angiotensin II per gram rat kidney, which increased upon sodium deprivation of the animals, and was markedly decreased by angiotensin-converting enzyme (ACE) inhibitors (MENDELSOHN 1979). More recently, KAWAMURA et al. (1985) isolated kidney cortex granules by density gradient fractionation and identified both angiotensin I and II in a granule fraction rich in renin. The angiotensin I content reported was about fivefold higher than that of angiotensin II (40 versus 9 pmol/g kidney cortex), with the concentration of the octapeptide being about 2% of that reported by MENDELSOHN (1979) for the whole kidney. The occurrence of angiotensin I and II in renal lymph is dealt with in Sect. 3.4.

The distinct angiotensin II-like activity in the JGA of rats was discovered independently by CELIO and INAGAMI (1981), TAUGNER and HACKENTHAL (1981), and BROOKS et al. (1982). In the following, the generally accepted details of this topic are described more closely. Subsequently, two opposing interpretations of the findings are examined, namely: (a) the intracellular synthesis followed by secretion of the "hormone" angiotensin II from a prohormone (angiotensinogen) by means of an enzyme (renin) and an activator (converting enzyme) (CELIO and INAGAMI 1981) and (b) the pinocytotic uptake of angiotensin II from the cortical interstitium followed by the transfer of the octapeptide to the lysosome-like secretory granules of epithelioid cells (TAUGNER et al. 1981, 1984d, 1985b).

Using paraffin sections, CELIO and INAGAMI (1981) as well as TAUGNER and HACKENTHAL (1981) showed that angiotensin II coexists with renin in the same epithelioid cells of the rat afferent glomerular arteriole (Fig. 4.13). When renin-positive cells occurred in the media of the efferent arteriole (Fig. 4.13), in the glomerular mesangium, or further

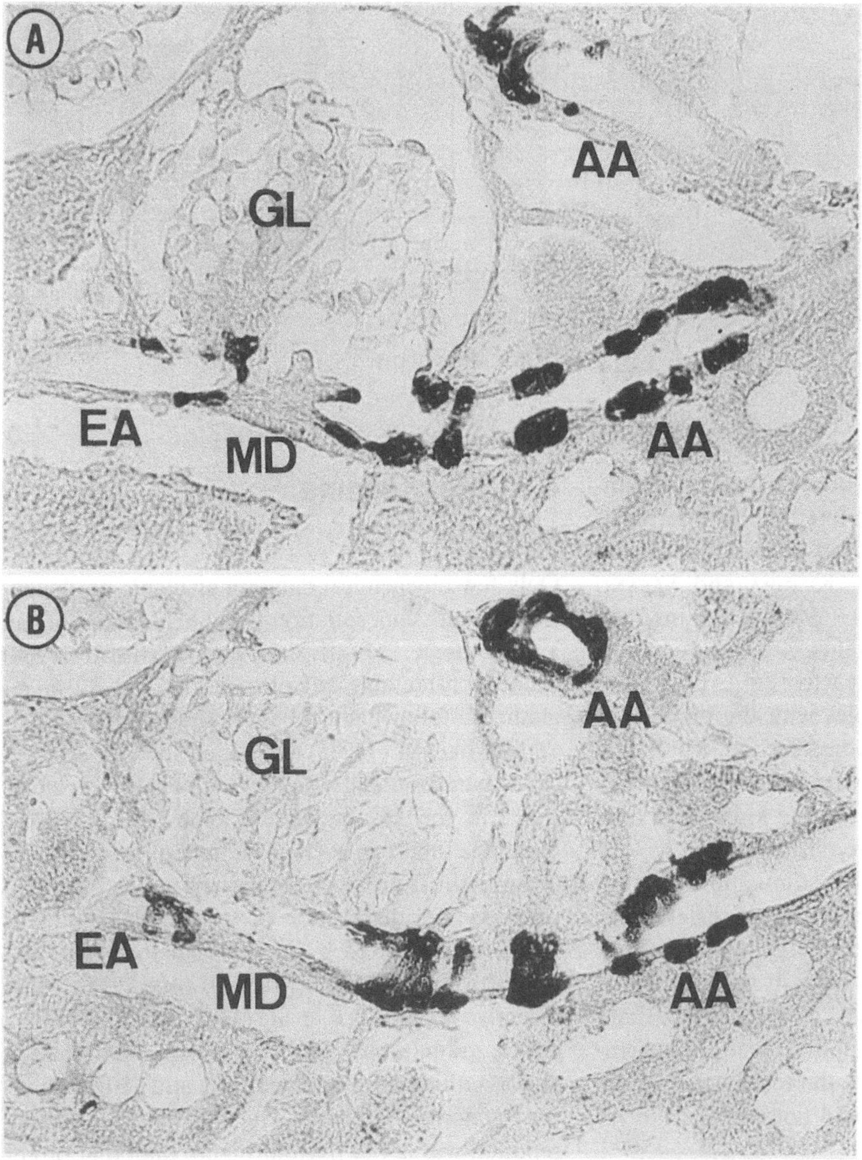

Fig. 4.13. Paired paraffin sections from rat kidney, immunostained for renin **A** and angiotensin II, respectively **B**. Groups of renin-positive cells in the afferent *(AA)* and efferent glomerular arterioles *(EA)* are also angiotensin II-positive. *GL*, glomerulus; *MD*, macula densa. Adrenalectomy 3 days before perfusion fixation with Bouin's solution containing 0.1% glutaraldehyde. Antiserum dilution, 2:25000 in **A** and 1:1000 in **B**, ×380

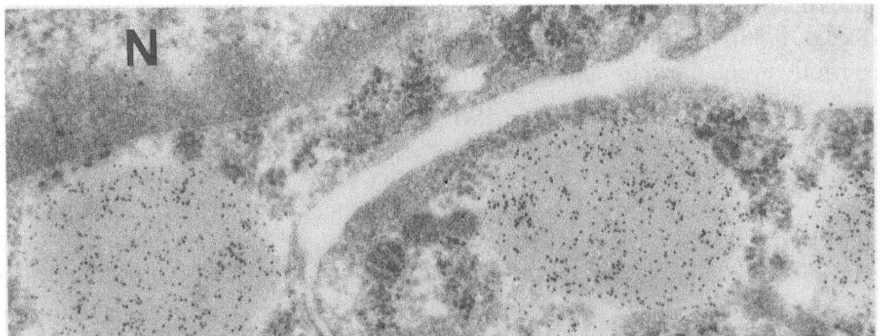

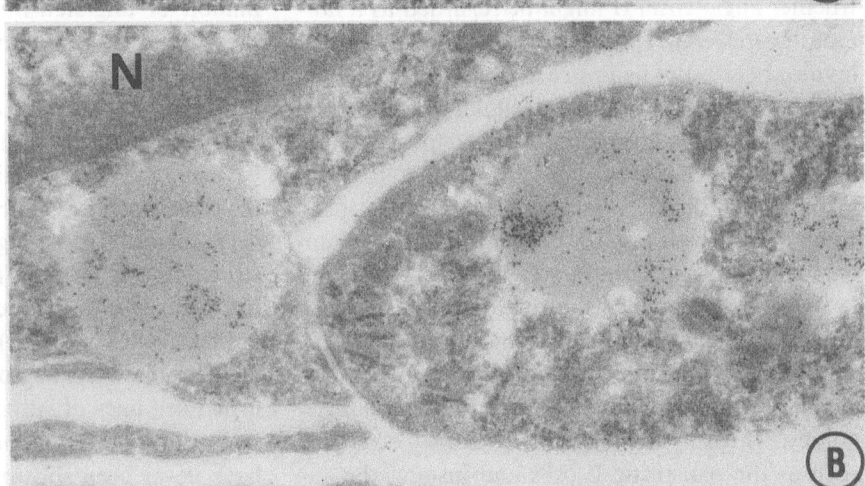

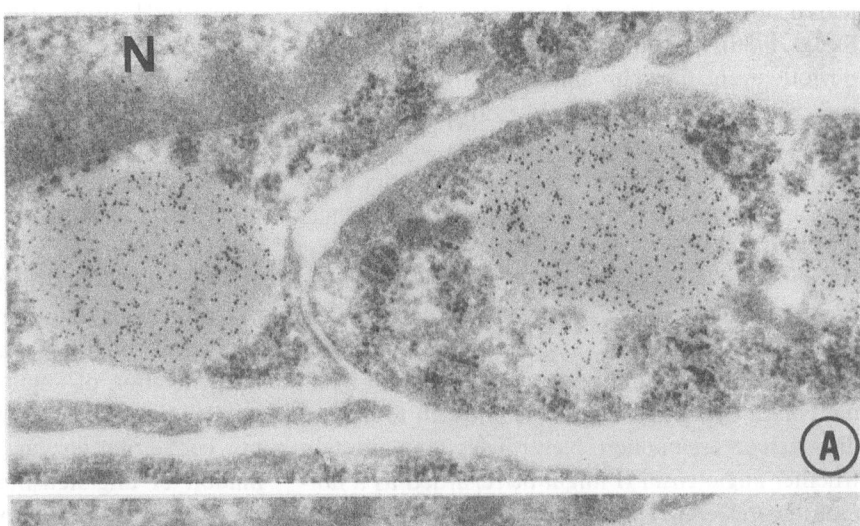

Fig. 4.14 A, B. Paired thin sections from rat kidney, immunoreacted for renin **A** and angiotensin II **B**. The same epithelioid cell granules are both renin and angiotensin II positive. Note that the immunostaining for angiotensin II tends to produce clustering of the label. *N*, nucleus. Adrenalectomy 3 days before perfusion fixation with 1% glutaraldehyde; Lowicryl-embedding, protein A-gold method. Dilution of the antisera, 1:100, ×42800

upstream in the media of the interlobular artery, they were regularly angiotensin II positive as well (TAUGNER and HACKENTHAL 1981; TAUGNER et al. 1982a). When stimulation of the RAS was followed by metaplastic transformation of smooth muscle cells, the newly recruited renin-positive cells proved also to be angiotensin II reactive, with the length of the renin-positive portion of the afferent arteriole increasing - and after cessation of stimulation decreasing - in parallel with the angiotensin II-positive portion of the vessel (TAUGNER et al. 1982a).

The question of whether the angiotensin II found in the juxtaglomerular epithelioid cells was of systemic origin or generated within the kidney was addressed with immunohistochemical experiments on ischemic and nonischemic kidneys of rats with renovascular hypertension (TAUGNER et al. 1982d). It was found that the immunoreactivity for angiotensin II increased in the ischemic organ, but was drastically decreased in the nonischemic kidney. As the plasma level of angiotensin II was elevated in the hypertensive animals, these results were suggestive of a local-interstitial-formation of angiotensin II by the intrarenal RAS.

Paraffin and semithin sections had already given the impression of angiotensin II being localized in granular structures (CELIO and INAGAMI 1981; TAUGNER and HACKENTHAL 1981). Subsequently, immunocytochemical experiments revealed that renin and angiotensin II indeed coexist in the same mature secretory granules of epithelioid cells (Fig. 4.14; CANTIN et al. 1984; TAUGNER et al. 1982c, 1984d; BÜHRLE et al. 1984).

In none of the experiments quoted so far was angiotensinogen, angiotensin I, or converting enzyme visualized in epithelioid cells and their secretory granules.

Given the coexistence of renin and angiotensin II in the matrix of epithelioid cell secretory granules and the fact that the content of these granules is released upon stimulation (TAUGNER et al. 1984a, b; SKØTT 1986), it is quite probable that with each exocytotic event, a certain amount of angiotensin II is cosecreted with renin as suggested by CANTIN et al. (1984). As the angiotensin II quanta liberated by the exocytosis of rat epithelioid cell secretory granules are unknown, it remains a matter of speculation as to whether the exocytosed octapeptide may substantially increase the angiotensin II concentration in the interstitium of the JGA. The relatively low concentrations of angiotensin II coincident with high angiotensin I concentrations found in renal lymph by INAGAMI et al. (1986) seem to contradict such an assumption. Nevertheless, the question whether angiotensin II might be regarded as a hormone secreted by epithelioid cells deserves special interest because of its principal importance.

The arguments for an intracellular synthesis and storage of angiotensin II in the secretory granules of epithelioid cells are mainly those of Inagami and his group. These arguments have recently been summarized with regard to an intrarenal renin-angiotensin system (INAGAMI et al. 1986). CELIO and INAGAMI (1981) pointed out that the exclusive production of the bioactive peptide hormone angiotensin II in the circulating blood from its prohormone angiotensinogen would be an exception in endocrinology. Thus, by analogy with classical examples such as the pancreatic B cells and insulin, the authors argued in favor of an intracellular generation of the octapeptide. While angiotensin I could not be detected in the JGA of normal rats, the epithelioid cells of animals treated with converting-enzyme inhibitors were reported to show angiotensin I-like immunoreactivity (NARUSE et al. 1982). KAWAMURA et al. (1985) found that the fractions of isolated granules from rat kidney cortex with the highest content of renin also showed the highest concentrations of angiotensin I and angiotensin II immunoreactive substances, identified by high-pressure liquid chromatography (HPLC). In contrast to the immunohistochemical results of several groups, the quantities of angiotensin I found in these granule fractions were higher than those of angiotensin II even in animals not treated with captopril. Intracellular mechanisms of angiotensin formation have been reported to exist in several models of cultured cells. Thus, neuroblastoma and pheochromocytoma cells, adrenocortical tumor as well as so-called juxtaglomerular cells in culture were shown to contain renin, angiotensin I, and angiotensin II, some of them also converting enzyme (for references see NARUSE et al. 1985; INAGAMI et al. 1986). The colocalization and possible cosecretion of several components of the RAS may also be inferred from results obtained with renin-positive cells of other organs, in particular the submandibular gland, the pituitary, the adrenal gland, and the testis (for reference see BARKA 1980; INAGAMI 1982; PANDEY et al. 1984; NARUSE et al. 1986).

However, as stated previously, various arguments also speak against the assumption of an intracellular generation of angiotensin II in juxtaglomerular epithelioid cells. Firstly, neither angiotensinogen nor converting enzyme, both required for the formation of angiotensin II, could be traced within these cells by immunohistochemical or immunocytochemical methods (CELIO and INAGAMI 1981; TAUGNER et al. 1982a, c; CANTIN et al. 1984). Secondly, biochemical findings with cultured cells can be extended to epithelioid cells in situ only with major reservations. Primary cultures of epithelioid cells known to discharge about 99% of their renin store within 24 h after isolation are not able to replace this loss by synthesis (KURTZ 1986a, own unpublished observations). On the other hand, "juxtaglomerular" cells grown in culture, and established cell lines of vascular and nonvascular origin, seem to generate predominantly inactive renin and have never been shown to harbor a granular renin store comparable to that of epithelioid cells in vivo.

Of special importance is the controversial question concerning the presence of angiotensin I in juxtaglomerular epithelioid cells. In one of our early studies on the distribution of the components of the RAS (TAUGNER et al. 1982c), immunocytochemical staining of juxtaglomerular cells of rats not only with antisera against angiotensin II but also with antisera against angiotensin I was observed. The cross-reactivity of these immunocytochemical angiotensin I-positive antisera with angiotensin II in the conventional displacement reaction of labeled angiotensin I by angiotensin II (ABRAHAM 1969) had been only about 0.01%. However, in the direct binding assay, cross-reactivities of up to 20% were found for these immunocytochemically angiotensin I-reactive antisera (HACKENTHAL and TAUGNER 1983). These data were interpreted to demonstrate that falsely positive immunohistochemical reactions may be obtained if the specifity of the reactions is judged only from the displacement reaction and the conventional preabsorption.

When the antisera were selected according to the above-mentioned criteria, angiotensin I could not be traced in rat juxtaglomerular epithelioid cells, not even after a long-term treatment with captopril (50 mg/kg/day for 3 weeks). Likewise, CANTIN et al. (1984) could only visualize angiotensin II, but not angiotensin I in these cells at both the light- and electron-microscopic levels. This discrepancy is even more surprising as the angiotensin I content in the renin granule fraction of KAWAMURA et al. (1985) in control as well as in low- and high-sodium rats was constantly found to be four to five times higher than that of angiotensin II.

As referred to earlier, there is still another possibility, apart from the intracellular generation, of explaining the existence of angiotensin II and, if actually present, also that of angiotensin I in epithelioid cells. Epithelioid cells have been shown to take up extracellular tracers such as HRP or cationized ferritin by pinocytosis and to transfer them to their secretory granules which are endowed with several features of lysosomes (TAUGNER et al. 1982c, 1985b, 1986a; TAUGNER and HACKENTHAL 1988). As HRP is considered a marker of fluid phase pinocytosis, it can be assumed that also the angiotensins - depending on their concentration in the interstitium of the JGA - may be transferred to the secretory granules in such an unspecific way. However, epithelioid cells have also been shown to be equipped with angiotensin receptors and to exhibit numerous coated pits, especially in membrane domains close to exocytotic events (cf. Chap. 6). It is therefore conceivable that angiotensin II gains access to the epithelioid cells via a specific receptor-mediated endocytosis. In this context, it would be interesting to know whether there is a down-regulation of the receptor density by internalization of the receptor-ligand complex in epithelioid cells (BÜHRLE et al. 1987a; cf. Chap. 8). As to angiotensin I, it may be assumed that, in accordance with its smaller affinity to the receptor, a correspondingly higher interstitial concentration would be required for the decapeptide to be internalized in immunohistochemically traceable amounts. Favorable conditions for such an event probably exist after long-term captopril treatment (NARUSE et al. 1982), although the situation after converting-enzyme inhibition is extremely complicated, among others by the existence of converting-enzyme independent pathways of angiotensin II generation (BÜHRLE et al. 1987b; cf. Sect. 2.6).

The effects which angiotensin II cosecreted with renin might have in the area of the JGA are discussed in the context of the intrarenal RAS. It may be tempting to link this fraction of angiotensin II with the afferent vasoconstriction thought to be involved in the TGF. It should not be overlooked, however, that those cells containing - and thus probably secreting - the larger part of angiotensin II, namely the fully transformed epithelioid cells, are the cells which are least contractile. It has therefore been suggested that the immediate effect of the angiotensin II cosecreted with renin into the extracellular space surrounding the epithelioid cells might rather consist in the inhibition of further secretory events, thus constituting an "ultrashort" negative feedback loop on renin secretion (TAUGNER et al. 1984a).

In the dispute over origin and possible local effects of the angiotensin II cosecreted with renin, one circumstance which should warn against drawing general conclusions from the data so far available is liable to be overlooked: the clear-cut angiotensin II-like reactivity of the juxtaglomerular ep-

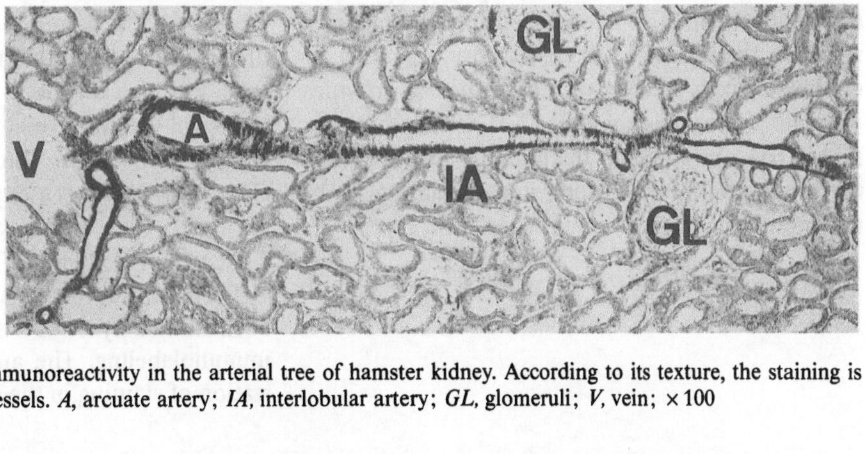

Fig. 4.15. Angiotensin II-like immunoreactivity in the arterial tree of hamster kidney. According to its texture, the staining is attributed to the media of the vessels. *A*, arcuate artery; *IA*, interlobular artery; *GL*, glomeruli; *V*, vein; ×100

ithelioid cells is probably a species-dependent phenomenon, especially significant in rats. In the Chinese hamster, angiotensin II-like reactivity was not only found in epithelioid cells of the JGA, but also in plain smooth muscle cells of arcuate and interlobular arteries as well as afferent arterioles (Fig. 4.15; TAUGNER et al. 1983a, b). Weaker immune reactions after administration of the anti-angiotensin II sera clearly effective in rats were observed only occasionally in mice (unpublished observations). In spite of identical tissue handling and immunohistochemical processing, the results were negative in rhesus monkies, dogs, cats, pigs, rabbits and guinea pigs (HACKENTHAL and TAUGNER 1987). Angiotensin II-positive reactions were reported by CELIO (1982) in humans. In our studies, weak positive reactions were observed in one out of eleven cases. The remaining ten tissue samples of human kidneys examined were angiotensin II-negative. Of these, three kidneys were from children with Bartter's syndrome and one from a partially infarcted kidney, all having enlarged JGAs (HACKENTHAL et al. 1987; TAUGNER et al. 1988c). These negative findings do not rule out the existence of immunohistochemically subthreshold angiotensin II concentrations in the secretory granules of epithelioid cells of the species involved. In this case, however, we may well be dealing with angiotensin II quanta, which would hardly be effective upon secretion when compared with the continuous generation of angiotensin II in the cortical interstitium.

In his histochemical experiments with aminopeptidase A (APA, angiotensinase A), KUGLER (1981, 1982a) found species differences between rats and mice, which may be relevant in the present context. While in rats, APA was mostly found in the vicinity of cell membranes, in mice, the enzyme could also be traced intracellularly, especially in lysosome-like structures. We are tempted to speculate that angiotensin II is internalized in both rats and mice in the lysosome-like secretory granules of epithelioid cells by way of pinocytosis. If enzymes capable of cleaving angiotensin II were present in higher concentrations in the granules of mice, then the result would be a lower - in some animals subthreshold - concentration of angiotensin II.

4.3 Angiotensin-Converting Enzyme

ACE (EC 3.4.15.1) is a peptidyl-dipeptide-carboxyhydrolase which transforms angiotensin I into angiotensin II, the effector peptide of the RAS. The enzyme is identical with kininase II and inactivates bradykinin by removal of the C-terminal dipeptide (ERDÖS 1975; SOFFER 1976). In addition, similarities to enkephalin-degrading enzymes have been shown (SWERTS et al. 1979). ACE was first detected in the blood (SKEGGS et al. 1956a) and later in the kidney (ERDÖS and YANG 1967) and various other tissues (GANTEN et al. 1975). Using biochemical and immunohistochemical methods, the enzyme was localized predominantly in endothelial cells, e.g., of the pulmonary and liver vasculature as well as in epithelial cells of the renal proximal tubule, the intestine and choroid plexus (BAKHLE et al. 1969; CALDWELL et al. 1976; IGIC et al. 1975, 1977; ARREGUI and IVERSEN 1978; WIGGER and STALCUP 1978; RYAN and RYAN 1980; RIX et al. 1981).

ACE has been purified to homogeneity from the kidneys of various species (see ONDETTI and CUSHMAN 1982 for ref.), including man (WEARE et al. 1982). It is a glycoprotein with a carbohydrate content of 10%-30% and a molecular weight of 130-160 kd. ACE contains a zinc atom, which is essential for its activity. Chloride and other monovalent anions increase the activity of the enzyme through lowering of K_m (see ONDETTI and CUSHMAN 1984 for review).

RYAN et al. (1975), using antibodies conjugated with microperoxidase, showed that pulmonary ACE is located on the luminal aspect of vascular endothelial cells in direct apposition to the circulation. By membrane isolation and immonocytochemical techniques, it was subsequently established that ACE is a membrane-bound ectoenzyme not only localized on the membranes of endothelial cells, but also on those of the brush border membranes of the above-mentioned epithelia, thus interacting with their luminal fluids or contents (Fig. 4.16; CALDWELL et al. 1976; WARD et al. 1975, 1976a, b; ODY and JUNOD 1977; RIX et al. 1981; TAUGNER et al. 1982c).

The distribution of ACE in kidney vessels has been studied by CALDWELL et al. (1976) using direct immunolabeling. The authors report on the localization of staining in the endothelial cells of the glo-

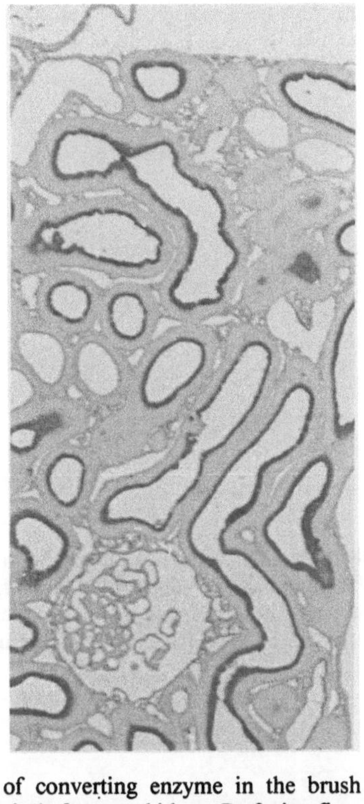

Fig. 4.16. Localization of converting enzyme in the brush border of the proximal tubule from rat kidney. Perfusion fixation with 0,4% glutaraldehyde in PBS; ×200. The high concentration of the enzyme in this extremely membrane-rich location may offer an explanation for the clear-cut immunohistochemical reactivity of the brush border - in contrast to that of the endothelium - even after fixation with somewhat higher concentrations of glutaraldehyde or formaldehyde (cf. TAUGNER and GANTEN 1982; DEFENDINI et al. 1983)

merular tuft and the delicate intertubular capillary network. HALL et al. (1976) found positive immunofluorescence reactions with tubuli and also with cells in the glomeruli. CELIO and INAGAMI (1981), using indirect peroxidase staining, found ACE immunoreactive material on the luminal surface of endothelial cells lining the arteries of the kidney, the glomerular arterioles, and some capillary loops.

A detailed study of ACE localization in kidney vessels was performed by TAUGNER and GANTEN (1982). With the indirect labeling PAP technique, specific immunostaining was found in the endothelial layer of all arteries and arterioles of kidney cortex, namely in the arcuate and interlobular arteries as well as in the pre- and postglomerular arterioles (Fig. 4.17). Positive reactions were also regularly seen in the outer zone of the renal medulla, corresponding to the descending (arterial) vasa recta. Kidney veins, in contrast, and the ascending (venous) vasa recta were ACE-negative. Positive reactions after administration of the ACE antiserum were seen in most glomeruli. The reaction product, however, was confined to the endothelium of only a few capillary loops, preferentially those in connection with the glomerular stalk (Fig. 4.18). With the exception of the first ramifications of the efferent arteriole (Fig. 4.17), the peritubular capillaries showed no ACE-specific staining.

These immunohistochemical results are not inconsistent with biochemical findings showing ACE activity in isolated glomeruli (WARD et al. 1977) and glomeruli with adherent afferent arterioles (DAHLHEIM et al. 1970; THURAU et al. 1970; GRANGER et al. 1972). However, with regard to the hypothesis of angiotensin II being produced in and secreted from epithelioid cells in effective quanta, it appears important to emphasize that the renin-positive cells in the area of the JGA were shown to be immunohistochemically ACE-negative (CELIO and INAGAMI 1981; TAUGNER and GANTEN 1982). Using radioenzymatic techniques, VELLETRI and BEAN (1982) found over 40% of the ACE activity localized within the tunica media of the aorta. In the renal arterial tree, however, the smooth muscle cells were - as the epithelioid cells - immunohistochemically ACE negative (TAUGNER and GANTEN 1982).

The question arises, whether in a number of organs and especially in the kidney, the generation of angiotensin II can take place outside the bloodstream. In connection with the concept of an intrarenal RAS (THURAU and MASON 1974), it would be important to know whether ACE, besides being present in the luminal membranes, is also localized in the abluminal membranes of the endothelial and proximal tubular epithelial cells of the kidney. In the renal arterial tree, ACE-specific reaction product could be seen in the luminal as well as in the abluminal aspect of endothelial cells (Fig. 4.17; cf. TAUGNER and GANTEN 1982). However, the tissue preservation after the mild fixation procedures tolerated by ACE as an immunhistochemical antigen was not sufficient to provide definitive answers to the above-mentioned question. TAKADA et al. (1982), KOKUBU et al. (1983), and BRUNEVAL et al. (1986), using the PAP method for preembedding staining of human kidney specimens, found the immunocytochemical reaction product not only close to the brush border but also in the area of the basolateral membranes of the proximal tubule.

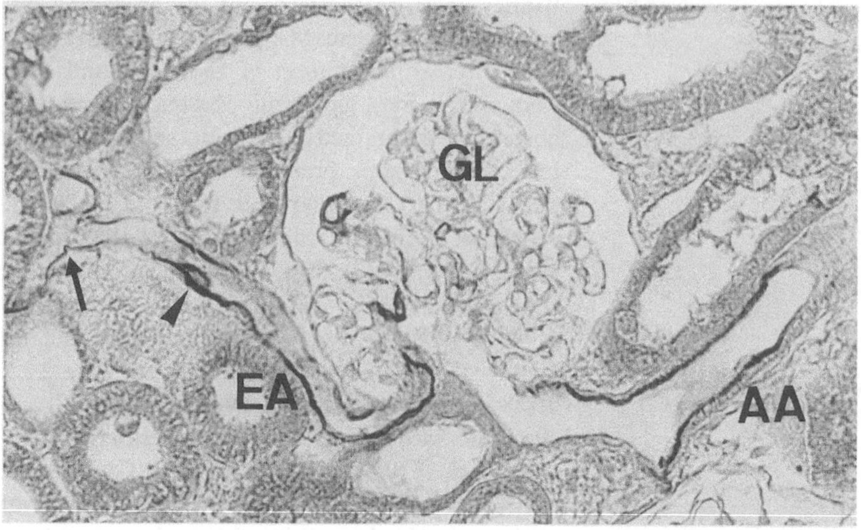

Fig. 4.17. Converting enzyme immunostaining of endothelial cells in the hilar vessels of rat renal corpuscle. Positive reactions are seen in the afferent arteriole *(AA)*, in the efferent arteriole *(EA)*, and in a few glomerular capillaries. In addition, the first branches of the peritubular capillaries seem to react positively *(arrow)*. The *arrowhead* points to an endothelial cell, in which reaction product is seen on both (the luminal and the abluminal) sides of the nucleus. *GL*, glomerulus. Antiserum dilution, 1:1000, PAP method; ×400

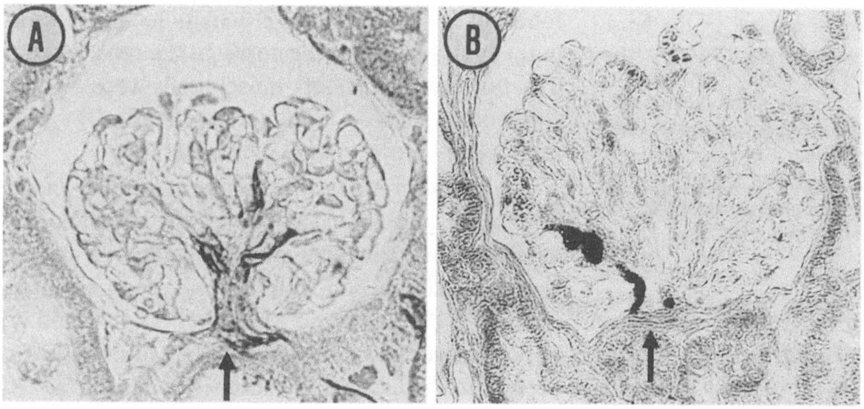

Fig. 4.18 A, B. Immunoreactivity for converting enzyme and renin, both related to the glomerular stalk *(arrows)*. In **A**, converting enzyme-positive glomerular capillaries, in **B** renin-positive mesangial cells, are seen. Paraffin sections, 7 μm. **A** control rat; **B** diabetes insipidus rat after a 4-week period of water restriction; ×400 and 300, respectively

Taken together, these experiments seem to support the assumption that abluminal membranes may play an important role within the scope of the intrarenal RAS (cf. Chap. 9).

When considering the possible intrarenal functions of ACE at locations detected by immunohistochemistry, it should be kept in mind, however, that ACE has a rather low specificity with respect to its substrates, i.e., it not only cleaves angiotensin I and bradykinin, but also a wide variety of other synthetic or natural peptide substrates (see Ondetti and Cushman 1984), indicating that there might be other, yet undiscovered functions of this enzyme, perhaps also in the kidney. The absence of ACE, on the other hand, does not permit to conclude that cleavage of angiotensin I to angiotensin II is unlikely to occur at a given site. This argument is based on observations on the presence, in the kidney, of enzymes other than the "classical" ACE, but nevertheless capable of converting angiotensin I into angiotensin II. Thus, tonin, originally detected in the rat submaxillary gland (Boucher et al. 1972, 1974; HACKENTHAL et al. 1972), is also present in the kidney (BOUCHER et al. 1974; LEDOUX et al. 1982). This enzyme, besides generating angiotensin II directly from angiotensinogen, can also convert angiotensin I to angiotensin II (BOUCHER et al. 1972). Since tonin is biochemically and immunologically not related to ACE, it escapes immunohistochemical detection with ACE antibodies. The same is true for another as yet poorly defined carboxypeptidase found in human kidney, which converts angiotensin I into angiotensin II by sequential cleavage of the two carboxy-terminal amino acids, but does not further hydrolyze angiotensin II (CHANGARIS et al. 1986). Finally, an ACE with a particularly high concentration in isolated glomeruli has been identified in rat kidney, which, although suppressed by ACE inhibitors, is clearly different from ACE in its pH optimum and in other properties, such as inhibition by diisopropyl fluorophosphate (DFP) (suggestive of a serine protease nature) and lack of inhibition by ethylenediaminetetraacetic acid (EDTA) as well as lack of chloride dependence (BURGHARDT et al. 1972; DAHLHEIM et al. 1983). The functional significance of this type of "converting enzyme" and its biochemical identity remain to be clarified.

The possible existence of still other, renin-independent pathways for the intrarenal generation of angiotensin II will be dealt with in Sect. 4.6.

4.4 Angiotensinogen

Angiotensinogen, or renin substrate, is the precursor of the angiotensin peptides. It is a glycoprotein with a molecular weight of 53-57 kd (SKEGGS et al. 1964; TEWKSBURY et al. 1978; HILGENFELDT and HACKENTHAL 1979; BOUHNIK et al. 1981) and carries a single angiotensin I moiety on its aminoterminal end. The amino acid sequence has been derived from the cloned cDNA sequence for rat (OHKUBO et al. 1983) and human (KAGEYAMA et al. 1983) angiotensinogen. Normal plasma concentrations of angiotensinogen in various species are in the range from 0.5 to 3 μM. Since this is also the range for the apparent Michaelis-Menten constants for the renin-angiotensinogen reaction (POULSEN and JACOBSEN 1986), the plasma concentrations of angiotensinogen, in addition to the primary role of renin, are also a determinant for the rate of angiotensin generation in the circulation. The half-life in the circulation and the distribution of circulating angiotensinogen have, so far, only been studied in the rabbit by LEWICKI et al. (1983). From the kinetics of disappearance of radiolabeled angiotensinogen, distribution into the extravascular compartment with a half-time of about 50 min and an elimination half-life of 9 h have been calculated. Extravascular distribution is an important factor in considerations of interstitial generation of angiotensin from angiotensinogen, particularly in the kidney (see below).

Angiotensinogen present in the circulation is probably exclusively synthesized and secreted by the liver. Conditions for secretion have been examined in the isolated liver (NASJLETTI and MASSON 1971, 1972; MURAKAMI et al. 1980; STUZMANN et al. 1986), liver slices (FREEMAN and ROSTORFER 1972; HERMANN et al. 1980; CLAUSER et al. 1983), and isolated hepatocytes (WEIGAND et al. 1977; STUZMANN et al. 1986; RADZIWILL et al. 1986). Summarizing the results of these and several other studies performed in vivo and in vitro, the rate of synthesis and secretion appears to be controlled by the RAS via a feedback stimulation by angiotensin II. In addition, hormonal factors, such as glucocorticoids, thyroxin, and estrogens contribute in an unspecific fashion to the control of angiotensinogen secretion (cf. HACKENTHAL et al. 1987).

Following bilateral nephrectomy, the rate of angiotensinogen synthesis and secretion from the liver increases severalfold within 8-12 h (CARRETERO and GROSS 1967; BING and POULSEN 1969; NASJLETTI and MASSON 1971, 1972). The mechanisms by which this occurs are still unknown. The presecnse of an angiotensinogen-stimulating factor in plasma of nephrectomized rats, which has been postulated by HASEGAWA et al. (1973), could not be confirmed in a recent study by KLETT et al. (1986).

In addition to angiotensinogen synthesis in the liver, there has been increasing evidence for some extrahepatic angiotensinogen synthesis. Unequivocal demonstration of a local synthesis of angiotensinogen is essential for the verification of independent tissue RASs, since different enzymes are able to release angiotensin I or angiotensin II from renin substrate, but angiotensinogen is the only known precursor of the angiotensins. Extrahepatic sites of angiotensinogen synthesis under discussion are, among others, the brain (PHILLIPS et al. 1979; CAMPBELL et al. 1984), the arterial tree (DESJAR-

DINS-GIASSON 1981; for references see SWALES et al. 1983; DAHLHEIM et al. 1983), and the kidney (CAMPBELL and HABENER 1986; FRIED and SIMPSON 1986; INGELFINGER et al. 1986; OHKUBO et al. 1986).

The presence of angiotensinogen in the kidney has never been demonstrated by *biochemical techniques*. However, in some studies, the generation of angiotensins in vitro, in the absence of exogenous angiotensinogen, has been taken as indirect evidence for the presence and local formation of an angiotensin-precursor, most likely angiotensinogen in the kidney. FINKIELMAN and NAHMOD (1969) reported that suspensions of washed rat glomeruli produced a pressure activity, which, in a rat blood pressure bioassay and in chromatographic behavior, was indistinguishable from angiotensin I. Since these authors could not detect angiotensinogen in their preparation, they concluded that this angiotensin may have originated from a different precursor. MORRIS and JOHNSTON (1976) found that subcellular fractions of kidney cortex generate angiotensin I when incubated with renin. This observation was taken to support the concept that angiotensinogen may be present in, and react with, renin in the granulated cells of the JGA, resulting in the intracellular formation of angiotensin I. As angiotensinogen from other sources, e.g., from the pinocytotic apparatus of tubular cells, might have been involved in the observed reactions, it remained uncertain as to whether they were suited to verify the production of angiotensinogen in epithelioid cells. Similarly, observations on the presence of angiotensin I in subcellular structures of renal cortical tissue have also been considered to indicate the local generation from angiotensinogen. Thus, KAWAMURA et al. (1985) isolated kidney cortex granule fractions by gradient fractionation, which contained renin as well as angiotensin I and II (identified by HPLC and radioimmunoassay). They concluded that all three components were present in the same granules and suggested that the whole cascade of angiotensin formation from angiotensinogen may be present within epithelioid cells. However, the absolute quantities detected in these experiments were extremely small, i.e. about 9 fmol angiotensin II and 40 fmol angiotensin I per gram kidney cortex were measured. Furthermore, the fourfold higher angiotensin I content as compared with angiotensin II is difficult to reconcile with results from immunocytochemical studies on the presence of angiotensin I oder II in the kidney (see Sect. 4.2).

Only a few *immunohistochemical experiments* with regard to the localization of angiotensinogen at the cellular level have been reported. In the liver, renin substrate was found to be located only in hepatocytes and never in endothelial or in Kupffer's cells. It is not yet certain whether there is a preferential localization of angiotensinogen in hepatocytes located around the central veins of liver lobules (MORRIS et al. 1979a; TAUGNER and HACKENTHAL, unpublished observations) or not (RICHOUX et al. 1983). In the kidney, angiotensinogen-reactive cells have only been found in the proximal tubule (RICHOUX et al. 1983; TAUGNER and HACKENTHAL, unpublished observations). As specifically stained granular formations were abundant in the apical regions of the involved tubular cells, it was suggested that - in analogy to the tubular renin shown in Figs. 4.7 and 4.8 - they represent filtered angiotensinogen taken up from the proximal tubular fluid via pinocytosis.

Juxtaglomerular epithelioid cells have never been found to be angiotensinogen reactive, in immunohistochemical (TAUGNER et al. 1982c; RICHOUX et al. 1983) as well as in immunocytochemical experiments (TAUGNER and HACKENTHAL, unpublished observations).

It should be recalled that immunohistochemical experiments are facing two limitations. Positive reactions permit definite conclusions as to the site of synthesis of a protein only when its uptake from extracellular sites can be safely ruled out. Negative results, on the other hand, might be hampered by the relative insensitivity of the available immunohistochemical methods. Safe conclusions can be expected from the *mRNA-hybridization techniques*. First reports on the presence of mRNA coding for angiotensinogen in the kidney came from OHKUBO et al. (1986) and FRIED and SIMPSON (1986). OHKUBO et al. (1986), using cDNA and rat genomic DNA hybridization, found angiotensinogen mRNA in brain, adrenal, ovary, lung, and kidney, with about equal distribution between medulla and cortex, whereas in the RNA-DNA dot blot hybridization studies of FRIED and SIMPSON (1986) the mRNA for angiotensinogen was almost exclusively confined to the renal medulla. In addition, the effects of uninephrectomy suggested that regulation of kidney angiotensinogen mRNA levels does occur.

CAMPBELL and HABENER (1986) reported the localization of angiotensinogen mRNA in 12 different tissues, besides the liver, aorta, and mesentery, among others, also the kidney. The regulation of the mRNA levels was found to be organ specific. In kidney tissue, hormone treatment with dexamethasone, ethinylestradiol, and triiodthyronine was followed by a 3.1-fold increase of angiotensinogen mRNA. However KALINYAK and PERLMAN (1987) observed no increase in renal angiotensinogen mRNA following dexamethasone stimulation.

Similar to OHKUBO et al. (1986), INGELFINGER et al. (1986) found angiotensinogen mRNA sequences in rat renal cortex and medulla in apparently equal portions, the relative signal intensity of the kidney to liver mRNA amounting to 3:100. According to densitometric Northern blot analysis, cortical and medullary angiotensinogen mRNA increased on a low-sodium compared with a high-sodium diet by a factor of 3.5 and 1.5, respectively. These studies again suggest the existence of intrarenal angiotensinogen generation, which is subject to organ specific control. In view of the intrarenal RAS discussed in Chap. 9, it will be most interesting to identify the exact intrarenal distribution of the angiotensinogen mRNA found by extraction procedures.

In this context of an intrarenal generation of angiotensinogen a possible relationship between angiotensinogen and erythropoietin, as suggested by GOULD et al. (1980), should be discussed. FYHRQUIST et al. (1984 a, b) noticed immunological similarities between both proteins and proposed that angiotensinogen is the precursor of erythropoietin in the kidney. This proposal seemed to receive support from immunocytochemical studies on the presence of both erythropoietin and angiotensinogen-immunoreactive material in hemangioblastoma tumor cells (ROSENLÖF et al. 1985). However, the amino acid sequence of erythropoietin has now been derived from the cloned cDNA sequence (LEE-HUANG 1984; JACOBS et al. 1985) and was found to be entirely unrelated to the amino acid sequence of angiotensinogen (OHKUBO et al. 1983; KAGEYAMA et al. 1984); thus the proposal of FYHRQUIST et al. (1984a, b) that angiotensinogen is the biochemical precursor of erythropoietin has to be dismissed. On the other hand, the immunological cross-reactivity between both proteins has recently been confirmed with highly purified components by ROSENLÖF (1986); however, the nature and significance of this observation remain unclear.

4.5 Angiotensinases

It is well-known that angiotensin I and angiotensin II are not only generated inside the kidney, but also cleaved and inactivated by ectoenzymes associated with the luminal membranes of endothelial and tubular epithelial cells (LEARY and LEDINGHAM 1969; BAILIE et al. 1971; AIKEN and VANE 1972; OPARIL and BAILIE 1973; PULLMAN et al. 1975; WARD et al. 1976; BAILIE and OPARIL 1977; PETERSON et al. 1977, 1979; CARONE et al. 1980; LOJDA and GOSRAU 1980; SUDO 1981). In contrast, there is little information about the so-called angiotensinases in the area of the vascular pole of the renal corpuscle which may affect the concentration of the angiotensins in the interstitial fluid of the JGA.

Several enzymes are thought to take part in the breakdown of angiotensin I and II. APA (GLENNER and FOLK 1961) is able to remove the N-terminal Asp from the octapeptide (GLENNER et al. 1962; HESS 1965; NAGATSU et al. 1965 and 1970; KHAIRALLA and PAGE 1967), the resulting heptapeptide angiotensin III displaying only a fraction of the vasoconstrictive ability of angiotension II. The cleavage of the N-terminal Arg from angiotensin III, in which aminopeptidase M (APM; cf. DELANGE and SMITH 1971; KUGLER 1982 b) can participate, then leads to the almost complete inactivation of the heptapeptide. In addition other enzymes, e.g., carboxypeptidase or endopeptidases, may be involved in the inactivation of the angiotensins (KHAIRALLAH and HALL 1977). Among these enzymes, only the localization of APM and APA has been studied with histochemical techniques.

In histochemical experiments, AMP was found to be localized predominantly in the brush border of the proximal tubule. The glomerulus and the JGA showed no reaction product (KUGLER 1981).

APA, on the other hand, has been localized by histochemical reactions not only in the brush border of the proximal tubule (GLENNER and FOLK 1961; LOJDA and GOSRAU 1980; KUGLER 1981), but also in glomerular cells, principally podocytes (LOJDA and GOSRAU, 1980; KUGLER 1981, 1983; SCHERBERICH et al. 1986; cf. also GRANGER et al. 1972 and SUDO 1981). The glomerular APA activity was shown to increase after short-term treatment with captopril or furosemide in rats, possibly by increasing the angiotensin I and II levels, i.e., the substrate supply (KUGLER and SCHIEBLER 1984).

In the JGA region, some species differences have been observed. APA-positive reactions were seen in the membranes of Goormaghtigh cells primarily of subcortical and intermediate nephrons and in the juxtaglomerular portion of both glomerular arterioles in rats. In the mouse, the epithelioid cells of the afferent arteriole showed a particularly marked reaction (KUGLER 1981, 1982a, c). In the golden hamster, the perivascular tissue of kidney vessels, including the juxtaglomerular arterioles, was APA-positive (KUGLER 1983). In ultracytochemical studies KUGLER (1982a) demonstrated that APA is associated mainly with the cell membranes of podocytes and glomerular endothelial and juxtaglomerular epithelioid cells in the mouse and with the membranes of Goormaghtigh cells in the rat. Interestingly, the reaction product was mainly localized in the area of cell contacts, i.e., of gap junctions. In mouse juxtaglomerular epithelioid cells, APA activity was also observed intracellularly in "lysosomal structures" which, as pointed out in Sect. 5.3, may be assumed to be altered secretory granules. The author suggests that the species differences encountered in the APA activity of epithelioid cells may explain the differences in the angiotensin II content of these cells in mice and rats (KUGLER 1982a; cf. Sect. 4.2).

Other possible implications of the APA activities in the region of the JGA will be discussed in the context of the intrarenal RAS (cf. Chap. 9).

4.6 Nonrenin Angiotensin-Forming Enzymes

Several proteases other than renin have been shown in vitro to generate angiotensin I or II from angiotensinogen. These include trypsin (AREKAWA et al. 1980), kallikrein (MARUTA and ARAKAWA 1983), cathepsin G (TENNESEN et al. 1982; WINTROUB et al. 1984), cathepsin D (DAY and REID 1976; HACKENTHAL et al. 1978), and tonin (HACKENTHAL et al. 1972; BOUCHER et al. 1972, 1974, 1977).

Whereas trypsin-catalyzed generation of angiotensin II is unlikely to occur and to have any physiological significance in the kidney, this may be different for kallikrein, which can generate angiotensin II from angiotensinogen (MARUTA and ARAKAWA 1983) and is present in high concentrations in the kidney, mostly at tubular sites, but also in the vicinity of epithelioid cells (BARAJAS et al. 1986; cf. Sect. 7.3). The mRNA for kallikrein and kallikrein-related enzymes from the rat submandibular gland has recently been sequenced, and a close sequence homology of kallikrein and tonin (of 85%) was found (ASHLEY and MACDONALD 1985 a). Since tonin can liberate angiotensin II directly from angiotensinogen (see below), this property may be shared by kallikrein or related enzymes. However, this enzymatic effect has only been described with hog pancreas kallikrein and no direct comparison can be made to renal kallikrein and its quantitative potential to generate angiotensin II in the kidney. Furthermore, the reported pH optimum for the reaction was pH 4-5 (MARUTA and ARAKAWA (1983). It therefore remains to be established whether this reaction occurs at all in the kidney, and, if so, whether it has any biological significance.

The same holds true for the angiotensin II-generating capacity of cathepsin G. This enzyme has been detected in neutrophil leukocytes (TONNESEN et al. 1982; WINTROUB et al. 1984). It is a serin protease with a molecular weight of 26-29 kd, which has functional properties very similar to those of tonin (see below). There are as yet no indications that cathepsin G, which is stored in and released from granulocyte lysosomes, participates in local angiotensin II formation in the kidney or other vascular beds, although a role in the control of vascular permeability has been suggested (WINTROUB et al. 1984).

The lysosomal acid protease cathepsin D, which is phylogenetically and biochemically closely related to renin (e.g., CORDES, 1984; MORRIS and CATANZARO 1986) has been found to generate angiotensin I from angiotensinogen (DAY and REID 1976; HACKENTHAL et al. 1978 a; MORRIS and REID 1978). Cathepsin D is present in the kidney in high concentrations and is colocalized with renin in the granules of epithelioid cells (TAUGNER et al. 1986c; cf. also Sect. 5.3). However, the rate of angiotensinogen cleavage by cathepsin D is about 100 000-fold lower than that by renin, and the pH-optimum for this reaction is around pH 5.0 (HACKENTHAL et al. 1978 a). It is therefore questionable, whether this enzyme can play any role in the generation of angiotensin I in the kidney.

Tonin is a serin protease capable of forming angiotensin II directly from angiotensinogen (HACKENTHAL et al. 1972; BOUCHER et al. 1974; GRISE et al. 1980). It has been isolated from the rat submaxillary gland and may also be present in other organs, including the kidney (BOUCHER et al. 1974, 1977). As mentioned already in Sect. 4.3, tonin can also convert angiotensin I to angiotensin II, a reaction not inhibited by converting-enzyme inhibitors (BOUCHER et al. 1972, 1974). Recent enzyme kinetic studies with purified components (GUTKOWSKA et al. 1984) have shown that tonin has a slightly higher affinity to angiotensinogen as a substrate when compared with renin (0.7 versus 2.8 μM), but renin has a turnover number about 3700 times that of tonin under optimum conditions. Since the molar concentration for tonin in the kidney is not known, no estimation of the angiotensin II-generating capacity of renal tonin in comparison to renin can be made. Furthermore, the reaction rate of tonin with angiotensinogen in plasma is much lower than with purified angiotensinogen (TREMBLAY et al. 1981; HACKENTHAL et al. 1972), probably due to the presence of a high-molecular-weight protease inhibitor in plasma (TREMBLAY et al. 1981). Therefore, it is not possible at present, to evaluate the potential of tonin (which may be identical with cathepsin G) to contribute to intrarenal angiotensin II generation. Some indirect evidence seems to suggest the possibility that tonin, or a functionally related enzyme, may play such a role. For example, HOFBAUER et al. (1973) observed in the isolated perfused rat kidney that the vasoconstriction elicited by infusion of angiotensin I, but not the vasoconstriction in response to the infusion of rat angiotensinogen, could be inhibited by the simultaneous infusion of the converting-enzyme inhibitor teprotide, indicating that angiotensin II may have been formed from angiotensinogen by a converting-enzyme independent pathway (such as tonin). Recently, BÜHRLE et al. (1987a) made analogous observations in electrophysiological studies on epithelioid cells of the mouse afferent arteriole, with respect to the depolarizing effects of tetradecapeptide renin substrate and converting-enzyme inhibitors (cf. Chap. 8). However, tonin-dependent generation of angiotensin II within the kidney has not directly been shown to occur. Earlier reports on the presence of tonin in rat kidney tissue (BOUCHER et al. 1974) as evaluated by demonstration of a tonin-like enzymatic activity and by immunohistochemical localization in distal tubular cells (LEDOUX et al. 1982), must be reevaluated in the light of recent findings that the tonin gene appears not to be expressed in the rat kidney (ASHLEY and MACDONALD 1985b). Still another aspect of the properties of tonin has been reported by GARCIA et al. (1981), who observed an effect of tonin on the contractility of vascular smooth muscle, which was not dependent on its enzymatic nature and appeared to be mediated by potentiation of the effect of norepinephrine. The authors discussed a role of tonin in the pathogenesis of hypertension.

In summary, no direct evidence is yet available for a non-renin-dependent pathway of angiotensin II generation in the kidney, although this pathway cannot be excluded to exist and to have a physiological or pathophysiological function.

Chapter 5

Synthesis and Traffic of Renin in Epithelioid Cells

5.1 Granulopoiesis in Epithelioid Cells and Packaging of the Secretory Product

In cell types with regulated secretion, the concentration and/or packaging of the secretory product occurs in specialized Golgi-dependent condensing vacuoles or in the transmost Golgi cistern (for review see FARQUHAR and PALADE 1981; FARQUHAR 1985). BARAJAS and LATTA (1965) and BARAJAS (1966) showed that in accordance with the second of these two possibilities, the secretory granules[1] of juxtaglomerular epithelioid cells originate from the dilated, pinched-off rim of the innermost Golgi cistern (Fig. 5.1). Corresponding to this origin, also noted by CHANDRA et al. (1965), the nascent or so-called protogranules exhibit rhomboid or fusiform profiles. After stimulation of renin synthesis, rhomboid protogranules with a paracrystalline core are conspicuous (Fig. 5.3 A; BARAJAS and LATTA 1965; BARAJAS 1966). In individuals without stimulation of the RAS, fusiform protogranules with an amorphous content predominate (Fig. 5.1; TAUGNER and METZ 1986). Besides, species differences have been observed not only in the electron density, but also in the shape of protogranules: in humans and monkeys they exhibit a polygonal shape, whereas in the rat protogranules tend to be more ovoid (BARAJAS 1966; ROSEN and TISHER 1968).

Immunocytochemical experiments revealed that these early rhomboid or fusiform protogranules are distinctly renin positive (Figs. 4.11, 5.2). There is, in fact, a jump in renin concentration between stacked Golgi cisterns and protogranules which seems to be bigger than any other concentration difference during the course of granulopoiesis to mature granules (LINDOP and DOWNIE 1984; LACASSE et al. 1985; LINDOP and GARDINER 1986: TAUGNER and METZ 1986). This early condensation of the secretory product supports the assumption by BARAJAS (1966) that renin is related to the paracrystalline core in protogranules. Immunocytochemical experiments with antibodies against the prosegment of renin suggest that protogranules may contain mainly prorenin (cf. Sect. 5.2).

According to the fundamental description of BARAJAS (1966), further maturation of epithelioid cell secretory granules may proceed through coalescence of several protogranules within smooth-surfaced sacs, in which the former protogranules at first retain their individual paracrystalline pattern (Figs. 5.2, 5.3). Paracrystalline cores may also persist in conglomerate or large polymorphous granules with irregular contours, especially after stimulation of the RAS (Fig. 5.4).

Besides by way of protogranule coalescence, mature secretory granules with their characteristically homogeneous internum were suggested to develop by continuous growth from individual protogranules (BARAJAS 1966). Although there is an abundance of small smooth and coated vesicles in the Golgi region of epithelioid cells (GORGAS 1978 a) and there are indications that some of these vesicles may fuse with juvenile or intermediate granules (Fig. 5.6 B), the available methods do not permit the classification of the vesicles with respect to their content and target organelles. Therefore, the question of this alternative way of mature granule formation remains unresolved (Fig. 5.7). This also applies to the possibility that - in analogy to the "maturation model" of lysosomes (HELENIUS et al. 1983, BROWN et al. 1986) - a small fraction of renin granules may originate from the fusion of carrier vesicles with endosomes (TAUGNER and METZ 1986).

There are, on the other hand, indications that coalescence phenomena after fusion of rhomboid granules with paracrystalline cores may be less significant than granulopoiesis by fusion of fusiform

[1] In the following, the entire organelle, i.e., internum and the limiting membrane, will be referred to as "granule" or "protogranule."

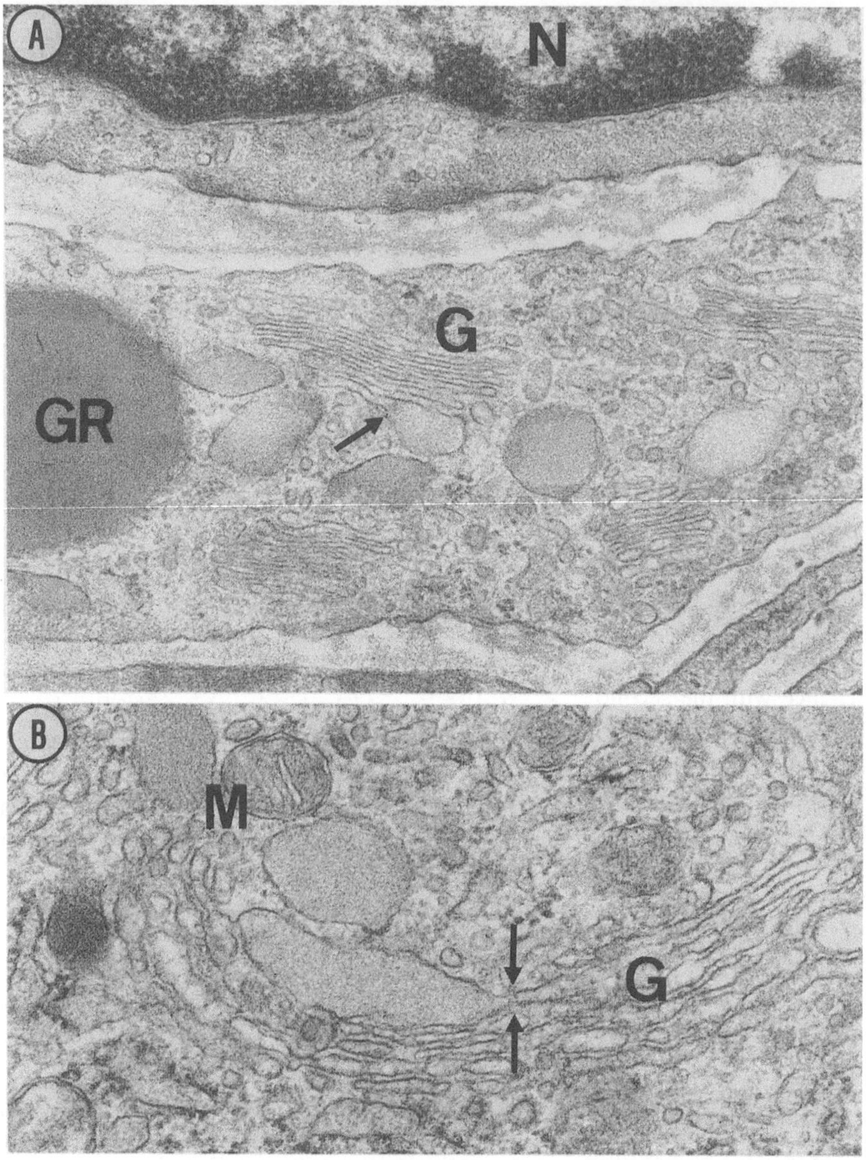

Fig. 5.1 A, B. Golgi regions *(G)* of mouse juxtaglomerular epithelioid cells showing nascent protogranules pinching-off from the transmost Golgi cistern *(arrows)*. Additionally, several juvenile secretory granules ranging from rhomboid through fusiform to roundish shapes are seen. *GR*, mature granule; *M*, mitochondria; *N*, nucleus. Note the difference in electron density between stacked Golgi cisterns, juvenile granules, and mature granules; ×45000 and ×58000, respectively

or roundish juvenile granules with a homogeneous internum: roundish juvenile and intermediate granules with an amorphous matrix are frequently seen in the Golgi region of epithelioid cells (Fig. 5.1), their occurrence outweighing that of rhomboid and polymorphous granules with paracrystalline cores, even in stimulated animals. This seems to contradict the finding that fusion processes between roundish juvenile granules are rarely observed in the section plane as compared with coalescence phenomena between rhomboid or polymorphous granules. However, this dissimilarity may be misleading inasmuch as the time scale of both events is likely to be quite different for several reasons. Above all, it is suggested that the intermixture of amorphous matrices occurs rapidly as compared with the relatively slow coalescence of paracrystalline contents, delaying the rounding of rhomboid

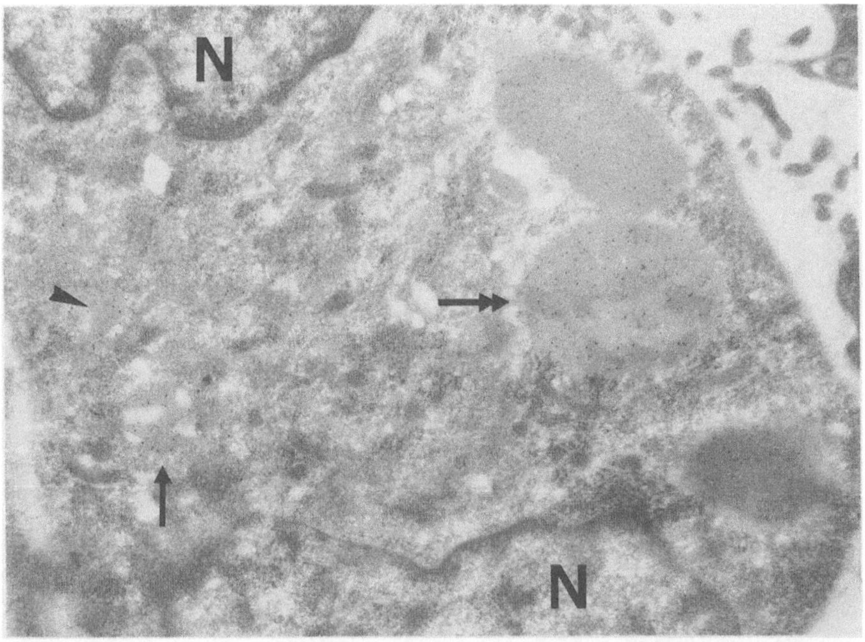

Fig. 5.2. Mouse epithelioid cells, immunoreacted for renin with labeled protogranules in the process of coalescence to intermediate or conglomerate granules *(arrow, double arrow).* Note that the electron-dense, renin-positive matrix of the former protogranules can still be distinguished in the forthcoming organelle, embedded in a renin-negative and less electron-dense matrix. *Arrowhead,* immunoreactive nascent protogranule; *N,* nucleus; ×21100

and polymorphous granules. It was therefore concluded that, diverging from their respective incidence in the section plane, the slower process, namely the fusion of rhomboid and polymorphous granules with paracrystalline cores, contributes proportionally less to the development of mature renin granules than does the more rapid fusion process between roundish granules displaying an amorphous matrix. According to this view, roundish juvenile and intermediate granules with a homogeneous matrix may represent plain juvenile granules which, after stimulation of renin synthesis, are accompanied by an increasing number of granules with paracrystalline cores due to an overload in processing and packaging of the secretory product (TAUGNER and METZ 1986). A distinct overcharge of these processes, such as in Bartter's syndrome or in animals after adrenalectomy or chronic administration of captopril, appears to lead to the occurrence of paracrystalline inclusions already in the RER cisterns (Figs. 14.5, 14.6; KANETA et al. 1981; HACKENTHAL et al. 1987; TAUGNER et al. 1988c).

The hypothetical concept shown in Fig. 5.7 presents, from left to right, the various putative routes of granulopoiesis from juvenile to intermediate and mature granule types. Below, the routine route via round granules with a homogeneous internum is shown, above, the route via rhomboid and polymorphous granules which is preferentially seen during stimulation of renin synthesis. Intermediate routes encountered between both these extremes are likely to occur at every stage of granulopoiesis, with the paracrystalline core becoming more and more lost. The significance of the disappearance of the paracrystalline cores and the rounding-off of the granules is unknown. Parameters such as the intragranular milieu and/or the activities of the various enzymes within the granule matrix have to be considered. However, as shown in Sect. 5.2, there is evidence that the maturation of the secretory product, i.e., the conversion of prorenin to mature renin, occurs during this stage of granulopoiesis.

The parallelism between renin granules and lysosomes is dealt with in Sect. 5.3. Here, only additional arguments derived from the examination of maturing granules will be emphasized (TAUGNER and METZ 1986). Pertaining to this are, in the first place, observations indicative of a remarkable turnover of membrane material in juvenile and intermediate granules: the preferential uptake of exogeneous tracers, the frequent microautophagic activity (Fig. 5.6, inset) probably responsible for the vesicu-

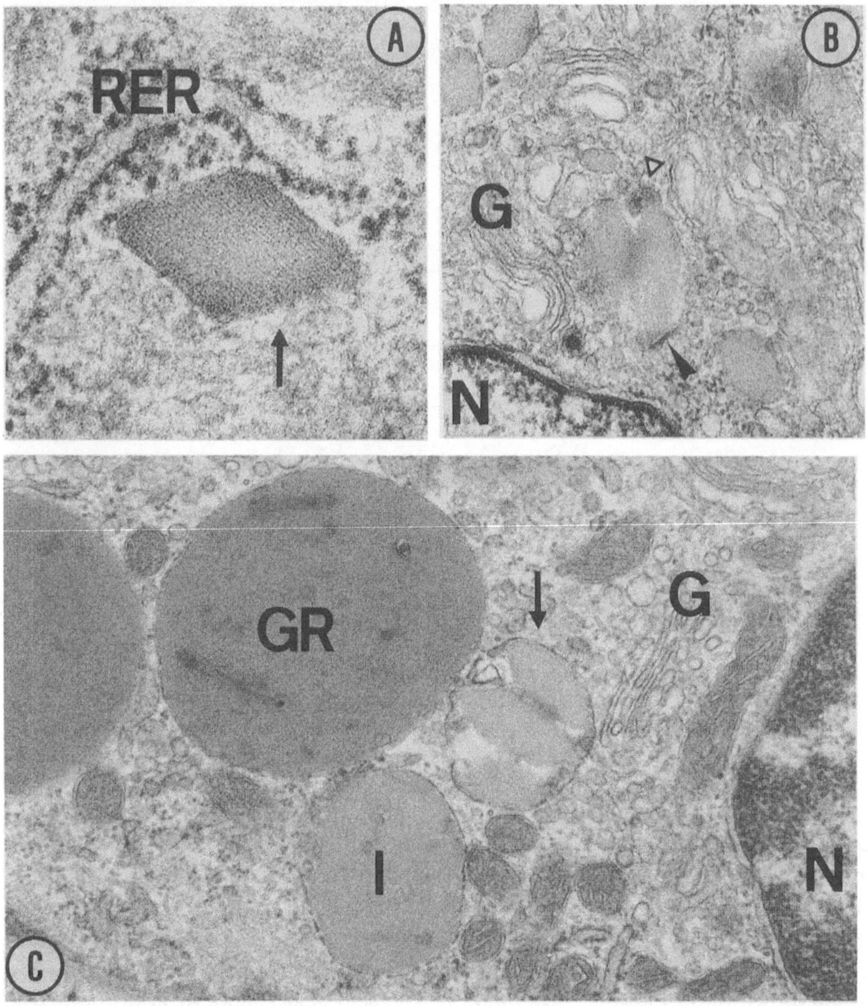

Fig. 5.3 A–C. Early stages of granulopoiesis in mouse epithelioid cells. **A** Rhomboid protogranule with paracrystalline contents *(arrow)*. *RER*, cistern of the rough endoplasmic reticulum; × 100000. **B** Coalescence of two fusiform protogranules exhibiting coated bud *(white arrowhead)* and surface plaque *(black arrowhead)*, *G*, Golgi complex; *N*, nucleus; × 38000. **C** Profile of juvenile granule showing the coalescence of three protogranules *(arrow)*. *I*, more advanced stage of juvenile granule with vesicular inclusions at the site of former protogranule fusion; *GR*, mature granule; *G*, Golgi complex; *N*, nucleus; × 36000

lar inclusions already observed, among others, by Barajas (1966), Lee et al. (1966), and Tsuda et al. (1971) (Figs. 5.3 C, 5.4 A, 5.5), and finally the accumulation of polar lipids after the administration of lysosomotropic substances not only in mature, but also in juvenile and intermediate, granules. A possible function of microautophagy at this early stage of granulopoiesis might be the incorporation of endogenous cytoplasmic inhibitors of the thiolpeptidase cathepsin B already present in maturing granules, where cleavage of the prosegment, i.e., the activation of renin, is thought to take place (Taugner et al. 1985 a, 1986 a, b, 1987 a). The preferential uptake of exogeneous tracers by juvenile granules was taken to indicate an accordingly high fusogeneity of their membranes, allowing for the exocytotic extrusion of inactive renin from immature granules (Taugner et al. 1986 a, 1987 a; Hackenthal and Taugner 1987).

Another similarity of renin granules with lysosomes relates to the decoration of maturing granules with bristle coats. Aside from coated buds located in particular around the apex of nascent and already pinched-off protogranules (Fig. 5.3 B), fuzzy bristle coats were also observed in relation to intermediate and large polymorphous granules (Taugner and Metz 1986). These bristle coats tend to occur in flattened, strikingly electron-dense mem-

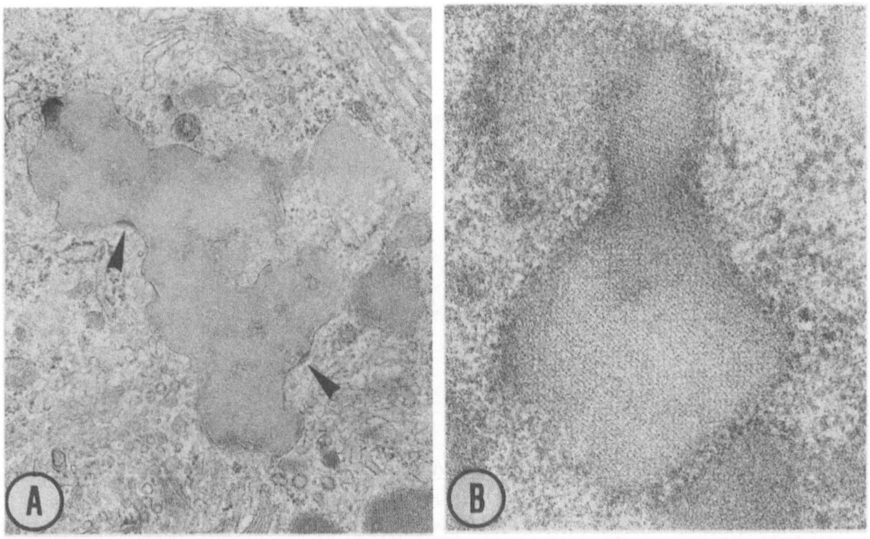

Fig. 5.4 A, B. Conglomerate epithelioid cell secretory granules of the mouse kidney. **A** Large conglomerate granule formed by fusion of several protogranules. Note the vesicular inclusions at the sites of previous fusions. *Arrowheads,* surface plaques; ×25500. **B** Polymorphous granule with paracrystalline patterns. Perfusion fixation with 2.5% glutaraldehyde and 2% OsO_4 in phosphate-buffered saline; ×97000

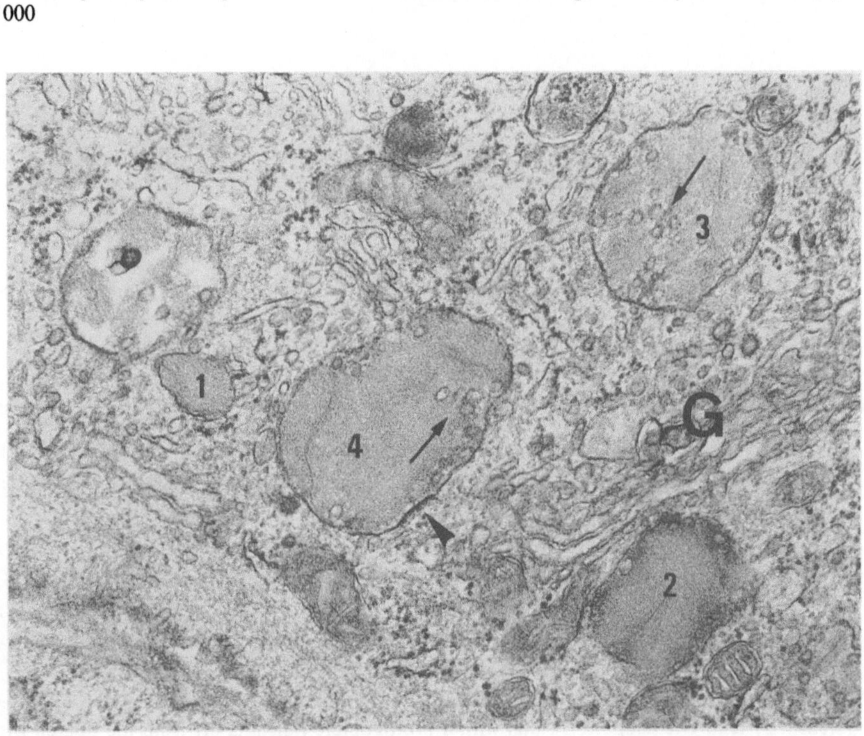

Fig. 5.5. Protogranule *(1)* and profiles of juvenile epithelioid cell secretory granules of different size and complexity, showing at least two fused protogranules *(2–4).* Vesicular inclusions are seen at the periphery of the granules and at putative previous fusion sites *(arrows). Arrowhead,* surface plaque; ×40000

brane areas of maturing granules (Figs. 5.3, 5.5, 5.6) and are evidently equivalents of surface plaques. Surface plaques are not present in membranes of typical secretory granules, but are well known in membranes of endosomes (WILLINGHAM and PASTAN 1984) and those of MVBs (HOLTZMAN 1976). The functional significance of these peculiar structures is, as yet, unknown. Their bristle coats show similarities to coated pits and coated vesicles, and they also seem to react with anticlathrin antibodies (TOUGARD et al. 1985). ORCI et al. (1984 a, b) recently proved that cleavage of the prosegment of insulin takes place in a partly clathrin-coated converting compartment. In this context, it appears remarkable that the prosegment of renin is cleaved in juvenile granules decorated with bristle coats.

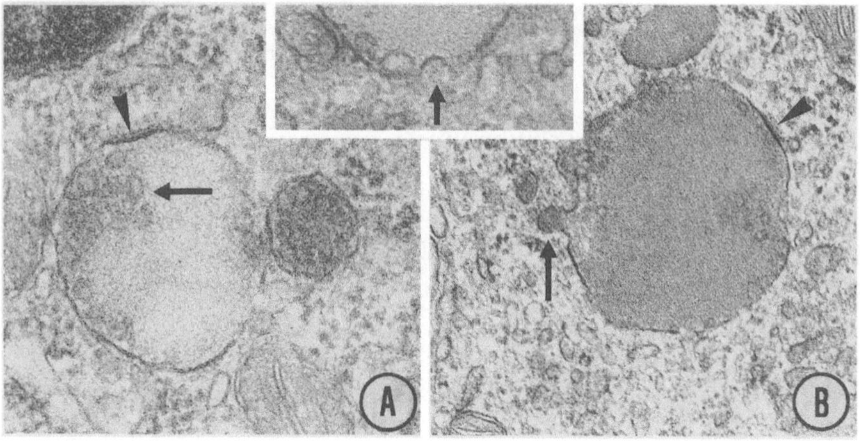

Fig. 5.6 A–B. Intermediate secretory granules of epithelioid cells with features uncommon to granules of other secretory systems. **A** Intermediate granule from rat epithelioid cell showing vesicular inclusions *(arrow)* and a surface plaque with bristle coat *(arrowhead)*; ×74000. **B** Mouse epithelioid cell secretory granule apparently in the process of fusion with a vesicle *(arrow)*; *arrowhead*, surface plaque; ×45700. *Inset:* Intermediate granule from mouse epithelioid cell with vesicular inclusions and an indentation of its outer membrane, suggesting a micropinocytotic (microautophagic) event *(arrow)*; ×65400

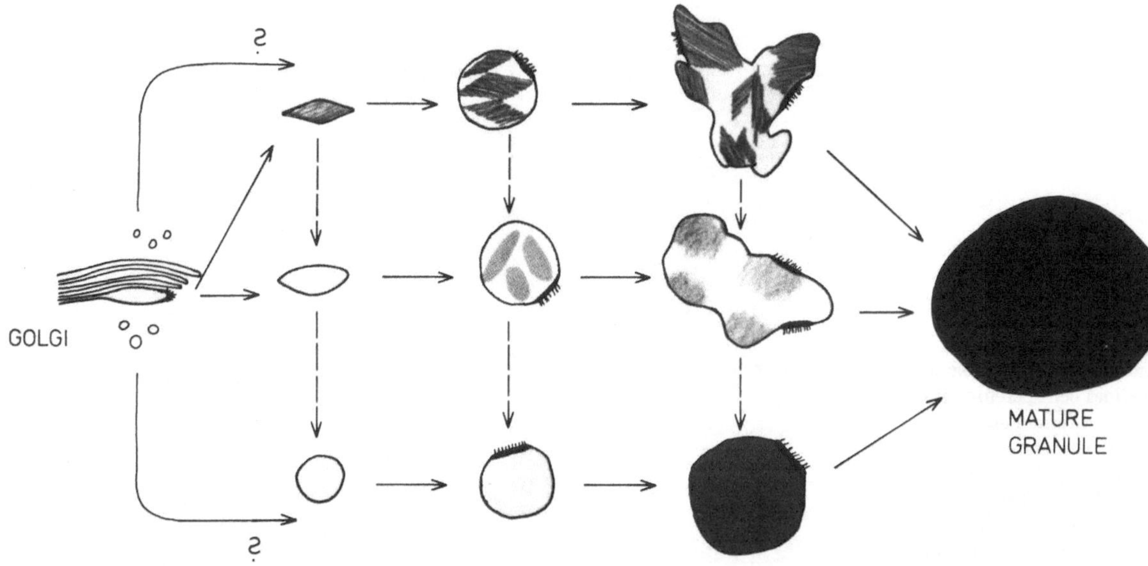

Fig. 5.7. Hypothetical concept of epithelioid cell granulopoiesis under control conditions and after stimulation of the renin-angiotensin system. Granulopoiesis starts with a protogranule pinching-off from the dilated rim of the transmost Golgi cistern (Barajas 1966). Under control conditions *(below)*, the fusiform protogranule is believed to round off and fuse with other amorphous juvenile or intermediate granules. After stimulation of the RAS *(above)*, the protogranules tend to exhibit rhomboid profiles and paracrystalline contents; after fusion with other juvenile or intermediate granules, the incomplete coalescence of the still paracrystalline content of the granule imposes polymorphous contours on the growing organelles. Finally, large round mature granules with a homogeneous internum are formed. Roundish intermediate granules with amorphous matrix can also be seen in stimulated animals. It is therefore suggested that the transition from rhomboid, fusiform, and polymorphous granules with paracrystalline cores to round granules with homogeneous interna may occur at every stage of granulopoiesis *(dashed arrows)*. It is furthermore assumed that the coalescence of the paracrystalline cores of rhomboid or polymorphous granules proceeds more slowly than the intermixture of the amorphous internum of round granules so that the more rapid course of granulopoiesis via roundish granules as compared with that via rhomboid or polymorphous granules occurs much more frequently than the incidence of the respective section profiles would suggest. The possible role of the Golgi-derived carrier vesicles in the granulopoiesis of epithelioid cells cannot be assessed with the methods available at present

5.2 Renin Synthesis and Activation

Analogous to the other proteases, renin is synthesized in epithelioid cells of the kidney as an inactive zymogen, called prorenin. In contrast to many other zymogens of proteases, prorenin is partially processed intracellularly to the active form and released from the kidney as both active and inactive renin. In the plasma of many species including man, a major portion of renin is in an inactive form which can be activated by various experimental procedures possibly similar to those by which kidney-inactive renin may be activated. This raises many interesting questions concerning the molecular properties of plasma-inactive renin, the activation process, the origin of plasma-inactive renin, the existence of control mechanisms for its release from the kidney or other organs, and the possible physiological function of inactive renin in the circulation. These questions will be discussed in Sect. 7.7. Here, the biochemical and immunocytochemical evidence for the maturation of immature renin in epithelioid cells will be reviewed.

Numerous studies on the characterization or purification of kidney renin have shown that, in addition to active renin with a molecular weight of 35-40 kd, also inactive, high-molecular-weight forms of renin are present in the kidney of man (DAY et al. 1975; BARRETT et al. 1977; SLATER and HABER 1978; GALEN et al. 1979; INAGAMI and MURAKAMI 1980; ATLAS et al. 1982; HSUEH et al. 1983; CHANG et al. 1984; MCINTYRE et al. 1984) hog (BOYD 1974; INAGAMI et al. 1977; LEVINE et al. 1978; MURAKAMI et al. 1980; TAKII and INAGAMI 1982), dog (FUNAKAWA et al. 1978; POTTER et al. 1979; IKEMOTO et al. 1980, 1982 a, 1983; TANAKA et al. 1985), rabbit (LECKIE and MCCONNELL 1975; GROSS and BARAJAS 1978), and rat (RUBIN et al. 1980; MORRIS and JOHNSTON 1976; INAGAMI et al. 1977), with apparent molecular weights ranging from 43 to 180 kd. Upon activation by acidification and/or treatment with proteases, the renin activity of the preparations was increased with or without changes in apparent molecular weight. In the mouse kidney, renin seemingly occurs only in the active 38-40 kd molecular form (POULSEN and NIELSEN 1981; IWAO et al. 1982). This rather complex situation has largely been clarified in several independent lines of investigation:

(1) In contrast to the large proportion of inactive renin with widely varying molecular weight found when renin was prepared from homogenates of kidney cortex, only active renin with a molecular mass of 38-40 kd and a smaller portion of inactive renin with molecular weights ranging from 43 to 50 kd was found when isolated renin granules were used as the source of renin. This has been observed in rats (MORRIS and JOHNSTON 1976; MURAKAMI et al. 1980; TAKAORI et al. 1981; IKEMOTO et al. 1982 b), and dogs (KAWAMURA et al. 1980). The existence of abundant high-molecular-weight inactive or partly active renin in homogenates, as opposed to mostly active renin with a small proportion of inactive renin with molecular weights never exceeding 50 kd in isolated granules, can now be explained by the demonstration of renin-binding protein(s) in the cytosolic fraction of renal cortical homogenates of man (TAKAHASHI et al. 1985), hog (TAKAHASHI et al. 1983; MURAKAMI et al. 1980), rat (TAKAORI et al. 1981; IKEMOTO et al. 1982 b), dogs (IKEMOTO et al. 1982 a; TANAKA et al. 1985), and mice (IWAO et al. 1982), thus confirming earlier suggestions of the existence of renin-binding proteins in the kidney by BOYD (1974), LECKIE and MCCONNELL (1975). From these and many additional studies it can be concluded that a significant source of high-molecular-weight inactive renin found in kidney homogenates originates from the association of mature, active low-molecular-weight renin with renin-binding proteins. This binding appears to depend on the interaction of sulfhydryl groups, as sulfhydryl oxidants favor formation of the complex. The complex, which has variably been found active, partly active, or completely inactive, can be dissociated by low pH and by treatment with various proteases. A high-molecular-weight "renin-converting enzyme" has been identified in dog kidneys, which is thought to participate in the interconversion of free and inhibitor-bound renin (TANAKA et al. 1985). The molecular weight of the binding proteins has been reported to vary between 10 and 40 kd. Most important, however, is the finding by IKEMOTO et al. (1982 b) that the renin-binding protein in the rat kidney is exclusively of tubular origin. The authors suggested that the binding protein does not contribute to the process of biosynthesis, processing, and secretion of renin in juxtaglomerular cells, but may play a role in tubular binding of filtered renin and thus perhaps in tubular functions. In other words, if

we assume that this observation can be extended to other species as well, this type of high-molecular-weight inactive renin originating from the association of renin and a binding protein is most likely an experimental artifact, because this association can only occur when mature, active renin stored in granules of epithelioid cells comes into contact with the cytosolic fraction of tubular cells during homogenization of the kidney cortex.

(2) Translation of renin mRNA extracted from the mouse kidney in a cell-free translation system was first described by POULSEN et al. (1980). The translation product, i.e., pre-prorenin, was shown to be a single polypeptide chain with a molecular weight of 50 kd, whereas mature, active renin was consistently found to have a molecular weight of 38-40 kd. The same value of 50 kd had previously been reported for the cell-free translation product of mouse submaxillary mRNA (POULSEN et al. 1979). These observations have been confirmed by DYKES et al. (1980) and MORRIS et al. (1981) with renin mRNA derived from mouse kidney and submaxillary gland, respectively, as well as with human kidney renin mRNA by PARMENTIER et al. (1983) (see also the review by POULSEN and JACOBSEN 1983). The molecular weight of 50 kd for the cell-free translation product is in fact identical to the molecular weight of completely purified prorenin from hog kidney (TAKII and INAGAMI 1982) as well as from human kidney (MCINTYRE et al. 1984). The difference in molecular weight to be expected from the presence of the presegment (about 20 amino acids) in the cell-free translation product and its absence in the purified prorenin preparation, is probably compensated by the carbohydrate residues attached to prorenin, but absent in the freshly translated pre-prorenin, which, in hog kidney renin, comprises about 3% of the molecule (CORVOL et al. 1977).

(3) Several laboratories have succeeded in cloning the complete cDNA prepared from renin mRNA in several species, such as man (MORRIS et al. 1983, 1984; IMAI et al. 1983; HARDMAN et al. 1984; HOBART et al. 1984; MYAZAKI et al. 1984), mouse (ROUGEON et al. 1981; MASUDA et al. 1982; PANTHIER et al. 1982, 1984; MORRIS et al. 1983; FIELD et al. 1984; CATANZARRO et al. 1985), and rat (NAKAMURA et al. 1985). This not only permits the deduction of the amino acid sequence of renin including its pre- and prosegment, but also gives more insight into the pathway of maturation of renin and the identification of potential glycosylation sites. Except for the mouse, where two distinct renin genes have been identified which are differently expressed in different tissues (MULLINS et al. 1982; FIELD et al. 1984), renin is coded in other species by a single gene. With regard to renin maturation and the relationship between intrarenal processing of the enzyme and the occurrence of prorenin in the circulation, the best studied example is the human renin gene and its expression in the kidney (see MORRIS 1986). Interestingly, the human renin gene has several potential promoter elements. Promoter 1 is used predominantly in the kidney, giving rise to a renin precursor with a hydrophobic signal peptide comprising 23 amino acids, which directs the nascent polypeptide chain to the endoplasmic reticulum for further processing in the Golgi complex (MORRIS 1986). Another promoter sequence, yielding a larger renin precursor, is probably used in the mouse submaxillary gland (FIELD et al. 1984; PANTHIER et al. 1984) and a third promoter element, if used, would give a much shorter expression product, lacking the presegment and also a portion of the prosegment. Interestingly, this latter expression product would carry a hydrophilic peptide sequence at its aminoterminal end, which could direct the enzyme to a cytosolic compartment where it could serve intracellular functions without being secreted (e.g., in the intracellular production of angiotensin in certain neurons in the brain). As derived from biosynthesis studies in vitro (HIROSE et al. 1985) and the known structure of the human gene, the primary translation product in the kidney is a 406 amino acid polypeptide, from which the 23 amino acid signal peptide is split off. The resulting prorenin polypeptide has a molecular mass of 42 kd, from which the prosegment comprising 43 amino acids with a molecular mass of 5.1 kd is removed. If the molecular mass of the carbohydrate moieties of 3-4 kd is added, a molecular weight of 41-42 kd can be calculated for mature, active renin. This figure is in close agreement with the reported molecular weight of purified human renin (SLATER and STROUT 1981). The presence of prorenin in the human kidney with the amino acid sequence of the aminoterminal portion predicted from the cDNA structure has recently been verified by ATLAS et al. (1985), BOUHNIK et al. (1985), and DAY et al. (1986) by means of antibodies raised against synthetic peptides representing the aminoterminal portion of the prosegment.

Taking these and further information from the literature together, it can be assumed that in the human kidney (and most likely also in kidneys of other species) the first product of ribosomal protein synthesis is pre-prorenin (about 45 kd), from which, following transfer into the endoplasmic reticulum, the signal peptide is cleaved, yielding unglycosylated prorenin (about 42 kd), which is confined to the interior of the Golgi complex. Subsequently, carbohydrate residues are attached, resulting in an increase in the molecular weight to about 45-46 kd. Following sequestration and packaging of prorenin in immature vesicles, the so-called protogranules, the prosegment is split off, yielding mature renin with a molecular mass of about 40 kd. The process of activation of prorenin to renin can be mimicked experimentally by exposing prorenin to various proteases under certain conditions (see Sect. 7.7). With immunocytochemical methods, both cathepsin B and cathepsin D have been demonstrated to be contained in renin granules (TAUGNER et al. 1985a; 1986c), suggesting that these two enzymes may participate in the intracellular processing of renin. Interestingly, many studies in vitro have shown that low pH leads to the reversible activation of prorenin without change of the molecular size (see DERKX et al. 1987 and Sect. 7.7 for further details). This "acid activation," in which probably proton-dependent conformational changes expose the active site of the enzyme, facilitates the subsequent irreversible activation of prorenin by enzymatic cleavage of the prosegment, and, as a result, the molecular weight is reduced to about 40 kd as expected. Since renin granules are very similar to lysosomes (see Sect. 5.3) they can be assumed to have an acidic interior, and the same mechanism of acid activation and facilitated proteolytic cleavage as observed in vitro may also be effective in the intravesicular maturation of renin. In this context, the observations of THURAU et al. (1972) and GILLIES et al. (1982) are of interest; these authors demonstrated that the apparent content of inactive, i.e., acid-activable renin, in rat glomeruli decreases when the delivery of sodium chloride to the kidney or to microperfused tubules is increased and vice versa. This may suggest the existence of a control mechanism for the maturation process in the kidney.

Recent immunocytochemical evidence suggests that the intracellular activation of renin proceeds in multiple steps by the sequential cleavage of distinct portions of the prosegment during granulopoiesis: KIM et al. (1985) synthesized three peptides, Pro 1, Pro 2A, and Pro 3, covering almost the entire span of the human prosegment from the NH_2- to the COOH-terminus. Antibodies against each of these three peptides and an antibody against mature human renin were used to compare the fate of the corresponding portions of the immature enzyme (TAUGNER et al. 1986a, b; 1987a). With anti-Pro 1, i.e., the antibody which recognizes the NH_2-termi-

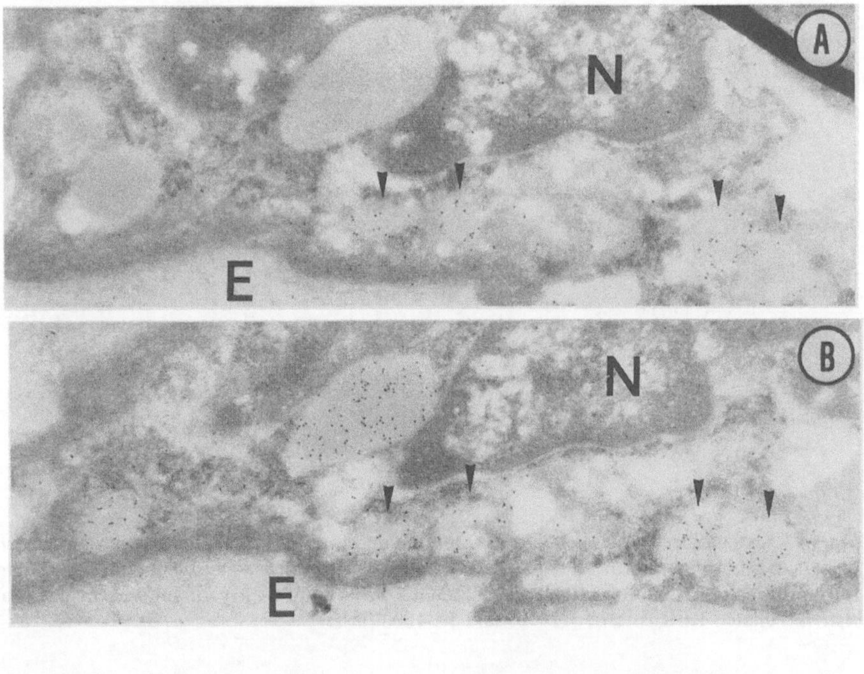

Fig. 5.8. Serial thin sections from human kidney immunoreacted for renin **B** and the middle portion of human renin prosegment **A**. Note that - in contrast to renin - only the protogranules *(arrowheads)* are prosegment positive. *N*, nucleus; *E*, extracellular space; ×32700

nus of human prorenin, no clear immunolabeling of epithelioid cell organelles could be obtained. After application of anti-Pro 2A, i.e., the antibody directed against the middle portion of the prosegment, only juvenile secretory granules were labeled (Fig. 5.8). With anti-Pro 3, the antibody against the COOH-terminus of the prosegment, labeling of all juvenile, some intermediate, and a few mature granules was obtained (Figs. 5.9, 5.10). In contrast to these results, the immunoreactivity of mature renin increased from juvenile to mature granules, the latter being labeled without exception (Figs. 5.8, 5.10). As the immunoreactivity of mature renin increased during granulopoiesis while that of the prosegment decreased to subthreshold levels, it was suggested that cleavage of the prosegment, i.e., the activation of renin, takes place in juvenile granules parallel to the condensation of the enzyme and that the secretion of active renin proceeds by the exocytosis of mature granules, the secretion of inactive renin by that of juvenile granules (cf. Chaps. 6, 7). In addition, the more or less apparent differences between the findings with the three prosegment-directed antibodies were taken to indicate segmental cleavage with subsequent breakdown of the respective prosegment components by hydrolytic enzymes.

According to HIROSE et al. (1985), human plasma prorenin is recognized only by anti-Pro 3, indicating that inactive renin may be a truncated version of intact prorenin, lacking a large portion of the NH_2-terminus of the prosegment. In the course of granulopoiesis in epithelioid cells, the immunoreactivity against anti-Pro 2A disappears somewhat earlier than that against anti-Pro 3 (TAUGNER et al. 1987a). Thus, a similar or even identical truncated version of prorenin may exist in late juvenile or intermediate granules, the exocytosis of which may be the source of inactive renin in human blood plasma.

Finally, it appears possible that maturation of renin proceeds even beyond the stage of the fully active 40 kd species. In the mouse submaxillary gland, an enzymatically active renin species consisting of two polypeptide chains with molecular weights of 31 and 5 kd, respectively, has been identified (MISONO et al. 1982; MORRIS et al. 1983; PRATT et al. 1983; 1986); the chains are held together by a disulfide bridge. Recently, synthesis of a fragmented form of renin with two subunits of molecular weights of 23.4 and 18.6 kd has been observed in the human kidney (PRATT et al. 1986). A similar subunit structure of purified human renin has been described by DO et al. (1987). It remains to be clarified, however, whether this complex, in which the two subunits are not linked by disulfide bridges, but rather held together by aggregation, is enzymatically active and released from the kidney or whether it is an artifact. One must also consider the possibility of this two-subunit complex representing the first step in the controlled intrarenal degradation of excess renin (cf. Sect. 5.3.1; Chap. 10).

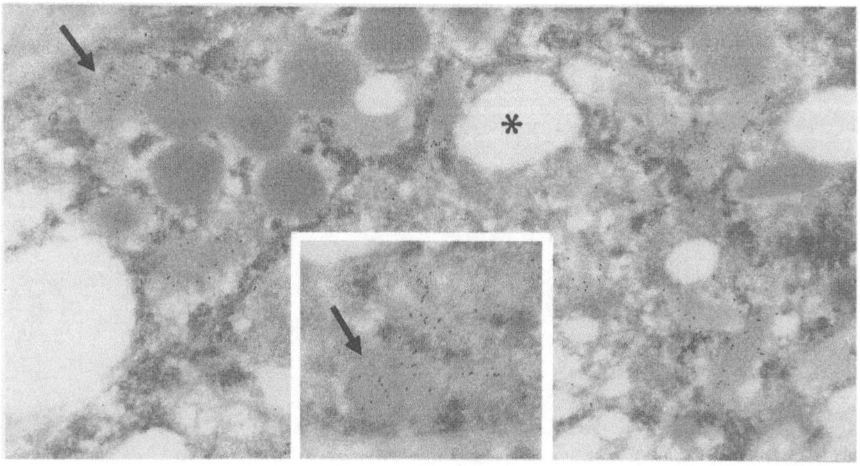

Fig. 5.9. Juxtaglomerular epithelioid cell from human kidney after incubation with an antiserum against the COOH-terminus of human renin prosegment. In contrast to the large electron-dense mature granules, the somewhat brighter rhomboid or fusiform protogranules and two intermediate granules *(arrows)* are clearly immunoreactive with one exception *(upper right corner)*. *Asterisk*, dilated RER cistern; ×29100 and ×34000 *(inset)*

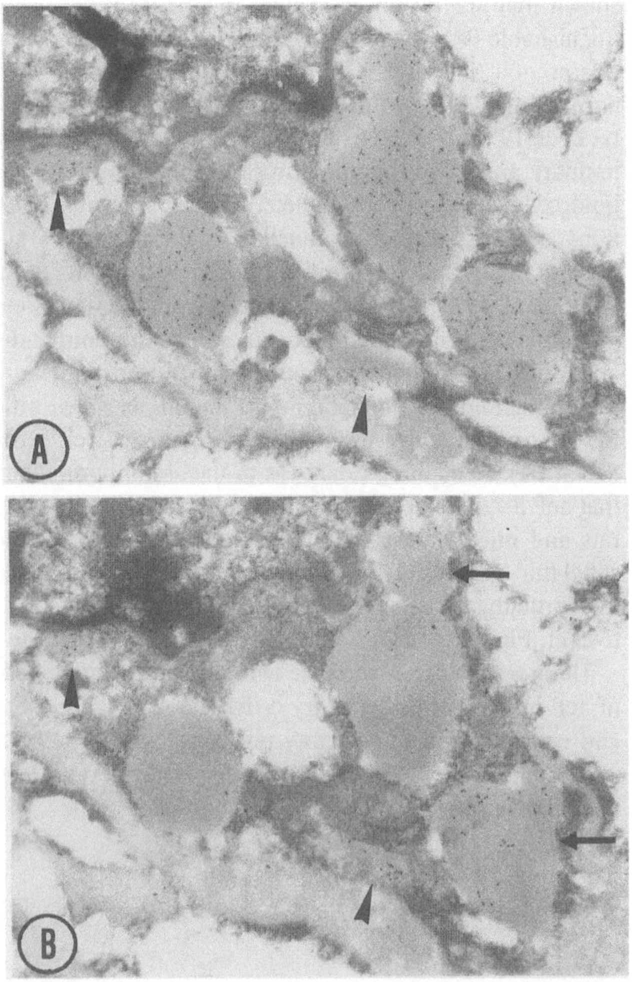

Fig. 5.10. Paired serial sections from human kidney, immunostained for renin **A** and Pro 3, the COOH-terminus of human renin prosegment **B**. **A** Renin-positive reaction of all juvenile *(arrowheads)* and mature secretory granules. **B** After application of the anti-Pro 3 serum, besides protogranules (*arrowheads*) only part of the mature and intermediate granules located close to protogranules are labeled (*arrows*); ×25300

5.3 Epithelioid Cell Secretory Granules Viewed as Lysosomes

On the basis of transmission electron microscope (TEM) studies, two types of Golgi-derived organelles have been assumed to exist within epithelioid cells of the juxtaglomerular apparatus: secretory (or specific) granules and lysosomes, also referred to as residual bodies or nonspecific granules (for review see ROUILLER and ORCI 1971). Apart from distinguishing features, however, there are indications of some relationship between secretory granules and lysosomes in epithelioid cells which appears to go far beyond that observed in other secretory cell systems. This becomes evident when the essential characteristics of lysosomes are considered (DE DUVE 1983). Lysosomes are membrane-bound pleomorphic organelles which contain acid hydrolases and degrade foreign (endocytosed) and cellular (autophagocytosed) material. In the following, renin granules are compared with lysosomes with regard to these and some other criteria, in particular their reaction to lysosomotropic substances.

5.3.1 Lysosomal Enzymes as Constituents of Renin Granules

RUYTER (1964) discovered acid phosphatase (acP) activity within epithelioid cells of the juxtaglomerular afferent arteriole, manifested in the form of fine black granules, which he assumed to be identical with lysosomes. Subsequently, LEE et al. (1965, 1966), FISHER (1966), and FISHER et al. (1966) have shown that acP occurs in cellular organelles indistinguishable from secretory granules at the electron microscopic level (cf. GOMBA et al. 1967; ROSEN and TISHER 1968; GOMBA and SOLTESZ 1969; TSUDA et al. 1971; cf. ECKERT and KUNDE 1974). A preliminary answer to the question of whether epithelioid cells contain two completely distinct populations of Golgi-derived organelles - acP-negative renin granules and acP-positive lysosomes (MATSUHASHI et al. 1977) - could be deduced from the first immunocytochemical experiments using antirenin antibodies. These experiments clearly showed that the epithelioid cells of laboratory animals besides MVBs contain only renin-positive organelles (cf. Sect. 4.1.1). Remaining doubts as to the assumption that acP is a regular constituent of renin granules in rats and mice were eliminated by acP staining of serial thin sections (TAUGNER et al. 1985b), showing that virtually all mature renin granules are acP reactive (cf. Fig. 5.11).

These results are at variance with the localization of acP in most other secretory cells. The packaging and concentration of secretory products, processes which start in the Golgi region, are believed not only to allow economic storage, but also to protect the secretory material from the actions of acidic hydrolases following the sorting of lysosomal enzymes from secretory proteins. It is true that in some cells secretory granules exhibit acP activity during their maturation, smaller granules usually being more reactive than larger ones (NOVIKOFF and ESSNER 1962; cf. HAND and OLIVER 1981). In a few cases, such as mammotrophs (SMITH and FARQUHAR 1966) and pancreatic B cells (ORCI et al. 1971), acP may even be found in mature secretory granules, although the reaction product occurs only in a few granules or is localized around (rather than inside) the granule core. One of the possible explanations for these findings involves incomplete sorting, so that lysosomal enzymes may be carried over haphazardly into secretory granules. As all renin granules are acP positive until maturity, acP rather seems to be a constitutive element of these granules.

Besides acP, a number of other enzymes characteristic of lysosomes have been demonstrated in epithelioid cells, e.g., β-glucuronidase (GOMBA and SOLTESZ 1969), arylsulfatase (GOMBA et al. 1970), *N*-acetyl-β-glucosaminidase (SOLTESZ et al. 1979), α-glucosidase (FRANSEN 1987), cathepsin B (TAUGNER et al. 1985a), and cathepsin D (TAUGNER et al.

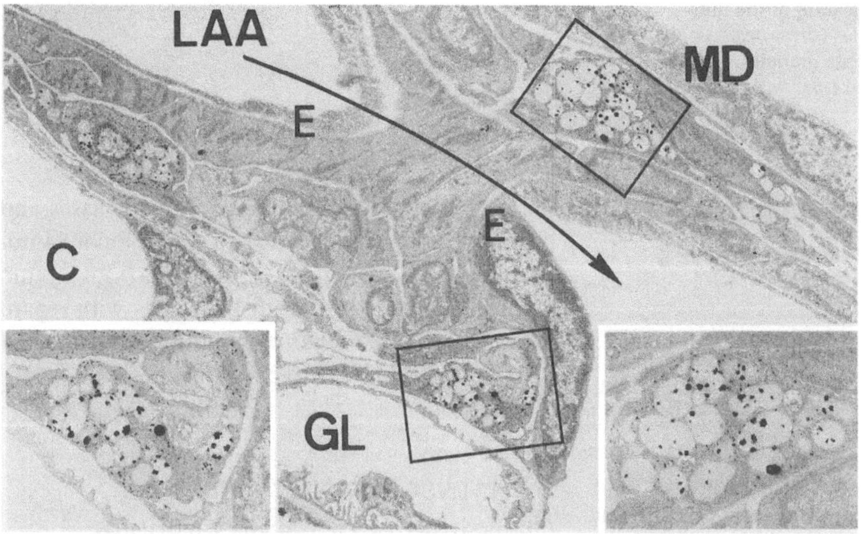

Fig. 5.11. Acid phosphatase in epithelioid cell secretory granules. Afferent arteriole in a rat kidney at its entry into the glomerules *(arrow)*. The arteriole runs in an arched manner, so that the section plane traverses the vessel wall tangentially. In this example, the media of the afferent vessel is formed by undifferentiated "plain" vascular smooth muscle cells in addition to granule-containing epithelioid cells. Upon reaction of the tissue for acid phosphatase (acP) with glycerophosphate as a substrate, in contrast to the adjacent smooth muscle cells, all granulated cells are labeled. The two areas *enclosed by rectangles* are shown at higher magnification in the *insets*. Individual patches of reaction product are confined mainly to the matrix of mature secretory granules. Serial sectioning revealed that virtually all renin granules are acP positive. *LAA*, lumen of the afferent arteriole; *E*, endothelium of the afferent arteriole; *GL*, urinary space of the glomerulus; *MD*, macula densa; *C*, peritubular capillary; ×3400, ×17000, ×16800

1986c). There is good reason to suggest that, like acP, also these enzymes are constitutive elements of renin granules. For cathepsin B and D, their coexistence with renin in epithelioid cell secretory granules has been demonstrated by serial thin sectioning (Figs. 5.12, 5.13).

Whereas the function of the other acidic hydrolases in epithelioid cell granules is so far completely unknown, the presence of the proteinases cathepsin B and D appears to be relevant for the fate of the epithelioid cell secretory product, renin. The thiolpeptidase cathepsin B is a lysosomal enzyme thought to be involved in the generation of bioactive products from precursor proteins (KATUNUMA and KOMINAMI 1983, 1985; STEINER et al. 1983; STEINER 1985). Cathepsin B is found in both juvenile and mature renin granules (TAUGNER et al. 1985b) and the prosegment of renin, clearly traceable in protogranules, is cleaved off in the course of granulopoiesis (TAUGNER et al. 1986b, 1987a). As a consequence, cathepsin B was suggested to be involved in the activation of renin prior to extrusion, capable of rapidly adjusting the proportion of active to inactive renin currently available for secretion (cf. Sects. 5.1, 5.2 and Chap. 6). Less specific proteases such as cathepsin D, on the other hand, could provide a mechanism by which the overall quantity of secretory product available in epithelioid cells is controlled at the stage of intracellular storage (TAUGNER et al. 1986a).

The outstanding large granular renin store is subject to wide variations according to the requirements of the RAS. Thus, a sudden decrease in the stimulation level of the system would require some efficient mechanisms for a drastic reduction of the accumulated secretory product. By contrast to other secretory cell systems (cf. FARQUHAR 1977), classical crinophagy does not seem to occur in epithelioid cells. Another possibility for the posttranslational down-regulation of renin stores not required for secretion would be the intragranular breakdown of renin by lysosomal enzymes. In the following

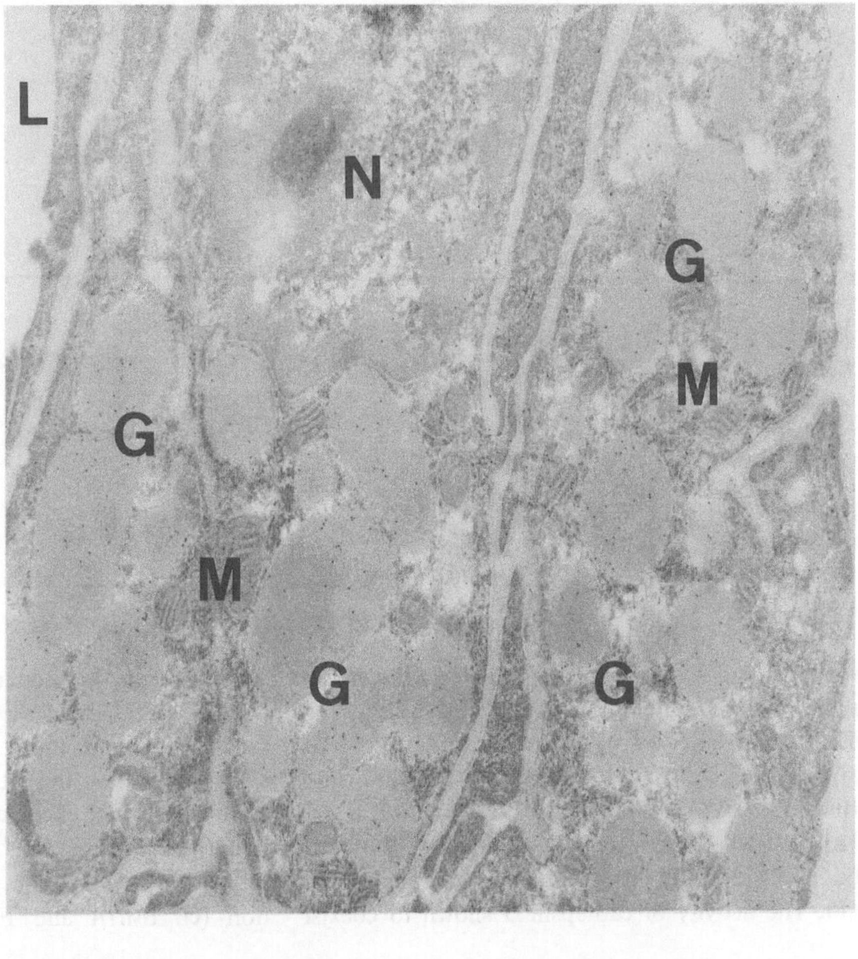

Fig. 5.12. Cathepsin B-positive epithelioid cell secretory granules *(G)* in the juxtaglomerular part of the rat afferent arteriole. Note that all the granules are labeled without exception. *M*, mitochondria; *N*, nucleus; *L*, lumen of the afferent arteriole. Perfusion fixation with 1% glutaraldehyde, embedded in London White resin, protein A-gold method; ×31600

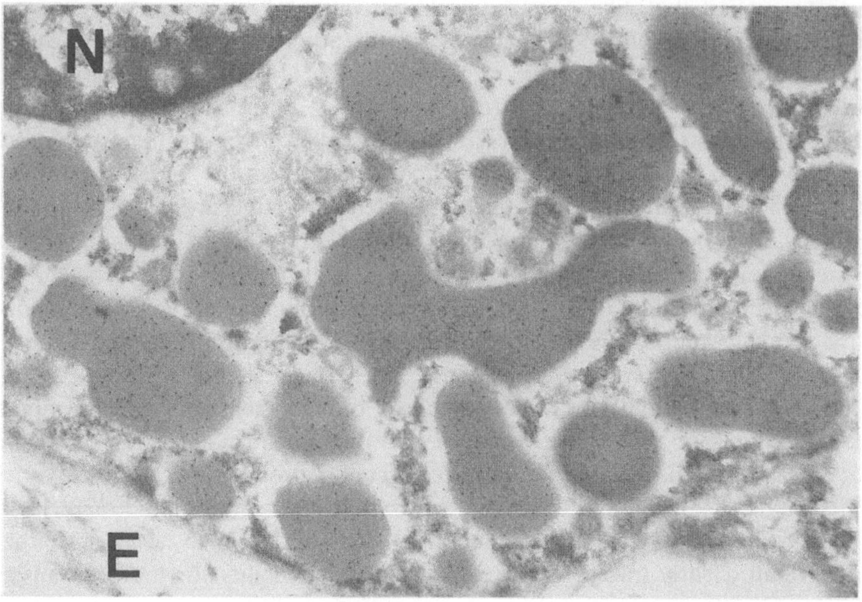

Fig. 5.13. Cathepsin D-positive epithelioid cell secretory granules in the juxtaglomerular part of the rat afferent arteriole. *N*, nucleus; *E*, extracellular space; ×25000

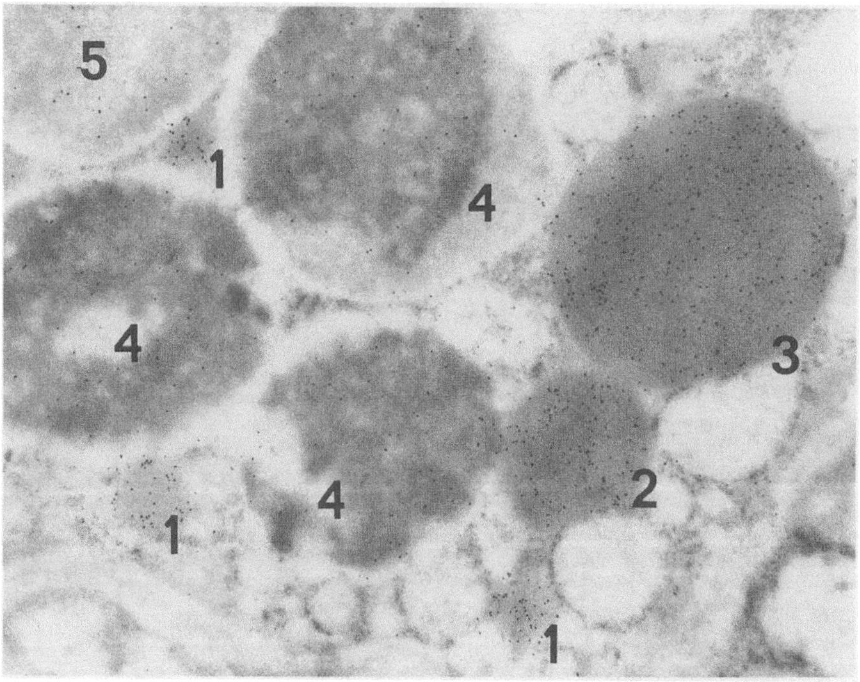

Fig. 5.14. Human juxtaglomerular epithelioid cell with different types of secretory granules immunoreacted for renin. In contrast to protogranules *(1)*, intermediate granules *(2)*, and mature secretory granules *(3)*, the oversized granules with an inhomogeneous, mostly flocculent internum (*4* and *5*) show only weak immunoreactivity. Biopsy specimen from a patient with Bartter's syndrome; ×27200

sections, indirect evidence is presented for a pH in renin granules comparable to that in lysosomes (cf. TAUGNER et al. 1985b). Therefore, cathepsin D, as an unspecific aspartyl protease (SCHWARTZ and BIRD 1977; OGUNRO et al. 1979; KATUNUMA and KOMINAMI 1983; YOKOTA et al. 1985), might indeed be involved in the intragranular degradation of renin. The activity of cathepsin D shown to coexist with renin in the secretory granules of epithelioid cells (Fig. 5.13) may be influenced by a variety of factors similar to those discussed for typical lysosomes, including the intragranular milieu. This, of course, does not exclude that cathepsin D or other hydrolytic enzymes might be transferred ad hoc into renin granules upon abrupt reduction of secretion (cf. SMITH and FARQUHAR 1966; HOPKINS

1969; FARQUHAR 1977). As cathepsin D could neither be traced in protogranules nor in Golgi-derived vesicles, the time and mode of its delivery into the granules during the course of granulopoiesis is open to question.

Systematic biochemical experiments with isolated renin granules or quantitative immunocytochemical experiments at different levels of stimulation would be necessary in order to verify the hypothesis that renin can be degraded within secretory granules to inactive fragments. Indications for such processes might be deduced from the observation that some epithelioid cells exhibit atypical secretory granules with a remarkably low degree of immunoreactivity to antirenin antibodies (Fig. 5.14). The properties of these granules which are larger than unaltered mature granules and, in contrast to these, display a heterogeneous, sometimes flocculent internum do not fit into the well-known scheme of granulopoiesis. As altered large granules seem to occur more frequently when the stimulation level of the RAS is reset to lower values, they might represent examples for the intragranular breakdown of renin. Another possibility for the reduction of the stored secretory product upon inhibition of the system would be the sequestration of granules together with other cell organelles during the metaplastic retransformation of epithelioid cells (cf. Chap. 10).

5.3.2 Autophagic Phenomena in Renin Granules

Autophagy is termed the uptake and degradation of cellular constituents by lysosomes. Microautophagy relates to cytoplasmic material including glycogen particles, macroautophagy to the fate of larger organelles, such as mitochondria and profiles of endoplasmic reticulum. There are several reasons to suggest that renin granules have both micro- and macroautophagic capabilities.

Juvenile and intermediate renin granules often contain small vesicles (BARAJAS 1966; LEE et al. 1966; SKELTON et al. 1967; TSUDA et al. 1971), initially located near the granule membrane. After mergence of several juvenile granules, the vesicles may be arranged in rows, apparently in those places where fusion of the involved membranes had occurred previously (Figs. 5.4 A, 5.5). These observations and occasional indentations (Fig. 5.6) suggest that the vesicular structures seen in juvenile granules may develop by internalization processes, i.e., by *microautophagy*. Microautophagic events are equivalents of endocytosis in the area of the plasmalemma. In analogy to the biogenesis of MVBs (ERICSSON 1964; MERKER 1965; FRIEND and FARQUHAR 1967; HIRSCH et al. 1968), they may therefore be considered as a response of the outer granule membrane to the repeated incorporation of membrane by fusion with Golgi-derived and endocytic vesicles (cf. Sect. 5.3.4). One of the possible functional roles of microautophagy might consist in the incorporation of endogenous cytoplasmic inhibitors of cathepsin B, suggested to participate in the activation of renin (cf. Sect. 5.2) or other thiolpeptidases. It should be pointed out, however, that there are other possibilities for the appearance of vesicular structures in renin granules, for instance, by rearrangement of previously incorporated and "solubilized" lipidic components.

Several observations indicate that *mature renin granules* have both *micro- and macroautophagic capabilities*. LEE et al. (1966) observed that mature epithelioid cell granules occasionally contain cytoplasmic structures and myelin figures, resembling those frequently found in lysosomes (for review see ROUILLER and ORCI 1971). Findings such as those shown in Fig. 5.15 suggest that these alterations are related to the *microautophagic* activity of the granules (cf. TAUGNER et al. 1985 b). The vesicle-like inclusions in the electron-dense matrix of mature granules (50–200 nm in diameter) are mostly larger than those of juvenile granules and contain conspicuous cytoplasmic material (Fig. 5.15 A). We may therefore be led to assume that they are formed by budding from the granule membrane (Fig. 5.15 A, B). The developing microautophagic vesicles may then migrate to the interior of the granules (Fig. 5.15 C, D) where some of them apparently lose their membranes, giving the impression of vesicle content lying freely within the granule internum. In the vicinity of the inclusions, some conspicuous small, stacked, membrane-like fragments are often seen (Fig. 5.15 A, D). These intragranular myelin figures may grow to the extent that, in freeze fractures, they appear as large smooth surfaces devoid of intramembrane particles, otherwise typical for polar lipid bilayers. This and the effects of lysosomotropic substances (Sect. 5.3.3) suggest that the intramatrical myelin-like structures are the corollary to a remarkable lipid turnover in epithelioid cell secretory granules.

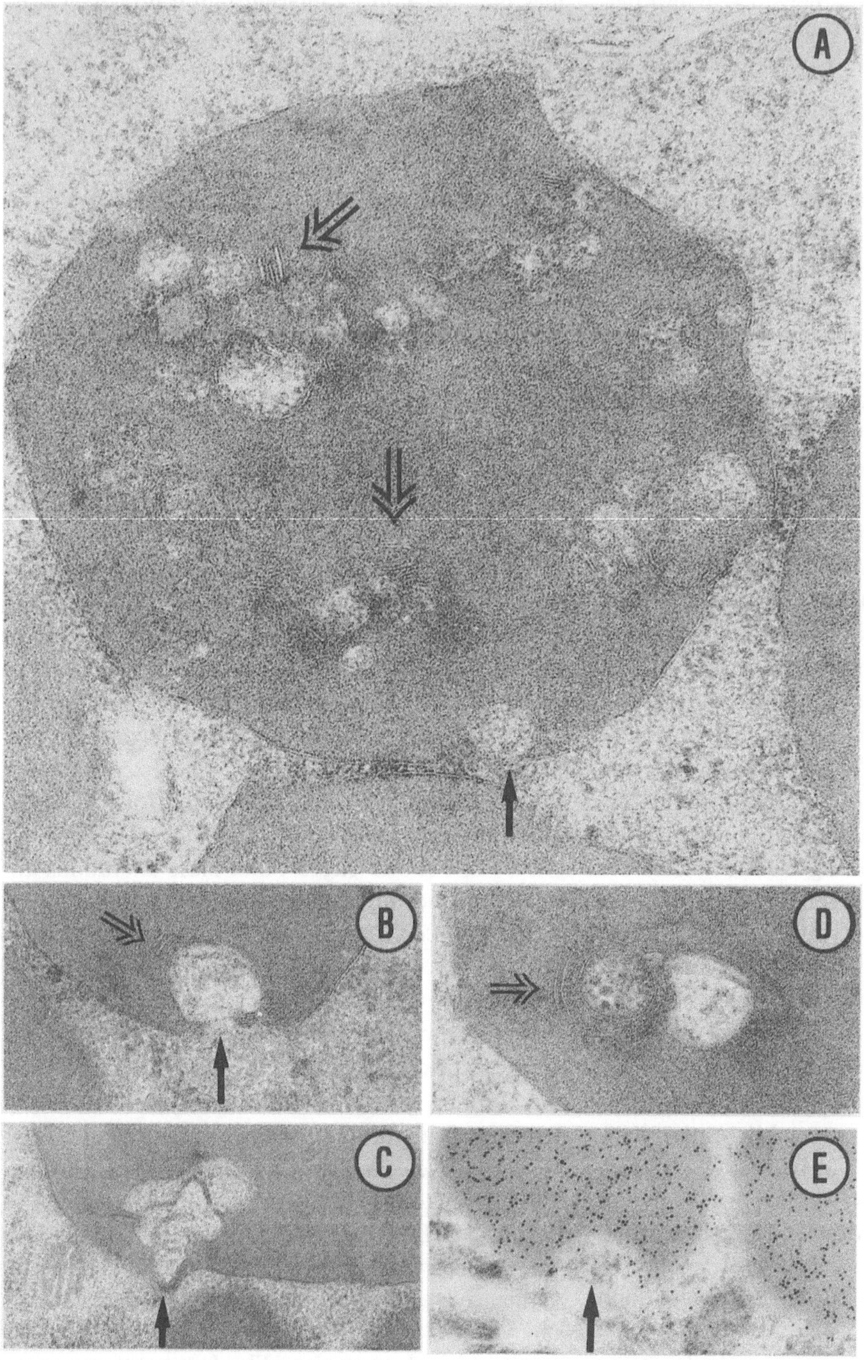

Fig. 5.15 A–E. Microautophagy by eptihelioid cell secretory granules. **A** Mature secretory granule from a mouse epithelioid cell with several vesicle-like cytoplasmic inclusions within the otherwise electron-dense, homogeneous internum. Some of the vesicles are membrane-bound and the content of others appears to be continuous with the granule matrix. The *arrow* points to an apparently nascent vesicle in the process of budding from the granule membrane. Discrete myelin-like figures are observed in the vicinity of the vesicles *(contoured arrows)*. Uncontrasted section, ×68000. **B–D** Presumed successive stages of microautophagy, i.e., incorporation of cytoplasmic material by vesicles budding from the granule membrane (*arrows* in **B** and **C**). Note the myelin-like material in **B** and **D** *(contoured arrows)*. Uncontrasted sections from mouse epithelioid cells; **B**, ×68000; **C**, ×42000; **D**, ×82500. **E** Incipient vesicle formation *(arrow)* by an immunoreactive secretory granule in a rat epithelioid cell. Indirect immunostaining with antirenin serum (1:100), followed by protein A-gold; ×41500

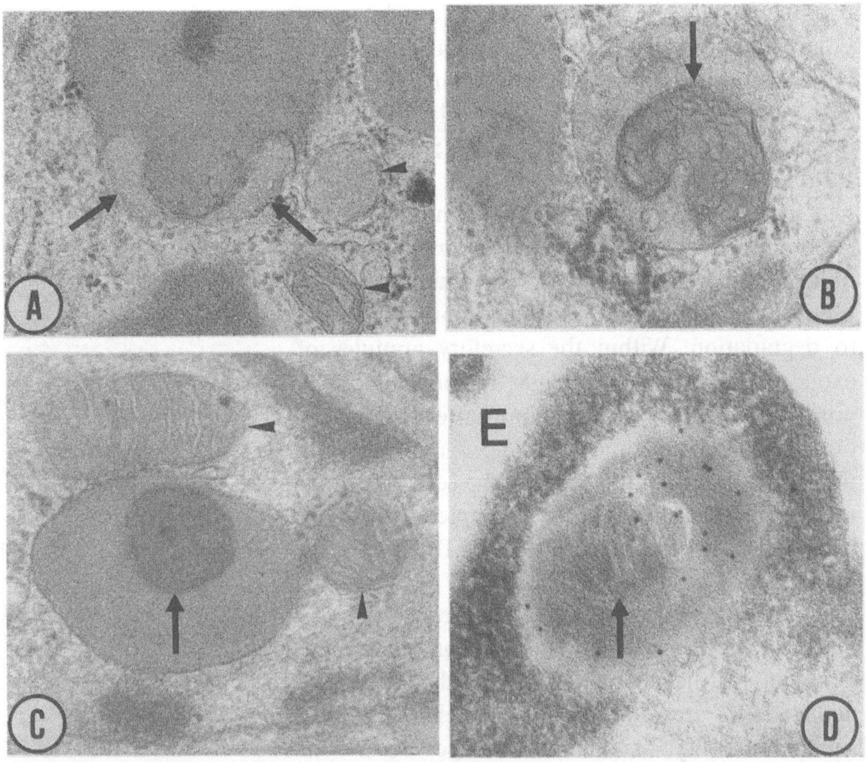

Fig. 5.16 A–D. Macroautophagy by mouse epithelioid cell secretory granules resulting in the uptake of mitochondria *(arrows)*. Note the matrix granule of the autophagocytosed mitochondrion in **C**. **A, D** From animals substituted with DOCA and NaCl from the 4th to the 6th day after bilateral adrenalectomy. **B, C** From hydronephrotic kidneys, 6 weeks after unilateral ureteral ligation. **A–C** Conventional fixation and embedding; **D** fixation with 1% glutaraldehyde, embedding in London White resin, immunocytochemical staining for renin with the protein A-gold method; × 50000 **A–D** and × 72000 **E**, respectively

Cytoplasmic inclusions have been observed in epithelioid cell granules of several species, occurring both in control animals and in animals with stimulated RAS. In adrenalectomized animals, in which renin secretion had additionally been increased by furosemide injection and prefinal bleeding, such inclusions were found most frequently (Taugner et al. 1984b). However, these preliminary results do not permit any definite connection between the functional status of epithelioid cells and the frequency of microautophagic events.

As the matrix of epithelioid cell granules is frequently seen to contain bi- and multilayered membrane-like structures, the unambiguous demonstration of *macroautophagy* by these granules meets with difficulty. However, granule profiles like those of Fig. 5.16 can hardly be interpreted other than by the assumption that renin granules are able to incorporate - and probably digest - mitochondria. As, apart from MVBs, renin-negative organelles with the appearance of secondary lysosomes could not be identified in our material, it is tempting to postulate that in epithelioid cells secretory granules take over the task of both micro- and macroautophagy (Taugner et al. 1988a).

5.3.3 Renin Granules and Lysosomotropic Substances

Lysosomotropic (or acidotropic) substances like NH_4Cl, chloroquine, or chlorphentermine are weak bases generally believed to enter the acidic matrix of cell organelles and especially that of lysosomes via nonionic diffusion (De Duve 1983). Inside lysosomes, they are trapped by protonation and may interfere with lipid catabolism, either by direct drug-lipid interactions (cf. Lüllmann-Rauch 1979) or by inhibition of lipid-processing enzymes (Seglen and Gordon 1980; Harder et al. 1981; cf. De Duve 1983; Lafont et al. 1984; Dean et al. 1984). Another possibility is the stimulation of lipogenesis by the lysosomotropic agents (Chen et al. 1986). As a consequence, there is a gradual accumulation of polar lipids, mainly phospholipids in the form of lamellated material within the lysosomes (Fig. 5.17), which are eventually converted into lamellated inclusion bodies (HRUBAN et al. 1972; LÜLLMANN-RAUCH 1975, 1979).

A prerequisite for this accumulation of polar lipids in the matrix of lysosomes is the internalization of membrane material by micro- and macroautophagy, which, under normal conditions, is subject to degradation. Within the secretory granules of exocrine and endocrine glands, neither autophagy nor the local pileup of polar lipids after administration of lysosomotropic substances are generally thought to occur. In this respect, the secretory granules of epithelioid cells show a completely different behavior (TAUGNER et al. 1985b; TAUGNER and METZ 1986).

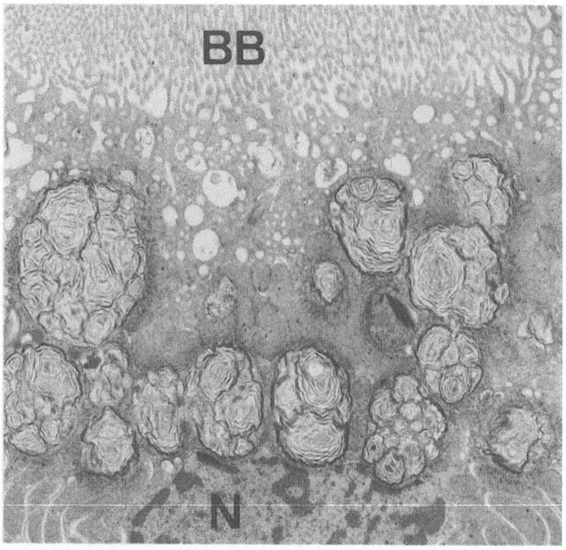

Fig. 5.17. Transformation of proximal tubular lysosomes into large lamellated bodies after long-term administration of chlorphentermine (40 mg/kg per day) in a rat. *BB*, brush border; *N*, nucleus; ×7200

Figure 5.18 shows murine (A) and rat (B) epithelioid cell profiles after long-term chlorphentermine treatment of the animals. Some of the organelles depicted are transformed into lamellated bodies similar, although smaller, to those shown in Fig. 5.17 for tubular lysosomes. In organelles subject to only partial changes, direct continuities between the unaffected matrix, crystalloid inclusions with hexagonally arranged tubular subunits, and lamellated structures are observed (inset). Because of their number, size, and location in epithelioid cells, these inclusion bodies had to be considered as altered renin granules (TAUGNER et al. 1985b). This assumption was corroborated by the observation that the unaltered matrix residues of such organelles are renin positive (Fig. 5.19; TAUGNER and METZ 1986). Apart from mature granules, chlorphentermine induced myelin-like inclusions were also found in juvenile and intermediate granules. Similar results were obtained in rats and mice by the long-term administration of chloroquine.

Summing up, the morphological alterations of renin granules induced by the acidotropic substances chlorphentermine and chloroquine suggest that these granules closely resemble lysosomes also with respect to their acidic pH and the intramatrical turnover of internalized membrane material.

According to Table 5.1, the long-term administration of chlorphentermine was also followed by a decrease of kidney renin content and plasma renin concentration. These effects might be related to the inhibition of intracellular fusion events or to interference with enzymes involved in renin processing (for review see DEAN et al. 1984; MELLMANN et al. 1986). The short-term effects of lysosomotropic substances on renin release in vitro have been studied by SKØTT (1987). The author found a delayed inhibition by low concentrations of NH_4Cl and chloroquine, probably as the result of an increase in the intragranular pH.

Table 5.1. Kidney renin and plasma renin concentration after long-term administration of chlorphentermine in mice

	Kidney renin (μg AI/mg per hour)	Plasma renin concentration (μg AI/ml per hour)
Control animals[a]	1.91 ± 0.46	3.43 ± 0.41
	$p<0.02$	$p<0.05$
Chlorphentermine-treated animals[a, b]	0.61 ± 0.1	2.0 ± 0.7

AI, angiotensin I
[a] Male NMRI mice; body wt., 25–30 g; $n=4$
[b] 50 mg chlorphentermine/kg i.p. for 4 weeks
(GORGAS, HACKENTHAL, and TAUGNER, unpublished observations)

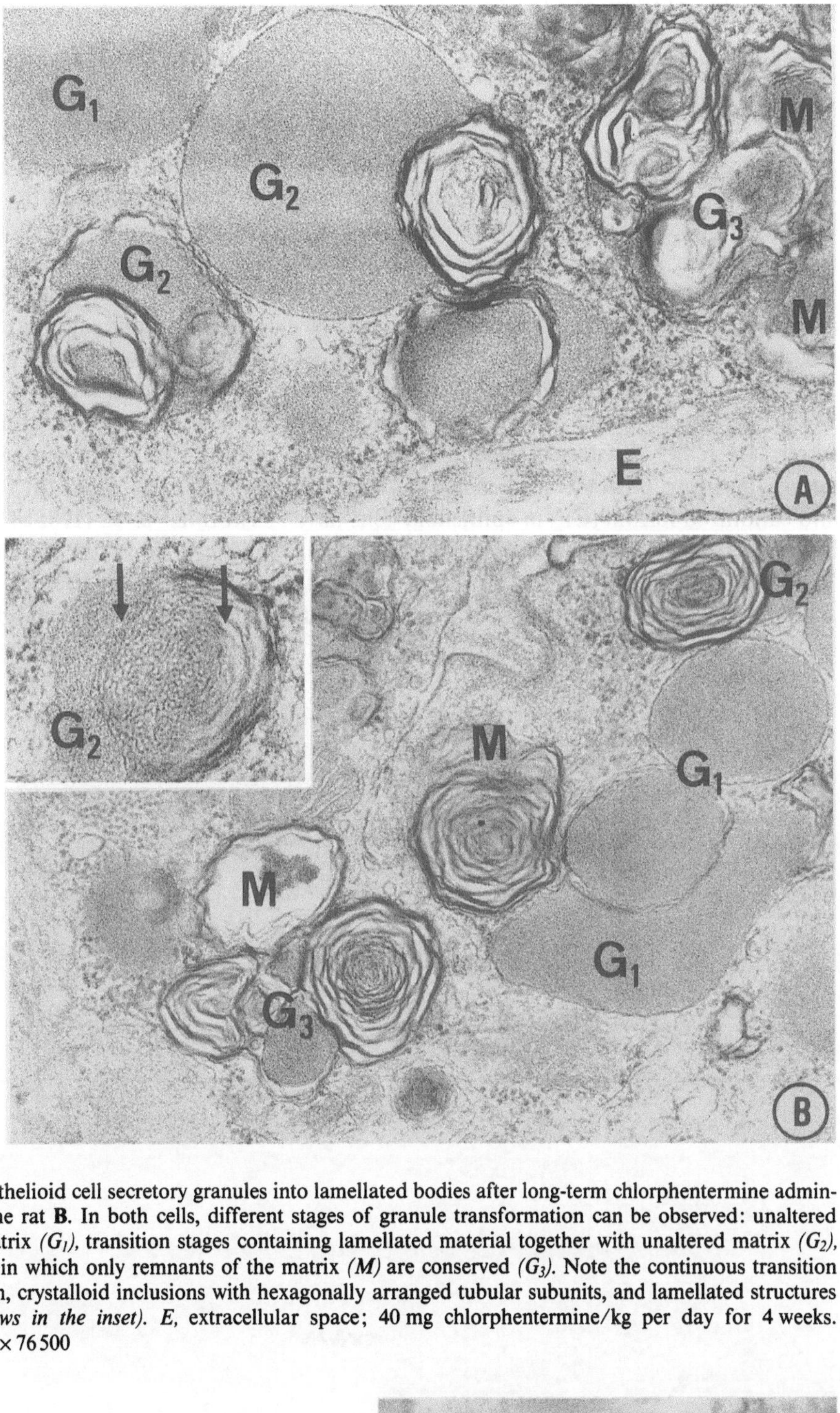

Fig. 5.18. Transformation of epithelioid cell secretory granules into lamellated bodies after long-term chlorphentermine administration in the mouse **A** and the rat **B**. In both cells, different stages of granule transformation can be observed: unaltered granules with homogeneous matrix *(G_1)*, transition stages containing lamellated material together with unaltered matrix *(G_2)*, and markedly altered granules, in which only remnants of the matrix *(M)* are conserved *(G_3)*. Note the continuous transition between homogeneous internum, crystalloid inclusions with hexagonally arranged tubular subunits, and lamellated structures in some of the granules *(arrows in the inset)*. *E*, extracellular space; 40 mg chlorphentermine/kg per day for 4 weeks. **A**, ×52500; **B**, ×43600; *inset*, ×76500

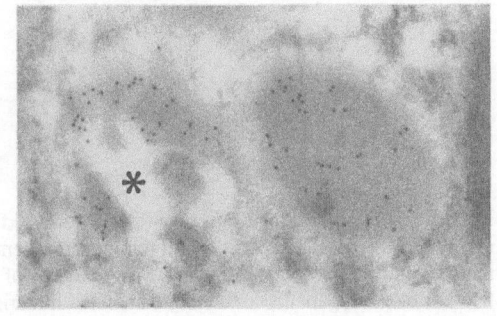

Fig. 5.19. Mouse epithelioid cell secretory granules after long-term chlorphentermine treatment, immunoreacted for renin. One of the granule profiles shows a homogeneously electron-dense internum. In the other, the matrix is interrupted by renin-negative electron-translucent areas *(asterisk)*. These discontinuities are suggested to result from the elution of lipid-soluble substances (i.e., myelin-like material) during the embedding procedures (cf. Fig. 5.18). Embedding in London White resin, protein A-gold technique; ×45200

5.3.4 Exogenous Tracers and Renin Granules

Several earlier reports concerning the uptake of dyes (for references see CANTIN et al. 1977b) and electron-dense tracers (LATTA and MAUNSBACH 1962b; DUNIHUE and BOLDOSSER 1963) indicated additional similarities between epithelioid cell secretory granules and lysosomes (TAUGNER et al. 1982c). Since negative findings (CANTIN et al. 1977b) casted doubt as to whether endocytosed tracers are really transferred to renin granules, the uptake and fate of horseradish peroxidase (HRP) and cationized ferritin (CF) in epithelioid cells have been investigated in some detail (TAUGNER et al. 1985b; TAUGNER and METZ 1986).

CF is considered a membrane marker which, after endocytosis, is transferred to the lysosomal compartment and stacked Golgi cisterns, thus tracing part of the recycling of the granule membrane back and forth, from and to the cell membrane (for review see FARQUHAR 1981). CF may also be found within the granules of secretory cells. However, here it is usually introduced at the earliest step of granulopoiesis, reaching more elaborate granules only after maturation and assembly of these early labeled forms. Fusions of CF-labeled endocytotic vesicles with mature granules are thought to be rare or nonexistent (FARQUHAR 1981, 1982; FARQUHAR and PALADE 1981). In this respect, the findings of TAUGNER et al. (1985b) in epithelioid cell granules differed from the results of most of the earlier studies. Although CF appeared in endosomes/pinosomes of epithelioid cells, in MVBs, and also in secretory granules, the labeling of Golgi cisterns,

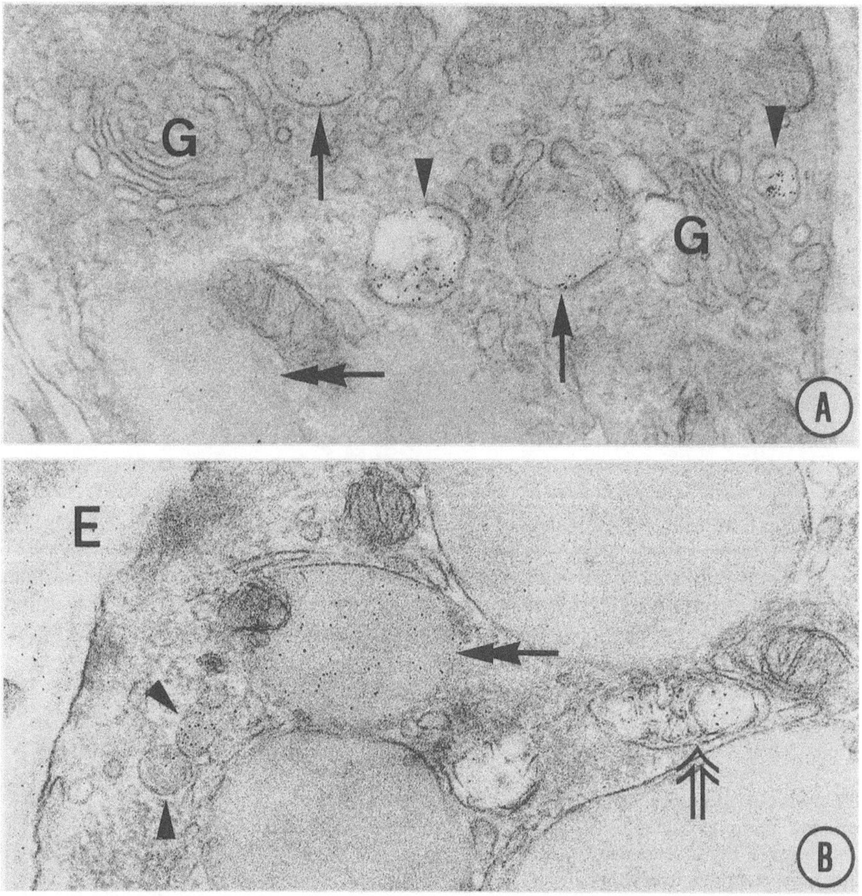

Fig. 5.20 A, B. Endocytosis and transfer of cationized ferritin into epithelioid cell secretory granules of the mouse. At 35 min after i.v. infusion, the tracer is found in pinosomes/endosomes and MVBs *(arrowheads)* as well as in intermediate *(arrows)* and mature renin granules *(double arrows)*. The cisterns of the Golgi complexes *(G)* are not labeled. *Contoured arrow:* ferritin-positive MVB-like organelle, containing lamellated material. *E,* extracellular space; ×53000

Golgi-derived vesicles, and early (rhomboid) protogranules was never observed (Fig. 5.20). This seemed to imply that CF reached the secretory granules of epithelioid cells without entering the Golgi-dependent part of the secretory pathway.

Similar results were obtained with HRP. HRP ist considered to be a content marker, generally directed toward the lysosomal compartment. Correspondingly, in epithelioid cells, as in other cell types, the tracer first became apparent in endosomes/pinosomes. This prelysosomal stage was followed by a longer-lasting phase not only in MVBs, but also in juvenile and intermediate renin granules (Figs. 5.21, 5.22, 5.23). With increasing exposure, more and more mature secretory granules seemed to be labeled (Figs. 5.22, 5.24). Like CF, HRP also appeared to reach the secretory granules without entering the Golgi complex.

These findings deviate substantially from those which are known of most other secretory cells, where, as a rule, HRP has no access to the secretory pathway (Oliver and Hand 1981). Only in a few secretory cells does HRP appear in the transmost cisterns of the Golgi stack (Pelletier 1973; Mata 1976; Farquhar et al. 1975; Orci et al. 1978; Broadwell and Oliver 1983), and even less frequently in forming granules (Pelletier 1973; Farquhar et al. 1975). For the labeling of mature secretory granules by HRP bypassing the Golgi apparatus, as has been found in epithelioid cells, there seems to be only one parallelism, namely that of basophilic leukocytes and bone marrow cells (Dvorak et al. 1972).

In the tracer studies reported, CF was not associated with the membranes of mature granules, but appeared in their matrix. Although preferentially binding to membranes, CF is not an ideal membrane marker, because its binding mainly depends on charge interactions; it is thus susceptible to competitive displacement by acidic groups of high

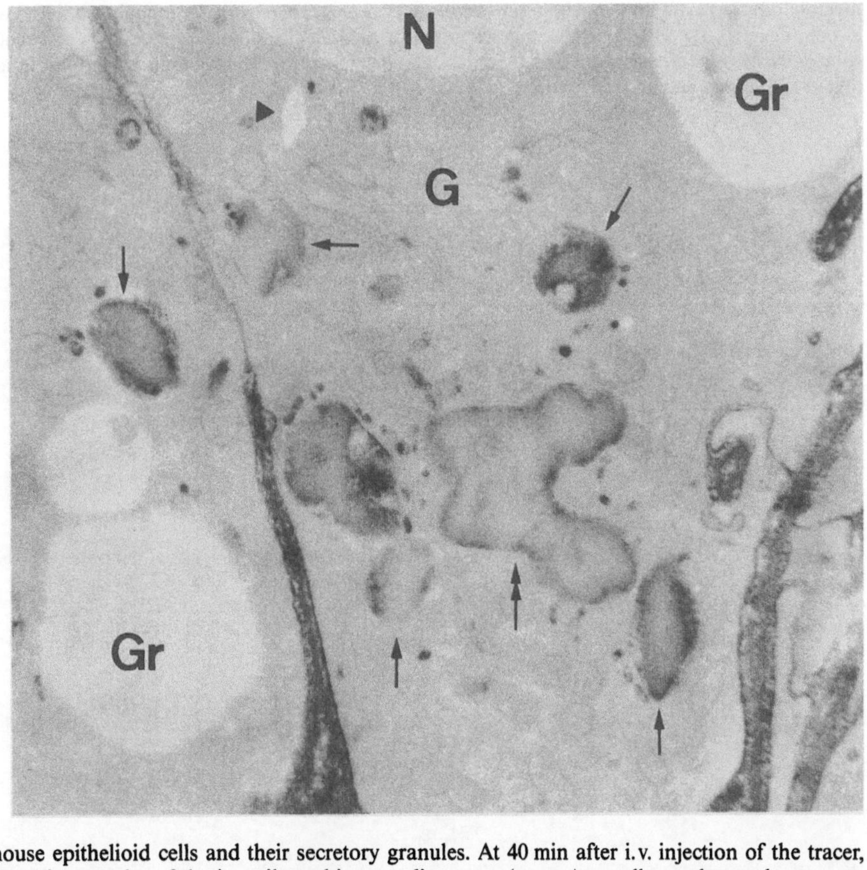

Fig. 5.21. Uptake of HRP by mouse epithelioid cells and their secretory granules. At 40 min after i.v. injection of the tracer, besides endocytotic vesicles, also renin granules of the juvenile and intermediate type *(arrows)* as well as polymorphous granules *(double arrow)* are labeled. The Golgi complex *(G)* and a nascent rhomboid granule *(arrowhead)* are free of reaction product. At this time, also most of the mature secretory granules *(Gr)* ar not labeled. *N*, nucleus. Uncontrasted section; ×27200

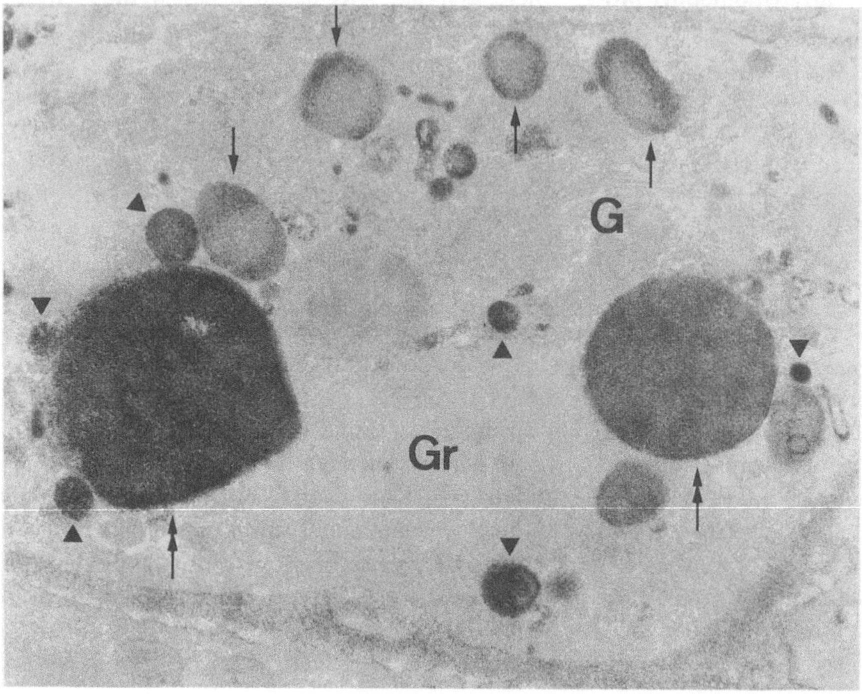

Fig. 5.22. Uptake of HRP by mouse epithelioid cells and their secretory granules. Two hours after i.v. injection of the tracer, juvenile *(arrows)* as well as some mature secretory granules *(double arrows)* are labeled. No reaction product is seen in the Golgi complex *(G)*. *Gr*, HRP-negative mature granule; *arrowheads*, pinocytotic vesicles or vacuoles (so-called pinosomes/endosomes). Note that in contrast to endocytotic organelles juvenile secretory granules exhibit inhomogeneous, peripherally accentuated staining. Uncontrasted section; ×34000

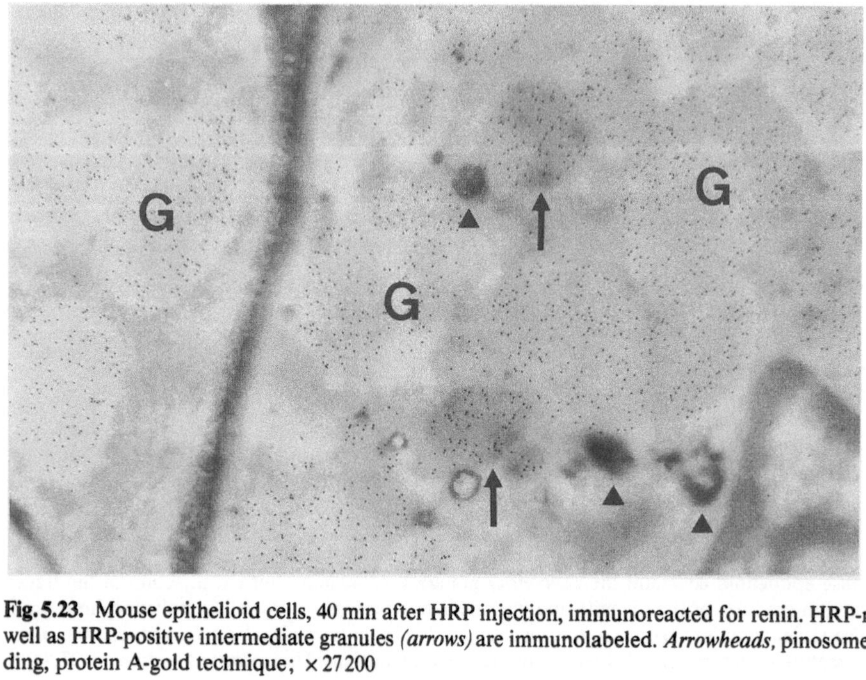

Fig. 5.23. Mouse epithelioid cells, 40 min after HRP injection, immunoreacted for renin. HRP-negative mature granules *(G)* as well as HRP-positive intermediate granules *(arrows)* are immunolabeled. *Arrowheads*, pinosomes/endosomes. Lowicryl embedding, protein A-gold technique; ×27200

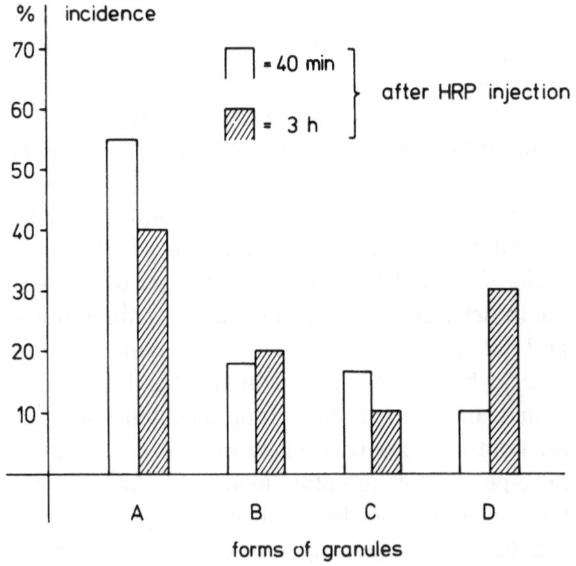

Fig. 5.24. Histogram, showing the relative frequency of labeling of the different forms of epithelioid cell secretory granules, 40 min and 3 h after the intravenous injection of horseradish peroxidase *(HRP)*, respectively. *A*, fusiforme juvenile granules (so-called protogranules); *B*, roundish juvenile granules; *C*, intermediate granules; *D*, mature granules. Note that after 3 h relatively fewer protogranules and more mature granules seem to be labeled. (METZ and TAUGNER, unpublished observations)

charge density (FARQUHAR 1981). It is conceivable that the matrix of mature renin granules contains strongly acidic macromolecules, e.g., lipoproteins (cf. TAUGNER et al. 1985b). HRP, on the other hand, is not an ideal content marker (cf. HOLTZMAN 1981; TARTAKOFF 1980). Hence, the question remains as to whether and where the segregation of endocytosed contents and membrane components occurs in epithelioid cells.

The biological importance of the transfer of endocytosed extracellular material into lysosomes probably lies in the digestion of macromolecules. If secretory granules of epithelioid cells were to be classified as lysosomes, then the same might be said for renin granules. However, components of the extracellular fluid gaining access to lysosomes and renin granules could possibly also change the milieu of their respective matrices or represent signal molecules. Angiotensin II, which in some species has been shown to coexist with renin in epithelioid cell granules, might be a pertinent example of such a possibility (cf. Sect. 4.2).

5.3.5 Conclusions and Outlook

The observations reported clearly demonstrate that the secretory granules of epithelioid cells are endowed with structural and functional properties which are otherwise characteristic of lysosomes. These include autophagic phenomena, the reaction to acidotropic agents, the uptake of exogeneous tracers, and the coexistence of renin with a variety of lysosomal enzymes. In addition, there are some ultrastructural similarities between juvenile renin granules and MVBs, such as the occurrence of surface plaques (cf. Sect. 5.1) and similarities between the exocytosis of renin granules from epithelioid cells (cf. Chap. 6) and the extrusion of lysosomes from endothelial cells of venous sinuses (DE BRUYN and CHO 1986). Another facet of the correspondence between epithelioid cell secretory granules and lysosomes is the close relationship between the lysosomal enzyme cathepsin D and renin, both belonging to the family of aspartyl-proteinases, and probably originating from a common ancestor (NEURATH 1984). These similarities altogether lend support to the hypothesis that renin granules are modified lysosomes, which, during metaplastic transformation of vascular smooth muscle cells to epithelioid cells, have acquired the capability to process, condense, store, and release renin upon appropriate stimulation.

According to TAUGNER and METZ (1986), in juxtaglomerular cells of rats and mice it is difficult to identify renin-negative organelles with the ultrastructural characteristics of secondary lysosomes, except for multivesicular bodies, which probably originate from endosomes. It is therefore tempting to speculate that renin granules are indeed lysosomes which have acquired the specialized function of renin handling, but have also retained, to some extent, the original lysosomal abilities, such as the uptake and degradation of cellular or foreign material by autophagocytosis or endocytosis, respectively. In this view, epithelioid cells would not contain any secretory granules in the strict sense.

It must be pointed out, however, that this proposal to define secretory granules of epithelioid cells as modified lysosomes rests on incomplete experimental evidence and on assumptions which still remain to be verified. Thus, it has not yet been possible to determine the pH value inside renin granules and only the occurrence, but not the concentration, of acid hydrolases in their matrix has been ascertained. In addition, the sorting mechanisms in epithelioid cells are completely unknown: is a mannose-phosphate receptor involved? Are all the lysosomal enzymes found in mature granules subject to condensation by protogranules in parallel with renin or do they enter the secretory granules via small carrier vesicles, the equivalents of primary lysosomes in other cells?

The concept of a lysosomal nature of renin granules received strong support from a recent study by FAUST and coworkers (1987) who demonstrated that renin carries the mannose-phosphate moiety, which represents the specific sorting signal for lysosomal enzymes.

Last not least, we have to indicate that in human epithelioid cells, as opposed, e.g., to those of rats and mice, there are obviously many more atypical granules. These are commonly termed "nonspecific granules" (BIAVA and WEST 1966b; BIAVA 1967; for review see ROUILLER and ORCI 1971). The question is whether these granules are true (renin-negative) lysosomes or perhaps modified renin granules (cf. Chap. 14).

For a different approach to the understanding of renin granules and their peculiarities, it may be worthwhile remembering that secretory granules could be viewed in general as evolutionary homologues of lysosomes (DE DUVE and WATTIAUX 1966) and that, therefore, similarities between these organelles might be expected. Along these lines, the question arises of whether secretory granules of epithelioid cells represent but one of the more extreme members of a continuous series of secretory granules, which usually differ markedly from lysosomes, but less so in exceptional cases, such as renin granules.

Chapter 6

Morphology of Renin Release from Epithelioid Cells

In their comprehensive review, ROUILLER and ORCI (1971) discussed three mechanisms of renin secretion from epithelioid cells: (a) emiocytosis, i.e., exocytosis, of renin-containing granules, (b) a lysosome-dependent mode of excretion, and (c) the intracytoplasmic solubilization of the stored secretory product. Since then, many studies have been published addressing the biochemical events concerning renin secretion, but remarkably few related to its morphology. As a consequence, the direct release of renin from the cytoplasm without passing through the stage of granular storage has still been taken into consideration (FRAY and LUSH 1984; KRIZ and KAISSLING 1985).

This lack of information is not only a result of the fact that the secretory process itself is transient and therefore difficult to visualize. More important are several difficulties which tend to impair the morphological study of renin secretion: the laboriousness of the search for epithelioid cells in ultrathin sections and even more so in freeze fractures of kidney cortex, where small groups of renin-secreting cells are dispersed within a large volume of tissue; focal renal ischemia followed by severe cell damage upon experimental stimulation of renin secretion (CAIN and KRAUS 1969, 1970, 1971; TAUGNER et al. 1984a, b); several peculiarities of renin secretion which compromise any comparison with other secretory events, among them the ambiguous role of lipidic structures; and, finally, the fact that cultures of true epithelioid cells, i.e., cells loaded with mature storage granules, have not been available up to now for release experiments.

PETER (1976), working with thin sections, described deep, channel-like invaginations of the plasma membrane in granulated cells in rats. He concluded that renin secretion represents an unusual type of exocytosis, with the plasmalemma invaginating toward the granules, thus providing increased sites for extrusion, instead of the granules moving toward the cell surface prior to release. According to RYAN et al. (1982), the incidence of these deep invaginations increases during stimulus-induced degranulation and decreases during regranulation of sheep epithelioid cells. TAUGNER et al. (1978a), on the other hand, using thin sections and freeze-fracture replicas, found preliminary evidence for an exocytotic release of renin from granules located at the periphery of epithelioid cells in *Tupaia belangeri*. In order to improve the probability of finding secretory-active epithelioid cells in the section or fracture plane, TAUGNER et al. (1984b) used mice a few days after adrenalectomy and stimulated renin secretion by furosemide administration and prefinal bleeding. In this way, it was possible to gather substantial evidence for the exocytosis of mature renin granules, later confirmed in control animals of this and other species. In addition, there have been observations indicating the exocytosis of juvenile granules.

According to TAUGNER et al. (1984b), the exocytosis of mature epithelioid cell secretory granules is not preceded by bulging of the cell membrane or by an extensive fusion between the granule and cell membranes. However, circumscribed contact areas between these membranes were observed, accompanied by small indentations of the plasmalemma (Fig. 6.1 A, B). Characteristically, the granule core was altered locally at the site of the incipient exocytosis: Sheets of membrane-like material seemed to produce small protrusions of the granule membrane (Fig. 6.1 C), often accompanied by electronlucent areas or vacuoles within the granule (Fig. 6.1 D). Some of these vacuoles were adjacent to multilayered or vesicular membrane-like profiles, whereas others appeared to contain them. Later, close to the fusogenic focus, parts of the granule content seemed to escape into the extracellular space (Fig. 6.1 E, F).

The interpretation of these observations, which – with the apparent exception of lysosome extrusion discussed in Sect. 5.3 – markedly deviate form those in other secretory systems, has so far been highly speculative. Granule extrusion is generally

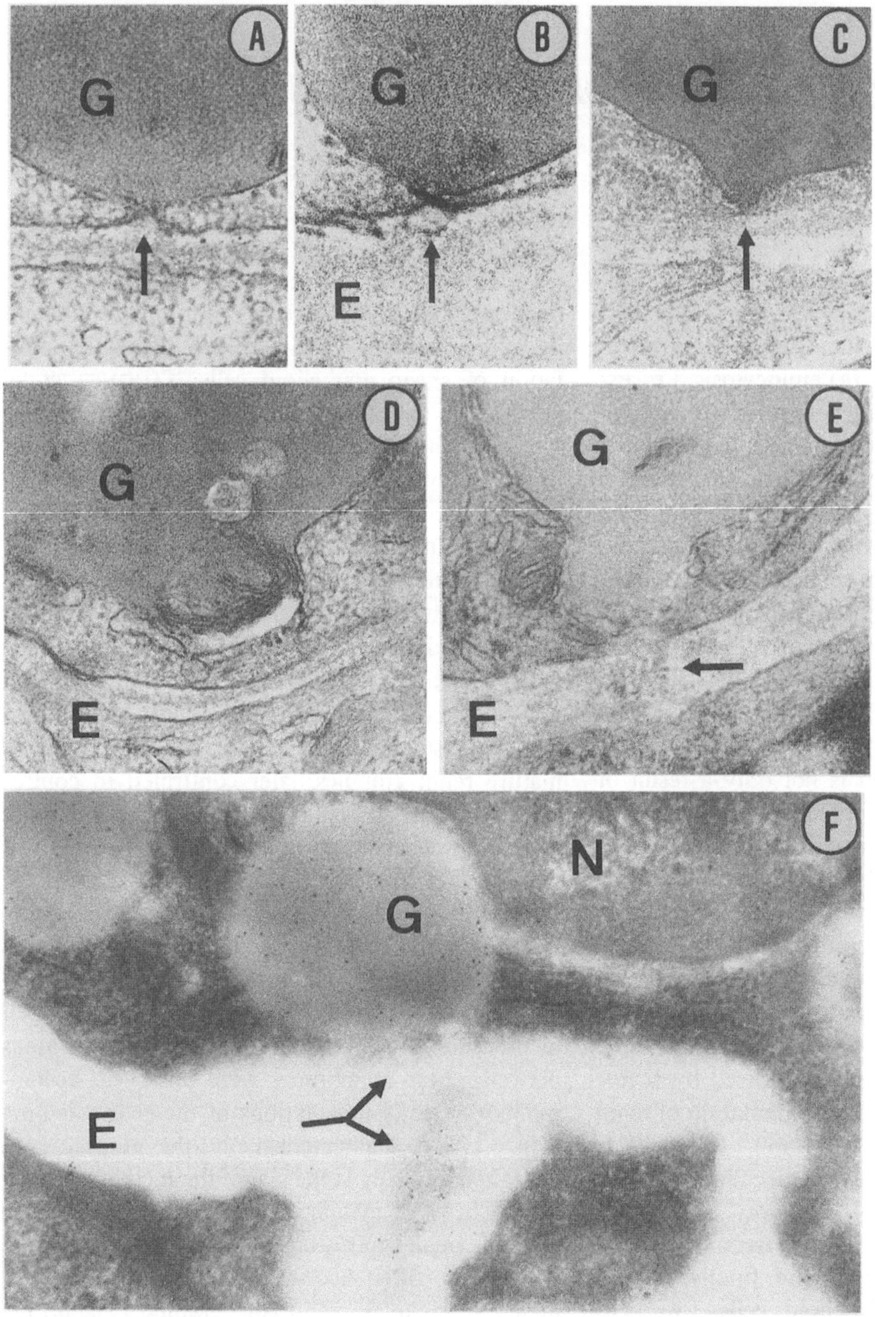

Fig. 6.1 A–F. Incipient exocytosis of mature secretory granules *(G)* from mouse juxtaglomerular epithelioid cells. The profiles of the granules are arranged according to the supposed successive stages of alteration and extrusion of their contents. **A** Punctiform area of contact between the cell and granule membranes *(arrow)* with small indentation of the plasmalemma; × 61 200. **B** Incipient local alteration of the granule content close to the area of contact between the cell and granule membranes *(arrow)*. *E*, extracellular space; × 91 800. **C** Small protrusion of the granule membrane *(arrow)* with somewhat more advanced local alterations of the granule internum; × 61 200. **D** Membrane-like material in local granule protrusion facing the extracellular space *(E)*; × 53 500. **E** Local protrusion of secretory granule facing the extracellular space *(E)* with fuzzy material *(arrow)* extruding through an incipient exocytotic opening. Noncontrasted section; × 60 800. **F** Somewhat larger exocytotic opening of renin-positive secretory granule *(G)* with extruded immunoreactive material *(arrows)* in the extracellular space *(E)*. *N*, nucleus. Fixation with 1% glutaraldehyde, embedding in London white resin. Protein A-gold method after application of antirenin serum (dilution, 1:100); × 120 000

known to be dependent upon mechanisms that, in the preexocytotic phase, lead to the juxtaposition followed by fusion of the granule and cell membranes. In fully transformed epithelioid cells, mature granules are often separated only by a narrow, homogeneously electron-translucent rim of cytoplasm from the cell membrane (Fig. 2.4). Accordingly, elements of the cytoskeleton do not seem to be involved in the fusogenic approach step of the membranes in these cells. Observations such as those shown in Fig. 6.1 may rather be interpreted to indicate that upon adequate stimulation circumscribed chemiosmotic effects induce a local protrusion in "stimulated" secretory granules facing the extracellular space, thus initiating exocytosis by limited fusions of the granule and cell membranes. As the local increase in granule volume was regularly associated with a spatially confined change in the granule matrix, which invariably appeared less electron-dense, a transmembrane water flow from the cytoplasm into the granule interior was suggested to occur, driven by an osmotic gradient. Since the adjacent cytoplasmic compartment often exhibited an increased electron density, and the cell membrane appeared to be indented, the osmotic gradient was tentatively assumed to extend to the extracellular space in the vicinity of the incipient exocytosis. It was further speculated that the preexocytotic increase in osmolality might be the result of a rapid chemical reaction initated at the fusogenic focus, which subsequently spread from the stimulated site across the entire granule matrix. The membrane-like material becoming visible upon stimulation in the granule core was thought be to related to the unmasking of membrane lipids previously incorporated through autophagic events (cf. Sect. 5.3.2). Figures 6.4–6.6 reveal that myelin figures tend to make their mark also on the subsequent stages of granule extrusion, rendering the interpretation of the ultrastructural details encountered difficult.

Independent from proper apposition of the involved membranes, water movement and vesicle or granule swelling on the basis of osmotic gradients have been suggested to be important preexocytotic fusogenic determinants in model systems such as

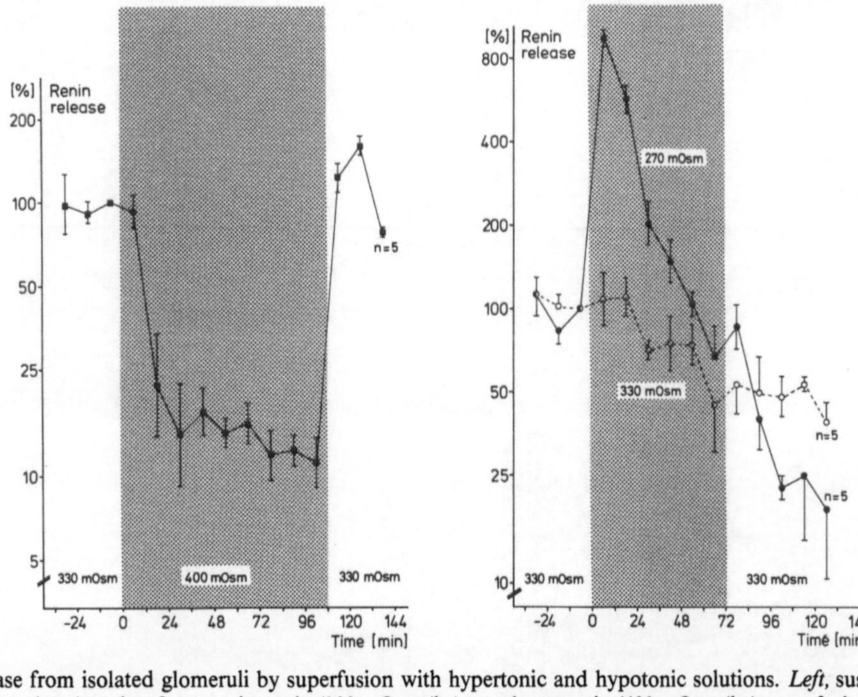

Fig. 6.2. Change of renin release from isolated glomeruli by superfusion with hypertonic and hypotonic solutions. *Left,* sustained depression of renin release by changing from an isotonic (300 mOsm/kg) to a hypertonic (400 mOsm/kg) superfusion solution. The absolute values of renin release in the last control period were used for normalizing the data of the individual experiments. The 100% value was 10.09 ± 2.13 (SD) μGU/300 glomeruli per 12 min. *Bars* indicate ± SE. *Right,* transient stimulation of renin release by exposure to a hypotonic (270 mOsm/kg) superfusion solution. The *dotted line* represents control experiments. The absolute values of renin release in the last control period were used for normalizing the data of the individual experiments. The 100% value was 3.84 ± 1.39 (SD) μGU/300 glomeruli per 12 min. *Bars* indicate ± SE (from Skøtt and Taugner 1987)

phospholipid vesicles and planar bilayers as well as in biological objects (COHEN et al. 1982; PREVOST and GALLEZ 1984; AKABAS et al. 1984; ZIMMERBERG and WHITACKER 1985; for review see FINKELSTEIN et al. 1986). On the basis of functional studies, osmotic effects have also been discussed in connection with renin release (FREDERIKSEN et al. 1975; BAUMBACH and SKØTT 1982; NAVAR et al. 1978, 1981). Parallel functional and morphological studies were performed by SKØTT (1986a) and SKØTT and TAUGNER (1987), investigating the effects of osmolality changes on epithelioid cells contained in superfused afferent arterioles. Slightly hypotonic solutions lead to a drastic increase of renin release from the preparations (Fig. 6.2, right). Under the electron microscope, a distinct swelling of the secretory granules often found in close proximity not only to each other, but also to the cell membrane, was observed (Fig. 6.3 B). As a consequence, numerous contacts and circumscribed fusions between the granule and cell membranes occurred (Fig. 6.3 C). It was therefore concluded that with the probability of fusion also the frequency of exocytotic events and therewith renin release increased

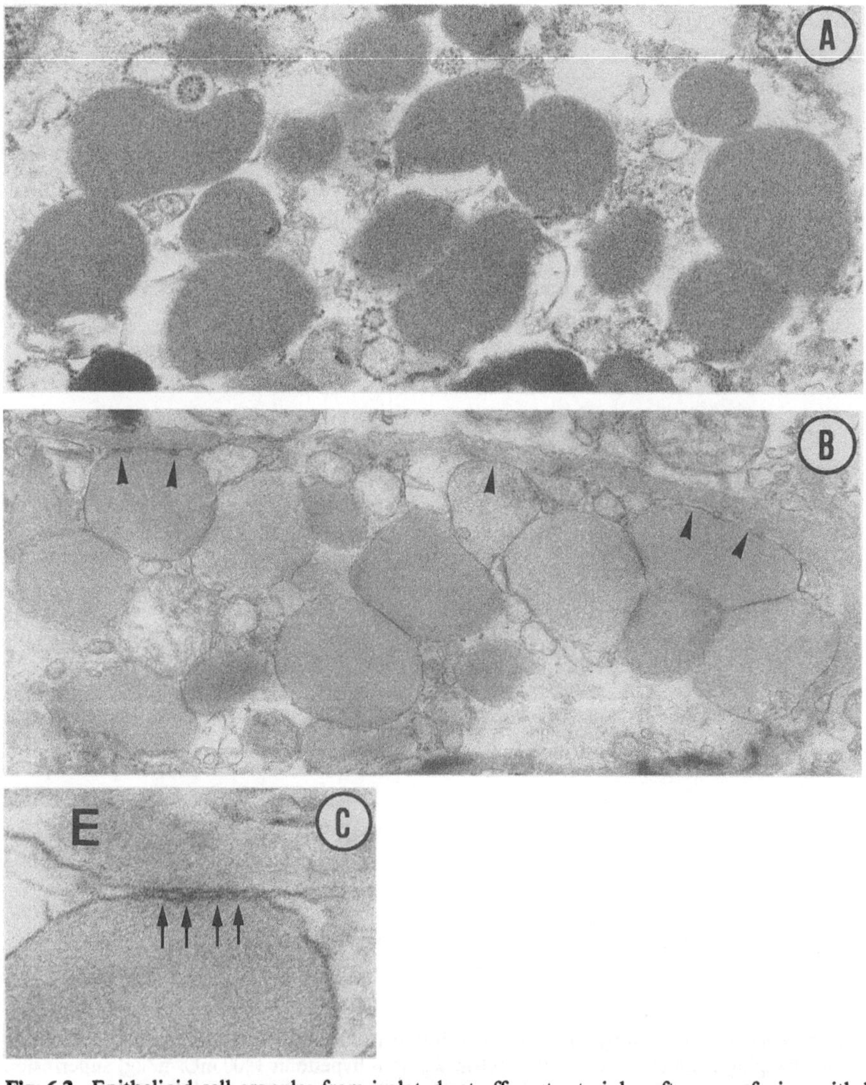

Fig. 6.3. Epithelioid cell granules from isolated rat afferent arterioles after superfusion with hypertonic **A** or hypotonic solutions **B, C**. **A** After 1 h superfusion with hypertonic (375 mOsm/kg) Ringer's solution, the secretory granules are of uniform high electron density. There is only moderate dilation of the RER and swelling of mitochondria; × 27000. **B, C** After superfusion with hypotonic (270 mOsm/kg) Ringer's solution for 4 min, the secretory granules are enlarged and electron translucent. Note the areas of contact or fusion between the membranes of adjacent granules and between the granule and cell membranes (*arrowheads* in **B** and *arrows* in **C**). *E*, extracellular space; × 27000 and × 130000, respectively (from SKØTT and TAUGNER 1987)

upon hypotonic stimulation. Hypertonic incubation of the isolated preglomerular arterioles led to opposite effects on granule morphology and renin release (Figs. 6.2 left/6.3 A). If one compares results obtained in isolated prepartions with those in the intact kidney after perfusion fixation, there seems to be one common denominator: osmotic swelling of the granules accompanied by membrane fusion and exocytosis. However, with the intact JGA, a circumscribed increase of the intragranular osmolality in the course of stimulus-secretion coupling is inferred to induce a local protrusion of the granules, whereas in vitro the entire epithelioid cells have been exposed to a hypotonic milieu, inducing a general swelling of the granules mediated by the cytosol.

In contrast to the incipient event, later phases of exocytosis of mature granules are found much more often. Hence, these later phases of the extrusion process appear to develop slowly in comparison with the preceding steps. Most typically, the modified granule matrix was found to lie within a membrane-bound, omega-shaped recess about to expel its contents into the extracellular space, the size of the emerging cavity being comparable to that of a mature granule (Fig. 6.4). In some cases, the omega-like cavity was particularly large, reminiscent of a compound exocytosis (Fig. 6.5). Large pools of dense material resembling congolomerate granules extending well into the cell, but also having continuity with the extracellular space, were observed by LEE et al. (1966) and HARTROFT (1968).

In the region of exocytotic events, large coated membrane areas are frequently encountered in the plasmalemma of the secretory active cell as well as in the membranes of neighboring cells (Figs. 6.4, 6.6). Therefore, these coated pits appear to be induced by the exocytosed granule content. Under the influence of appropriate ligands, coated pits may develop in conjunction with the clustering of membrane receptors, followed by their internalization, and, often, recycling (WILLINGHAN and PASTAN 1984). As epithelioid cells are equipped with angiotensin II receptors and the concentration of angiotensin II in the extracellular space is probably increased after the extrusion of renin, angiotensin II appears to be a candidate for the induction of coated pits in epithelioid cells.

Several arguments could be directed against the assumption that the exocytosis of mature granules is the principal event in renin secretion. They would refer to the rarity of such exocytotic events in sections or freeze-fracture replicas, to the fact that extensive fusion sites between the cell and granule membranes are not observed after in vivo fixation of the intact kidney, as well as to the peculiar

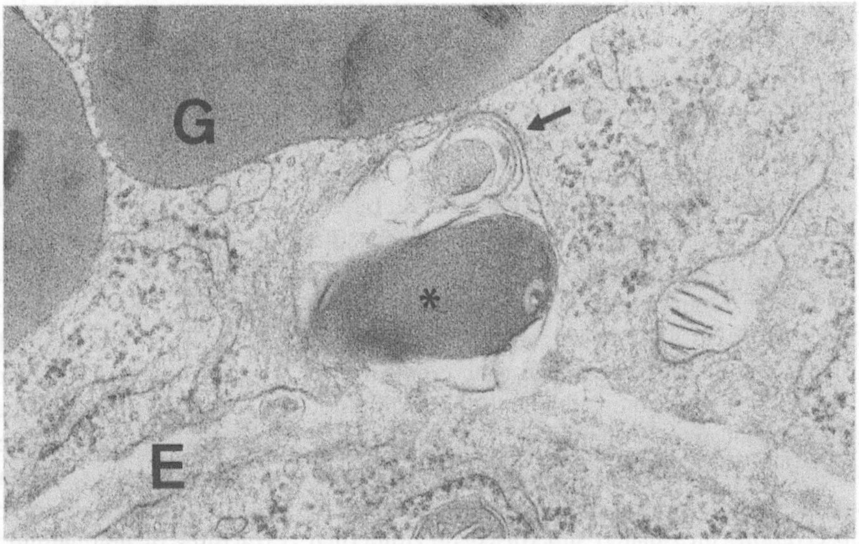

Fig. 6.4. Omega-shaped exocytotic cavity continuous with the extracellular space *(E)* after fusion and fission of the cell and secretory granule membranes in mouse epithelioid cell. The resulting cavity is partially filled by unaltered contents, apparently immobilized during extrusion *(asterisk).* In addition, whorls of multilayered membrane-like material are seen in the cavity adjacent to a coated region of the cell membrane *(arrow).* Note the membrane-like coat around part of the former granule internum. *G,* mature secretory granule; × 60800

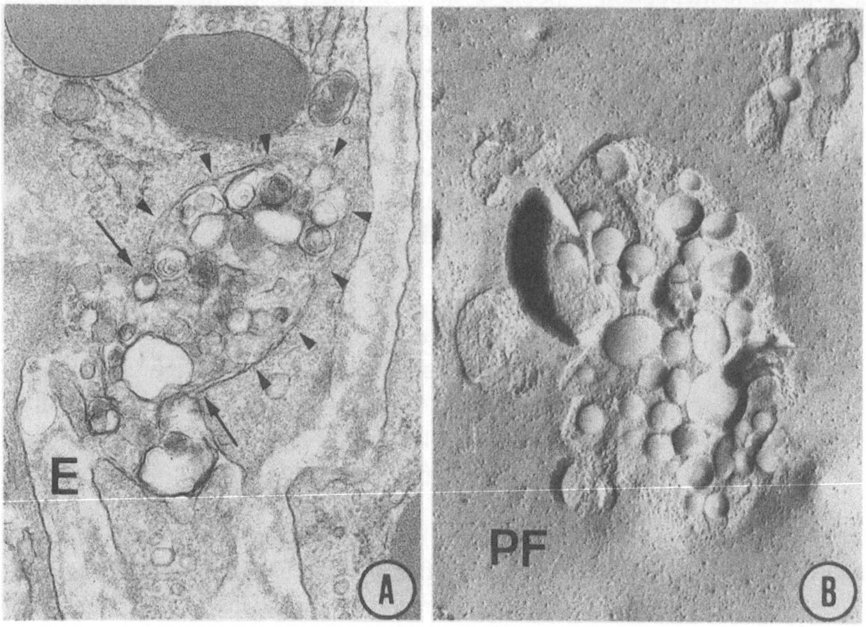

Fig. 6.5 A, B. Images suggestive of compound exocytosis (i.e., exocytosis involving more than one secretory granule) from mouse epithelioid cells. **A** Vesiculogranular material in the process of extrusion from a large recess *(arrowheads)* into the extracellular space *(E)* as delineated by the arrows, × 34000. **B** Vesiculogranular masses apparently protruding from a large opening in the membrane of an epithelioid cell. Freeze-fracture replica. *PF*, P-face of the cell membrane; × 56000

changes of the granule matrix during and shortly after its extrusion. Finally, alternative models of renin secretion have to be considered.

(1) Although ROUILLER and ORCI (1971) have observed phenomena reminiscent of exocytosis, they considered it doubtful that emiocytosis is the major mode of renin secretion, because of the rarity of such events. However, a calculation (using the following data: plasma renin in mice 300 ng AI/ml per hour[1]; kidney renin, 2 × 700 μg AI per hour[2]; half-life of plasma renin, 30 min; number of JGAs, 2 × 12500; granulated cells per JGA, 10–40; mature granules per epithelioid cell, 10–100) indicates that the frequency of exocytotic events required to keep the plasma renin level constant would be only of the order of 20–500/min in control animals. Therefore, the rare occurrence of extrusion events in thin sections is indeed compatible with the assumption that the exocytosis of mature granules is the essential pathway for the secretion of active renin from epithelioid cells. These arguments are supported by observations on isolated superfused preglomerular arterioles from rat kidney: According to SKØTT (1986b) the secretion of renin is an intermittent phenomenon. He showed that under control conditions there is an episodic release of active renin with a discharge of 45.2 ± 3.3 nano-Goldblatt units per event ($n = 114$) and a frequency of one episode/5 min from a single afferent arteriole. Referring his functional data to the morphological observations of TAUGNER et al. (1984b), SKØTT suggested that each of these secretory episodes represents the exocytotic extrusion of one mature secretory granule.

(2) For the release of the granule content, preexocytotic fusion of the granule and cell membranes is required. In epithelioid cells, extended fusion sites have not been encountered (LEE et al. 1966; TAUGNER et al. 1984b). However, although at early stages of exocytosis there is only a small and short-lived orifice toward the extracellular space, this hole then appears to be enlarged rapidly by circular extension of the small and transient fusion site. The lack of intramembrane particle-free areas in freeze fractures of epithelioiid cell membranes, which in some systems are considered indicators of membrane fusion, may be explained by the swift fusion processes and also by the fact that renin granules do not produce bulging of the cell membrane, a

[1] Angiotensin I generated from excess angiotensinogen by renin contained in 1 ml plasma/h.

[2] Angiotensin I generated from excess angiotensinogen by renin contained in 1 mouse kidney/h.

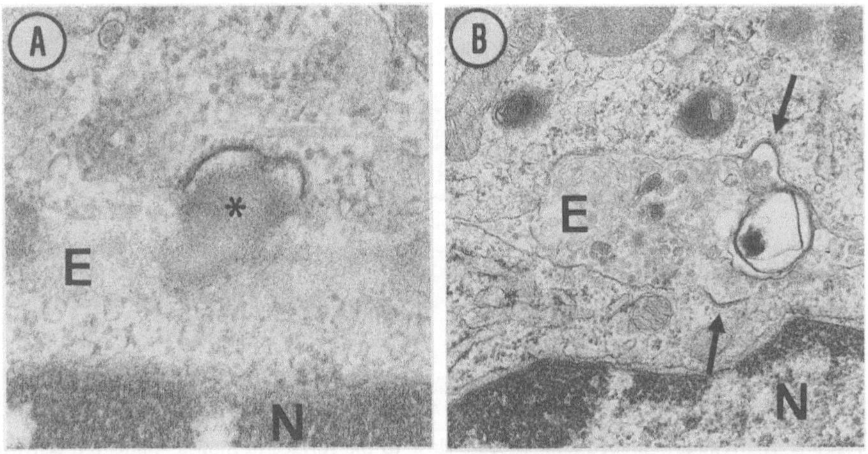

Fig. 6.6. Putative postexocytotic events in rat **(A)** and mouse **(B)** epithelioid cells. **A** Coated region in the membrane of an epithelioid cell, forming a small cavity continuous with the extracellular space *(E)* and partly filled with electron-dense amorphous material *(asterisk).* The structures depicted are interpreted as remnants of an exocytotic event. *N*, nucleus of endothelial cell; × 62300. **B** Heterogeneous granule content in a widened extracellular space *(E)*, suggested to result from previous exocytotic events. The *arrows* point to coated pits in the membranes of the adjacent epithelioid cells. *N*, nucleus; × 31600

prerequisite for particle clearing in glutaraldehyde-fixed, glycerinated specimens (TAUGNER et al. 1984b).

(3) As outlined in Sect. 5.3, there is probably a turnover of membrane material in renin granules similar to that in lysosomes. As a consequence, in thin sections, vesicular profiles or myelin figures, and in freeze fractures large smooth areas indicative of this material, may be found within the otherwise homogeneous matrix of "resting" granules (TAUGNER et al. 1984b, 1985b). The tendency of the formerly amorphous granule matrix to transform or rearrange itself into membrane-like structures is suggested to increase during or shortly after extrusion of the granule content. This may result in the demarcation of part of the granule core, in the inclusion of freed and essentially unaltered granule content by membrane-like structures (Fig. 6.4), or in the appearance of vesicular or multilayered membrane-like material in the extracellular space (Figs. 6.5, 6.6B). As a consequence, "membrane-bound" amorphous granule matrix may appear to lie freely within the extracellular space (Fig. 6.4; cf. ROUILLER and ORCI 1971). More often, however, altered granule content in the form of vesicular or multilayered structures is seen in or close to positions formerly held by depleted granules (Figs. 6.5A, 6.7A). At later stages of granule extrusion, only small amounts of membrane-like material are seen in round electron-lucent areas, suggested to be remnants of extruded mature granules.

(4) PETER (1976) and RYAN et al. (1982) have proposed a renin secretion mechanism for epithelioid cells of rats and sheep, respectively: this is based on the depletion of mature granules deeply embedded within the cytoplasm by the formation of long, channel-like plasmalemmal invaginations, which ultimately gain access to the granules. The results of our experiments in mouse and rat kidneys do not support this hypothesis. We have observed two features reminiscent of such invaginations, which, however, in our opinion, have to be interpreted differently. One possibility is that the membrane-bound saccule that persists after granule depletion collapses first in the region of its neck, thus resembling an invagination (Fig. 6.7A). Secondly, processes of epithelioid cells often closely approach each other or the perikaryon, thus forming folds or clefts which mimick invaginations in transverse sections (LINDOP 1987; cf. Fig. 6.7B). Since these clefts most often contain basal lamina-like material, it is improbable that they are related to renin secretion. In addition, similar clefts with reflexive gap junctions have been found in Goormaghtigh, mesangial, and smooth muscle cells (IWAYAMA 1971; own unpublished observations).

Our observations are also incompatible with the repeatedly suggested model of renin secretion by intracytoplasmic solubilization (TSUDA 1969), recently revived by ZAVAGLI et al. (1983) on morphological and by FRAY and LUSH (1984) on functional grounds. With conventional as well as with immu-

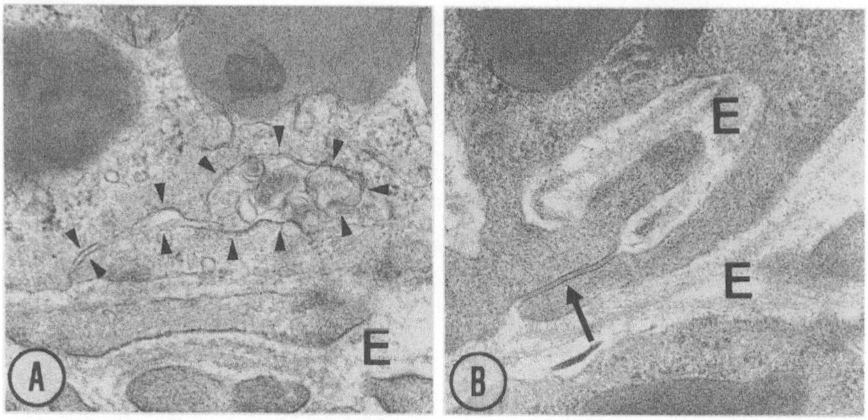

Fig. 6.7. **A** Irregularly-shaped large saccule close to the surface of mouse epithelioid cell with membrane-bound connection *(arrowheads)* to the extracellular space *(E)*, altogether interpreted to be the remnants of an exocytotic event. Note the inhomogeneous contents of the resulting saccule, suggested to represent parts of the former granule matrix; × 36500. **B** Mouse epithelioid cell processes joined by reflexive gap junction *(arrow)*. The "lumen" of the apparent "saccule" containing basal lamina material is in fact an integral part of the extracellular space *(E)*, × 48600

nocytochemical methods, neither the intracellular deterioration of granules nor the solubilization of their content could be observed (unpublished observations). Erroneous interpretations might be the result of poor fixation, of tangentially sectioned granule profiles, or of the occurrence of spherical, crystalloid masses in strongly dilated RER cisterns after stimulation of renin synthesis by adrenalectomy or administration of captopril (Kaneta et al. 1981, 1983; Zaki et al. 1982; cf. Hackenthal et al. 1987c). Biochemical evidence that renin release comes from a compartment other than the cytoplasm of epithelioid cells has been supplied by Baumbach and Skøtt (1982).

In connection with the intrarenal renin-angiotensin system (RAS) discussed in Chap. 9, it has to be assumed that renin is secreted into the interstitium surrounding the granulated cells. The question arises whether the exocytotic events occur more often in the direction of the lumen of the preglomerular arteriole or in an abluminal direction. It has already been mentioned in Sect. 2.2.2.2 that the arrangement of secretory granules in the cytoplasm of epithelioid cells shows no polarity. Furthermore, an evaluation of the approximately 15 exocytotic events observed revealed no preferential direction of the secretory event: exocytosis occurred with a similar frequency in both the luminal and the abluminal direction. For the further pathway of the secreted renin, it is important that the intima of the preglomerular vessels is equipped with interendothelial tight junctions, and that a considerable transmural pressure gradient exists from the lumen in the direction of the surrounding interstitium (cf. Sects. 2.8 and 3.3). It is therefore suggested that the bulk of the newly secreted renin reaches the general circulation via peritubular capillaries and, to a minor extent, via lymph vessels. Granulated cells of atypical localization which would allow the exocytosed renin to gain access to the glomerular filtrate are described in Sect. 3.3.

In the blood plasma of several species, besides active renin, there is also a comparatively large amount of inactive renin, in man probably in the form of the truncated version of prorenin described by Hirose et al. (1985). Since an activation of the enzyme within the circulating blood does not seem to be of any relevance, it must be assumed that both active and inactive renin are secreted from the kidney. In Sect. 5.2, it is reported that in humans prorenin, in contrast to active renin, is only found in juvenile and intermediate, but not in mature, granules of epithelioid cells. Hence, the question arises, whether - besides mature granules - also immature secretory granules containing prorenin may be extruded by these cells.

In humans as well as in rats and mice, juvenile granules were frequently found in a peripheral position adjacent to the cell membrane. On closer inspection, punctiform fusions between the membranes of intermediate granules and the plasmalemma have been observed (Fig. 6.8). Granular or

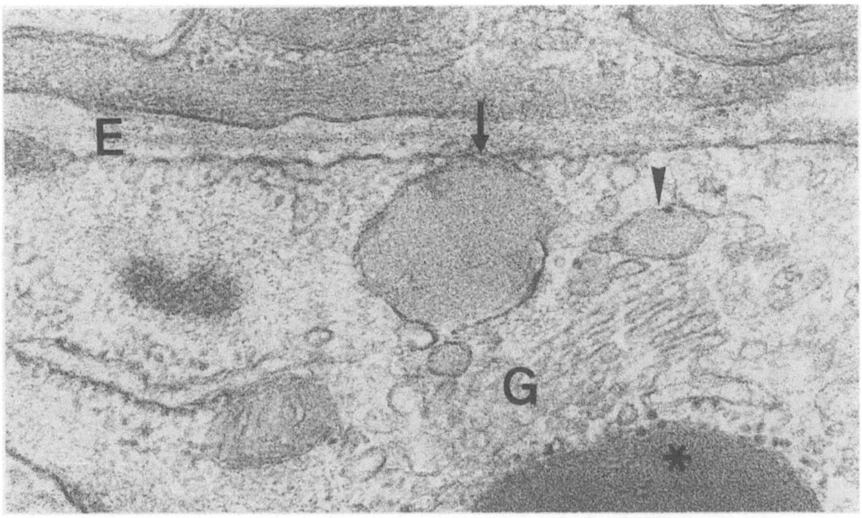

Fig. 6.8. Punctiform fusion between the membrane of an intermediate secretory granule and the cell membrane *(arrow). Arrowhead,* protogranule; *asterisk,* mature secretory granule; *G,* Golgi complex; *E,* extracellular space. Epithelioid cell in the juxtaglomerular portion of the afferent arteriole from mouse kidney; × 60700

smooth vacuolar profiles distinctly smaller than mature granules, but larger than any endocytotic event, had already been reported by TAUGNER et al. (1984b) to fuse with the plasma membrane or to open into the extracellular space. Figures 6.9 and 6.10 show pertinent examples.

Images such as these were most frequently seen in epithelioid cell processes, where Golgi elements are located close to the cell membrane (Fig. 6.9; cf. also Fig. 5.1 A). Their comparatively high fusogeneity is also in favor of an exocytosis of juvenile and intermediate granules (TAUGNER and METZ 1986).

A different model for the secretion of inactive renin has been proposed by LACASSE et al. (1985) assuming the exocytosis of vacuoles derived from RER cisterns and the nuclear envelope. In our material, such vacuoles have not been observed, and RER or Golgi cisterns continuous with the extracellular space have only been encountered in association with severe cell damage due to focal ischemia upon excessive stimulation of the RAS (TAUGNER et al. 1984b).

Still another model for the release of inactive renin from epithelioid cells refers to some direct ("constitutive") secretion of prorenin from the Golgi compartment circumventing the granular renin store (GALEN et al. 1984; for review see MORRIS 1986). This proposal, based mainly on biochemical experiments, will be discussed in more detail in Sect. 7.7. Immunocytochemical methods did not permit us to classify the many small carrier vesicles seen in epithelioid cells with respect to their contents and target organelles. Therefore, the possibility of prorenin secretion circumventing the granular renin store, including juvenile and intermediate granules which have been shown to contain prorenin (TAUGNER et al. 1986b, 1987a), cannot entirely be ruled out on morphological grounds. It is also conceivable that renin granules may not only be emptied by bulk exocytosis, as shown above, but also via small transport vesicles. Questions such as these can probably be resolved with morphological methods only in connection with those of the sorting processes which are unknown to this date in the epithelioid cells.

In our opinion, the results reported here on the ultrastructure of renin secretion in combination with the immunocytochemical findings in Sect. 5.2 justify the conclusion that the secretion of inactive renin occurs via the exocytosis of juvenile granules, while that of active renin takes place by the exocytosis of mature granules. If this assumption is correct, the secretion of active renin should match the exocytosis of mature granules, and the secretion of inactive renin the exocytosis of juvenile granules under varying experimental conditions (cf. Sect. 7.7).

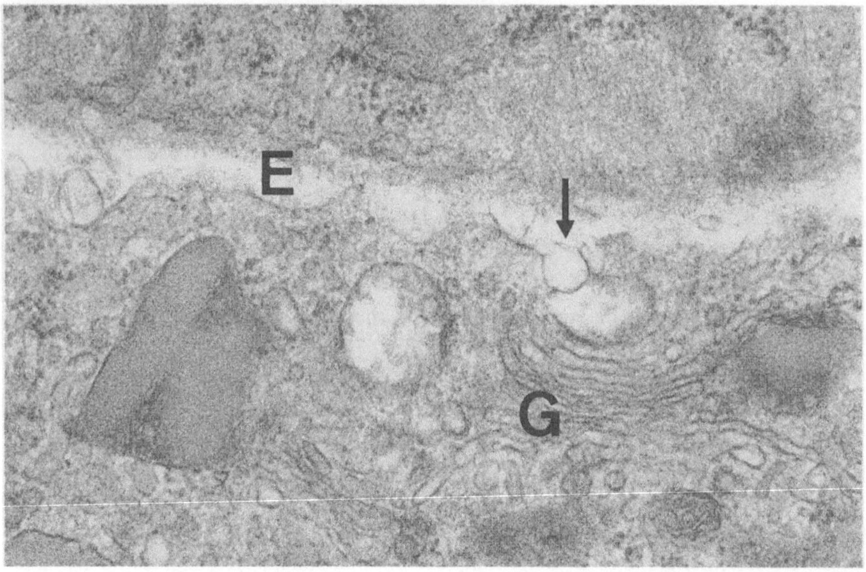

Fig. 6.9. Vacuolar structures between the Golgi region *(G)* and the surface of a mouse epithelioid cell. One of the organelles *(arrow)* protrudes into the extracellular space *(E)*. The electron-dense organelles to the left and right of the Golgi region are juvenile secretory granules; × 52200

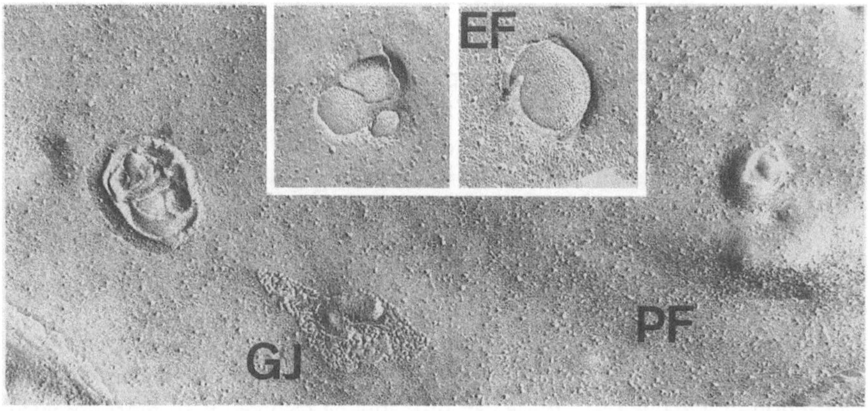

Fig. 6.10. Smooth-surfaced flat and vesicular structures exposed by freeze-fracturing the membrane of a juxtaglomerular mouse epithelioid cell. *PF, EF,* P- and E-face, respectively; *GJ,* gap junction. Freeze-fracture replica from mouse kidney; × 62200 and × 74000 *(insets)*

Chapter 7

Physiology, Pathophysiology, and Biochemistry of Renin Secretion

In some contrast to the preceding and subsequent sections of this monograph, we will not be able, within the limits of this section, to cover the major part of the pertinent literature, which amounts to several thousand publications. Rather, we are forced to use an eclectic approach to the literature in an attempt to characterize the present state of knowledge in a way which permits us to identify promising directions of future research into this subject. We do apologize to all those whose valuable contributions could not be cited, which can be found, however, in previous reviews covering various aspects of renin secretion (SKINNER et al. 1964; VANDER 1967; THURAU and MASON 1974; DAVIS and FREEMAN 1976; GANTEN et al. 1976; HOFBAUER et al. 1976; ZANCHETTI et al. 1976; PEART 1978; REID et al. 1978; FRANCO-SAENZ et al. 1980; KEETON and CAMPBELL 1980; DIBONA 1982; TORETTI 1982; HACKENTHAL et al. 1983; FREEMAN et al. 1984; TAUGNER et al. 1984; GIBBONS et al. 1984; HACKENTHAL and TAUGNER 1986; KURTZ 1986).

In this section, first some consideration is given to the experimental models which have been used to study renin secretion, followed by a discussion of the more integrated control mechanisms of renin secretion, i.e., the renal baroreceptor mechanism, renal sympathetic tone, and the still controversial tubulovascular interactions, including the macula densa mechanism. Subsequently, the surprisingly numerous hormonal signals which, at least experimentally, affect renin secretion, will be described. No special consideration is given to data obtained with pharmacological agents, unless they provide insight into physiological or biochemical mechanisms. An exhaustive review of the pharmacology of renin secretion has been given by KEETON and CAMPBELL (1980). A major focus of discussion, however, will be on the intracellular events during renin secretion, because it appears, from the still scanty or indirect evidence, that profound or even principal differences to other secretory mechanisms may exist.

7.1 Experimental Models

Experimental models, in which mechanisms of renin secretion have been studied, encompass all levels of structural and functional integration, such as isolated renin granules, suspensions or primary cultures of renal cortical cells or cell populations enriched in epithelioid cells, isolated afferent arterioles, isolated glomeruli, kidney cortex slices, the isolated perfused kidney, the kidney in situ, and, finally, the intact animal or human being. Obviously, each experimental approach has its own characteristic advantages, disadvantages, and limitations, which cannot be discussed here in detail. However, results and conclusions may vary significantly with the model and the experimental conditions used. Therefore, a few of these aspects will be briefly discussed in order to provide a methodological background against which to evaluate the available data.

7.1.1 Studies in the Intact Organism

With respect to renin secretion, the intact conscious animal or man is obviously best suited to study integrated regulatory mechanisms of blood pressure control, salt and water homeostasis, as well as other regulatory mechanisms, in which the renin-angiotensin system (RAS) is thought to be involved, whereas they are less well suited for studies addressing single mechanisms which participate in the control of renin release. This is not only because renin secretion cannot be measured without instrumentation in this situation, but also because the complexity of reactions to a given experimental maneuver does not make it possible to differentiate between direct and indirect effects on the renin-containing cells of the afferent arteriole.

7.1.2 The Kidney In Situ

Numerous studies have been performed with the kidney in situ, i.e., by instrumentation of kidney vessels and the ureter in the open abdomen during acute experiments. In this preparation, it is possible to measure, separately, arterial and venous plasma renin as well as renal blood flow, enabling renin secretion rates to be calculated. The error introduced into the calculation by neglecting secretion of renin into the renal lymph is very small, since the total amount of renin secreted into this pathway is usually in the order of 1% of total secretion or less, despite high concentrations of renin in renal lymph (cf. Sect. 4.6). A more significant, albeit mostly unrecognized, source of error in the calculation of renin secretion is the renal "extraction" of renin from the circulation. Renal clearance of arterial renin may significantly contribute to apparent net arteriovenous differences, as evidenced by "negative" secretion rates under conditions of acute suppression of renin release (MURRAY et al. 1982). Furthermore, the advantage in this preparation of providing the kidney with an optimal "perfusion medium" may be balanced by the disadvantages of unwanted systemic influences via changes of perfusion pressure, sympathetic nerve activity to the kidney, hormonal signals of unknown identity and magnitude, of possible indirect tubular influences, and of systemic effects of agents infused into the renal artery. Some of these disadvantages have been overcome by denervation of the kidney, by controlling perfusion pressure, by subjecting the kidney to total ischemia for 1-2 h resulting in a "non-filtering" kidney suitable for exclusion of tubular effects, by administration of receptor antagonists and/or by complete pharmacological vasodilatation of the kidney to exclude vascular effects, or by a combination of various measures. Obviously such interventions are likely to produce new sources of artifacts or misinterpretations. Nevertheless, such preparations have been used successfully to study the baroreceptor mechanism of renin release, especially by the group of DAVIS (cf. DAVIS and FREEMAN 1976), and to establish the effects of prostaglandins on renin release (e.g., by NIES, GERBER, and colleagues). The dog kidney in situ was also a particularly successful preparation in the studies of DIBONA and his associates to characterize the role of renal nerve activity in renin secretion and its influence on kidney function (for review see DIBONA 1982).

Another disadvantage of the in situ kidney preparation used in acute experiments is the need for anesthesia and acute surgery. Both anesthesia and surgical trauma drastically increase renal nerve activity, renin secretion rate, and plasma renin activity. They may alter the basal reactivity of the secretory mechanisms, and obscure more delicate reactions of epithelioid cells to experimental maneuvers. These disadvantages have been overcome in studies with chronically instrumented, awake, and trained dogs. Such experiments, which include chronic implantation of renal arterial and venous catheters, pressure transducers, flowmeters, infusion lines, and inflatable cuffs around the renal artery or the suprarenal aorta to control renal perfusion pressure, have been performed most elegantly by the groups of BARGER, FARHI, and coworkers, and KIRCHHEIM, GROSS, and associates, in studies on the precise role of renal perfusion pressure and its interplay with renal nerve activity (see below).

Recently, the hydronephrotic rat or mouse kidney has been established as a tool to examine various aspects of the renal circulation and renin secreting cells (STEINHAUSEN et al., 1983, 1986b). Hydronephrosis is induced by ureteral ligation. Within 6-8 weeks most of the tubular structures are lost, while the vascular system including the glomerulus are well preserved. This preparation permits to visualize individual vessels of all calibers and to measure local changes in flow rate and vessel diameter (STEINHAUSEN et al., 1986a, b, 1987). The preparation has also been used to examine renin release mechanisms without interference from tubular structures.

7.1.3 The Isolated Perfused Kidney

The isolated perfused kidney has been developed into a useful tool to examine several aspects of kidney function (ROSS 1978; MAACK 1980; EPSTEIN et al. 1980), including mechanisms of renin secretion. In the isolated kidney (mostly taken from the rat) neuronal and humoral interferences are absent, perfusion pressure or flow can be controlled, and renin secretion can be measured directly. For perfusion, balanced salt solutions supplemented with energy sources, such as glucose, pyruvate, and amino acids, are used. In order to prevent or delay the development of interstitial edema, and to mimic blood with respect to viscosity and oncotic pressure, various oncotic substances are added to the perfusion medium, such as albumin (SCHUREK and ALT 1981), gelatine (HOFBAUER et al. 1976), or hydroxyethylstarch (SCHWERTSCHLAG and HACKENTHAL 1982; HACKENTHAL et al. 1987). Oxygenation is usually accomplished by physical saturation of the perfusion fluid.

In kidneys perfused with albumin (60-70 g/l), physiological functions are best preserved in comparison to other oncotic substances (ROSS 1978; MAACK 1980; SCHUREK and ALT 1981). However, these kidneys can usually only be operated on in a recirculating system since in an open, single-pass system, the quantities of albumin required are hardly affordable. Recirculation of the perfusion medium, on the other hand, has the disadvantage that arteriovenous differences must be measured to calculate renin secretion rates, and that various metabolites, degradation products, or biologically active compounds released from the kidney will accumulate in the perfusate and modify basal conditions in an unknown fashion.

In all types of isolated perfused kidneys, morphological and functional deterioration of medullary structures occurs within a few hours (ROSS 1978; MAACK 1980; ALCORN et al. 1981; GRONOW and COHEN 1984), although large differences in the extent and time course of damage between various perfusion media exist. However, in studies on renin secretion, this can be considered as a secondary problem, particularly in the single-pass system, since in the area of interest, i.e., in the pre- and immediate postglomerular vascular tree of the kidney, signs of structural or functional damage are barely detectable for several hours of perfusion, probably because oxygen supply in this region is sufficient, as evidenced by the ultrastructural appearance of the otherwise highly vulnerable epithelioid cells after several hours of perfusion (unpublished observations).

Since, in the isolated kidney, epithelioid cells are still integrated in their natural cellular setting, this preparation is suited for studies on the interaction of epithelioid cells with their microtopographical environment, such as macula densa cells or endothelial cells, or to examine the role of renal perfusion pressure, renal nerve stimulation, etc. However, this preparation has also proved to be exceptionally useful to identify hormonal mediators of renin secretion and to obtain indirect information on intracellular events in epithelioid cells during modulation of renin secretion (see below).

7.1.4 Kidney Slices

Because of the ease of preparation and handling, kidney cortex slices from the rat, dog, or cat kidney have been used extensively in studies or renin secretion. Although with this experimental preparation valuable information on mechanisms of renin secretion have been obtained, it is likely to attract more methodological reservations and criticism than others for the following reasons: Kidney slices have a large surface area of damaged cells. These, as well as cells deteriorating in the interior of the slice, release lytic enzymes which may degrade experimental agents added to the bath or substances released from the cells, such as renin (CHO and MALVIN 1979). Cells damaged in the inner layers of slices will release K^+, which, before diffusing into the incubation medium, may attain high local concentrations, sufficient to disturb the function of epithelioid cells. Kidney slices have been shown, as the consequence of tissue damage and mechanical irritation, to release high amounts of various prostaglandins during the first 1-2 h of incubation, which stimulate and interfere with renin release influenced by experimental maneuvers (KNAPP et al. 1977; WHORTON et al. 1977). In addition, agents added to the incubation medium must diffuse through several layers of cells to reach epithelioid cells. Likewise, renin released from these cells has to diffuse to the surface of the slice, and, on its way, is particularly prone to proteolytic degradation by enzymes released from deteriorating tubules.

Finally, interactions of granulated cells with other cells, such as macula densa or endothelial cells, and thus indirect effects of experimental agents added are not excluded in this preparation.

Another experimental aspect, pertinent not only for kidney slices, but also for isolated glomeruli, afferent arterioles, and isolated cells, is the absence of intravascular pressure and consequently of a transmural pressure gradient across the media of arterioles carrying epithelioid cells. Intravascular pressure is an important determinant in the control of renin release, and pressure-dependent mechanisms interact in a delicate fashion with other control mechanisms for renin release. Pressure gradients are also likely to determine the direction of interstitial fluid flow (cf. Sect. 3.3). It therefore remains to be clarified how and to what extent responses of renin release to experimental maneuvers observed in isolated preparations lacking intravascular pressure reflect the behavior of epithelioid cells in the intact organ.

7.1.5 Isolated Glomeruli

Isolated glomeruli with attached segments of the afferent arteriole have been used less frequently for studies on renin secretion. They are more difficult to prepare than slices, and stronger mechanical forces have to be used for their isolation, usually causing loss of most of their renin content before the start of the actual experiment, as documented, for example, by MORRIS et al. (1976). Another methodological problem is the need for continuous agitation of isolated glomeruli in suspension, which affects their metabolic activity (e.g., increased synthesis of prostaglandins) and viability. This latter problem has been overcome by BLENDSTRUP et al. 1975 and BAUMBACH et al. 1976, who used the method of iron-particle infusion prior to the preparation of glomeruli in order to keep the isolated glomeruli in place in a small, perfused glass tubing by strong electromagnets. The functional properties with respect to renin secretion of this elegant preparation have not yet been fully evaluated in other laboratories. Agitation of glomeruli in suspension can also be avoided by a superfusion technique, as described by MORRIS et al. (1976) and BEIERWALTES et al. (1981, 1982).

7.1.6 Isolated Cells

In view of the wealth of information accumulated in numerous studies on cell suspensions and cell cultures in many fields of cell biology, it is clearly desirable to establish such preparations also for epithelioid cells to examine cellular events in the control of renin release, and many attempts have been made to that end. Necessarily, the preparation of isolated cells requires their dissociation from the tissue texture, which is usually accomplished by calcium-free perfusion or incubation, followed by collagenase digestion of the intercellular matrix, mechanical sieving, and washing to the cells. In striking contrast to the isolation of other types of secretory cells, incubation of renal tissue in a calcium-free medium with or without the addition of EDTA or EGTA is known to increase renin release from epithelioid cells dramatically, possibly as a consequence of the inverse ("paradox") relationship between internal calcium and renin release (cf. Sect. 7.5). Collagenase treatment, which always includes exposure to some tryptic activity contained in the collagenase preparations, is likely to affect the receptor properties on the cell surface. In several studies, in which renal cortical cell suspensions with little enrichment of granulated cells have been used, the methodological problems associated with this technique became apparent, such as poor responsivness to various stimuli or spontaneous loss of renin during cell preparation and initial incubation, usually amounting to 95% or more of the initial renin content.

In addition, unexpected differences to results obtained with other preparations, such as slices, may occur. For example, in a study by LYONS and CHURCHILL (1975a) with rat kidney slices, an increase of the sodium concentration in the incubation medium from 50 to 150 m*M* resulted in a dramatic increase of renin release, whereas in the presence of ouabain renin release decreased with increasing sodium concentration. The same authors (LYONS and CHURCHILL 1975b) reported that in a rat cortical cell suspension, incubated under experimental conditions comparable to the slice experiments, the same increase of sodium concentration induced a decrease of renin release, which was not affected by ouabain. Also, the addition of furosemide produced different reactions in the two preparations.

These comments are not intended to disqualify the use of isolated cell preparations for studies of renin secretion. On the contrary, by proper handling of isolated cells and selecting gentle conditions for incubation, and by recognizing the limitation of this method, they can provide useful information. For example, by superfusion of cells embedded in a layer of Sephadex gel, it has been possible to characterize the stimulation of renin secretion by dopamine (WILLIAMS et al. 1983; DRURY et al. 1986).

The most fascinating aspect, however, of studies with isolated cells is the chance to obtain homogeneous suspensions of pure granulated cells in sufficient quantity to measure intracellular reactions associated with renin release, and to maintain these cells in culture. This goal has not yet been reached, but some investigators have come close to it, in particular the group of KURTZ and BAUER (cf. KURTZ 1986), who provided exciting data on changes of intracellular messengers during stimulation or inhibition of renin release (cf. Sect. 7.5).

In all of these studies, however, as a result of the initial loss of renin during preparation the cellular renin content was far below that predicted for epithelioid cells. In the best preparation so far described (cf. KURTZ 1986), this amounted to about 0.1%-0.3% of the expected renin content. This low renin content of granulated cells does not represent a limiting factor for the assay of renin, which can be made extremely sensitive. Rather, the concern stems from the consideration that in normal, undisturbed epithelioid cells in vivo, renin is not in a homogeneous state, both with respect to biochemical maturation from prorenin to active renin (cf. Chap. 5) and with regard to compartmentalization, e.g. its presence in granules differing in maturity. Taking these aspects into account, it is conceivable, or even likely, that the small amount of renin remaining in isolated cells following isolation and plating is in a state not representative for cells in vivo and may respond to various stimuli quite differently from the "normal" epithelioid cell. For example, one could speculate that in these cells all the renin available for rapid secretion has been lost and that renin release depends largely on de novo synthesis and/or maturation of the secretory product. If this were true, one would obviously not be able to identify signals for secretion, but rather those for renin synthesis, processing, and/or exocytosis of immature granules. Eventually, this situation may turn out to be of major advantage in studies addressing the control of renin synthesis rather than secretion.

Another, partly methodological aspect, which has so far not been evaluated, is the possibility of major species differences in the reaction of renin secretory mechanisms to various maneuvers. For practical reasons, most of the numerous studies have been done in the rat and rat kidney preparations, followed by studies in the dog kidney. Very few studies have been done in the rabbit, or in the cat. Surprisingly, one can hardly find any studies done in mice, guinea pigs, hamster, or minipigs. In view of the well-documented species differences in the absolute and relative numbers of epithelioid and intermediate cells (cf. Chap. 12), a comparative search for corresponding functional differences would be of interest.

7.2 Renal Baroreceptor Mechanism

Based on the pioneering work of GOLDBLATT and associates (GOLDBLATT et al. 1934), the observations of TOBIAN (1960, 1962) of an inverse relationship between the long-term changes of "stretch" across the afferent arteriole, and "granularity" of juxtaglomerular cells and subsequent studies on the role of renal perfusion pressure in the control of renin secretion, SKINNER et al. (1964) proposed the existence of a renal "baroreceptor," which signals the kidney to increase renin release at lower and to reduce renin release at higher perfusion pressure. SKINNER and colleagues also emphasized the primary importance of renal perfusion pressure in comparison to other mechanisms, for the control of renin secretion, a view which has only recently been fully appreciated and confirmed (see below).

Since these early studies, many investigators have attempted to establish the existence of such a "baroreceptor" and to identify its mode of operation. Among the difficulties encountered in these experiments was the lack of control of other factors that may influence renin release, such as changes of renal nerve activity, autoregulatory mechanisms, or tubular effects. These difficulties have partly been overcome by using the isolated kidney (see below) and by developing the nonfiltering, denervated kidney model in the dog (BLAINE et al. 1970, 1971a, b; WITTY et al. 1971). In this latter model, in which the pressure response was further isolated from other interfering mechanisms by removing the contralateral kidney and by adrenalectomy, the existence of a renal pressure receptor controlling renin secretion could clearly be established. Studies in the isolated perfused rat kidney also demonstrated an inverse relationship between renal perfusion pressure and renin release (HOFBAUER et al. 1974, 1976; ZSCHIEDRICH et al. 1975; FRAY 1976, 1978a; TOKUMORI et al. 1983).

For the quantitative description and functional characterization of the relationship between perfusion pressure and renin secretion, studies in the

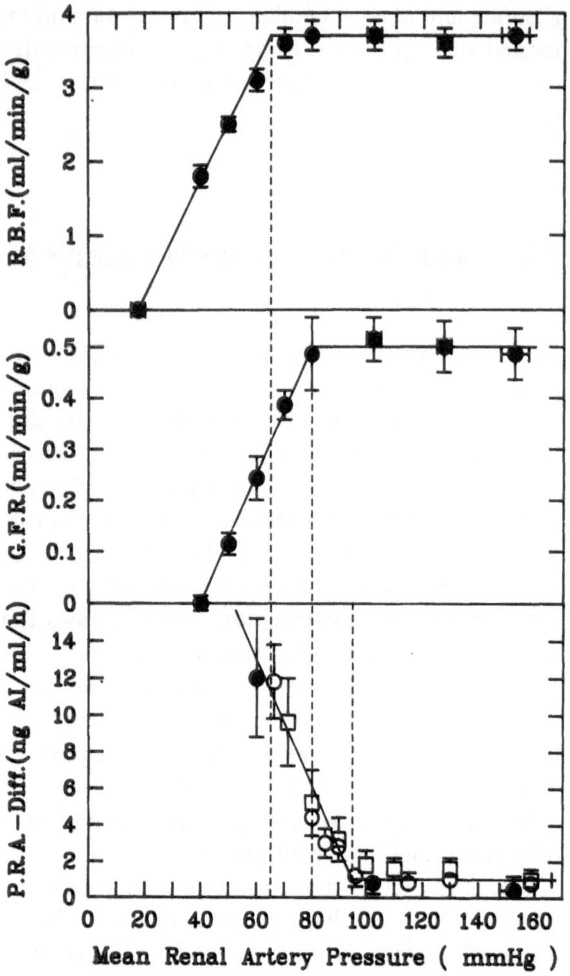

Fig. 7.1. Autoregulation of renal blood flow *(RBF)*, glomerular filtration rate *(GFR)* and renin release *(PRA-Diff.)* in the chronically instrumented conscious dog. *A I*, angiotensin I. Experimental details as described by KIRCHHEIM et al. (1987). See text for discussion

conscious, unrestrained dog proved to be most valuable. In these studies, stepwise reduction of renal perfusion pressure resulted in little change of renin secretion as long as perfusion pressure remained above a certain threshold level, which, in the dog, is around 90 mm Hg, i.e., close to the normal resting blood pressure. When renal perfusion pressure was lowered below this threshold, a steep increase of plasma renin activity or renin secretion was observed (KIRCHHEIM et al. 1981; FARHI et al. 1982; FINKE et al. 1983; KIRCHHEIM et al. 1985). This situation is illustrated in Fig. 7.1. A similar threshold pressure for renin secretion of about 100 mm Hg has recently been described in anesthetized rats (IMAGAWA et al. 1985). When basal renin secretion in the dog is increased by a low-salt diet, the response to changes of perfusion pressure is shifted upwards with a proportional increase at all pressure levels, but without changes of the threshold pressure point for the increased secretion of renin (FARHI et al. 1983; GIBBONS et al. 1984). A qualitatively similar relationship between salt intake and the baroreceptor-mediated renin release has been shown by FRAY et al. (1977) in the rat.

The interrelationship of the renal baroreceptor and renal nerve activity influencing renin secretion is more complex. Previous investigations by ROCCHINI and BARGER (1979), KIRCHHEIM et al. (1980), and GROSS et al. (1981) in the conscious dog had demonstrated that mild sympathetic stimulation, produced by carotid sinus hypotension, increased renin secretion, when perfusion pressure of the kidney was kept constant. If the perfusion pressure was allowed to rise with systemic pressure, the sympathetic nerve-mediated stimulation of renin secretion no longer occurred due to the simultaneous inhibition by activation of the renal baroreceptor. In accordance with these results, THAMES and DIBONA (1979) and KOPP and DIBONA (1984) observed "sensitization" to the effect of mild renal nerve stimulation, which, per se, did not alter renal hemodynamics or tubular functions, by lowering renal perfusion pressure in anesthetized dogs. Similarly, ABE et al. (1977) described an increased sensitivity to pressure reduction during norepinephrine infusion. A quantitative description of this relationship was again established in the conscious, unrestrained dog: In these studies an increased renal nerve activity was either simulated by infusion of epinephrine (FARHI et al. 1982, 1983) or actually induced by producing carotid sinus hypotension (FINKE et al. 1983; KIRCHHEIM et al. 1986). In both experimental situations a significant upward shift of the threshold pressure for the steep increase of renin secretion was measured, i.e., the stimulus-response curve relating perfusion pressure to renin release was shifted to the right. In the upper pressure range, i.e., above threshold pressure, only a small elevation of the plateau of renin secretion was noticed.

Indirectly, these observation can explain, why in previous studies on the quantitative aspects of this inverse relationship between perfusion pressure and renin release in anesthetized animals a characteristic threshold pressure was not observed. For example, in the anesthetized dog, COWLEY and GUYTON (1972) found almost proportional inhibition of renin release with increasing perfusion pressure over the whole pressure range. In these studies intrarenal arterial pressure had been increased by infusion of norepinephrine, which must have led to a stepwise upward shift of threshold pressure.

If it is assumed that infusion of norepinephrine into the renal artery is functionally equivalent to increased sympathetic activity of renal nerves, this explanation is also valid for many other studies in anesthetized animals, in which the precise quantitative relationship between perfusion pressure and renin secretion was obscured by the consequences of anesthesia and acute surgery, such as increased sympathetic discharge, increased prostaglandin synthesis, increased levels of circulating catecholamines, and elevated basal renin secretion.

The local cellular and biochemical mechanisms translating changes in perfusion pressure into changes in renin secretion have not yet been identified. Prostaglandins, such as PGE_2 and PGI_2, are potent stimulators of renin secretion (see below) and are released from afferent vascular structures upon renal nerve stimulation (cf. DiBona 1982) as well as upon reduction of renal perfusion pressure (Jackson et al. 1982a,b; Vikse and Kiil 1985). Therefore, prostaglandins have contracted considerable attention as possible mediators of the baroreceptor response. In the anesthetized rat, the reduction of renal perfusion pressure below 100 mmHg by suprarenal aortic constriction results in a dramatic increase of plasma renin activity, which could almost completely be blocked by indomethacin (Imagawa et al. 1985). In contrast, in studies by Villareal et al. (1984), also performed in anesthetized rats, meclofenamate failed to block or attenuate the increase in plasma renin activity in response to mild or severe reduction of renal perfusion pressure in the nonfiltering, denervated, as well as the intact kidney model. In the majority of studies in anesthetized dogs, inhibition of prostaglandin synthesis by indomethacin or meclofenamate significantly attenuated the increase of renin secretion during reduction of perfusion pressure (Data et al. 1978; Berl et al. 1979; Blackshear et al. 1979; Seymour and Zehr 1979). In the conscious dog, however, no effect of prostaglandin inhibition on the stimulus-response curve of the renal baroreceptor was observed (Freeman et al. 1982; Anderson 1982; Anderson et al. 1983). Thus, it may be concluded that the baroreceptor-mediated renin secretion can, under normal conditions, i.e., in the conscious animal, operate without participation of prostaglandins, whereas during anesthesia and increased sympathetic activity prostaglandins may contribute in a modulating fashion to the net response of renin secretion to alterations of perfusion pressure. As pointed out by Kirchheim et al. (1985), it cannot be excluded that the upward shift of threshold pressure for renin release by mild renal nerve stimulation may, at least in part, bemediated by a previously described α-adrenoceptor-dependent and prostaglandin-mediated stimulation of renin secretion (Olson et al. 1983).

Little attention has been given so far to the role of calcium in the baroreceptor control of renin secretion. In the isolated rat kidney, Ettienne and Fray (1979) and Fray (1980) found that the stimulatory effect of lowering perfusion pressure was attenuated by increasing the calcium concentration of the medium, and that the inhibitory effect of high perfusion pressure was blocked by verapamil, a calcium-channel antagonist, or in the absence of extracellular calcium. These observations have been incorporated into a stretch receptor hypothesis for renin secretion (Fray 1976, 1978a,b; Fray et al. 1983), which postulates that stretch of the vascular wall facilitates Ca^{2+} entry into epithelioid cells. Since intracellular calcium appears to be inversely related to renin secretion (cf. Sect. 7.6), an increased rate of calcium entry would then induce inhibition of renin secretion. Observations of a calcium dependency of renal vascular autoregulation (Ogawa and Ono 1986, 1987; Loutzenhiser et al. 1987) would be in agreement with this role of calcium. Recently, activation of calcium channels of vascular endothelial cells by stretch has been described by Lansman et al. (1987). The presence of mechanotransducing ion channels in vascular endothelial cells may not only help to explain the participation of the endothelium in vascular responses to hemodynamic changes, but perhaps also the response of epithelioid cells to changes of perfusion pressure.

The functional significance of the baroreceptor control of renin secretion has mainly been interpreted as an emergency reaction in response to a dramatic fall of arterial pressure, e.g., by hemorrhage. Recently, however, it became clear that, in the conscious dog, the threshold pressure for renin secretion is close to the mean resting arterial pressure (Finke et al. 1983; Kirchheim et al. 1985) and that the pressure-response curve of renin release corresponds quantitatively to the autoregulatory behavior of renal blood flow (RBF), with the threshold pressure of the latter being but a few millimeters of mercury above threshold for renin secretion (see Fig. 7.1). This has led KIRCHHEIM et al. (1981, 1985) to formulate the hypothesis of autoregulation of RBF, renin secretion, and glomerular capillary pressure by descending myogenic dilatation in response to lowering perfusion pressure. This hypothesis is an extension of the autoregulation hypothesis put forward by KIIL and coworkers (EIDE et al. 1973; KIIL 1975; LANGGARD et al. 1981) and the model of autoregulation of RBF by OIEN and AUKLAND (1983). According to KIRCHHEIM, gradual reduction of perfusion pressure induces myogenic vasodilation, which spreads from larger vessels, such as arcuate arteries, via interlobular arteries to afferent arterioles, eventually (at threshold pressure) reaching the site of epithelioid cells, which respond with a dramatic increase of renin secretion. In view of the close association of resting mean arterial pressure and threshold pressure for renin secretion, this hypothesis was further extended by assigning this mechanism an important role in the maintenance of normal blood pressure. It has been shown that, during daily fluctuations of resting arterial pressure in the individual dog, a significant portion of single pressure values fall below the threshold pressure of renin release established for this individual dog. This prompts an immediate rise of renin release, an increased rate of angiotensin II generation, and thus restoration of suprathreshold arterial pressure (KIRCHHEIM et al. 1986). Considering the fact that the baroreceptor mechanism is extremely sensitive below threshold pressure, i.e., that a reduction of perfusion pressure by 2-3 mm Hg results in a twofold increase of renin release (cf. Fig. 7.1), considering further the close association of arterial pressure and the threshold pressure for renin secretion, the baroreceptor mechanism of renin secretion must be considered as one of the most important physiological mechanisms in the control of renin secretion and in the short term maintenance of systemic blood pressure.

7.3 Tubular Signals, Renin Secretion, and Vascular Reactions - The Macula Densa Hypothesis

Intrigued by the unique anatomical relationship between the distal tubule and the vascular pole of the glomerulus, GOORMAGHTIGH (1939, 1945a) was the first to speculate on the possibility of the macula densa serving as a sensor for the tubular control of renin secretion. Later, VANDER and MILLER (1964) condensed this view and the observation of an inverse relationship between dietary salt intake and plasma renin into the "macula densa theory of renin secretion." According to this theory, the composition and/or flow rate of the fluid to the early distal tubule is sensed by the macula densa cells. This information is transformed into an intercellular message to which the neighboring epithelioid cells of the JGA respond with changes of renin secretion. Based on the general observation of an inverse relationship between salt intake and plasma renin, VANDER and MILLER (1964) proposed that an increased tubular load of salt is associated with a decrease in renin secretion and vice versa. In developing the concept of an intrarenal RAS that participates in the control of glomerular filtration rate (GFR), THURAU and colleagues (THURAU et al. 1967, 1970; THURAU and MASON 1974) arrived at the opposite conclusion, i.e., that an increased distal tubular salt concentration would stimulate renin secretion.

Although the available data favor the view of an inverse relationship, this controversy has, for several reasons, not yet been definitely settled. One of the reasons is the intimate association which had been constructed between the macula densa hypothesis of renin secretion and the macula densa mechanism for the control of GFR, better known as tubuloglomerular feedback (TGF). In fact, the RAS and its local activation has long been considered the principal effector system of TGF, and only recently has the TGF mechanism been recognized as a parallel or concomitant event to the macula densa mediated renin release (for review see BRIGGS and SCHNERMANN 1986).

Another source of existing disagreements and controversies may be differences in the methodolgical approach: the TGF-related mechanisms have been studied preferentially at the single nephron level (cf. SCHNERMANN and BRIGGS 1985), whereas renin secretion, for methodological reasons, has mostly been studied in the whole kidney, in which functional isolation of a tubular mechanism relevant for renin release is difficult to accomplish. Furthermore, many investigators attempting to characterize the macula densa mechanism of renin release did not clearly differentiate conceptually between long-term changes of salt intake or salt balance, on the one hand, and experimentally induced acute changes of intratubular salt concentration or flow rate, on the other hand. It is important to realize that it is unknown at present to what extent or whether at all these two conditions are related to each other with respect to the consequences for renin secretion and the mechanisms involved (cf. Section 11.1). In the following, the evidence in support of the existence of an acutely operating tubular mechanism for renin release and some aspects of its possible mode of operation will be discussed as such, and in relation to TGF.

7.3.1 Acute Tubulovascular Interactions

Since the existence of a TGF mechanism was first demonstrated by THURAU and SCHNERMANN (1965) and SCHNERMANN et al. (1970), it has now been clearly established in many studies that an increase in flow rate or sodium chloride concentration in the early distal tubule induces changes of vascular resistance resulting in a decreased ultrafiltration pressure, and hence a decreased GFR (for review see SCHNERMANN and BRIGGS 1985; AHRENDSHORST 1987; BRIGGS and SCHNERMANN 1987). The existence of this TGF mechanism has also been demonstrated at the level of the whole kidney: Infusion of hypertonic saline into the kidney results in vasoconstriction and reduction of GFR (e.g., GERBER et al. 1979; SCHNERMANN et al. 1979). This effect can be modified by salt depletion, furosemide, DOCA-salt treatment, or other interventions in the same direction as can be observed in single nephron studies (for references see BRIGGS and SCHNERMANN 1986) and is best explained by whole kidney activation of the TGF mechanism.

As pointed out above, the single-nephron technique was used in only a few studies on the putative macula densa control of renin release. MORGAN and GILLIS (1977), for example, reported that venules draining oil-blocked tubules, i.e., tubules with zero flow and probably low tubular sodium chloride concentrations and transport, had lower renin concentrations than venules draining unblocked nephrons. These results have been interpreted to demonstrate a direct relationship between tubular sodium chloride and renin secretion, as proposed by THURAU et al. (1967) and more recently by WRIGHT (1984). Results pointing into the opposite direction were recently described by LEYSSAC (1984, 1986). He observed that an increased tubular flow rate is associated with a decrease of renin concentration both in proximal tubular fluid and in plasma obtained from efferent arterioles at the welling point. These observations were thought to reflect a decrease in renin secretion from the corresponding afferent arteriole, assuming that a major portion of newly secreted renin gains access to the lumen of this vessel (LEYSSAC 1978a,b). However, other micropuncture studies suggest that newly secreted renin reaches the circulation by way of the cortical interstitium and the peritubular capillaries (e.g., MORGAN and DAVIS 1975; MORGAN et al. 1982). Furthermore, morphological studies at the ultrastructural level have demonstrated that along the afferent arteriole a barrier of tight junctions between endothelial cells exists, making it unlikely that renin released from epithelioid cells into the interstitium can directly reach the lumen of the afferent arteriole (cf. Sect. 2.2.8 and Chap. 6). Also, consideration of the possible direction of interstitial fluid flow on the basis of existing pressure gradients speaks against entry of renin into the circulation at the level of the afferent arteriole (cf. Sect. 3.3, and ROSIVALL and TAUGNER 1986 for discussion).

Single-nephron studies by THURAU and colleagues (THURAU et al. 1970, 1982; DAHLHEIM et al. 1970; GREGER et al. 1972) and by GILLIES and MORGAN (1978, 1982) have introduced still another aspect of the relationship between the macula densa and epithelioid cells. When the delivery of sodium chloride to a given distal tubule was increased by increasing distal flow rate or sodium chloride concentration, the content of active renin in the associated glomerulus, obtained by microdissection or biopsy, increased dramatically. It was concluded that this phenomenon represents intracellular activation of inactive renin by a tubule-dependent mechanism. It should be emphasized that in these studies the release of renin could not be measured.

It therefore remains to be clarified whether this observation can be generalized to the level of the whole kidney, and whether this activation process in accompanied by an increased rate of renin release. Since there is only very little information available on the control and the time course of the intracellular activation of renin by external stimuli (cf. Sect. 7.8), these interesting observations made by THURAU, GILLIES, and MORGAN should certainly be extended in order to clarify this facet of tubulovascular interaction.

Studies in the whole kidney can necessarily only provide indirect evidence for the existence of a distal tubular or macula densa control of renin secretion. For example, infusion of hypertonic saline into the dog kidney resulted in a decrease in renin release (e.g., SHADE et al. 1972; GERBER et al. 1979); the same infusion into the nonfiltering kidney had little effect (SHADE et. al. 1972), which was ascribed to the nonfunctioning macula densa. GERBER et al. (1981 a, b) used the denervated, β-blocked, papaverine-treated dog kidney originally described by WITTY et al. (1971) to "dissect" the macula densa from other mechanisms. When a reduction of renal blood flow and tubular load of sodium chloride was induced by aortic constriction, renin secretion was increased, and the increase could be prevented by indomethacin. It was concluded that this effect reflects activation of the macula densa mechanism and is mediated by prostaglandins. It remained uncertain, however, whether the baroreceptor mechanism had indeed been blocked. Furthermore, subsequent studies from the same laboratory have demonstrated that acute renal artery constriction by itself induced an increase in vascular prostaglandin (PGI_2) synthesis and release (JACKSON et al. 1982), and PGI_2 might have stimulated renin release without participation of a macula densa mechanism. In intact or adrenalectomized rats acute saline infusion has been shown to enhance GFR, chloride delivery to of Henle's loop and chloride reabsorption. The concomitant decrease in plasma renin was thought to result from macula densa dependent inhibition of renin secretion (WELCH et al. 1985).

In the isolated perfused rat kidney, large variations of the sodium chloride concentration in the perfusion fluid resulted in significant changes of urinary sodium chloride, but had no effect on the rate of renin secretion (FRAY 1976). ROSTAND et al. (1985) observed a significant negative correlation between renin secretion and absolute tubular chloride reabsorption when the latter was varied by gradual replacement of chloride by other anions or by furosemide. Surprisingly, partial replacement of chloride by nitrate significantly enhanced renin secretion without affecting GFR. In the same study, kidneys made nonfiltering by increasing the perfusate albumin concentration to 14% had a 40-fold higher rate of renin secretion, which was no longer affected by the perfusate chloride concentration. Although these and several similar studies (e.g., LINAS 1984; FRANCISCO et al. 1982; WELCH et al. 1986) can be taken to suggest the existence of a tubular mechanism in the control of renin secretion, many uncertainties and inconsistencies remain to be clarified (see below).

Another approach to this problem has been described by ITOH and CARRETERO (1985), who measured renin release from dissected afferent arterioles from the rabbit kidney, with or without macula densa segments still attached. They found spontaneous renin release from arterioles with macula densa segments to be lower than that from free arterioles. Furosemide, which is thought to suppress the macula densa mechanism by inhibiting chloride reabsorption, had no effect on free arterioles but increased renin release about fourfold from macula densa attached arterioles. This observation was considered to be the first direct evidence that furosemide-induced renin release is mediated by the macula densa. In the same preparation, adenosine was found to suppress renin release from free arterioles, but not from macula densa associated arterioles (ITOH et al. 1985). The authors concluded that macula densa cells exert a tonic inhibitory control on renin secretion from epithelioid cells, and that adenosine may be the inhibitory signal released from macula densa cells that inhibits renin release through activation of adenosine A_1 receptors (see below). It remains questionable, however, whether in the nonperfused segments of the distal tubules attached to the afferent arteriole furosemide-sensitive transport of sodium chloride still exists. Furthermore, the dissected preparations always contained, in addition to the macula densa, neighboring segments of the distal tubule. It therefore remains to be clarified whether the observed effects are related to a specific function of the macula densa cells (cf. Sect. 7.3.2). It should also be considered that adenosine A_1-receptors have as yet not been identified on vascular structures (see below). Finally, differences in renin release from macula densa attached segments and free arterioles may still have another explanation: As described in detail previously, in the rat kidney, epithelioid cells in the wall of the afferent arteriole close to the glomerulus (and the macula densa) exhibit distinct ultrastructural differences when compared with the intermediate renin containing cells further upstream (cf. Sect. 2.2). It is suggested that these morphological differences are reflected in functional dissimilarities. In fact, studies by SKØTT and BAUMBACH (1985) and BAUMBACH and SKØTT (1986) with afferent arteriolar segments differing in length have shown that responsiveness of renin release to adenosine and isoproterenol is restricted to the more proximal parts of the afferent arterioles, in which intermediate cells are more abundant.

Perhaps the most convincing argument in favor of a macula densa mechanism for the control of renin secretion is provided by a study of SKØTT and BRIGGS (1987). These authors measured renin release from microdissected JGAs including the glomerulus, in which the distal tubule including the macula densa segment was perfused with solutions differing in sodium chloride concentration. A reduction of the sodium chloride concentration by 96 mM (the osmolality was maintained by mannitol) resulted in a prompt increase of renin release from the associated afferent arteriole.

7.3.2 Location of the Tubular Sensor

As to the anatomical location of the sensor cells for TGF and/or the control of renin secretion, it is generally assumed, on the basis of their anatomical association, that the tubular signal is received by the macula densa and transmitted to the media cells of the afferent arteriole. The macula densa region itself is not directly accessible to micropuncture to verify this assumption, but can functionally be considered as part of the thick ascending limb of Henle's loop, for which active, furosemide-sensitive chloride reabsorption is characteristic (cf. HERBERT and ANDREOLI 1984; GREGER 1985).

On the other hand, histochemical data suggest that the macula densa differs from adjacent thick ascending limb cells in several biochemical aspects, such as the presence of carbonic anhydrase (LÖNNERHOLM and WISTRAND, 1984), the absence of Tamm-Horsfall glycoprotein (HOYER et al. 1979, SIKRI et al. 1981) and the absence of epidermal growth factor (BARAJAS and POWERS, 1984; SALIDO et al. 1986a; BARAJAS et al. 1988). The possible functional implication of these properties of macula densa cells have been discussed by BARAJAS et al. (1988), and it remains to be established whether these and other pecularities are associated with TGF or the control of renin secretion.

With regard to the location of the sensor, attention is called to the observations that not only the macula densa but also adjacent tubular segments can contact the Goormaghtigh cell field (see section 2.6), that the distal connecting tubule frequently contacts the afferent arteriole (BARAJAS et al. 1988) and the unmodified thick ascending limb comes into close topographical association with the efferent arteriole (BARAJAS and POWERS, 1984).

In view of these multiple tubulo-vascular proximities it remains open where exactly the sensing mechanism is located along the ascending limb and the early distal tubule. With respect to the control of renin secretion, the functional implications of the topographical association of tubular structures with epithelioid cells will be discussed later in this section (see also Sect. 2.6).

The nature of the signal perceived by the cells of the macula densa (or other segments of the distal tubule) is still debated. Recently, evidence is increasing that the chloride anion is the tubular signal rather than sodium, and transport of chloride by the sensor cells may be an essential prerequisite for the operation of the TGF and perhaps also for the macula densa mechanism of renin secretion (ROSTAND et al. 1985; SCHNERMANN and BRIGGS 1985).

7.3.3 The Macula Densa Signal

As pointed out before, uncertainty also still exists with respect to the nature of the message transmitted from macula densa cells to both epithelioid and smooth muscle cells to modulate their functional state. The earlier concept, in which locally generated angiotensin II was thought to be the extracellular messenger for the vascular response in TGF, has been abandoned by many investigators in favor of a concept, in which the RAS has an important modulating function, but is not the causal link in the expression of TGF (see below and reviews by PLOTH and ROY 1982a, b; NAVAR and ROSIVALL

1984; BRIGGS and SCHNERMANN 1986). Instead, various other humoral agents came into consideration as mediators of both the renin and the vascular response. Of these, prostaglandins and adenosine have been studied most extensively.

7.3.3.1 Prostaglandins

A prostaglandin-dependent macula densa mechanism for renin secretion has been concluded by GERBER et al. (1981 a, b) from the observation that in denervated, β-blocked, and papaverine-treated dog kidneys suprarenal aortic constriction induced an increase in renin release which could be prevented by indomethacin. As pointed out already, participation of the baroreceptor was not completely excluded, and vascular prostaglandin synthesis is an alternative for this effect. Participation of prostaglandin in the macula densa mechanism has also been inferred from studies in sodium-depleted dogs by FRANCISCO et al. (1982) and by LINAS (1984) in the isolated perfused rat kidney. More recent studies, however, failed to demonstrate a major dependency of the macula densa mechanism on prostaglandin synthesis (e.g., VIKSE and KIIL 1985; OSBORN et al. 1984). This is in accordance with the view that also in the mediation of TGF local prostaglandin formation is a modulating factor rather than a causal link in the sequence of events between signal recognition and vascular response (SCHNERMANN et al. 1979; BRIGGS and SCHNERMANN 1986; GERBER and NIESS 1986).

At present, several investigators focus on the adenosine hypothesis first proposed by OSSWALD et al. in 1978, in which transport-dependent generation of adenosine from ATP is the message released from macula densa cells to reach both epithelioid and vascular smooth muscle cells. In view of the importance of this aspect, the adenosine hypothesis will be discussed in the following.

7.3.3.2 The Adenosine Hypothesis

Adenosine is generally known as a potent vasodilator and has been implicated in the local regulation of organ blood flow. In the heart, for example, locally formed adenosine is thought to be the key factor in the metabolic control of coronary blood flow (cf. BERNE et al. 1973). According to this hypothesis of coronary autoregulation, interstitial concentrations of adenosine rise by exercise-induced increases of ATP hydrolysis. Increased levels of adenosine then decrease coronary vascular resistance, and oxygen supply is adjusted to the increased demand as a result of an increase in coronary blood flow.

As pointed out by THURAU (1964), a similar type of metabolic control of RBF would be unsuitable for the maintenance of GFR, since renal metabolism is determined mainly by sodium reabsorption and a decrease of RBF would decrease GFR, decrease sodium reabsorption, decrease oxygen expenditure, decrease the metabolic signal, further decrease RBF and GFR, and so on. Therefore, a modification of the metabolic hypothesis has been proposed (THURAU 1964; GUYTON et al. 1964), in which a tubular metabolic signal *decreases* rather than increases GFR and RBF by mediating constriction of the afferent arteriole.

Adenosine has later been propagated as a candidate for this metabolic signal (OSSWALD et al. 1978, 1980, 1982; SPIELMAN and THOMPSON 1982), because in the kidney, in contrast to other organs, intraarterial infusion of adenosine may produce vasoconstriction and a decrease in GFR (THURAU et al. 1967; TAGAWA and VANDER 1970; OSSWALD et al. 1980; SPIELMAN and THOMPSON 1982). This vasoconstrictor effect of adenosine was initially thought to be mediated by angiotensin II, since the vasoconstriction could be abolished or attenuated by the infusion of saralasin, a competetive antagonist of angiotensin II, or by converting enzyme inhibition (SPIELMAN and OSSWALD 1979; HALL et al. 1985; HALL and GRANGER, 1986; SPIELMAN et al. 1987). However, such a mechanism is difficult to reconcile with the observation that adenosine infusion into the renal artery suppressed renin secretion (OSSWALD et al. 1978; TAGAWA and VANDER 1970; SPIELMAN and THOMPSON 1982; SPIELMAN 1984), which should result in a decrease of intrarenal generation of angiotensin II. In order to accommodate these conflicting aspects, complex reaction schemes have been proposed which allow for an increased generation rate of angiotensin II in the face of suppressed renin secretion (OSSWALD et al. 1978, 1982). This possibility is not only unlikely in itself, but has been excluded by the recent demonstration that angiotensin II measured in hilar lymph is not increased, as would be predicted from this hypothesis, but rather suppressed during intrarenal adenosine infusion (SPIELMAN 1984). Furthermore, recent studies in the isolated rat kidney have shown that the vasoconstrictor effect of adenosine is totally independent of angiotensin II (HACKENTHAL et al., unpublished data; ROSSI et al. 1987).

Recently, OSSWALD (1984) proposed a modified hypothesis of the role of intrarenally formed adenosine, in which an increased generation rate and release of adenosine induces, in parallel, preglomerular vasoconstriction as well as suppression of renin secretion. In this hypothesis, angiotensin II is no longer considered as an essential link in TGF, and the effect of adenosine on renin secretion is thought to be independent of the vascular and filtration effects of adenosine (AREND et al. 1984).

It must be emphasized, however, that although adenosine seems to be an important regulator of renal hemodynamics and GFR and inhibits renin release in vivo (see review by SPIELMAN 1984), the evidence for the participation of adenosine in a macula densa mechanism of renin secretion and in TGF is fragmentary, and several observations even seem to contradict this hypothesis: In the cited studies by OSSWALD and SPIELMAN, vasoconstriction upon infusion of adenosine was only transient and was always followed by sustained vasodilatation. In the isolated perfused rat kidney, concentration-dependent stimulation rather than inhibition of renin release by adenosine and the adenosine-analogs phenylisopropyladenosine (PIA) or *N*-ethylcarboxamide adenosine (NECA) has been observed (HACKENTHAL et al. 1983, 1987; MURRAY and CHURCHILL 1984, 1985. Recently, CHURCHILL and BIDANI (1987) reported that NECA increases plasma renin in rats. Occasionally, a biphasic response to some agonists was observed, e.g., cyclohexyladenosine or 2-chloroadenosine at low concentrations inhibited and at higher, micromolar concentrations stimulated renin release both in the isolated kidney (MURRAY and CHURCHILL 1985) and in kidney slices (CHURCHILL and CHURCHILL 1985). As an explanation, the existence of two types of adenosine receptors (A_1 and A_2 receptors) present on epithelioid cells has been proposed, based on the assumed receptor preferences of the agonists used and on the general concept of adenosine receptor classification: Adenosine A_1-receptors are activated by low concentrations of adenosine and are coupled to adenylate cyclase via the inhibitory G_i-coupling protein, wheras adenosine A_2-receptors require higher (0.1-1 μM) concentrations of adenosine and are coupled to adenylate cyclase via a stimulatory G_s-coupling protein (for survey see WOLFF et al. 1981). Since it is known from many studies that an increase or decrease of intracellular adenosine 3′, 5′-cyclic monophosphate (cAMP) is associated with stimulation or inhibition of renin secretion, respectively, the proposal of dual receptor activation would explain the observed biphasic effect. The presence of both A_1- and A_2-receptors on glomeruli and attached arterioles has recently been described by FREISSMUTH et al. (1987).

Direct evidence for an adenosine A_1-receptor-mediated inhibition of renin release was recently obtained by HACKENTHAL et al. (1987c) in the isolated rat kidney. Infusion of the specific A_1-receptor agonist cyclopentyl-adenosine concentration-dependently inhibited renin release and induced vasoconstriction. Conversely, infusion of the A_1-receptor antagonist dipropylcyclopentyl xanthine (DPCPX) into the isolated kidney induced vasodilatation and an increase in renin release, indicating that, in the isolated rat kidney, endogenous adenosine exerts an inhibitory control on renin release. Similarly, ROSSI et al. (1987) recently reported that cyclohexyladenosine inhibits renin release from rat kidney slices and that this inhibition was reversed when slices were taken from pertussis-toxin treated rats. From these data the authors conclude that adenosine inhibits renin secretion via A_1-adenosine receptors.

In a previously discussed study, ITOH et al. (1985) examined the effect of adenosine in dissected afferent arterioles. Inhibition of renin release was observed in free afferent arterioles but not in arterioles to which the macula densa segment was still attached. They concluded that adenosine is responsible for a macula densa mediated suppression of renin secretion. It is difficult to understand, however, how in a preparation of isolated afferent arterioles adenosine, once it has been released from macula densa cells, could have escaped diffusion and dilution into the incubation medium. Also at variance with the conclusions of ITOH et al. (1985) are the data presented by SKØTT and BAUMBACH (1985): Using isolated glomeruli carrying afferent arterioles of different length, they observed inhibition of renin release by adenosine only in long afferent arterioles, but not in glomeruli with short fragments of the afferent arteriole. They concluded that only those epithelioid cells that are located at some distance from the glomerulus, respond to adenosine, and that adenosine is therefore unlikely to be the tubular (macula densa) signal for suppression of renin release.

Preliminary evidence for a direct effect of adenosine on epithelioid cells was reported by KURTZ (1987) who, in isolated epithelioid cells, observed inhibition of renin release by adenosine, which was accompanied by an increase of intracellular guanosine 3′,5′-cyclic monophosphate (cGMP).

Finally, the adenosine hypothesis of TGF and the control of renin secretion, as described by OSSWALD and SPIELMAN, requires that both the vascular smooth muscle cells and the epithelioid cells of the afferent arteriole carry adenosine A_1-receptors, coupled in an inhibitory fashion via G_i-protein to adenylate cyclase. Except for the recent report by FREISSMUTH et al. (1987), such receptors have not yet been identified on either cell type, nor is there evidence for the presence of adenosine A_1-receptors on vascular smooth muscle cells in general (WOLFF et al. 1981; STILES 1986).

This confusing situation with regard to the role of adenosine in the control of renin secretion and TGF may eventually be clarified by taking other renal effects of adenosine into account. For example, adenosine affects presynaptic release of noradrenaline from sympathetic terminals of blood vessels and in the kidney (EKAS et al. 1983; BURNSTOCK and KENNEDY 1985) and may thus modulate vascular tone. At least some of the vascular effects (and perhaps the renin secretion effect) of adenosine may also be mediated by the vascular endothelium (e.g., D'ORLEANS-JUSTE et al. 1985; RUBANYI and VANHOUTTE, 1985).

7.3.4 Concluding Remarks

In summary, there is now considerable evidence to suggest that a tubular mechanism controlling renin secretion exists - one which operates in parallel with the TGF response, i.e., an increase in sodium chloride delivery to the distal tubule induces both a reduction in GFR and a decrease in renin release. Contrary to earlier views, angiotensin II generated by the released renin does not mediate the TGF response, but modulates vascular reactivity. As the mediator of both the vascular and the epithelioid cell response to macula densa activation, adenosine has to be considered as a candidate, although many inconsistencies still exist.

In this context it should be noted that the exact effector site of the tubuloglomerular feedback mechanism is still a matter of controversy, i.e., it is debated whether preglomerular vasoconstriction, glomerular vasoconstriction, total glomerular contraction, postglomerular vasodilatation, or a variable mixture of these components is most important (cf. Chap. 9). Furthermore, most investigators, in their effort to establish and define the macula densa mechanism for renin secretion, have based their studies on the simplifying assumption that renin is more or less exclusively contained in the small segment of the late distal afferent arteriole directly adjacent to the glomerulus and the macula densa. This view, however, does not reflect the topographical reality. As has been described extensively in Sect. 2.2, renin positive cells extend much further upstream and may even occur in interlobular arteries. Thus, a significant portion of renin-secreting cells is probably outside the action of any putative message from macula densa cells. In addition, the possibly macula densa controlled portion of epithelioid cells varies considerably between species and also within a given species depending on the state of stimulation of the RAS. For example in mice on a low sodium diet, only a negligible portion of the total population of renin secretion cells can potentially receive such a humoral signal. These and other considerations do not support the view that a macula densa mechanism plays a major role in the systemic regulation of plasma renin activity under physiological conditions. On the other hand, such a mechanism could be important in local intrarenal effects of the RAS, which will be discussed in chapter 9.

The "classical" macula densa concept has also to be revised for another reason. As described in Sect. 3.1, the area of morphological and perhaps also of functional contact between tubular and vascular structures is much larger than defined by the "juxtaglomerular" apparatus. A most striking illustration of this situation is the recent demonstration by BARAJAS et al. (1986) that between 70% and 90% of the afferent arterioles come into close proximity (i.e., less than 3 μm) of distal connecting tubules. Interestingly, it is this segment of the distal tubule that synthesizes and secretes kallikrein into both the tubular lumen and the interstitium (see BARAJAS et al. 1986). This observation is even more intriguing, as kallikrein is thought to play a role in the activation of prorenin to renin (see SEALEY 1980), and direct stimulation of renin secretion by kallikrein and kinins has been described (OKAMURA and INAGAMI 1984; BEIERWALTES et al. 1985a, b).

Finally, another aspect should be recalled which concerns the original basis of the macula densa hypothesis, i.e., the inverse relationship between long-term dietary salt intake and plasma renin, kidney renin content, or renin secretion. It should be emphasized again that there is no evidence at all which suggests the participation of the macula densa or other tubular structures in the expression of this relationship. It must be conceded, however, that alternative explanations for this relationship do not exist at the present time.

7.4 Renal Nerves and Renin Secretion

Numerous studies have demonstrated that β-adrenergic agents, such as isoproterenol, stimulate renin secretion, and that this stimulation can be abolished by β-adrenergic antagonists (see review by KEETON and CAMPBELL 1980). In fact, β-adrenergic agonists are the most reliable experimental stimuli of renin secretion in all types of preparations and in all experimental conditions examined. Obviously, this situation suggests that renal sympathetic nerve activity and circulating catecholamines may be an important factor in the control of renin secretion via activation of β-adrenoceptors.

The intrinsic innervation of the kidney, as described mainly by BARAJAS (1978, 1979), is predominantly noradrenergic and reaches all segments of the nephron and the vasculature. The area of the JGA receives a particularly rich supply of adrenergic nerve terminals (BARAJAS 1979; GORGAS 1978a, b). There is no evidence for renal cholinergic nerves (ZAMBRASKI et al. 1978). The existence of a specific dopaminergic innervation of the kidney has been suggested. Since, however, the major part of renal dopamine appears to be of nonneural origin, the possible participation of dopamine in the control of renin release will be discussed later (cf. Sect. 7.5). For a comprehensive discussion of the innervation of the JGA see Sect. 3.5.

Stimulation of renal nerve activity may have profound effects in the kidney, most prominent of which are vasoconstriction, a reduction of GFR, an increased tubular sodium reabsorption from the proximal tubule, the ascending limb of Henle's loop and the collecting ducts, and alterations of renin release (for survey and references see reviews by DIBONA 1982, 1985; JOHNS 1985). Since each of these effects can be assumed to influence renin release indirectly, it was difficult, in earlier studies, to identify and describe selectively the neural component of renin secretion. More recently, experiments with carefully graded direct renal nerve stimulation, or mild reflex renal sympathetic nerve activation produced by carotid sinus hypotension, have provided conditions which enable the neural component of renin secretion to be characterized with little or no interference from other renal effects of sympathetic stimulation. Thus, in the conscious, chronically instrumented dog, Kirchheim and colleagues (GROSS et al. 1981; KIRCHHEIM et al. 1981) stimulated renal sympathetic nerve activity (about 60% increase) by carotid occlusion. When a rise in renal perfusion pressure was prevented by an aortic cuff, renin secretion was increased without changes of RBF. This rise could be prevented by β-adrenoceptor blockade or by previous surgical denervation of the kidney. In the anesthetized, acutely instrumented dog, low frequency (1.0-Hz) electrical renal nerve stimulation was shown to increase renin release and decrease urinary sodium excretion, without affecting RBF and GFR (OSBORN et al. 1983). By using various α- and β-adrenoceptor antagonists, the authors were able to dissociate these two effects. Renin secretion was shown to be a β-adrenoceptor-mediated effect and tubular sodium reabsorption an α-adrenoceptor-mediated event. In the nonfiltering kidney preparation, HOLDAAS et al. (1981a, b) also observed selective β-adrenergic-mediated renin release upon renal nerve stimulation. By further reducing the electrical stimulation rate to 0.5 Hz, renin secretion could be stimulated without alterations of RBF, GFR, or sodium reabsorption (OSBORN et al. 1982a, 1984). Furthermore, subthreshold stimulation at 0.25 Hz which, per se, did not affect renin secretion, augmented the renin secretion response to nonneural stimuli (e.g., aortic constriction) (KOPP and DIBONA 1984; OSBORN et al. 1984). Further studies demonstrating that the stimulatory effect of renal sympathetic stimulation on renin secretion is mediated by β-adrenoceptor activation have been described by HOLDAAS et al. (1981a, b), JOHNSON et al. (1984), and JOHNS (1985).

The observation made by GROSS and colleagues (1981) that in the resting, conscious unrestrained dog on a normal sodium diet infusion of a β-adrenoceptor antagonist reduced renin secretion rate and plasma renin activity to about half the normal values suggests that renal sympathetic nerves exert a tonic stimulatory effect on renin secretion via β-adrenoceptor activation also under physiological conditions.

Concerning the classification of the β-adrenoceptor involved in renin secretion, the majority of studies in several species suggest that β_1-adrenoceptors mediate this effect (e.g., KOPP et al. 1980; KOPP and DIBONA 1983; CHURCHILL et al. 1983; JOHNSON et al. 1984; ZAMBRASKI et al. 1984; JOHNS 1985; MILAVEC-KRIZMAN et al. 1985), although participation of β_2-adrenoceptors has also been suggested (OLSON et al. 1982; NAKANE et al. 1980; see also KEETON and CAMPBELL 1980).

With respect to the mechanisms of these β-adrenoceptor-mediated effects, some controversy existed as to whether stimulation of renin release by β-adrenoceptor activation is indeed a direct effect on juxtaglomerular cells via the known adenylate cyclase cAMP pathway, or may be mediated by prostaglandins. Thus, CAMPBELL et al. (1979), FEUERSTEIN and FEUERSTEIN (1980), and SUZUKI et al. (1981), on the basis of their experiments in the intact rat or cat, advocated an essential mediator function of prostaglandins in β-adrenoceptor-mediated stimulation of renin secretion. Subsequent studies, however, have clearly demonstrated that, if prostaglandins contribute to β-adrenoceptor stimulation of renin secretion at all, this is a modulating and parallel, but not an essential, mediator function of prostaglandins (HENRICH and CAMPBELL 1984; KOPP and DIBONA 1983; OLSON et al. 1983; ZAMBRASKI et al. 1984).

The topographical distribution of both α- and β-adrenaoceptors in the JGA is discussed in Sect. 3.5. Direct functional evidence for the presence of β-adrenoceptors on granulated cells is provided by studies with isolated renin-secreting cells (KURTZ et al. 1984; KURTZ 1986). In these studies, isoproterenol stimulated renin release and increased intracellular concentrations of cAMP. Heterogeneity of renin-containing cells with respect to the response to β-adrenergic stimulation has been observed in biochemical studies by BAUMBACH and SKØTT (1986) with dissected afferent arterioles and with glomeruli to which fragments of the afferent arteriole were still attached. In these preparations, β-adrenergic stimulation of renin secretion was confined to glomeruli with long afferent arterioles and to microdissected arterioles, but was not detectable in short afferent arterioles. BAUMBACH and SKØTT concluded that only granulated cells at some distance from the glomerulus, further upstream along the afferent arteriole, respond to β-adrenoceptor activation with an increase of renin release. Again, as suggested for the response to adenosine, this difference in the response to β-adrenoceptor activation may reflect differences of the type of renin-containing cells involved, i.e., epithelioid or intermediate cells. The functional significance of this observation remains to be clarified.

In summary, although uncertainties still exist in some minor aspects, the neural control of renin secretion via activation of β-adrenoceptors located on granulated cells can be considered a definitively established mechanism.

In contrast, the role of α-adrenoceptors in the control of renin secretion is still controversial. The discussion of this problem receives continuous support from diverging observations on the effects of α-adrenergic activation on renin release, i.e., stimulation, inhibition, or no effect, and from recent electrophysiological data.

If we first consider stimulatory effects of α-adrenoceptor activation on renin release, these have mainly been observed in studies in anesthetized animals, in which the effects of electrical renal nerve stimulation or of intrarenal infusion of α-adrenergic agonists and antagonists on renin secretion have been examined. Thus, in a study by OSBORN and colleagues (1982b) in the dog kidney, mild renal nerve stimulation (0.5 Hz) increased renin secretion, an effect that was partly prevented by phenoxybenzamine, but not by prazosine or phentolamine. The authors concluded that a nonadrenergic mechanism must be involved. In some contrast to this study are the observations by BLAIR (1983), who, in the acutely denervated dog kidney, found methoxamine, an α_1-adrenoceptor agonist, to increase renin secretion, an effect abolished by prazosin but not by propranolol. Similar observations were reported by TAKAHASHI et al. (1984) and HISA et al. (1984). One possible explanation for this α-adrenergic effect, not requiring direct stimulation of epithelioid cells, is provided by the work of OLSON and colleagues

(1981), who also examined the effect of phenylephrine in the dog kidney. In these experiments the stimulatory effect was totally abolished by indomethacin, whereas the decrement in urinary sodium excretion and the vasoconstriction by phenylephrine were not affected. This interpretation of an α-adrenergic stimulation of prostaglandin release and the subsequent response of renin secretion to the increased concentrations of prostaglandins in supported by similar data obtained by HISA and SATOH (1983), who directly measured an increased release of PGE_2, in addition to the increase of renin release, upon infusion of subpressor doses of phenylephrine into the renal artery in dogs.

In some studies, however, the α-adrenergic stimulatory effect could not be abolished by inhibition of prostaglandin synthesis (HESSE and JOHNS 1985; TAKAHASHI et al. 1984; HISA et al. 1984). This was usually only observed at higher intensities of renal nerve stimulation or higher concentrations of α-adrenoceptor agonists, which profoundly affected renal hemodynamics, glomerular filtration, and sodium reabsorption. DIBONA (1985) therefore considers this type of renin response as nonneurally mediated, i.e., secondary to other α-adrenoceptor-dependent effects. This has recently been confirmed by BLAIR et al. (1986), who demonstrated that the response of renin secretion to low-frequency renal nerve stimulation, which did not change GFR or RBF, but increased sodium reabsorption, was entirely dependent on renal β-adrenoceptors, with no contribution by α-adrenoceptor activation.

The opposite observation, i.e., inhibition of renin secretion by α_1-adrenergic agents, has almost exclusively been observed in experiments with rat kidney cortex slices (CAPPONI and VALLOTON 1976; MORRIS et al. 1979; MATSUMURA et al. 1985; SASAKI et al. 1986) or the isolated perfused rat kidney (VANDONGEN et al. 1979). Most investigators using kidney slices have tacitly or explicitly assumed that indirect modulation of renin release via vascular or tubular effects can be excluded in this preparation and that any effect on renin release can therefore be considered as a direct effect on juxtaglomerular cells. However, this has not been ascertained in any of these studies, and it is quite conceivable that, e.g., the strong vasoconstriction induced by high concentrations of α_1-adrenoceptor agonists applied to these preparations can interfere with renin release, e.g., by changing the surface area of relevant cell structures or the diffusion pattern of released renin without direct effects of these agents on juxtaglomerular cells. Indirect effects of α_1-agonist via vasoconstriction have, at least partly, been excluded in experiments by VANDONGEN and colleagues (1979) in the isolated rat kidney. These authors prevented vasoconstriction by the simultaneous infusion of hydralazine and found that methoxamine as well as phenylephrine still inhibited renin release. In this case, however, a macula densa or distal tubule mediated inhibition of renin release via the α-adrenergic stimulation of tubular sodium reabsorption may be an explanation. In our studies, methoxamine failed to produce inhibition of renin release from the isolated perfused rat kidney (HACKENTHAL et al., unpublished obervations).

In summary, it is safe to say that renal nerve stimulation induces an increase fo renin secretion, in which the direct neural component is exclusively mediated by β-adrenoceptors located on epithelioid, renin-secreting cells. An α_1-adrenoceptor-dependent increase of renin secretion occurs only with higher intensity of stimulation, and is probably mediated by the hemodynamic and prostaglandin-stimulating effects of α_1-adrenoceptor activation. An α_1-adrenoceptor-mediated inhibition of renin secretion has consistently been observed in kidney slices. Its mediation remains unclear, although an indirect rather then a direct effect of high "pharmacological" concentrations appears possible. However, a strong argument in favor of the existence of inhibitory α-adrenoceptors on epithelioid cells is provided by two observations: KURTZ and coworkers (1984, 1986) described the inhibitory effect of norepinephrine on renin release from isolated juxtaglomerular cells, which was accompanied by an increased calcium influx, and BÜHRLE and colleagues (1984, 1986a), in elctrophysiological studies, observed that α_1-adrenoceptor agonists depolarized juxtaglomerular cells, a reaction known to be associated with inhibition of renin release (cf. Chap. 8).

In conclusion, the issue of α_1-adrenoceptor representation on epithelioid cells and their participation in the control of renin secretion remains unsettled.

Another aspect associated with the neuronal control of renin secretion is the possible modulation of presynaptic function at the site of the terminals. There is some indication from the work of VANDONGEN and GREENWOOD (1975) and PEDRAZA-CHAVERRI et al. (1986) that α_2-adrenoceptor agonists may inhibit renin release via presynaptic inhibition, but further experiments are necessary to define this pathway.

Perhaps more important for the understanding of a possible inhibitory neuronal pathway is the recent observation that neuropeptide Y (NPY), which is known to be costored with noradrenaline in terminals supplying epithelioid cells, is a potent inhibitor of renin secretion (HACKENTHAL et al. 1987). This aspect is discussed in more detail in Sect. 7.5.

7.5 Hormonal Signals Influencing Renin Secretion

In this section, hormonal control mechanisms of renin secretion will be discussed. The term hormone is used here in a rather broad sense, encompassing also so-called tissue hormones, such as prostaglandins and histamine, as well as potential neurotransmitters or neuromodulators, such as acetylcholine and NPY. The control of renin release by adrenaline and noradrenaline via activation of β- or α-adrenoceptors has been described in Sect. 7.4 and will not be discussed here again. This also pertains to the role of adenosine in the control of renin secretion, which is discussed in the context of the macula densa mechanism of renin secretion (Sect. 7.3).

7.5.1 Dopamine

It is still debated whether the kidney receives a significant dopaminergic innervation (cf. Sect. 3.5). Based on fluorescence-histochemical studies by BELL et al. (1978) and DINERSTEIN et al. (1979), the presence of a major dopaminergic nerve supply of the kidney is strongly advocated by LEE (1982). In these studies, dopaminergic nerve terminals were found to be preferentially located near the vascular pole of the glomerulus. Further support for the existence of renal dopaminergic nerves can be seen in the observations by MORGUNOV and BAINES (1981), STEPHENSON et al. (1982), and KOPP et al. (1983) of an increased urinary output of dopamine upon electrical stimulation of renal nerves. In contrast, ADAMS and ADAMS (1985), examining the handling of L-dopa and dopamine by the isolated kidney and reviewing the data in the literature, questioned the existence of a major dopaminergic neural pathway to the kidney. In any case, there appears to be agreement that the kidney is capable of producing dopamine in large quantities from circulating L-dopa, most of which is then excreted with the urine (MORGUNOV and BAINES 1981; LEE 1982; STEPHENSON et al. 1982; ADAMS and ADAMS 1985). The exact site of dopamine production, however, is not known.

Dopamine is thought to modulate both renal hemodynamic and tubular functions. When infused into the renal artery, it induced vasodilatation, natriuresis, and diuresis, in both rats and dogs (cf. LEE 1982; MCGRATH et al. 1985 for references). The diuretic effect is thought to be independent of hemodynamic changes (PELAYO et al. 1983; MCGRATH et al. 1985). At higher concentrations of dopamine administered to the kidney or kidney vessels, vasoconstrictor effects can also be observed. Thus STEINHAUSEN et al. (1986) examined the effect of dopamine on different vascular segments in the model of the split hydronephrotic kidney, permitting the direct visualization of individual vessel segments. In accordance with the observations and conclusions of other investigators on the vascular effects of dopamine, STEINHAUSEN et al. (1986) found that lower concentrations of dopamine (10^{-7}–10^{-5} *M*) caused selective vasodilatation of preglomerular arterioles, without significant effect on postglomerular vessels. At higher concentrations, i.e., above 10^{-5} *M*, postglomerular vasoconstriction in addition to preglomerular vasodilatation occurred.

The analysis of the mechanisms involved in the renal effects of dopamine and their biological significance is difficult, not only because different types of dopamine receptors have been identified in the kidney, such as D_1-receptors linked to activation of adenylate cyclase in arterial vessels and proximal tubules, or D_2-receptors in isolated glomeruli, but also because exogenous dopamine may stimulate receptors other than those specific for dopamine, e.g. α- and β-adrenoceptors (cf. FELDNER et al. 1984; STOOF and KEBABIAN 1984). Furthermore, there is evidence for presynaptic receptors of the D_2-type on the terminals of postganglionic sympathetic neurons, whose activation inhibits the release of norepinephrine (STOOF and KEBABIAN 1984). Recently, a significant attenuation of agiotensin II-induced contraction of isolated glomeruli and mesangial cells has been described, which is independent of prostaglandins (BARNETT et al. 1986). The authors suggest that dopamine may play a role in the regulation of glomerular filtration.

In view of the complexity of the hemodynamic effects of dopamine in the kidney, it is not surprising to note that the studies on the effect of dopamine on plasma renin activity or renin release do not give a clear picture on the role of endogenous dopamine in the control of renin secretion.

In man, infusion of dopamine in high concentrations resulted in an elevated plasma renin activity (WILCOX et al. 1974). IMBS et al. (1975) observed an increased plasma renin activity and renin secretion rate in anesthetized dogs without changes of renal blood flow and sodium excretion. This effect was attenuated by the dopamine antagonist haloperidol, but not by propranolol, suggesting a specific dopaminergic effect. Similarly, in the conscious dog, MIZOGUCHI et al. (1983) observed a stimulation of renin secretion by dopamine without changes of blood pressure. The increase of renin secretion was attenuated by sulpiride or haloperidol, but not by propranolol. In order to exclude an indirect effect of dopamine via arteriolar dilatation, the authors compared the effect of dopamine with that of other renal vasodilators, such as papaverine and acetylcholine. Since these two vasodilators did not produce increases in renin release comparable to those of dopamine despite similar effects on RBF, the authors concluded that dopamine stimulates renin release by a direct dopaminergic effect on granulated cells. In contrast, SOWERS et al. (1980, 1981) and GORDON et al. (1983) concluded from observations of an increased plasma renin activity following infusion of metoclopramide in rats that renin secretion is under inhibitory dopaminergic control. Studies in vitro more consistently demonstrated stimulation of renin release by dopamine, e.g., in rat kidney cortical slices (HENRY et al. 1977), the isolated perfused rat kidney (QUESADA et al. 1979), and isolated cortical cells in a superfusion system (WILLIAMS et al. 1983). However, also in these studies the results were conflicting with respect to the specificity of the effect of dopamine: whereas in the experiments of HENRY et al. (1977) with kidney slices and in the isolated rat kidney (QUESADA et al. 1979) the stimulatory effect of dopamine could be prevented or significantly attenuated by propranolol, but not by dopamine antagonists, WILLIAMS et al. (1983) postulated a dopamine-specific effect from the observation that renin release from isolated cells was also stimulated by dopamine agonists with little affinity to β-adrenoceptors, such as bromocryptine and α-flupenthixol, and that the effect of dopamine was partially inhibited by the dopamine antagonist sulpiride. Still another aspect was introduced by the studies of HOLDAAS et al. (1982). In the anesthetized dog, these authors found dopamine to stimulate renin release only when preglomerular vessels are dilated, such as during reduction of renal perfusion pressure, but not during preglomerular vasoconstriction.

In summary then, it appears likely that dopamine stimulates renin release by an intrarenal mechanism. However, from the available data it is not possible to delineate a clear picture of how this or these mechanism(s) operate, whether this is an effect predominantly mediated by β-adrenoceptors, by presynaptic inhibition or by a direct dopamine receptor-mediated effect on epithelioid cells, let alone the question of whether dopaminergic nerves are involved or what the physiological significance of these obervations may be.

7.5.2 Prostaglandins and Related Products

Since VANDER first demonstrated in 1968 that intrarenal infusion of prostaglandins, such as PGE_2 or PGI_2, increases renin release from the kidney, the stimulatory effect of prostaglandins on renin release has been confirmed in numerous subsequent studies performed in different preparations in vivo and in vitro. A direct effect of prostaglandins on epithelioid cells to stimulate renin release can therefore be considered established (see reviews by CAMPBELL and KEETON 1980; FRANCO-SAENZ et al. 1980; FREEMAN et al. 1982, 1984). Furthermore, infusion of arachidonic acid, the common precursor of various prostaglandins, has been shown in vivo and in vitro to stimulate renin release, an effect that could be blocked by inhibitors of cyclooxygenase, such as indomethacin, meclofenamate, or ibuprofen (for references see FREEMAN et al. 1984). This demonstrates that the enzymatic machinery for the synthesis of prostaglandins is present in the kidney in locations from where the latter can reach the granulated cells to stimulate renin release. The key enzyme cyclooxygenase has been localized by immunofluorescence and by direct biochemical estimation to endothelial cells of the renal arteries, to arterioles, glomeruli, and interstitial cells. No fluorescence was detected in epithelioid renin-producing cells (SMITH and BELL 1978; cf. also FREEMAN et al. 1984; LINAS 1984).

Since the kidney synthesizes a number of different prostaglandins both in the cortex and the medulla, the question of which of the prostaglandins is responsible for the stimulation of renin release in vivo has been raised. Several investigators have attempted to answer this question by comparing the potency of exogenously administered prostaglandins, such as PGE_2, $PGF_{2\alpha}$, PGI_2, 6-keto-PGE_1, PGD_2, 13,14-dihydro-PGE_2, and others (GERBER et al. 1979; IMANISHI et al. 1980; HACKENTHAL et al. 1980; SCHWERTSCHLAG et al. 1980, 1982; JACKSON et al. 1981; WHORTON et al. 1981; MCGIFF et al. 1982), and, depending on the laboratory and the model studied, either PGE_2, PGF_2, or 6-keto PGE_1, a more stable metabolite of PGI_2, have been favored in this role. However, a rank order of potency on a molar basis does not necessarily reflect the order of biological significance, and locally occurring concentrations of a given prostaglandin may be more important. Since no qualitative differences in the effects of various prostaglandins on renin secretion have been found, the question of potency may, at the present time, be considered academic, having little relevance for the understanding of the control mechanisms of renin release. Since PGI_2 is thought to be the major vascular prostaglandin not only in the general circulation, but also in the kidney (WEEKS and COMPTON 1979), and since epithelioid cells are integral parts of the renal vasculature, this prostaglandin is now generally considered to be the physiological agent involved (FREEMAN et al. 1984). No direct information is available on the cellular mechanism by which prostaglandins stimulate renin secretion from juxtaglomerular cells. Since the vasodilator effects of PGE_2 and GI_2 are mediated by stimulation of adenylate cyclase and an increase of intracellular cAMP, the same mechanism has been assumed to mediate the effect of prostaglandins or renin release (MCGIFF 1981). While it was stated above that no qualitative differences exist between various prostaglandins concerning renin secretion, one possible exception should be mentioned: exogenous $PGF_{2\alpha}$ has occasionally been found to suppress renin release. Furthermore, $PGF_{2\alpha}$ can be generated intrarenally from PGE_2, and in the rabbit kidney, the concentration of the enzyme responsible, PGE-9-ketoreductase, has been claimed to depend on the salt balance of the animal. These observations have been incorporated into the hypothesis that the enzymatic conversion of PGE_2 into $PGF_{2\alpha}$ by PGE-9-ketoreductase may, depending on the salt balance of the organism, switch the prostaglandin-modulated renin release from stimulation to inhibition (cf. LARSSON and WEBER 1978). However, LIFSCHÜTZ et al. (1980) could not confirm these observations in the rabbit. Also, no further data on other species in support of such a mechanism are available, and others have even found that $PGF_{2\alpha}$ stimulates renin release (HACKENTHAL et al. 1980; SCHWERTSCHLAG et al. 1980). Therefore, this hypothesis has not received further attention.

In contrast, there has been a growing interest in determining the role prostaglandins play in the various intrarenal mechanisms which control renin secretion, i.e., the pressure-dependent mechanism, renal sympathetic activity, and the macula densa mechanism. As discussed in the respective section, there are data that both support or refute such a role of prostaglandins in each of the three mechanisms. The discrepancies in results probably reflect the complexity of interactions between vascular, tubular, and neuronal mechanisms which participate in the control of renin secretion as well as vascular tone, making it almost impossible to dissect a single pathway for renin stimulation. In addition, the experimental conditions also influence the rate of prostaglandin synthesis. In particular, all types of anesthesia (PETTINGER et al. 1975; SWAIN et al. 1975; TERRAGNO et al. 1977; ZIMMERMANN 1978) induce a dramatic increase of prostaglandin synthesis and release, stimulating renin release. This latter effect is, of course, preventable by inhibition of cyclooxygenase.

Another possible source of misinterpretation are the mutual interactions of intrarenal humoral systems: not only that prostaglandins stimulate renin release, angiotensin II also induces the release of prostaglandins from glomeruli (SCHLONDORFF et al. 1980), cultured distal tubule cells (GRENIER et al. 1981), isolated perfused kidneys (ELLIS and ITSKOWITZ 1980), and renal slices (SATOH et al. 1981). In addition, prostaglandins and angiotensin II have opposing effects on renal vascular tone (cf. CURRIE and NEEDLEMAN 1984). Similarly, kinins have been demonstrated to stimulate renal prostaglandin release (cf. NASJLETTI and MALIK 1981; CURRIE and NEEDLEMAN 1984) and are also thought to stimulate the secretion of renin. There is a complex functional interaction of intrarenal prostaglandins, the kallikrein-kinin system, and the RAS to control and modulate renal functions (cf. NASJLETTI and MALIK 1981; CURRIE and NEEDLEMAN 1984; MARIN-GREZ 1982). Vasopressin induces renal prostaglandin synthesis, thereby attenuating its own intrarenal vasoconstrictor action (for references see MILLER et al. 1986), but, in contrast to kinins, appears to inhibit renin release (cf. Sect. 7.5.10). This apparently confusing situation may eventually be resolved by taking into account the topographical relationship of the sites of generation and possible sites of action of the various interacting humoral signals, since demonstration of a stimulatory or inhibitory effect of exogenously administered agents does not necessarily imply that the same agent, when generated intrarenally, reaches its presumed site of action, i.e., the granulated, renin-secreting cells of the afferent arteriole.

The possible role of prostaglandins in mediating the effects of other control mechanisms of renin secretion, such as the pressure-dependent release, the macula densa mechanism, and activation of renal nerves has been discussed in the respective sections (cf. Sects. 7.2–7.4) and will not be discussed here again. Although the experimental data do not provide a clear picture, collectively these studies indicate that each of the mechanisms can operate without participation of prostaglandins, i.e., prostaglandins are apparently not essential mediators of the renal baroreceptor, the macula densa, or the β-adrenoceptor-dependent renin secretion. On the other hand, activation of either mechanism does stimulate the renal synthesis and release of prostaglandins, which may thus contribute to the overall response to the primary signal. Such an important modulatory function of prostaglandins may also be involved in the action of other signals affecting renin secretion.

Finally, it should be pointed out that in most of the studies on the role of renal arachidonic acid metabolism in the control of renin release, only the effects of the classical prostaglandins such as PGE_2 $PGE_{2\alpha}$, PGD_2, PGI_2, and 6-keto-PGE_2 have been examined, whereas other arachidonic acid metabolites have received little attention. Renal synthesis of thromboxane A_2 (TxA_2) has been demonstrated in vivo and in vitro, e.g., in isolated glomeruli (FOLKERT and SCHLONDORFF 1977; HASSID et al. 1979). Recently, DATAR et al. (1987) have examined the effect of indomethacin and a specific inhibitor of thromboxane synthesis on the furosemide-induced renin release in the anesthetized rat. Since inhibition of thromboxane synthetase in a narrow dose range potentiated furosemide-induced renin release, it appears possible that TxA_2 is a potent inhibitor of renin secretion. This vasoconstrictor, that appears to be released preferentially under pathophysiological conditions, such as ureteral obstruction, renal ischemia, or local inflammation, may be an interesting substance for further studies on the control of renin secretion. The same holds true for arachidonic acid metabolites generated either through the lipoxygenase pathway or the reduced nicotinamide adenine dinucleotide (NADH)-dependent cytochrome P-450-monooxygenase pathway. Both enzymes have been described in renal cortex. Their main enzymatic products are various hydroxyeicosatetraenoic (HETE) and hydroperoxyeicosatetraenoic (HPETE) acids that are thought to be mediators of the inflammatory response in glomerular disease and modulators of glomerular circulation via regulation of prostacyclin production (cf. CURRIE and NEEDLEMAN 1984). To date, no information on the effects of these noncyclic arachidonic acid products on renin secretion is available.

7.5.3 Kallikrein and Kinins

In this context, the kallikrein-kinin system deserves special attention for the following reasons: (a) The kallikrein-kinin system is frequently considered a system which functionally balances the RAS, since the effectors of the two systems, angiotensin II and bradykinin, have opposing effects on vascular resistance, blood pressure, and renal functions (see PISANO et al. 1978; REGOLI and BARABE 1980; MARIN-GREZ 1982; MARGOLIUS 1984). (b) The two systems are biochemically interdependent, since the enzyme which ultimately activates the RAS and the enzyme which inactivates the kallikrein-kinin system, i.e., angiotensin I-converting enzyme and kininase II, are identical. (c) Kallikrein has been identified and characterized by SEALEY et al. (1978; 1980) DERX et al. (1979), and YOKOSAWA et al. (1979) as belonging to a group of enzymes which activate inactive renin by limited proteolysis in vitro. There are also indications for a physiological function of kallikrein as an activator of prorenin. (d) Kinins have been shown in vivo and in vitro to stimulate renin secretion either directly or indirectly (see below). (e) Surprisingly, kallikrein itself has been reported to stimulate renin secretion (see below). (f) All these different aspects of a relationship between the RAS and the kallikrein-kinin system gained an additional dimension with the observation of BARAJAS et al. (1986) that the kallikrein-synthesizing and -releasing segment of the late distal tubule (connecting tubule) comes into close anatomical contact with the afferent arteriole in 60%-90% of all nephrons examined (cf. also Sect. 3.1). Since it has been established that kallikrein is secreted from these tubular cells not only into the tubular lumen, but also into the interstitium (YAMADA and ERDÖS 1982; MARGOLIUS 1984), it is conceivable that kallikrein occurs physiologically in the immediate vicinity of epithelioid cells of the afferent arteriole, and that this may be of functional importance for tubulovascular interactions. (g) Finally, it has been claimed by MARUTA and ARAKAWA (1983) that kallikrein itself is capable of generating angiotensin II from angiotensinogen. In view of the close genetic relationship between kallikrein and the serine protease tonin, which also generates angiotensin II from angiotensinogen (cf. Sect. 4.6), such an enzymatic property of kallikrein appears possible.

It is against this complex background that studies on the influence of the kallikrein-kinin system on renin secretion have to be viewed. First, observations on a putative direct effect of kallikrein on renin secretion without participation of kinins will be discussed.

Thus, SUZUKI et al. (1980; 1981) reported that superfused rat kidney cortical slices respond to the addition of kallikrein to the medium with an increase in renin release. More recently, in a study from the same laboratory (DOI et al. 1985), the authors suggested that the apparent increase in renin secretion is due to the protection of renin from enzymatic degradation by kallikrein and other proteases during incubation, rather than to an increased release. This latter interpretation has been questioned by BEIERWALTES et al. (1985a), who also observed an increase in renin release from isolated rat glomeruli by kallikrein; however, this effect did not seem to depend on the prevention of proteolytic degradation of renin by kallikrein, since it was abolished by aprotinin, a kallikrein inhibitor. In these studies, activation of prorenin by kallikrein had not been excluded as an explanation for the increased renin activity. The latter aspect has been more specifically addressed in a study by OKAMURA and INAGAMI (1984), who examined the release of active and inactive renin from superfused hog kidney cortical slices. Both active and inactive renin were released at a higher rate when isoproterenol, PGE_2, or dibutyryl-cAMP (dbcAMP) were added to the bath. However, upon addition of kallikrein, the amount of active renin rose, whereas inactive renin decreased. As the total net amount of active plus inactive renin was higher in kallikrein-treated slices than in controls, the authors concluded that the effect cannot be explained by activation of prorenin to active renin. Nevertheless, also in this study the effect of kallikrein was abolished by aprotinin, and alternative explanations for the results may be an activation of prorenin, combined with physical protection of active renin from degradation, and/or incomplete activation in the biochemical estimation of inactive renin. Taken together, the available evidence does not permit the conclusion that kallikrein indeed has a direct stimulatory effect on epithelioid cells to release renin.

Similar uncertainties exist with respect to the enzymatic products of kallikrein, i.e., bradykinin and kallidin. Studies in the conscious rat by KOBAYASHI et al. (1983) indicate that kinins may participate in the control of renin release, since the furosemide-induced increase in plasma renin activity could be prevented by aprotinin or soybean trypsin inhibitor (SBTI). Infusion of bradykinin resulted in an increase in plasma renin activity in anesthetized dogs (BROUHARD et al. 1979; FLAMENBAUM et al. 1979; YUN et al. 1982) or anesthetized rats (SUZUKI and SATOH 1984). These effects can be regarded as the consequence of the known stimulatory effect of bradykinin on renal prostaglandin synthesis and release (e.g., OLSEN 1978; MULLANE and MONCADA 1980; OKAHARA et al. 1981), since SUZUKI and SATOH found the effect of kinins on renin activity to be abolished in indomethacin-pretreated rats. However, indomethacin failed completely to suppress the renin-stimulatory effect in the anesthetized dog (BROUHARD et al. 1979; FLAMENBAUM et al. 1979; YUN et al. 1982); this was taken to suggest a direct effect of bradykinin on renin release without mediation by prostaglandins. Studies in vitro were also inconclusive: in the already cited study of OKAMURA and INAGAMI (1984), bradykinin or kallidin had no effect on renin release from hog kidney slices, whereas BEIERWALTES et al. (1985b) were able to induce a two- to threefold increase in renin release by bradykinin in superfused rat glomeruli.

In summary, several studies suggest a role of the kallikrein-kinin system in renin secretion, and, in view of the multiple relationship between the RAS and kallikrein-kinin system, it is intriguing to speculate on the possible physiological significance of such a role. However, before embarking on the discussion of reaction sequences and regulatory schemes, it seems essential to establish and characterize the participation of kallikrein and/or kinins in the control of renin release.

7.5.4 Histamine

The effects of histamine on renal hemodynamics and renin release have been examined in several studies both in vivo and in vitro. From these studies, where histamine itself and, in the attempt to characterize the receptor type involved, histamine analogs with specificity for either H_1- or H_2-receptors and the corresponding receptor antagonists have been used, no clear picture as to the role of histamine in the kidney has emerged so far. For example, in the dog or cat kidney, histamine was found to decrease vascular tone (CAMPBELL and ITSKOVITZ 1976). This effect was blocked by diphenhydramine, but not by metiamide, indicating a histamine H_1-receptor-mediated vasodilatation, whereas in the isolated perfused rabbit kidney histamine increased vascular resistance (ERCAN and TURKER 1975; CAMPBELL and ITSKOVITZ 1976). In the latter study the vasoconstrictor effect was converted to a vasodilator effect in the presence of a histamine H_1-receptor antagonist. It was concluded that both H_1- and H_2-receptors are present in renal vasculature, the former mediating vasoconstriction, the latter vasodilatation. Others (ANGUS and KORNER 1977) observed vasodilatation by histamine in the rabbit kidney in situ, which was abolished by either H_1- or H_2-receptor antagonists. BANKS et al. (1978) examined the renovascular effects of histamine in the dog kidney and came to the conclusion that both H_1- and H_2-receptors are primarily vasoconstrictive at postglomerular vessels, while vasodilating H_2-receptors exhibit both pre- and postglomerular distribution. Histamine H_2-receptors have also been identified by binding studies in isolated rat glomeruli (TORRES et al. 1978; CHANSEL et al. 1982), and the presence of histamine has been demonstrated in the postganglionic renal nerves of the dog kidney (RYAN and BRODY 1972). Consistently, intrarenal infusion of histamine was found to increase sodium excretion (SELKURT et al. 1979; GERBER and NIES 1983; BANKS et al. 1978).

The effect of histamine on renin release was first examined by SELKURT et al. (1979) in the anesthetized dog. Intrarenal infusion of histamine produced an increase in RBF, sodium excretion, and renin release without any change in arterial pressure and GFR. The participation of the macula densa and/or the renal baroreceptor mechanism in the renin response appeared possible. The type of histamine receptor involved was not examined. Subsequently, SCHWERTSCHLAG and HACKENTHAL (1982) used the isolated rat kidney perfused at constant pressure with a synthetic medium, to analyze the effect of histamine on renin release. From their studies with several histamine H_1 or H_2 agonists and antagonists, selective stimulation of renin release by H_2-receptor activation without participation of the renal baroreceptor and probably also without participation of the macula densa could be established. These results and conclusions were subsequently confirmed by GERBER and NIES (1983) and RADKE et al. (1986) in anesthetized dogs. In both studies, however, histamine H_2-receptor activation induced a significant increase of sodium excretion, leading to the suggestion that the effect on renin secretion was secondary to activation of a tubulo-vascular interaction, such as the macula densa mechanism. It therefore remains unclear whether histamine has a direct effect on epithelioid cells and whether this renin stimulatory effect is of physiological significance. Concerning the mechanism by which histamine stimulates renin release, it also has to be considered that histamine H_2-receptor-mediated vasodilatation is thought to be mediated by the endothelium, i.e., by inducing the release of endothelial-derived relaxing factor (EDRF) (FURCHGOTT 1983; VAN DE VOORDE and LEUSEN 1983; PEACH et al. 1985). The possible role of EDRF in the mechanism of renin release will be discussed later. Since histamine has been shown to stimulate prostaglandin release from the vasculature (JUAN and SAMETZ 1980), this could also explain the stimulation of renin release. However, SCHWERTSCHLAG and HACKENTHAL (1982) found the effect of histamine H_2-agonists not to be attenuated by indomethacin, indicating that prostaglandins were not involved.

7.5.5 Platelet-Activating Factor

The phospholipid, platelet-activating factor (PAF), is chemically defined as 1-alkyl-2-acetyl-*sn*-glycero-3 choline. It is biologically generated by immunological or other stimulation of neutrophils, basophils, macrophages, and endothelial cells and is coreleased with arachidonic acid (cf. LAI et al. 1983; PAGE et al. 1984; HANAHAN 1986; SCHLONDORFF and NEUWIRTH 1986). PAF induces platelet activation and aggregation, neutrophil activation, smooth muscle contraction, and increases vascular permeability. The vascular effects of PAF are poorly understood. Despite a dramatic hypotensive reaction and a decrease of total peripheral resistance following injection of PAF into the intact animal, the majority of studies in vitro suggest a direct vasoconstrictor effect on vascular smooth muscle cells. In some studies, however, vasodilatation has been observed (cf. LAI et al. 1983; SCHLONDORFF and NEUWIRTH 1986). It has been suggested that PAF is, in principle, a vasoconstrictor with a direct action on vascular smooth muscle and that vasodilatation or general hypotension ensues from indirect effects, i.e., from the release of EDRF and prostaglandins (KAMITANI et al. 1984; D'HUMIERES et al. 1986).

Given these uncertainties about the general vascular effects of PAF, it is not surprising that divergent results are also described with respect to the renovascular actions of PAF. In anesthetized dogs intrarenal infusion of PAF produced a dose-dependent decrease in blood flow, GFR, urine flow, and sodium excretion (SCHERF et al. 1986). In the isolated perfused rat kidney, either a dose-related decrease in renovascular resistance (YOKOTA et al. 1983; SCHWERTSCHLAG et al. 1986) or a biphasic response, i.e., vasodilation at lower (below 1 μM) and vasoconstriction at higher (10 μM) concentrations, have been observed (HACKENTHAL and TAUGNER 1986; HACKENTHAL, unpublished observations).

Observations on the generation and release of PAF from isolated glomeruli, glomerular cells, mesangial cells in culture, and the isolated rat kidney (PIROTZKY et al. 1984; SCHLONDORF et al. 1985, 1986) raised the question whether PAF may affect renin secretion. In spontaneously hypertensive rats, injection of PAF lowered systolic blood pressure, increased heart rate, and induced a twofold increase in plasma renin activity (HUBBARD et al. 1983). This latter effect was interpreted to result from a baroreceptor-mediated increase in sympathetic discharge to the kidney, secondary to the induction of systemic hypotension.

In a preliminary communication by SCHWERTSCHLAG et al. (1986), no change of renin release from the isolated perfused rat kidney upon infusion of PAF was reported despite concomitant release of PGE_2. In contrast, HACKENTHAL and TAUGNER (1986) and HACKENTHAL (unpublished observations) found a dose-dependent stimulation of renin release by PAF in the isolated perfused rat kidney. At the highest concentration, a threefold increase in renin release occurred despite a 40% reduction in perfusate flow. The stimulating effect of PAF on renin secretion was not attenuated, but slightly enhanced, by indomethacin, indicating that it was not secondary to the release of prostaglandins. The simultaneous infusion of a calcium channel antagonist (nimodipine, 1 μM) did not significantly alter the renin response, but clearly reduced the vasoconstrictor response. Recently, SCHWERTSCHLAG et al. (1987) confirmed the observation made by HACKENTHAL and TAUGNER (1986) that PAF stimulates renin release from isolated rat kidney despite simultaneous vasoconstriction.

These data, as well as the results of HUBBARD et al. (1983) and SCHWERTSCHLAG et al. (1986), are at variance with the inhibition of renin release from isolated renin-producing cells, as described by PFEILSCHIFTER et al. (1985). In this latter study, inhibition of renin release was accompanied by an enhanced calcium permeability and stimulation of phospholipase C, and was blocked by verapamil. Clearly, further studies are necessary to characterize the effect of PAF on renin secretion. The local generation and release of PAF in the immediate vicinity of epithelioid cells make this substance an interesting object for further study, especially in view of the question whether PAF may participate in the local feedback control of renin secretion, as suggested by PFEILSCHIFTER et al. (1985).

7.5.6 Vasoactive Intestinal Polypeptide

This 28 amino acid peptide, which has been found in neurons of the brain and the peripheral autonomous nervous system, is another example of the coexistence and corelease of multiple neurotransmitters or modulators. Vasoactive intestinal polypeptide (VIP) is costored with acetylcholine in cholinergic nerve terminals of the gastrointestinal tract (LUNDBERG et al. 1980), and VIP found in the circulation is thought to be released from this site. VIP immunoreactivity has also been found in nerve terminals of the mammalian kidney (HÖKFELDT et al. 1978; BARAJAS et al. 1983; REINECKE and FORSSMANN 1988); however, it was located mainly along the larger vessels without clear identification near juxtaglomerular cells, and its representation was sparse (BARAJAS 1983). Other investigators have failed to detect VIP in the kidneys of several species (see BARAJAS et al. 1983; for ref. PORTER et al. 1985).

The possible existence of VIP in the kidney and its vasodilator properties have prompted several studies, in which the effect of VIP on renin secretion was examined. Thus, stimulation of renin release from the kidney by VIP has been observed in man (CALAM et al. 1983a), rabbit (CALAM et al. 1983b; DIMALINE et al. 1983), and dog (PORTER et al. 1982a, b, 1985). The mechanism by which VIP stimulates renin release remains unclear. CALAM et al. (1983b) were able to prevent the rise in plasma renin activity in conscious rabbits upon infusion of VIP by indomethacin, but not by β-blockade. They concluded that the stimulation of renin release by VIP is dependent on prostaglandin synthesis and may be the consequence of lowering the renal perfusion pressure and activation of the intrarenal baroreceptor. However, as pointed out before, the baroreceptor mechanism does not depend on prostaglandin synthesis.

PORTER and colleagues (1985) attempted to identify the source of endogenous VIP which may influence renin release. Electrical stimulation of renal nerves in the anesthetized dogs did, as was to be expected, stimulate renin release, but failed to increase the VIP concentration in the renal vein. On the other hand, systemic plasma concentrations of VIP were raised two- to threefold above basal values by injection of neostigmine, and plasma renin concentration rose in parallel. This was also seen in denervated and β-blocked kidneys. The authors concluded that intrarenally generated VIP, if at all existent, is not directly involved in the neuronal control of renin release, but rather that circulating VIP can occur in sufficiently high concentrations to affect renin release by a direct action on granulated cells.

Clearly, further studies are necessary to clarify whether (a) VIP is present in the kidney (which does not receive postganglionic cholinergic innervation to any significant degree; cf. Sect. 3.5), (b) granulated cells respond directly to VIP, and (c) systemic VIP participates in the humoral control of renin release.

7.5.7 Acetylcholine

From the few studies in which the effects of acetylcholine or other parasympathicomimetic agents on renin release have been examined, no conclusive evidence for a role of acetylcholine (or parasympathetic nerves) in the control of renin release can be derived. Thus, no significant response of renin secretion to the infusion of acetylcholine into the renal artery has been observed in conscious (ABE et al. 1973) or anesthetized dogs (BUNAG et al. 1966; TAGAWA and VANDER 1969; OSBORN et al. 1978) despite a significant increase of renal blood flow and urinary sodium excretion in these experiments. In contrast, in conscious dogs with chronic renal hypertension, AYERS et al. (1969) found a marked stimulation of renin release. In the experiments of TAGAWA and VANDER (1969) in salt-depleted dogs, an inhibition of renin release was seen, when preinfusion renovascular resistance was high. The authors pointed out that in experiments of this type an existing modulation of renin release by acetylcholine may have been obscured by the concomitant vascular and/or tubular actions of acetylcholine. Therefore, more conclusive evidence might be expected from in vitro studies, in which at least some of these variables can be excluded. However, in studies by DEVITO et al. (1970) acetylcholine failed to influence the rate of renin release from rat renal cortical slices, whereas in studies with the isolated perfused rat kidney acetylcholine clearly stimulated renin release (HACKENTHAL and SCHWERTSCHLAG, unpublished observations). As shown in Fig. 7.2, infusion of acetylcholine increased renin release, renal perfusate flow, GFR, and urinary sodium excretion. Since these kidneys were perfused at constant pressure, a participation of the intrarenal baroreceptor seems unlikely. However, indirect influences on epithelioid cells through tubular effects of acetylcholine cannot be excluded.

Concerning the possible mechanism by which acetylcholine may stimulate renin release, one must

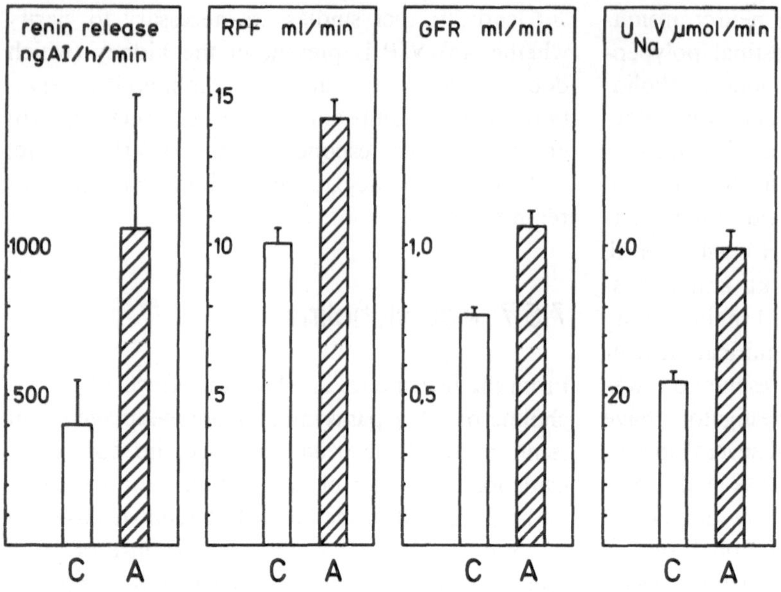

Fig. 7.2 Effects of acetylcholine in the isolated rat kidney. Rat kidneys ($n=4$) were perfused with a synthetic medium at constant pressure as described by SCHWERTSCHLAG et al. (1978). Acetylcholine *(A)* was infused for 5-min periods (10 µg/ml). *C*, control; *RPF*, renal perfusate flow; *GFR*, glomerular filtration rate; $U_{Na}V$, urinary sodium excretion (SCHWERTSCHLAG and HACKENTHAL, unpublished data)

consider, as pointed out before, that there is no evidence for postganglionic cholinergic innervation of the kidney (ZAMBRASKI et al. 1978; DIBONA 1985). Nevertheless, acetylcholine is a powerful vasodilator when infused into the renal artery (TAGAWA and VANDER 1969; ABE et al. 1973; ITSKOVITS and CAMPBELL 1976). In other vascular beds, the vasodilator effect of acetylcholine is thought to be mediated by an acetylcholine-induced release of EDRF from the vascular endothelium (see FURCHGOTT 1983; BUSSE et al. 1985; PEACH et al. 1985) because acetylcholine has no vasodilator properties in contractile vascular preparations, from which the endothelium has been removed by mechanical destruction. Although this has not been demonstrated for renal vessels, there is no reason to believe that these vessels are principally different from others with respect to acetylcholine. This would imply that EDRF mediates the observed vasodilator response to acetylcholine and may also be a stimulator of renin release from epithelioid cells. This aspect is discussed in more detail later in this chapter.

Interestingly, in the turtle *(Pseudemys scripta)*, infusion of acetylcholine produced a threefold rise of plasma renin activity (CIPOLLE et al. 1986) which could be prevented by propranolol. The authors concluded from these and further experiments that the response of renin release to acetylcholine, in this species, was secondary to the release of endogenous catecholamines. They also emphasize that these data and conclusions obtained in a primitive species cannot be transferred to mammals, where autonomic control mechanisms are organized differently. Thus, regardless of cellular mechanisms, most observed effects of acetylcholine on renin release may well represent a pharmacological effect.

7.5.8 Atrial Natriuretic Factor

DE BOLD and his colleagues reported in 1981 that rat cardiac atria produce and release a substance that lowers blood pressure and causes a dramatic diuresis and natriuresis. This initiated intensive research activities in many laboratories aimed at the biochemical identification of this factor and at the elucidation of its physiological and pathophysiological properties and functions (see recent reviews by CANTIN and GENEST 1985; MAACK 1985; NEEDLEMAN et al. 1985; ACKERMANN 1986; ATLAS and LARAGH 1986; LANG et al. 1987). Within a few years, the factor had been purified from atria of several species, its peptide nature and amino acid sequence elucidated, and the encoding gene cloned. Antibodies were raised for radioimmunological and immunohistochemical studies, receptor properties and distribution were studied, and a wealth of information on the physiological and pathophysiological properties of atrial natriuretic factor (ANF) has been accumulated. In the present context, only studies relevant for the understanding of the relationship between renal effects of ANF and renin secretion can be considered.

In cardiac atria, the main storage product of the intracellular processing of pre-pro-ANF appears to be a 126 amino acid peptide, from which the active circulating peptide, comprising the 28 carboxyterminal amino acids, is cleaved off. Several other peptides differing in length of the amino acid chain have been identified in plasma and tissues. Apparently, they do not differ significantly in their biological properties (except potency) and the term ANF will therefore be used throughout.

ANF is released from myocytes of the cardiac atria by volume and/or salt loading and by atrial distension. Upon intravenous infusion, it causes general arterial vasodilatation, which, in combination with a reduction in cardial output and the sequestration of plasma fluid into the interstitium (HUXLEY et al. 1987), reduces systolic blood pressure. Consequently, ANF is thought to play an important role in the control of blood pressure and the pathogenesis of arterial hypertension (cf. LANG et al. 1987). The hypotensive effect induces reflex sympathetic activation which may partly compensate the fall in blood pressure by increasing vascular resistance. It should be mentioned, however, that other investigators have questioned the biological significance of systemic vascular effects of ANF and its role in the control of blood pressure (e.g., HINTZE et al. 1985; LAPPE et al. 1985).

ANF causes inhibition of aldosterone release from zona glomerulosa cells of the adrenals, particularly when stimulated by angiotensin II (cf. DE LEAN et al. 1984; ELLIOT and GOODFRIEND 1986). ANF is the most potent diuretic and natriuretic agent known so far. Surprisingly, despite numerous studies on this subject, it is still debated whether this diuresis is primarily the consequence of renal hemodynamic changes or of direct tubular effects. When infused into the renal artery, ANF produces an increase in GFR with or without a concomitant increase in RBF. Some investigators feel that the increase in GFR and filtration fraction accompanied by a medullary wash out effect secondary to vasodilatation can fully account for the observed increase in sodium excretion (ATLAS et al. 1986; BOURGOIGNIE et al. 1986; CAMARGO et al. 1986; DUNN et al. 1986; SOSA et al. 1986), while others locate the main site of natriuretic action to the distal tubule (POLLOCK and ARENDSHORST 1986; SALAZAR et al. 1986; SONNENBERG 1986; SPRINGATE et al. 1987). These inconsistencies may partly be explained by different dose-regimens used, since in studies where the concentration of ANF was kept in the upper physiological or lower pathophysiological range of plasma concentrations, natriuresis and diuresis still occurred even when renal hemodynamic changes were minimal or absent (SEYMOUR et al. 1985; POLLOCK and ARENDSHORST 1986). Recently, SPRINGATE et al. (1987) examined the natriuretic effect of ANF in the chicken kidney, the unique portal circulation of which permits access to the peritubular space independent of arterial blood flow and filtration. Infusion of ANF into the portal system induced a fourfold increase in urine flow and a doubling of sodium excretion, providing clear evidence that ANF-induced diuresis and natriuresis is, at least in part, independent of changes in RBF or GFR.

However, a combined effect of ANF on both GFR and tubular structures to induce natriuresis is not only suggested by many studies in vivo, but also by the distribution of ANF-binding sites in the kidney. High-affinity binding sites have been localized in glomeruli (HEALY and FANESTIL 1986; LYNCH et al. 1986; KOSEKI et al. 1986a, b; BUTLEN et al. 1987; HAMADA et al. 1987) and the distal collecting duct (HEALY and FANESTIL 1986; KOSEKI et al. 1986a, b). This latter site of ANF-receptor localization is consistent with the observations by NONOGUCHI et al. (1987), who studied the influence of ANF on cGMP production, the presumed cellular mechanism of action of ANF in isolated tubule preparations. The authors found an increase in cGMP only in inner medullary collecting ducts, but not in other tubular segments. Similar observations have been described by TAKEDA et al. (1986) and ZEIDEL et al. (1987).

On the other hand, there are also observations which suggest a proximal site of action: BURNETT et al. (1984) found an increase of fractional lithium and phosphate clearance by ANF, indicating inhibition of proximal tubular sodium and water reabsorption, and COGAN (1985) noted a decrease in fractional proximal sodium reabsorption by ANF in anesthetized rats.

In summary, ANF seems to exert its natriuretic action partly by increasing GFR and filtration fraction, an effect observed most consistently at high, probably pathophysiological concentrations and by a direct tubular action already evident at physiological concentrations. This latter effect seems to be localized preferentially at the distal collecting duct, although other sites of action may also exist. Recent studies by HARRIS et al. (1987) may help to clarify this issue. In micropuncture studies in the rat kidney, the authors observed no direct effect of ANF on proximal tubular sodium and water reabsorption in the basal state. When, however, proximal sodium reabsorption was stimulated by angiotensin II, this increase was inhibited by peritubular ANF at physiological concentrations and abolished at higher concentrations of ANF.

The renovascular effects on ANF are still poorly understood. From studies in the anesthetized rat, DUNN et al. (1986) concluded that ANF has a selective renal vasodilator effect. Both whole kidney and single nephron GFR increased by about 20%, while RPF remained constant. The increase of glomerular capillary hydraulic pressure was ascribed by DUNN and colleagues to a combination of afferent arteriolar dilatation and efferent arteriolar constriction. These conclusions receive support from the studies of MARIN-GREZ et al. (1986) in the split hydronephrotic rat kidney. This preparation permits the direct visualization of renal vessels and measurement of their diameters. In this preparation, ANF produced preglomerular vasodilatation and postglomerular vasoconstriction. These observations in vivo are in striking contrast to results obtained in afferent and efferent arterioles microdissected from the rabbit kidney (EDWARDS and WEIDLEY 1987). In these studies, no effect of ANF on arteriolar diameter of either segment was observed, both at basal tone and when precontracted with angiotensin II or norepinephrine. Other vasodilators were fully active and the biological activity of the ANF preparation was ascertained by demonstrating an increase of cellular cGMP. In the isolated perfused rat kidney, MAACK and coworkers (1986) found ANF to slightly constrict renal vessels. However, when renovascular resistance was first increased by various vasoconstrictors, ANF was a potent vasodilator. Others, also examining the effects of ANF on the isolated perfused rat kidney, observed either no change (MURRAY et al. 1985) or a decrease in renovascular resistance (HACKENTHAL et al. 1985, 1987). In both studies, GFR and filtration fraction increased to a larger extent than perfusate flow. This could be explained by the differential effect on ANF on renal arterioles, i.e., dilatation of the afferent and constriction of the efferent arteriole, as described by DUNN et al. (1986) and MARIN-GREZ et al. (1986).

The effects of ANF on renin secretion are not clear, which is not surprising in view of the complexity of renovascular and tubular actions of ANF. From the hypotensive action of ANF, the accompanying reflex sympathetic activation, and the preglomerular vasodilatation, one would predict an indirect stimulation of renin release via activation of the renal baroreceptor and the renal sympathetic tone. However, in most studies in man or the intact animal, the systemic administration of ANF resulted in either no change or a slight decrease in plasma renin activity (MAACK et al. 1984; GARCIA et al. 1985; OBANA et al. 1985; SOSA et al. 1985; ATLAS et al. 1986; STRUTHERS et al. 1986; ANDERSON et al. 1987; FLÜCKIGER et al. 1987; HIRATA et al. 1987; MIYAMORI et al. 1987). Only in Goldblatt-hypertensive rats was an increase in plasma renin activity by ANF observed (VOLPE et al. 1985). The possibility of endogenous ANF participating in the control of renin release is suggested by the observation of NARUSE et al. (1985), that infusion of an antiserum against ANF decreases water and sodium excretion and increases plasma renin activity in anesthetized rats.

When ANF was infused into the renal artery (to minimize systemic effects) of anesthetized dogs, a reduction in renin release was observed (BURNETT et al. 1984), which, however, was no longer demonstrable when the kidney had previously been made nonfiltering by ureteral ligation and transient ischemia (OPGENORTH et al. 1986). The authors concluded from these observations that inhibition of renin release by ANF is mediated by the macula densa. This conclusion is supported by a subsequent study from the same laboratory (SALAZAR et al. 1986) which demonstrated that intrarenal infusion of ANF in concentrations that did not affect RBF and GFR nevertheless induced diuresis, natriuresis, and a depression of renin release. This interpretation would be consistent with the newly described inhibition by ANF of the sodium-retaining effect of angiotensin II in the proximal tubule (HARRIS et al. 1987). Similarly, DERAY et al. (1987) reported that in the nonfiltering canine kidney ANF did not attenuate hormone-induced renin release. VILLAREAL and coworkers (1986) also compared the influence of intrarenal infusion of ANF on the intact and the denervated, nonfiltering dog kidney. In both preparations ANF decreased renin secretion. The authors suggested that renin release is controlled by ANF through a combined action of ANF on the delivery of sodium chloride to the macula densa, on intravascular pressure, and possibly by a direct effect on epithelioid cells.

Contrary to most studies in vivo, ANF induced a concentration-dependent stimulation of renin secretion in the isolated perfused rat kidney, which was accompanied by an increase of renal perfusate flow and GFR (HACKENTHAL et al. 1985, 1987). Similar observations have been made by others (MURRAY, personal communication). The observed stimulation of renin release by ANF occurred in the concentration range from 0.3 to 30 n*M* (HACKENTHAL et al. 1985, 1987). It was not dependent on prostaglandin synthesis, presynaptic release of noradrenaline, or interference with the macula densa mechanism, since neither indomethacin, β-blockade, nor furosemide attenuated the response. Also, activation of the baroreceptor mechanism seems to be an unlikely explanation, since predilatation of the renal vasculature by infusion of hydralazine did not attenuate the renin response to ANF, whereas the vascular response was abolished at the lower and attenuated at the highest concentration of ANF. Interestingly, in the presence of nordihydroguaretic acid, an inhibitor of lipoxygenase activity and of the release of an EDRF (FURCHGOTT 1983), renin secretion was increased, and the subsequent infusion of ANF now inhibited renin secretion. It therefore appears possible that EDRF or a lipoxygenase product may be involved in the action of ANF in the kidney. The presence of receptors for ANF on vascular endothelial cells has been demonstrated, e.g., by SCHENK et al. (1985) and LEITMAN et al. (1986). At present, however, there is no ready explanation for the discrepancies between the effect of ANF on renin secretion in vivo and in the isolated perfused kidney. This is also true for results from studies on the release of renin from kidney cortex slices: OBANA et al. (1985) and HENRICH et al. (1987) observed a decrease, HIRUMA and colleagues (1986) a dose-dependent increase, and ANTONIPILLAI et al. (1986) no change in basal renin release by ANF. In the latter study, ANF augmented the angiotensin II-induced inhibition of renin release. Similarly, RODRIGUEZ-PUYOL et al. (1986) found no effect of ANF on renin release in isolated glomeruli over a wide dose range.

Finally, in studies of KURTZ with primary cultures of rat cortical cells enriched in renin-producing cells (KURTZ 1986; KURTZ et al. 1986), ANF inhibited renin release in low concentrations and increased intracellular cGMP. This observation is in accordance with the cellular mechanism of action of ANF. As established in many tissues, such as vascular smooth muscle, endothelial, tubular, glomerular, and zona glomerulosa cells, the primary intracellular event induced by ANF is the stimulation of particulate guanylate cyclase (see LANG et al. 1987 for references). However, there are also data pointing to different mechanisms of transmembrane signaling for ANF, which might be of interest for the interpretation of the modulation of renin release by ANF. Thus, inhibition of adenylate cyclase has been observed in aortic tissue, adrenal membranes, and hypophyseal membranes (ANAND-SRIVASTAVA et al. 1984, 1985, 1986). CHIU and SYBERTZ (1986) reported on the inhibition of calcium influx and intracellular calcium mobilization in cultured vascular smooth muscle cells by ANF, and TAYLOR and MEISHERI (1986) observed inhibition of norepinephrine-induced contraction and calcium influx. Furthermore, the natriuretic response to ANF has been associated with dopamine-receptor activation (MARIN-GREZ et al. 1985; WEBB et al. 1986).

In conclusion, the interference of ANF with the renin-angiotensin system is complex and not well understood. The majority of data indicate that ANF inhibits renin release from the kidney. It is not entirely clear whether this represents a direct effect on granulated cells. Other studies suggest a stimulatory function of ANF on renin secretion. Again, it is not clear whether this occurs by a direct action on renin-producing cells. In any case, it appears premature to consider the inhibition of renin secretion an essential component in the humoral regulation of extracellular fluid homoeostasis by ANF.

Finally, another aspect should briefly be considered: in many studies, a direct antagonism of angiotensin II and ANF has been described, e.g., in the control of aldosterone secretion, of vascular smooth muscle tone, and proximal tubular sodium transport. Whenever studied in detail, this appeared not only to be a functional, but rather a specific, antagonism at the level of signal transduction mechanisms (see above). The only exception to this "rule" appears to be the granulated cell, where ANF is thought to inhibit renin secretion, as does angiotensin, and where potentiation of the inhibitory effect of angiotensin II by ANF has been described (ANTONIPILLAI et al. 1986). This apparent discrepancy may be a clue to the understanding of the effect of ANF on renin-producing cells.

7.5.9 Angiotensin

In this section, the effects of angiotensin II on granulated, renin-producing cells will be discussed with respect to the basic effect on renin release, the possible role of angiotensin III, the origin of angiotensin II participating in the negative feedback control of renin release, the type of angiotensin II receptor, and the associated signal transduction mechanism. The vascular and tubular effects of angiotensin II, as well as the role of the RAS in the control of renal hemodynamics and tubular functions, will be discussed seperately in Chap. 9.

The existence of an inhibitory feedback mechanism of renin secretion by angiotensin II was first suggested by VANDER and GEELHOED (1965), who observed a decrease of renin release upon intravenous infusion of angiotensin II. The authors excluded an indirect effect of angiotensin II via activation of the intrarenal pressure receptor by demonstrating that inhibition of renin release still occurred when a rise of intrarenal perfusion pressure was prevented by a suprarenal aortic clamp. Since then, the inhibition of renin release by angiotensin II has been confirmed in numerous studies (cf. KEETON and CAMPBELL 1980) which provide abundant evidence that this is a direct effect on granulated cells. It can be observed in the denervated kidney (MCDONALD et al. 1975) and in the nonfiltering kidney (SHADE et al. 1973) and is independent of changes in renal hemodynamics (VANDER and GEELHOED 1965; BUNAG et al. 1967; BLAIR-WEST et al. 1971; MCDONALD et al. 1975). Angiotensin II inhibits renin release in the isolated perfused kidney (HOFBAUER et al. 1973; VANDONGEN et al. 1974; VANDONGEN and PEART 1974; ZSCHIEDRICH et al. 1975; HACKENTHAL et al. 1985), in kidney cortex slices (MORIMOTO et al. 1970; ROSSET and VEYRAT 1971; SARUTA and MATSUKI 1975; CAPPONI et al. 1977; NAFTILAN and OPARIL 1978; CHURCHILL 1980), and in isolated, dispersed cortical cells (MICHELAKIS 1971; DRURY et al. 1986; KURTZ et al. 1986). Although in some preparations, particularly in kidney slices, very high concentrations, i.e., in the micromolar range of angiotensin II, are necessary to observe the inhibitory effect, the response of renin secretion to angiotensin II may be extremely sensitive in other preparations. For example, renin release from the isolated perfused rat kidney responds to intrarenal infusion of angiotensin II at concentrations of 10^{-11}-10^{-10} *M*. These concentrations are in the upper physiological range in this species (OSTER et al. 1973); it therefore appears possible that renin secretion in vivo may be under tonic inhibitory control by angiotensin II. This question has been examined in several studies, in which the putative feedback control was pharmacologically interrupted by the administration of either converting enzyme inhibitors, such as SQ 20881 or captopril, or by angiotensin II receptor antagonists, such as saralasin. Thus, AYERS et al. (1974) observed a rapid increase in plasma renin activity following infusion of either saralasin or captopril in anesthetized dogs, in which the RAS had been stimulated by renal vasoconstriction. In the anesthetized rat, both the converting enzyme inhibitor SQ 20881 (teprotide) or saralasin induced an immediate and dramatic increase of plasma renin activity, whereas in the conscious rat only SQ 20881, but not saralasin, had this effect (BING 1973; BING and POULSEN 1974; OATES et al. 1974). This lack of increase in plasma renin activity upon saralasin infusion was explained by its partial agonistic activity exerted on epithelioid cells (BING and POULSEN 1974). In man, converting enzyme inhibitors also increased plasma renin activity concomitant with the decrease of angiotensin II plasma concentration and independently of hemodynamic effects of the inhibitor (KONO et al. 1981; LEBOFF et al. 1982). Similar to the cited studies in rats, the effects of saralasin on renin secretion in man are not as consistent as those of converting-enzyme inhibitor (see KEETON and CAMPBELL 1980), and stimulation of renin release by indirect vascular effects has been discussed (see HOLLENBERG and WILLIAMS 1979; KEETON and CAMPBELL 1980 for discussion). Nevertheless, the sum of all data convincingly demonstrates that endogenously formed angiotensin II is the effector of a negative feedback loop controlling renin secretion, which effectively operates in the state of a stimulated RAS and most probably also in the normal resting renin state. This control mechanism is often called the "short feedback loop," in contrast to the "long-loop" feedback control exerted through stimulation of aldosterone secretion by angiotensin II, subsequent retention of salt and water, and long-term suppression of renin secretion.

In only few studies have functional changes of the sensitivity of the short feedback loop been examined, e.g., WILLIAMS et al. (1978) described a lack of inhibition of renin secretion by graded infusions of angiotensin in patients with high or normal renin essential hypertension, whereas the same infusions caused a dose-related fall of plasma renin in normotensive subjects. LEBOFF et al. (1982) found an attenuation of the increase of plasma renin activity after a single dose of captopril also in patients with essential hypertension, as compared with normotensive patients, and suggested a decreased sensitivity of the negative feedback response in hypertensive patients. In the isolated rat kidney, the relative response of renin release to graded infusion of angiotensin II was found to be higher in kidneys taken from rats on a high-salt diet as compared with normal salt rats (see Fig. 7.3). The functional and morphological aspects of adaptation of renin synthesis and secretion to long-term changes of salt intake or other conditions will be discussed in Chap. 11.

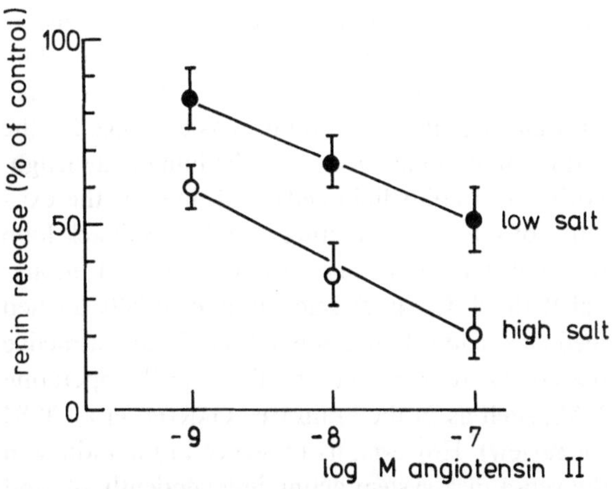

Fig. 7.3. Salt balance and angiotensin responsiveness of renin release. Rats were fed a high-salt diet (1000 mEq NaCl/kg diet) and received 150 mEq NaCl/l with drinking water *(open circles)* for 10 days. Low-salt rats were fed a diet containing approximatley 5 mEq NaCl/kg diet *(full circles)*. Kidneys were taken for perfusion as described by SCHWERTSCHLAG et al. (1978). Inhibition of renin release is expressed as percent of release in the respective control periods. (SCHWERTSCHLAG and HACKENTHAL, unpublished data)

With respect to the topographical origin of angiotensin II inhibiting renin release, several possibilities must be considered. First, angiotensin II generated in the systemic circulation pre- or intrarenally may reach angiotensin II receptors on granulated cells and, depending mainly on the plasma concentration of renin, modulate renin secretion in a short inhibitory feedback loop. This pathway must be considered as the major regulatory feedback system of renin release. Such a conclusion can be derived primarily from numerous studies (some of which are cited above) in which an increase of systemic angiotensin II concentration by the infusion of pressor or subpressor doses of angiotensin II has been shown to inhibit renin release. It can also be derived from studies where the consequences of an acute increase of endogeneously formed angiotensin II have been examined. For example, MURRAY et al. (1982) reduced renal perfusion pressure of one kidney by a clamp on the renal artery in the anesthetized dog. They observed an almost immediate cessation of renin release from the contralateral, untouched kidney. This inhibition of renin release occurred concomitantly with the rise of plasma renin activity originating from the clamped kidney, but was not related to systemic pressure changes. These observations can be interpreted to indicate that the increase of renin release from the clamped kidney raises plasma angiotensin II concentrations which reach the contralateral kidney and suppress renin release. Although systemically generated angiotensin II probably does not reach epithelioid cells directly from the vascular lumen of the afferent arteriole down to a tight barrier of endothelial tight junctions (cf. Sect. 2.8), the most likely access via peritubular capillaries and the cortical interstitium is nevertheless a fast effect; e.g., in the isolated perfused rat kidney, a significant drop of renin release upon infusion of angiotensin II can be measured within 10 s after the beginning of the infusion (own unpublished observations).

Another possible source of angiotensin II participating in the control of renin release is the interstitial space surrounding the vascular pole of the glomerulus. As pointed out in Chap. 6, the bulk of secreted renin is not released immediately into the vascular lumen, but rather into the interstitium, and reaches the systemic circulation most likely at the level of the peritubular capillaries. This implies that in the interstitial fluid in the immediate vicinity of epithelioid cells renin is present in high or even very high concentrations, as compared with concentrations in plasma. The demonstration of high concentrations of renin in renal hilar lymph, which also drains the interstitial space of the vascular pole, may serve as a confirmation of this conclusion. Since renal lymph has also been shown to contain angiotensinogen (cf. Sect. 3.4) as well as angiotensin I and II, this has been taken as evidence for the formation of angiotensin II in the immediate neighborhood of epithelioid cells, and thus for the existence of a locally operating negative feedback loop for renin release (as part of a most general intrarenal RAS, cf. Chap. 9). Such an interstitially located feedback control of renin release is an attractive possibility, reminiscent of other locally operating RAS, such as in the brain (see GANTEN et al. 1984 for review). However, its existence in the kidney in the sense of a system acting independently of, and perhaps even opposing, the systemic RAS has not yet been proven. The presence of angiotensinogen in renal lymph is no sufficient evidence for the local generation of angiotensin I and II in the interstitium close to epithelioid cells, since it is not known how and where during formation of renal lymph angiotensinogen enters this fluid. This aspect will also be discussed in more detail in Chap. 9. In conclusion, it is conceivable, but by no means certain, that angiotensin II is generated extravascularly in the interstitium surrounding epithelioid cells. However, even if such local generation exists, there is no evidence that this angiotensin II formation is controlled differently from systemic angiotensin II formation, except with respect to the time course of generation and the concentrations reached.

A third possibility for the generation of angiotensin II participating in the inhibitory control of renin release has been discussed by TAUGNER et al. (1982a, 1984d). This speculative possibility is based on the observation that, in rat kidney epithelioid cells, renin occurs together with angiotensin II in the same granules (CANTIN et al. 1984; TAUGNER et al. 1984) (cf. Sect. 4.2). It must therefore be assumed that with each discharge of renin from mature granules into the interstitium, angiotensin II is also released, which may then, by activating angiotensin II receptors on the surface of the same or neighboring cells, inhibit further release of renin. This mechanism, which has been called the "ultrashort feedback loop" in the control of renin release (TAUGNER et al. 1984) is, however, unlikely to be of major importance in the overall control of renin release, since in most other species examined angiotensin II could not be detected in epithelioid cells or their granules (HACKENTHAL et al. 1987).

With respect to the biochemical origin of angiotensin II, there is some evidence that additional angiotensin II may be generated within the kidney by a renin and/or converting-enzyme independent pathway, e.g., by serin proteases such as tonin. This aspect has been discussed in Sect. 4.6.

For some time, angiotensin III, i.e., des-asp-angiotensin II, which is generated from angiotensin II by unspecific aminopeptidases (cf. Sect. 4.5), had been considered as a separate hormonal entity with biological functions different from those of angiotensin II (see review by LEVENS et al. 1981). This was discussed particularly with respect to stimulation of aldosterone secretion, where angiotensin III has been claimed to be more potent than angiotensin II (e.g., CHIU and PEACH 1974; CAMPBELL et al. 1984). In the kidney, the potency of angiotensin III to inhibit renin release appears to be similar to that of angiotensin II (VANDONGEN et al. 1974; LOHMEIER et al. 1975), while the vascular response to angiotensin III is somewhat smaller than that to angiotensin II (LOHMEIER et al. 1975; TAUB et al. 1977). On the basis of receptor-antagonist studies, TAUB et al. (1977) came to the conclusion that both peptides act on a single renal vascular receptor. In conclusion, there is no convincing evidence at present for a specific role of angiotensin III in the control of renal functions including renin secretion.

The mechanism of action of angiotensin II on epithelioid cells has attracted considerable attention. VANDONGEN and PEART (1974) were the first to demonstrate, in the isolated rat kidney, that inhibition of renin release by angiotensin II is a calcium-dependent phenomenon. Since then, many investigators have addressed the question of the cellular mechanisms of action of angiotensin II on epithelioid cells. Since isolated epithelioid cells were not available for such studies, most investigators used an indirect approach to the problem. Nevertheless, an increasing body of evidence has now been accumulated, which suggests that indeed angiotensin II acts on epithelioid cells by either facilitating the entry of calcium into the cell and/or by mobilization of intracellular calcium. As the intracellular events during stimulation or inhibition of renin release will be discussed in more detail in Sect. 7.6, only data pertinent to the action of angiotensin II will be discussed here.

The general conclusion that the inhibitory effect of angiotensin II on renin secretion rests on its interaction with intracellular calcium is based on the following evidence: The inhibition is attenuated or abolished, when calcium is omitted from the perfusion medium of the isolated rat kidney (VANDONGEN and PEART 1974; HACKENTHAL et al. 1983; SCHWERTSCHLAG and HACKENTHAL 1983), or from the incubation medium of kidney cortex slices (CHURCHILL 1980; ANTONIPILLAI and HORTON 1985; MAY and PEART 1986). The inhibitory effect can also be attenuated by calcium-entry blockers such as verapamil in kidney slice experiments (PARK et al. 1981; NAFTILAN and OPARIL 1982), although this effect was not consistently observed by others with either verapamil (CHURCHILL 1980) or diltiazem (CHURCHILL et al. 1981). Furthermore, in the presence of inhibitors of calmodulin, the ubiquitous intracellular calcium-binding protein mediating many intracellular actions of calcium, the inhibitory effect of angiotensin II was also found to be attenuated, e.g., with trifluoperazine (TFP) in kidney slices (FRAY et al. 1983), or the isolated perfused rat kidney (SCHWERTSCHLAG and HACKENTHAL 1983), as well as with other calmodulin inhibitors, such as calmidazolium or the sulfonamide derivative W-7 in the isolated kidney (HACKENTHAL et al. 1983). Finally, angiotensin II has been shown to stimulate calcium uptake and release in isolated epithelioid cells, concomitant with inhibition of renin secretion (KURTZ et al. 1984), and, as discussed in more detail in Chap. 8, the effects of angiotensin II on the membrane potential of epithelioid cells are compatible with an increased influx of calcium (BÜHRLE et al. 1984, 1986a).

In conclusion, there is sufficient evidence for a role of calcium in mediating the inhibitory effect of angiotensin II on renin release. However, in many cell preparations, angiotensin II receptors have been shown to be linked to adenylate cyclase inhibition, e.g., in hepatocytes (JARD et al. 1981), vascular smooth muscle cells (ANAND-SRIVASTAVA 1983), or kidney cortex membrane preparations (WOODCOCK and JOHNSTON 1982). As an increase of intracellular cAMP has been associated with stimulated renin secretion, inhibition of adenylate cyclase may be expected to result in inhibition of renin release. This possibility has been tested by HACKENTHAL et al. (1985) by using the isolated kidney taken from pertussis toxin pretreated rats. Pertussis toxin inactivates the inhibitory G_i-protein coupling the receptor to adenylate cyclase, and this permits the identification of such a pathway (KATADA and UI 1983). In these experiments, the inhibiting effect of angiotensin II on renin release was significantly attenuated, suggesting that angiotensin II may inhibit renin release not only through a calcium-dependent mechanism, but also, perhaps to a small extent, through inhibition of adenylate cyclase.

7.5.10 Antidiuretic Hormone

Since the early work of BUNAG et al. (1967), VANDER (1968), and TAGAWA et al. (1971), in which the inhibitory effect of antidiuretic hormone (ADH) on plasma renin activity was established in the anesthetized or conscious animal, several studies have addressed the questions of whether this effect (a) is of physiological or pathophysiological significance, (b) is related to the vasoconstrictor or the antidiuretic properties of ADH, and (c) is a direct effect of ADH on epithelioid cells of the kidney. A critical evaluation of these studies reveals that none of these questions can be answered satisfactorily at present.

In most studies in the anesthetized animal, high doses of ADH (or arginine-vasopressin, AVP) had to be infused in order to observe a decrease in plasma renin activity (for references see KEETON and CAMPBELL 1980). In the conscious dog, however, infusion rates of AVP required for inhibition of renin release increased plasma concentration to a level which was in the upper pathophysiological range (e.g., MALAYAN et al. 1980; GREGORY and REID 1984; SCHWARZ and REID 1986). In cardiac patients receiving an infusion of AVP, HESSE and NIELSEN (1977) observed a 40% fall in plasma renin activity without any changes of blood pressure or colloid oncotic pressure, while others were unable to detect changes in plasma renin activity in man following infusion of AVP (e.g., GOODWIN et al. 1970; KHOKHAR et al. 1976). In all of these studies, indirect effects of AVP on renin release through changes in blood pressure, sympathetic tone, or extracellular fluid volume could not be excluded. More conclusive evidence for a physiological or pathophysiological role of AVP in the control of renin secretion can be expected from studies using AVP antagonists or antibodies under suitable conditions.

Differentiation between the vasoconstrictor and the antidiuretic activity of ADH in its effect on renin secretion has been attempted in studies with selective vasoconstrictor or antidiuretic analogs of AVP. In the isolated rat kidney, for example, AVP and D-amino-D-arginine-vasopressin (DDAVP), an analog almost devoid of vasoconstrictor activity, both produced inhibition of renin release at high concentrations, despite profound differences on renal hemodynamics (KONRADS et al. 1978). These authors concluded that the inhibitory effect of AVP on renin secretion is *not* related to its vasoconstrictor action. Similar observations were made in the intact rat by KNEPEL et al. (1982). MALAYAN and REID (1982) came to the same conclusion from studies with another nonpressor analog of AVP (deamino-threonine-D-arginine-vasopressin) in the anesthetized dog. In a later study from the same laboratory in conscious dogs, the authors came to the opposite conclusion, i.e., that acute inhibition of renin secretion by vasopressin is due to its vasoconstrictor rather than to its antidiuretic activity (SCHWARTZ and REID 1986). Similar inconsistencies exist with respect to the site and mechanism of action of ADH on renin secretion. In the conscious dog, the suppression of renin secretion by ADH could be prevented by renal denervation (GREGORY and REID 1984), indicating that this inhibition is a reflex response to general vasoconstriction. In the isolated perfused rat kidney, infusion of high concentration of AVP did not alter basal renin secretion, but decreased isoproterenol-stimulated renin secretion (VANDONGEN 1975). However, this effect on renin secretion may have been secondary to the drastic increase in renal perfusion pressure by AVP in these experiments. KONRADS et al. (1978), perfusing rat kidneys at constant perfusion pressure, also found no effect on basal renin secretion, but inhibition of isoproterenol-stimulated renin release both by AVP and DDAVV at high concentrations, i.e., 0.4–2 ng AVP/ml. In these experiments, inhibition never exceeded 30%. In some studies with rat kidney cortex slices, AVP in very high concentrations did not affect renin release (DE VITO et al. 1970; KNEPEL et al. 1982). In other studies spontaneous release of renin was decreased, an effect prevented by verapamil (PARK et al. 1981) but not by diltiazem (CHURCHILL et al. 1981). Finally, inhibition of renin release was observed in isolated epithelioid cells (KURTZ et al. 1986) by 1 μM AVP, a concentration exceeding normal plasma concentration in the rat by a factor of 100000. With respect to the cellular mechanism, by which AVP might inhibit renin release, only few data are available. In the cited studies of PARK et al. (1981), the attenuation of the inhibitory effect by verapamil points to a calcium-dependent action. This would be in accordance with the observation of KURTZ et al. (1986) in isolated cells, showing that AVP increased calcium entry, and with the electrophysiological studies of BÜHRLE et al. (1986a), also suggesting a calcium-mediated effect on epithelioid cells. However, as pointed out already, very high concentrations of AVP were necessary in these studies to demonstrate such effects. In summary then, the inhibitory effect of vasopressin on renin release is clearly established; its physiological significance, site, and mechanism of action, however, remain to be clarified.

7.5.11 Neuropeptide Y

The 36 amino acid neuropeptide Y (NPY) was originally identified as a neuromodulator in the central nervous system (TAKEMOTO 1982). Subsequently, it was also found peripherally in sympathetic varicosities surrounding the arterial vessels, where it is stored and released along with noradrenaline (LUNDBERG et al. 1982; EKBLAD et al. 1984). Most interestingly, immunohistochemical staining with NPY antibodies has been observed in adrenergic nerve terminals supplying the juxtaglomerular region (BALLESTA et al. 1984; TAUGNER et al., unpublished observations; REINECKE and FORSSMANN 1988; cf. Sect. 3.5). Its presence in the kidney has also been demonstrated by biochemical techniques (RAINE et al. 1984).

In many vascular beds, NPY potentiates the vasoconstrictor activity of noradrenaline and various α_1-adrenergic agents (LUNDBERG et al. 1986; EKBLAD et al. 1984; PERNOW and LUNDBERG 1986). However, in the heart (RUDEHILL et al. 1986) and in the kidney (RAINE et al. 1984; ALLEN et al. 1985), NPY is a vasoconstrictor on its own. Studies in the isolated rat kidney (HACKENTHAL and TAUGNER 1986; HACKENTHAL et al. 1987) revealed that NPY is also a potent inhibitor of renin secretion. This effect is independent of the vasoconstriction elicited by NPY and is probably mediated by inhibition of adenylate cyclase activity, as concluded from experiments with pertussis toxin (HACKENTHAL et al. 1987). This interpretation is in agreement with the observation that NPY lowers intracellular cAMP in vascular tissue via inhibition of adenylate cyclase (FREDHOLM et al. 1985). In addition, recent electrophysiological experiments by BÜHRLE et al. (1986a) demonstrated that NPY, although inhibiting renin release, did not alter the membrane potential of epithelioid cells, whereas other inhibitory signals for renin release, which act through facilitation of calcium entry, consistently depolarized these cells. In view of the immunocytochemical staining for NPY of nerve terminals in the immediate neigborhood of juxtaglomerular cells (cf. Sect. 3.5) and the partial depletion of kidney NPY stores by chemical sympathectomy (BALLESTA et al. 1984), it is conceivable that NPY is coreleased together with the adrenergic transmitter and may play a role in the control of renin release as a modulator of the stimulatory effect of noradrenaline. This peptide and its role in renin secretion clearly deserves more attention in future work on the control of renin secretion.

7.5.12 Other Hormonal Agents

Several other hormones or hormone-like substances that have vascular and/or renal effects have also been studied in regard to their interference with renin secretion. These include *parathyroid hormone* (PTH), *substance P, glucagon,* and *opioid peptides.*

In several studies in man and dogs, intravenous or intraarterial infusion of *parathyroid hormone* was followed by a rise in plasma renin activity (MCCREDIE et al. 1975; LINDNER et al. 1978; POWELL et al. 1978; SMITH et al. 1979). These observations have led SMITH and colleagues (1983) to examine the effects of intrarenal infusion of PTH in more detail in the anesthetized dog. They found that renin secretion rate in the infused kidney was not different from control kidneys nor from the contralateral noninfused kidney, despite higher excretion rates of both phosphate and sodium in the infused kidney. They concluded that PTH has no direct effect on renin secretion. In contrast, in the isolated perfused rat kidney, infusion of high concentrations of PTH induced a two- to threefold increase in renin release without altering renovascular resistance (SCHWERTSCHLAG and HACKENTHAL, unpublished observations). Thus, the role of PTH in the control of renin release remains to be clarified.

Substance P has also been localized in renal nerves by REINECKE and FORSSMANN (1988), and inhibition of renin release from denervated dog kidneys by intraarterially infused substance P was reported by GULLNER and colleagues (GULLNER and BARTTER 1979; GULLNER et al. 1979). Since alterations of hemodynamic or renal excretory parameters could not account for the observed changes in renin release, the authors concluded that substance P participates in the control of renin secretion by direct action on granulated cells. In the isolated perfused rat kidney, even high concentrations of substance P had no influence on renin release (HACKENTHAL, unpublished observations). No further studies on this subject are available.

Glucagon, the effect of which is mediated in various tissues by stimulation of adenylate cyclase and an increase of intracellular cAMP, has also been studied with respect to renin secretion. In man, intravenous infusion was found to increase plasma renin activity (FERNANDEZ-CRUZ et al. 1975). In the anesthetized dog, intrarenal infusion of high doses of glucagon increased heart rate, RBF, GFR, urine flow rate, and renin release (UEDA et al. 1978). The stimulation of renin release, but none of the other effects of glucagon, were augmented by the simultaneous infusion of theophylline (to inhibit phosphodiesterase activity, therety potentiating cAMP-mediated effects). The authors concluded from these data that glucagon directly stimulates renin release from epithelioid cells by activation of adenylate cyclase. In the isolated perfused rat kidney, also very high concentrations of glucagon were necessary to stimulate renin release (VANDONGEN et al. 1973), and an indirect mechanism by means of vascular effects could not be excluded. More recently, MIYAWAKI et al. (1985) reported on the stimulation of renin release from rat kidney slices by glucagon, which was not observed in slices taken from vitamin E-deficient rats. The physiological significance of these observations indicating a relationship between glucagon and renin release remains to be established.

This is also true for experiments on the participation of *opioid peptides* in the control of renin release. RABINOWE et al. (1983) infused β-endorphin into normal human subjects and found an increase in plasma renin activity. In unilaterally nephrectomized dogs, SZILAGYI et al. (1986) constricted the renal artery of the remaining kidney and found that the rise in blood pressure and plasma renin activity could be prevented by naloxone. In the chronic one-kidney Goldblatt hypertension model, continuous administration of naloxone prevented the development of renovascular hypertension and diminished the increase in plasma renin activity (SZILAGYI et al. 1986). The authors concluded that endogenous opioid peptides may participate in the control of renin release and in the pathogenesis of renovascular hypertension. These observations and conclusions, which would predict a stimulatory function of opioid peptides on renin secretion, are difficult to reconcile with those of KOYAMA and HOSOMI (1986), who, in the anesthetized dog, found that the increase of renin release and renovascular resistance induced by high-frequency renal nerve stimulation was inhibited by infusion of leu-enkephalin and augmented by naloxone. In view of the renal effects of opioid peptides (for reference see RIBSTEIN et al. 1983) and their participation in blood pressure control (SCHAZ et al. 1980; KOYAMA et al. 1983), the relationship between the endogenous opioid peptide system and the control mechanisms for renin secretion certainly deserves further attention.

7.5.13 Synopsis of Hormonal Signals

In the preceding section hormonal signals have been discussed individually but not compared in their potency or biological significance. Therefore, a brief comparative evaluation now seems appropriate.

Among the inhibitory signals, angiotensin II is by far the best established, most potent, and most specific agent. Its role in a very sensitive, continuously operating inhibitory feedback system has been demonstrated in numerous studies. The integration of this feedback inhibition in the function of the intrarenal RAS will be discussed in a subsequent section.

Vasopressin, often cited together with angiotensin II as an inhibitor of renin release, is clearly less potent, and a physiological or pathophysiological role in the control of renin secretion remains to be established. NPY has only recently been identified as a potent inhibitor of renin release; it is therefore difficult to evaluate its physiological significance in the control of renin secretion. In view of its localization in sympathetic nerve terminals in the vicinity of juxtaglomerular cells, NPY may eventually turn out to be an important modulator of the response of renin release to changes in renal sympathetic tone. As discussed extensively, the role of ANF in the control of renin secretion remains obscure, although plasma concentrations of ANF occurring under pathophysiological conditions have been demonstrated to influence renin secretion. The possible significance of ANF and its interplay with other mechanisms in the control of renin secretion will become more clear once the complex effects of ANF on renovascular and tubular functions are better known.

Surprisingly, there appears to be no humoral agent that stimulates renin release with a specificity comparable to that by which angiotensin II inhibits renin release (except noradrenaline and possibly dopamine, released by renal sympathetic nerve stimulation). Kinins, VIP, histamine, and acetylcholine, when administered exogenously, have all been shown to stimulate renin secretion. However, there is no evidence suggesting that this may be of physiological importance. In contrast, prostaglandins formed locally in the kidney have clearly been shown to participate in the control of renin secretion even under physiological conditions. However, their action seems to be restricted to a modulator function of other signals. Another more general aspect should briefly be mentioned: several of the humoral signals which influence renin secretion are known to exert their vascular effects at least partly via the endothelium by modulating the release of an EDRF. This is true for acetylcholine, histamine, and possibly kinins. For other humoral agents such as ANF and adenosine, the presence of membrane-bound receptors on endothelial cells has been demonstrated. It may therefore be of interest to examine the possible interaction of renin-secreting cells with the endothelium not only from the morphological point of view, as described in Chap. 2, but also with respect to their biochemical communication. Another aspect that has received little attention so far, is the possibility that renin-producing cells further upstream in the afferent arteriole respond differently from juxtaglomerular cells to various hormones.

7.6 Intracellular Control of Renin Release

7.6.1 Introductory Remarks

The ultimate biochemical evaluation of intracellular events associated with renin secretion obviously depends largely on the availability of pure populations of renin-producing cells. However, these cells comprise only about 0.01% of the total kidney cell population, and most attempts to isolate and culture these cells in sufficient quantity, which would permit the measurement of intracellular constituents, have not been successful - except for a few encouraging studies, particularly those of KURTZ, BAUER and PFEILSCHIFTER (KURTZ 1986; KURTZ et al. 1984, 1986a, b, c), who were able to obtain primary cultures of rat renal cortical cells rich in granulated cells. Others have used cell cultures derived from human granulated cells transfected with simian virus (PINET et al. 1985, 1986), from a human juxtaglomerular tumor (GALEN et al. 1984) or cultures of renal cortical cells of unclear identity and homogeneity (RIGHTSEL et al. 1982).

Besides the difficulties encountered in isolating epithelioid cells out of the mass of contaminating cells, their maintenance in culture beyond the stage of the primary culture has not yet been possible (except in transfected tumor cells). This may relate to the fact that mitotic events are very rare in epithelioid cells. It is also conceivable that epithelioid cells in culture dedifferentiate once they enter the mitotic cycle, which may be accompanied by loss of renin synthesis and thus detectability. The observation of PINET et al. (1985, 1986) that human juxtaglomerular tumor cells have lost the capability of processing prorenin to mature renin and of storing and condensing renin in secretory granules points in this direction.

In addition, even in the best preparations of isolated epithelioid cells described so far, the cells loose 95%-99% of their renin content during preparation and primary culture, which limits their usefulness as models for renin secretion (cf. also Sect. 7.1). Thus, although studies in cell preparations have contributed valuable and interesting data, the bulk of information on the role of intracellular messengers in the control of renin secretion has come from studies with kidney slices, isolated glomeruli, isolated kidneys, or in vivo studies, in which hormonal or pharmacological agents have been used to manipulate intracellular concentrations of putative messengers or other cell constituents, and, as the only detectable response of the cell, renin release was measured. Despite this indirect approach, these studies taken collectively have provided important preliminary insight into the early intracellular events during inhibition or stimulation of renin secretion. It appears clear now that cAMP is the messenger in a stimulatory pathway, and that intracellular calcium, in most but not all circumstances, is inversely related to renin secretion. Furthermore, a role for cGMP has been postulated, and the activation of protein kinase C may be an important intermediate step in signal transduction leading to inhibition of renin release. In the following, the evidence for these conclusions will be discussed on the background of knowledge of intracellular events during secretory processes in general, and by taking into account the close relationship between renin-secreting cells and vascular smooth muscle cells. This latter relationship is not only based on the vascular smooth muscle cell origin of epithelioid cells and their metaplastic interconversion (cf. Chap. 10), but also on the general experience that agents and maneuvers which stimulate renin secretion produce vasodilatation in the kidney.

7.6.2 Role of Adenosine 3′,5′-Cyclic Monophosphate

As described in section 7.4, stimulation of renin secretion by mild renal nerve stimulation has been identified as a *β-adrenoceptor*-mediated effect. Furthermore, numerous observations on the stimulation of renin release in various preparations by β-adrenergic drugs and its inhibition by β-adrenoceptor antagonists have left little doubt that epithelioid, renin-secreting cells possess β-adrenoceptors that are linked to a stimulatory pathway for renin secretion. With respect to the intracellular messenger involved in this stimulatory pathway, it should be recalled that in all tissues examined the activation of β-adrenoceptors invariably stimulates adenylate cyclase activity, which results in a rise in intracellular cAMP. In fact, the β-adrenoceptor adenylate cyclase complex in which the guanylnucleotide-binding G_s-protein complex serves as a coupling factor, is the best established signal transduction mechanism (see GILMAN 1985 for review). It is therefore highly unlikely that in epithelioid cells a different coupling mechanism exists, even more so, as there is no experimental evidence for a different signal transduction pathway of β-receptor activation. Furthermore, in isolated epithelioid cells, an increase in intracellular cAMP by β-adrenergic agents has been described (KURTZ et al. 1984; KURTZ 1986), providing direct evidence for the presence of β-receptors and their coupling to adenylate cyclase in renin-secreting cells. Additional evidence for a stimulatory intracellular pathway involving cAMP has come from studies in which intracellular cAMP was manipulated by receptor-independent mechanisms, e.g., by phosphodiesterase inhibitors, exogenous cAMP, dibutyryl-cAMP (db-cAMP), or forskolin.

Studies with various *inhibitors of phosphodiesterases,* such as theophylline or isobutylmethyl xanthine (IBMX), produced somewhat inconsistent results. When infused into the renal artery in anesthesized animals, either theophylline or IBMX produced no effect (JOHNS and SINGER 1973), or the expected increase in renin release was small and accompanied by profound changes in renal hemodynamics, which may have affected renin release indirectly (REID et al. 1972; NOLLY et al. 1974; SARUTA and MATSUKA 1975; UEDA et al. 1978).

In the isolated perfused rat kidney, theophylline or IBMX stimulated renin release (PEART et al. 1975; HOFBAUER et al. 1982; SCHWERTSCHLAG and HACKENTHAL 1982), whereas the more specific phosphodiesterase inhibitor Ro-20-1724 failed to alter renin release in this preparation (SCHWERTSCHLAG and HACKENTHAL 1982). In kidney slices, CHURCHILL and CHURCHILL (1982) also observed stimulation of renin release by IBMX. The interpretation of such studies is hampered by the later recognition of theophylline and of IBMX as an adenosine-receptor antagonist and by their lack of specificity with regard to cAMP or cGMP phosphodiesterase inhibition (WELLS and KRAMER 1981).

Another means of increasing intracellular cAMP is the exogenous application of either cAMP or dbcAMP. While cAMP does not easily enter cells due to low lipid solubility, dbcAMP has been shown to permeate into cells much more readily and, following conversion to monobutyryl-cAMP, to mimic intracellular cAMP (KAUKEL and HILZ 1972). It is therefore not surprising that in some studies the renal arterial infusion of cAMP failed to stimulate renin release (HOFBAUER et al. 1974; PEART et al. 1975; OKAHARA et al. 1977) while stimulating renin release in other studies (MICHELAKIS et al. 1968; HAUGER-KLEVENE 1970; WINER et al. 1970; GAAL and FORGACS 1975, 1977; GAAL et al. 1976). More consistent stimulation of renin release was observed when dbcAMP was infused into the renal artery in vivo (OKAHARA et al. 1977; CAMPBELL et al. 1979; LANGARD et al. 1981; BONDAR et al. 1984; DERAI et al. 1987), or the isolated perfused rat kidney (HOFBAUER et al. 1974, 1982), and applied to kidney cortex slices in vitro (YAMAMOTO et al. 1973; CHURCHILL and CHURCHILL 1982; SATOH et al. 1982).

As a more reliable and effective tool to increase intracellular concentrations of cAMP, the diterpene *forskolin* has been used in some studies. Forskolin is a fairly lipid soluble agent that directly activates adenylate cyclase by a receptor-independent mechanism (SEAMON and DALY 1981). Forskolin, first used in studies on renin secretion by SCHWERTSCHLAG and HACKENTHAL (1982), was found to be a potent stimulator of renin release from the isolated rat kidney. Subsequent studies with the isolated rat kidney (FRAY and PARK 1986; PARK and FRAY 1986) and with rat kidney slices (HENRICH and CAMPBELL 1986) confirmed these observations. Furthermore, KURTZ et al. (1984) not only observed a stimulation of renin release, but also the expected increase in cellular cAMP in renin-secreting cells in primary culture.

Collectively these results obtained with various agents that increase intracellular cAMP provide compelling evidence that an increase in cAMP is followed by an increase in renin release. However, this association is not obligatory, as alterations of renin secretion can be induced which are not accompanied by corresponding changes in intracellular cAMP (e.g., KURTZ et al. 1984). This aspect as well as the question of how an increase in intracellular cAMP is translated into an increase in renin release will be discussed in the context of the interaction of intracellular messengers (cf. Sect. 7.6.5).

Here, two other aspects related to cAMP and renin secretion should be briefly mentioned, that is, a possible role of cAMP in the control of renin synthesis and the putative function of prostaglandins as intracellular messengers in epithelioid cells.

In the elegant studies of KURTZ and colleagues

(KURTZ et al. 1984, 1986; KURTZ 1986a) primary cultures of isolated epithelioid cells of the rat kidney were used to examine intracellular events during renin secretion. Considering the fact that these cells have lost most of their natural renin content during preparation and primary culture, one might ask whether the observed changes in renin release in response to various experimental maneuvers do not only represent renin secretion in the conventional view but perhaps also reflect changes in the rate of renin synthesis. This question receives some justification from several observations discussed in Chap. 5, that in the intact kidney in vivo only a portion of total renin is thought to be available for immediate secretion, a portion which may no longer be present in these cells. If correct, this speculative view would imply that renin secretion in this situation may, at least partly, represent constitutive, i.e., nonregulated secretion, and hence renin synthesis. This could further imply that the observed changes in intracellular messengers would, in addition to controlling secretion, also control synthesis of renin.

Although the mechanisms participating in the control of renin synthesis will be discussed in more detail later (Sect. 7.8), one pertinent observation should be mentioned here. PINET and colleagues (1986) examined the conditions for renin secretion from human juxtaglomerular tumor cells transfected with SV 40 or SV 40 mutants. These cells are no longer capable of activating prorenin and storing active renin. Hence, their release of renin has been considered as constitutive secretion. In these cells, dbcAMP, forskolin, isoproterenol, and histamine - all known to stimulate renin secretion in normal epithelioid cells by an increase in intracellular cAMP - induced the release of prorenin. It is thus conceivable, although far from being established, that this effect represents stimulation of de novo renin synthesis. If true, this model would be extremely useful and convenient in studying the largely unknown mechanisms by which renin synthesis is regulated. Another possibility would be that cAMP stimulates exocytosis of juvenile granules containing prorenin. It should be noted that HSUEH (1984) concluded from his data on the effect of β-adrenergic agents on the release of active and inactive renin that cAMP may have a role in the activation process or in the control of renin synthesis.

Prostaglandins have for some time been taken into consideration as intracellular second messengers in renin-secreting cells. This notion originated from the observations of CAMPBELL and colleagues (1979) that the renal arterial infusion of dbcAMP in dogs stimulated renin release, which could be blocked by indomethacin. Similar observations were reported by FEUERSTEIN and FEUERSTEIN (1980). These and additional data led HENRICH (1981) to propose that prostaglandins mediate or act as stimulatory second (or third) messengers in renin secretion at a site distal to the β-agonist-induced increase in cAMP.

However, several subsequent studies failed to confirm the suppression of β-agonist-induced renin release by indomethacin (BERL et al. 1979; JACKSON et al. 1982; LINAS 1984; OSBORN et al. 1984; VIKSE et al. 1984, 1985). Also HENRICH and CAMPBELL, in a later study (1983), found β-agonist-stimulated renin secretion to be independent of the presence of prostaglandin synthesis inhibitors such as indomethacin. Furthermore, when intracellular cAMP was raised by other means than by β-agonists, e.g., by dbcAMP or glucagon, indomethacin also failed to block the associated renin release (VIKSE et al. 1984, 1985; BONDAR et al. 1984). Finally, SATOH et al. (1984) demonstrated that a concentration of indomethacin that completely inhibited renal prostaglandin synthesis was without effect on cAMP-mediated renin release. In summary, there is sufficient evidence to exclude a second messenger role of prostaglandins in renin secretion. Rather, a first messenger function similar to that in other cell types has to be assumed, as described in Sect. 7.5.

7.6.3 Role of Guanosine 3′,5′-Cyclic Monophosphate

Only a few studies have addressed the possible function of cGMP in renin secretion. Such a role appears possible on theoretical grounds since cGMP is considered as a second messenger mediating relaxation of vascular smooth muscle cells (MURAD 1986; RAPOPORT 1986; WALDMAN and MURAD 1987), and, as pointed out before, vasorelaxation is usually associated with stimulation of renin secretion. The stimulation of renin secretion in the isolated rat kidney by atrial natriuretic peptide (HACKENTHAL et al. 1987), by diazoxide (VANDONGEN and GREENWOOD 1975), or hydralazine (SINAIKO 1983; HACKENTHAL 1988), and by acetylcholine in the intact animal (TAGOWA and VANDER 1969) and in the isolated rat kidney (HACKENTHAL et al.; cf. Sect. 7.5) is in accordance with this view, because all these agents are thought to increase intracellular cGMP either directly or via release of EDRF in target organs such as vascular smooth muscle cells (for review see MURAD 1986). However, the infusion of cGMP or dbcGMP into the isolated rat kidney failed to influence renin release (PEART et al. 1975; HOFBAUER et al. 1976). Furthermore, CRAVEN and DERUBERTIS (1985), who compared renin release and cGMP content of isolated glomeruli under various conditions, found no indication for a significant role of cGMP in renin secretion. However, KURTZ and coworkers (1986a) observed a striking inverse correlation between cGMP content of isolated granulated cells and renin release in response to atrial natriuretic peptide and other agents known to affect intracellular cGMP. Since ANF did not alter intracellular free calcium concentrations, as measured by the Quin-2 fluorescence method, KURTZ and colleagues (1986a) concluded that the rise in cGMP inhibits renin secretion by a calcium-independent pathway. Nevertheless, the final pathway of inhibition seems to require a normal calcium concentration, as the effect was attenuated by the calcium-entry blocker verapamil (KURTZ 1986b). In summary, the role of cGMP as an intracellular messenger in the control of renin secretion is not clear. Since cGMP has been implicated as a stimulatory second messenger in stimulus-secretion coupling (NISHIZUKA 1983; GOMPERTS et al. 1986), its putative inhibitory function in renin-producing cells should be further examined. If the inhibitory function of cGMP on renin secretion could be confirmed, this would open a new aspect with regard to intracellular control mechanisms of renin secretion: As pointed out before, there is a general conformity of vascular smooth muscle relaxation and stimulation of renin release. This relationship is also maintained in alterations of intracellular messengers, as a rise in intracellular free calcium causes vasoconstriction and inhibition of renin release, whereas a rise in intracellular cAMP causes vasodilation and stimulation of renin release. Thus, if a rise in intracellular cGMP would inhibit renin secretion while relaxing vascular smooth muscle cells, this would be the crucial point where messenger pathways functionally dissociate in the two cell types.

7.6.4 Role of Calcium

Calcium is generally accepted as an essential intracellular messenger for stimulus-secretion coupling in almost every endocrine, exocrine, or neurosecretory process. Therefore, the general observation that renin secretion appears to be inversely related to intracellular calcium must be considered as a most unusual feature of epithelioid cells. This inverse relationship was first recognized by VANDONGEN and PEART (1974) and was later called the "calcium paradox" of renin secretion (HACKENTHAL and TAUGNER 1986). The evidence for this inverse relationship is based on numerous studies with a variety of different experimental approaches. In many experiments in which the *extracellular calcium concentration* was varied, an inverse relationship between extracellular calcium and renin release was observed, for example, in renal cortical suspensions (FRAY and LAURENS 1981; O'DEA et al. 1984), renal cortical slices (SARUTA and MATSUKI 1975; NAFTILAN and OPARIL 1982; GINESI and NOBLE 1984), isolated glomeruli (BAUMBACH and LEYSSAC 1977; BAUMBACH and SKØTT 1981, 1982; SKØTT 1986), isolated perfused rat kidneys (VANDONGEN and PEART 1974; FYNN et al. 1977; FRAY 1977, 1980; PEART et al. 1977; FRAY and LAURENS 1981; COGAN and CHATZILLES 1980; COHEN et al. 1980; HACKENTHAL et al. 1983), as well as the kidney in situ (KISCH et al. 1976; WATKINS et al. 1976). Also during chronic calcium loading in rats, a decrease in plasma renin was observed, and sodium depletion failed to elicit the known response of an increase in plasma renin activity (KOTCHEN et al. 1974).

Since acute changes in extracellular calcium concentration do not necessarily translate into corresponding changes in intracellular calcium, it is not surprising - and thus not contradicting the view of an inverse relation between intracellular calcium and renin release - when acute alterations in extracellular free calcium occasionally failed to affect renin release in man (KOTCHEN et al. 1974; LLACH et al. 1974; EPSTEIN et al. 1976), in the isolated rabbit kidney (VISKOPER et al. 1976; OPGENORTH and ZEHR 1982), and in isolated rat glomeruli (BAUMBACH and SKØTT 1982). In the latter study, some calcium dependency of renin release was observed in experiments performed during the summer, but not during the winter. However, there are also reports on a positive correlation of extracellular calcium and renin secretion which will be discussed later in the context of intracellular events, by which calcium may modulate renin secretion.

In several studies *calcium-channel antagonists* (or calcium-entry blockers) have been used to examine the role of extracellular calcium in mediating inhibitory effects on renin secretion. In isolated epithelioid cells, KURTZ (1986a) found verapamil to stimulate renin release. Verapamil, D-600, or diltiazem stimulated renin release from isolated glomeruli (BAUMBACH and SKØTT 1981) and kidney slices from rats, dogs or pigs (PARK et al. 1981; NAFTILAN and OPARIL 1982; HENRICH and CAMPBELL 1986). Renin secretion by the isolated perfused rat kidney was enhanced by verapamil (MARRE et al. 1982; LOUTZENHISER et al. 1985; HACKENTHAL et al., unpublished observations) and by the dihydropyridines nifedipine and nimodipine (MARRE et al. 1982; HACKENTHAL et al.; cf. Fig. 7.4).

Studies in vivo are in agreement with these observations. ABE et al. (1983) as well as IMAGAWA et al. (1986) infused verapamil or nifedipine, respectively, into the renal artery of anesthetized dogs and observed increases in renin secretion, urine flow, and sodium excretion without changes in blood pressure or total peripheral vascular resistance. To exclude indirect renovascular or tubular effects of calcium antagonists as the cause of changes in renin release, ROY and colleagues (1983) infused verapamil into the renal artery of nonfiltering, denervated, and papaverine-treated dog kidneys in situ and still observed an increase in renin release. Calcium-channel blockers have also been used to analyze the mechanism by which hormonal signals inhibit renin secretion. Thus, PARK and colleagues (1981) found verapamil to block the inhibitory effect of angiotensin II on renin release from dog and pig kidney slices. Similarly, D-600 blocked the inhibitory effect of angiotensin II on renin release from rat kidney slices (NAFTILAN and OPARIL 1982), and verapamil attenuated the inhibition of renin release by norepinephrine from the isolated rat kidney (LOGAN and CHATZILIAS 1980; see also Fig. 7.4). Although in some studies verapamil or diltiazem failed to prevent the inhibition of renin release from rat kidney slices by angiotensin II or vasopressin (CHURCHILL 1980; CHURCHILL et al. 1981), collectively the data obtained with calcium-channel antagonists in various preparations add to the evidence that calcium entry into epithelioid cells promotes inhibition of renin secretion, and that, at least in part, inhibitory hormones such as angiotensin II, vasopressin or norepinephrine may act through facilitation of calcium entry into the cell.

Direct evidence for this mechanism has been obtained by KURTZ and colleagues (1984), who measured the effect of angiotensin II and vasopressin on the influx of ^{45}Ca into granulated cells and observed an increase of the calcium influx which was inversely related to renin release.

As a counterpart to calcium-channel blockers, some dihydropyridines have been found to act as *activators of calcium channels,* notably Bay K8644 (GROSS et al. 1985; SCHRAMM and TOWART 1985). In the isolated rat kidney Bay K8644 concentration-dependently induced a slight vasoconstriction and a profound reduction of renin secretion (HACKENTHAL and TAUGNER 1986). The addition of Bay K8644 to rat kidney slices inhibited renin release either per se (CHURCHILL and CHURCHILL 1987) or only when the potassium concentration in the medium was raised to 15 mM (MATSUMURA et al. 1985, 1987). This latter observation is in accordance with the experience that Bay K8644 activates preferentially voltage-dependent calcium channels (YAMAMOTO et al. 1984; PREUSS et al. 1985), which would be facilitated by a slight depolarization by the elevated potassium concentration. Another calcium-channel agonist, CGP 28392, also required the presence of 15 mM potassium in the incubation medium of rat kidney slices to inhibit renin release, an effect which could be attenuated by the calcium-channel antagonists verapamil and nifedipine (MATSUMURA et al. 1987). Obviously, these data are consistent with an inhibitory function of intracellular calcium.

Also consistent with this interpretation are the observations on the inhibition of renin release from kidney slices by *high potassium* in the incubation medium (PARK and MALVIN 1978; CHURCHILL and CHURCHILL 1982a, b; CHURCHILL et al. 1983a, b; PARK et al. 1986). Potassium at extracellular concentrations of 40-60 mM is thought to depolarize epithelioid cells, activate voltage-dependent calcium channels, and permit accelerated entry of calcium, which then inhibits renin release. In support of this view are observations demonstrating that the potassium effect could be attenuated or prevented by low external calcium, i.e., at less than $10^{-8} M$ (PARK and MALVIN 1978; CHURCHILL and CHURCHILL 1982a) or by the calcium channel antagonist D-600 (CHURCHILL et al. 1983a). Analogous results

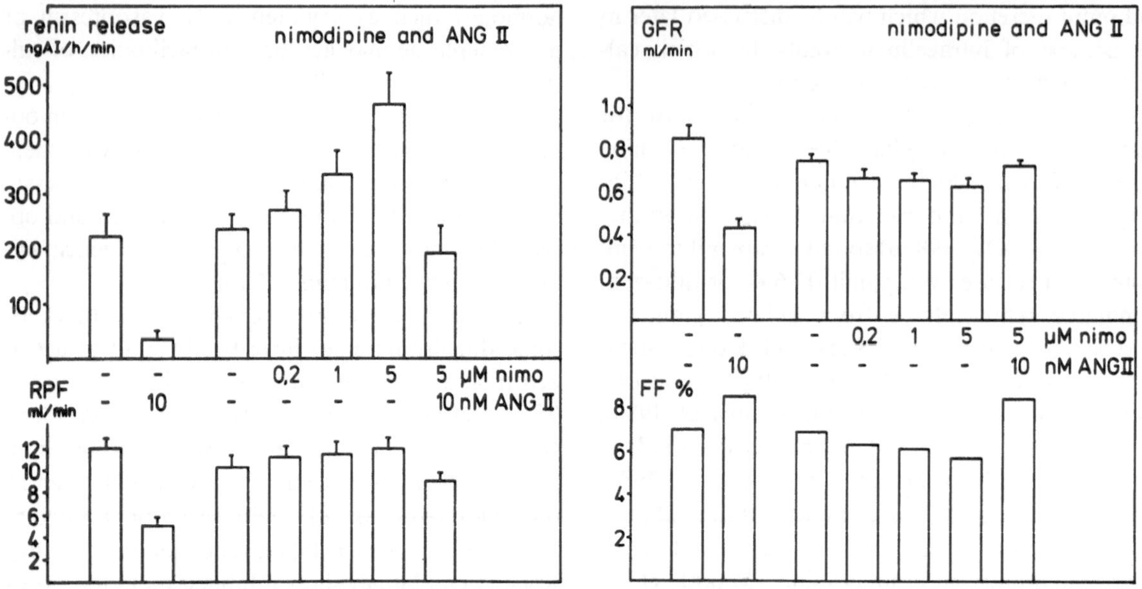

A B

Fig. 7.4 A, B. Effect of nimodipine *(nimo)* on renin release, perfusate flow *(RPF)*, glomerular filtration rate *(GFR)*, and filtration fraction (FF) and its interaction with angiotensin II *(ANG II)* in the isolated rat kidney. Isolated rat kidneys were perfused at constant pressure with a synthetic medium, as described by HACKENTHAL et al. (1987). Nimodipine was infused at the indicated concentration for 5-min periods. ANG II (10 n*M*) was infused in the absence or presence of nimodipine (5 μ*M*). **A** Nimodipine concentration-dependently stimulates renin release and blunts the inhibitory effect of ANG II, as well as its vasoconstrictor effect. **B** Nimodipine slightly reduces GFR and prevents the reduction in GFR by ANG II. *A I*, angiotensin I

were obtained when the Na-K-ATPase activity was inhibited by ouabain in superfused glomeruli (BAUMBACH et al. 1976), in kidney slices (LYONS and CHURCHILL 1974; PARK and MALVIN 1978; CHURCHILL 1979; CHURCHILL and CHURCHILL 1980, 1982b; LYONS 1980; CHURCHILL et al. 1983a), or by infusion of ouabain into the renal artery in dogs (CRUZ-SOTO et al. 1982, 1984). Inhibition of Na-K-ATPase, intracellular accumulation of sodium and an increased rate of Ca-Na exchange, with the consequence of an increased intracellular calcium concentration, was suggested as the mechanism involved (CHURCHILL 1979; CHURCHILL and CHURCHILL 1980). In support of this interpretation, a decrease in effectiveness of ouabain with decreasing extracellular calcium concentration was observed (CHURCHILL and CHURCHILL 1980).

As another means of raising intracellular calcium concentrations, the *calcium ionophore A 23187* has been used in several studies. However, the results were less consistent than expected. Infusion of A 23187 into the isolated rat kidney produced inhibition of renin secretion (FYNN et al. 1977) or variable effects, i.e., both stimulation and inhibition of renin secretion (SCHWERTSCHLAG et al. 1978). In the isolated cat kidney, HARADA et al. (1979) observed consistent stimulation of renin release, which could be blocked by propranolol or reserpine-pretreatment. In isolated glomeruli also variable responses to A 23187 were found (BAUMBACH and LEYSSAC 1977). In kidney slices, either inhibition (BAXTER et al. 1985; HENRICH and CAMPBELL 1986) or no effect was found, regardless of the presence or absence of calcium in the medium (GINESI and NOBLE 1984). In the latter study a decrease in inactive renin secretion in the absence of extracellular calcium was found. The significance of this finding remains to be clarified. Studies in vivo also failed to give a clear picture, as the infusion of A 23187 into the renal artery in dogs either failed to influence (BONDAR et al. 1984), inhibited (CADNAPAPHORNCHAI et al. 1987), or stimulated (OKABARA et al. 1980) renin release. A possible explanation for these discrepancies was provided by HACKENTHAL et al. (1983) who could demonstrate in the isolated perfused rat kidney that in indomethacin-treated kidneys the infusion of A 23187 produced a consistent inhibition of renin secretion, while in nonpretreated kidneys erratic effects of the ionophore on renin secretion were obtained. Obviously, the stimulation of prostaglandin synthesis by the ionophore, which had already been described for other tissues (e.g., KNAPP et al. 1977; OKANO et al. 1985), can interfere with the inhibitory effect of A 23187 on renin release.

Further indirect evidence for the putative inhibitory function of calcium in the pathway for renin secretion can be inferred from studies demonstrating that the inhibition of renin secretion by angiotensin II or vasopressin (ADH) is attenuated or prevented in the absence of extracellular calcium or when calcium influx is blocked by calcium-channel antagonists in kidney slice preparations (CHURCHILL 1980; PARK et al. 1981; NAFTILAN and OPARIL 1982; ANTONIPILLAI and HORTON 1985; MAY and PEART 1986) and in the isolated rat kidney (VANDONGEN and PEART 1974; HACKENTHAL et al., unpublished data; cf. Fig. 7.4). In other studies, however, the inhibitory effect of angiotensin II on renin release from kidney slices was not influenced by verapamil or diltiazem (CHURCHILL 1980; CHURCHILL et al. 1981).

As mentioned already, direct evidence for an increased calcium influx in primary cultures of granulated cells exposed to angiotensin II or vasopressin has been described by KURTZ et al. (1984, 1986a). The authors also demonstrated that renin secretion from these cells was inversely correlated with the rate of ^{45}Ca influx under a variety of conditions.

All these studies, although providing compelling evidence for a role of calcium as a second messenger in renin secretion, do not conclusively answer the question whether extracellular calcium is the exclusive source for an intracellular increase in cytosolic calcium. According to the established signal transduction mechanism for angiotensin II in vascular smooth muscle cells, occupancy of angiotensin II receptors activates a phospholipase C (or phosphodiesterase), which hydrolyzes phosphatidylinositol to yield two second messengers, inositol 1,4,5-trisphosphate (IP_3) and diacylglycerol. IP_3 initiates the release of calcium from intracellular stores, presumably vesicular structures related to the endoplasmic reticulum, resulting in a rise in cytosolic calcium (cf. EXTON 1985; ABDEL-LATIF 1986; BERRIDGE 1986; SEKAR and HOKINS 1986). The stimulation of phosphatidylinositol,4,5-biphosphate hydrolysis and this initial and transient increase in cytosolic calcium has also been demonstrated to occur in granulated cells exposed to angiotensin II by KURTZ et al. (1984, 1986b).

The role of intracellular calcium mobilization in renin secretion has also been studied by examining the effects of *TMB-8* (8-(*N,N*-diethylamino)dioctyl-3,4,5 trimethoxybenzoate). This agent has been reported to prevent the release of calcium from the endoplasmic reticulum (CHIOU and MALAGODY 1975) and has since widely been used as an "intracellular calcium antagonist" in various tissues. Indeed, renin release was found to increase in response to TMB-8 in rat kidney slices (ANTONIPILLAI and HORTON 1985; BAXTER et al. 1985; HENRICH and CAMPBELL 1986) and in the isolated perfused rat kidney (HACKENTHAL and TAUGNER 1986). As shown in Fig. 7.5, TMB-8 stimulates renin release from the isolated rat kidney in a dose-dependent fashion. It also attenuates the inhibition of renin release by angiotensin II (not shown), e.g., at 10^{-9} *M* angiotensin II, renin secretion was inhibited by 78% and 22% in controls and TMB-8 (0.5 μ*M*) infused kidneys, respectively. However, recent studies on the mechanism of action of TMB-8 have seriously challenged the putative mechanism of an intracellular calcium antagonism (e.g., BRAND and FELBER 1984; KOGIMA et al. 1985, 1986), and the interpretation of its effect on renin release remain unclear.

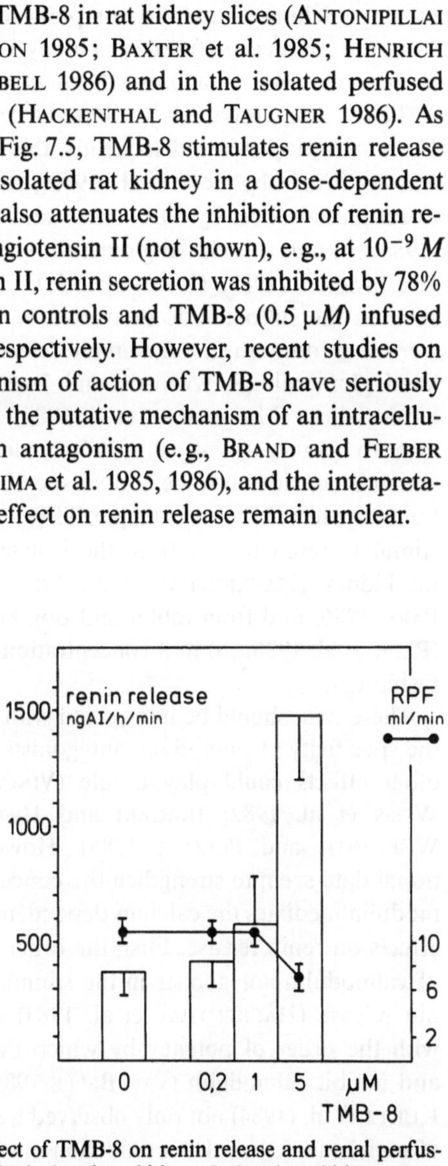

Fig. 7.5. Effect of TMB-8 on renin release and renal perfusate flow in the isolated rat kidney. Isolated rat kidneys were perfused at constant pressure with a synthetic medium (HACKENTHAL et al. 1987). TMB-8 was infused at the concentrations indicated for 5-min periods. *RPF*, renal perfusate flow; *A I*, angiotensin I

7.6.5 Role of Calmodulin

Many of the intracellular actions of calcium are mediated by calmodulin, e.g., in vascular smooth muscle, contraction is dependent on calcium-calmodulin activation of myosin light-chain kinase or of thin filaments (see MEANS et al. 1982; TOMLINSON et al. 1984 for survey). Therefore, several investigators have examined the effect of calmodulin antagonists on renin secretion. In accordance with an inhibitory function of intracellular calcium and its mediation by the calcium-calmodulin complex, renin release from the isolated perfused rat kidney is stimulated by a calmodulin antagonist, the neuroleptic agent TFP, as first described by SCHWERTSCHLAG and HACKENTHAL (1983). Subsequently, this observation was confirmed by FRAY et al. (1983), and extended to rat kidney slices by CHURCHILL and CHURCHILL (1983a, b) and to isolated glomeruli by KAWAMURA and INAGAMI (1983). Other calmodulin antagonists, such as the sulfonamide derivative W-7 and TFP, were also found to stimulate renin release from isolated glomeruli (KAWAMURA and INAGAMI 1983), kidney slices (MATSUMURA et al. 1984), and in rats in vivo (SHINYAMA et al. 1987). Calmidazolium, which is considered to be a more specific calmodulin antagonist than, for example, TFP, was found to stimulate renin release from the isolated perfused rat kidney (HACKENTHAL et al. 1983; FRAY and PARK 1986) and from rabbit and dog kidney slices (PARK et al. 1986a, b) in a concentration-dependent fashion.

These data should be interpreted with caution, as the specificity of calmodulin antagonists is low, and other effects could play a role (VINCENZI 1981; WEISS et al. 1982; INAGAKI and HIDAKA 1984; WÜLFROTH and PETZELT 1985). However, additional data seem to strengthen the concept that calmodulin mediates the calcium-dependent inhibitory effects on renin release. First, the order of potency of calmodulin antagonists in the stimulation of renin release (HACKENTHAL et al. 1983) agrees well with the order of potency by which they bind to and inhibit calmodulin (VAN BELLE 1981). Second, KURTZ et al. (1984) not only observed a stimulation of renin release but also a reduction in calcium permeability of isolated epithelioid cells by TFP. Third, various calmodulin antagonists were not only found to stimulate renin release, but also to block the inhibitory effect of angiotensin II on renin release (HACKENTHAL et al. 1983), which can be assumed to result from an increase in cytosolic calcium (KURTZ et al. 1984). This is in line with the observation of MATSUMURA and colleagues (1987) that the inhibitory effect of the calcium-channel agonist Bay K8644 on renin release from rat kidney cortex slices could be significantly attenuated by calmidazolium or TFP.

In conclusion, the consistent finding of a stimulation of renin release in various preparations by calmodulin antagonists is compatible with a role of calmodulin in the regulation of renin secretion, although direct evidence for this role is still lacking.

7.6.6 Role of Diacylglycerol and Protein Kinase C

As mentioned before, receptor occupation by Ca^{2+}-mobilizing hormones such as angiotensin II causes a rapid hydrolysis of phosphatidylinositol-4,5-bisphosphate with the production of IP_3 and the lipophilic modecule diacylglycerol. Diacylglycerol activates membrane-bound protein kinase C by enhancing the calcium sensitivity of this enzyme (BERRIDGE 1984, 1986, 1987; EXTON 1985; ABDEL-LATIF 1986). Activation of protein kinase C results in the phosphorylation of several distinct proteins with largely undefined functions and various cellular responses. Among these responses is also the regulation of exocytosis in various cell types in concert with other intracellular messengers, such as calcium and cAMP (cf. RASMUSSEN and BARRETT 1984; BAKER and KNIGHT 1986 for review).

As a means of activating protein kinase C independent of the calcium pathway, synthetic diacylglycerols or related phorbol esters such as 12-*o*-tetradecanoylphorbol 13-acetate (TPA) have been used in many studies including those on the control of renin release. TPA was first used by CHURCHILL and CHURCHILL (1984) to probe the role of protein kinase C in renin secretion. It was found that TPA inhibited renin secretion from rat cortical slices in a concentration-dependent manner (10^{-7}–10^{-5} *M*), with about 40% inhibition at the highest concentration. In the presence of 2 m*M* EGTA, which by itself stimulated renin release, TPA (10 μ*M*) reduced renin release to a level slightly above that in untreated controls. The authors concluded from these data that the TPA-induced inhibition of renin secretion was not due to the calcium-ionophoric properties of TPA, but rather to its activation of protein kinase C. KURTZ and colleagues (1986a) also found TPA to inhibit renin release in isolated granulated cells in the concentration range between 10^{-8} and 10^{-6} *M*. However, at the same time, a significant increase in calcium influx into these cells occurred, which could be attenuated by verapamil. It therefore cannot be excluded that inhibition of renin release by TPA was due to a calcium-ionophoric effect rather than to inhibition of protein kinase C. Experiments with other protein kinase C activators, such as oleylacetyl-glyceride (OAG), may perhaps clarify this issue.

If eventually the inhibitory role of protein kinase C in the control of renin secretion could be confirmed, this would add to the unique position of renin secretion among secretory processes in general, since in this case not only the role of intracellular calcium, but also that of protein kinase C activation would run counter to the situation predominating in "conventional" secretion. While the IP_3-induced intracellular calcium release is thought to mediate the immediate secretory response to various secretagogues, the continuous occupation of the secretagogue receptors and the resulting activation of protein kinase C seems to be responsible for the sustained secretory activity in the platelet-release reaction, insulin secretion, angiotensin II-stimulated aldosteron secretion, exocrine pancreas secretion, and acetylcholine secretion (for review and references see RASMUSSEN et al. 1984a, b, 1987; WILLIAMSON 1986; BERRIDGE 1987).

7.6.7 Interaction of Messengers

Several attempts have been made to identify intracellular messenger interactions and pathways in the intracellular control of renin secretion beyond the initial steps, i.e., stimulation of adenylate cyclase or calcium activation. Theoretically, three possibilities for the interaction between the two major second messengers cAMP and calcium can be described (CHURCHILL 1985): (a) calcium is the final signal, and cAMP modifies the calcium pathway at an intermediate step; (b) cAMP is the final signal, modified by calcium; (c) cAMP and calcium control renin secretion by independent pathways. Several investigators consider calcium to be the ultimate and more important signal (KURTZ et al. 1984; CHURCHILL 1985; KURTZ et al. 1986a; FRAY et al. 1987). In support of this view are experiments by CHURCHILL and CHURCHILL (1982), who observed that the stimulatory effect of isoproterenol of dbcAMP on renin release from rat kidney slices could be almost completely blocked by agents that raise intracellular calcium, such as ouabain, angiotensin II, and potassium depolarization. More conclusive data have been provided by KURTZ et al. (1984), who could demonstrate in cell cultures of granulated cells that a consistent inverse relationship between renin release and ^{45}Ca influx exists, regardless of the mechanism by which renin secretion was manipulated, whereas intracellular cAMP was only increased when renin secretion was stimulated by forskolin, prostacyclin, or isoproterenol, but not when stimulated by trifluoperazine or verapamil.

However, there are also observations in disagreement with this view. For example, PARK and MALVIN (1978) observed that renin secretion from rat kidney slices could still be stimulated when renin secretion had been inhibited by 60 m*M* potassium. Similarly, BONDAR and colleagues (1984) could stimulate renin release from kidney slices by dbcAMP in the presence of the calcium ionophore A 23187.

With regard to cGMP, the recent studies of KURTZ and colleagues (1986b) on the mechanism by which atrial natriuretic peptide might inhibit renin release, have demonstrated a close correlation between intracellular cGMP and renin release, whereas intracellular cAMP was only altered at the highest concentration of atrial natriuretic peptide.

Such observations are not only difficult to reconcile with each other but are also difficult to interpret against the background of knowledge of interactions between messengers in other cell types, e.g., vascular smooth muscle cells or various secretory cells. This is not surprising in view of the complexity of interactions of these messengers in a given cell type and the variations in these interactions between cell types. To mention only a few aspects, adenylate cyclase activity has been shown to be modulated by both calmodulin (e.g., KATADA et al. 1987) and by protein kinase C (JAKOBS et al. 1985; KATADA et al. 1985), and cAMP is known to lower intracellular calcium in many cell types, including vascular smooth muscle, by decreasing calcium influx or by promoting efflux (RASMUSSEN et al.

1984). Depending on the cell type, secretion is controlled by protein kinase C activity, calcium and/or cAMP. With regard to the role of cGMP in the control of vascular tone and of renin secretion, this messenger is thought to exert its vasorelaxant actions through inhibition of phosphotidylinositol-bisphosphate hydrolysis (RAPOPORT 1986), whereas KURTZ (1986c) concluded from his data that cGMP and phosphatidylinositol hydrolysis act in concert to inhibit renin release in epithelioid cells.

These and many other complex actions and interactions of messengers in stimulus-secretion coupling and in vascular smooth muscle contraction are still not clearly understood (see reviews by BAKER and KNIGHT 1986; HEAGERTY and OLLERENSHAW 1987; BERRIDGE 1987; RASMUSSEN et al. 1987), and much less is known of the identity and actions of messengers in renin secretion. It therefore appears premature to engage in extensive speculations on possible interactions of the putative intracellular messengers and subsequent biochemical events in renin-secreting cells. Rather, the possible nature of the inverse relationship between renin release and intracellular calcium will be discussed.

7.6.8 A Possible Explanation of the "Calcium Paradox" in Renin Secretion

As discussed in the previous sections, an apparently overwhelming number of studies describe an inverse relationship between intracellular free calcium concentrations and renin secretion. This has made the inhibitory function of calcium a generally accepted, although poorly understood fact. However, this inverse relationship between calcium and renin release is not observed as consistently as is usually claimed. In several studies, which cannot be simply dismissed on methodological grounds, a positive correlation has been observed which would be consistent with a "conventional" role of calcium in stimulus-secretion coupling of renin release. For example, intra-arterial infusion of calcium stimulated renin release from the dog kidney (IWAO et al. 1974). Dog, cat, or rat kidney cortex slices were found to release more renin when the calcium concentration in the medium was increased (MORIMOTO et al. 1970; MICHELAKIS 1981; SARUTA and MATSUKI 1975; GINESI et al. 1981). CHEN and POISNER (1976) as well as HINKO et al. (1984) reported that renin release from cat kidney slices was decreased when calcium was omitted from the medium and rose sharply upon reintroduction of calcium. LESTER and RUBIN (1977) and SCHWERTSCHLAG and colleagues (1978) made similar observations in the isolated perfused kidney of cats and rats, respectively. Furthermore, HARADA and colleagues (1979) observed an increase in renin release from the isolated perfused cat kidney upon infusion of the calcium ionophore A 23187. Similarly, when the calcium ionophore was infused into the renal artery in dogs, renin secretion increased, and the effect was not altered by indomethacin (OKAHARA et al. 1980). Pointing in the same direction is the observation of MAY and PEART (1986) that the calcium agonist Bay K8644 induced a transient increase in renin release from rat kidney slices. These observations do not contradict the more often observed inverse relationship. However, they demonstrate that this is not a straightforward phenomenon.

This situation has led TAUGNER et al. (1988b) to consider the possibility that the apparently conflicting data concerning the role of calcium in renin secretion may not depend primarily on differences in methodology but rather reflect the heterogeneity of renin-producing cells. Although in most studies on renin secretion, renin-producing cells have been considered as a homogeneous cell population, in reality these cells are profoundly different in fine structure, depending on their location along the afferent arteriole. This relates to the fact that renin-producing cells are modified smooth muscle cells, which can transform into granulated cells when the RAS is chronically stimulated (CANTIN et al. 1977), and retransform back into vascular smooth muscle cells, when renin secretion is suppressed for a prolonged time (cf. Chap. 10). This metaplastic transformation is a continuous process with all intermediate forms present at any given time. Thus, along the afferent arteriole, fully granulated secretory or epithelioid cells which are devoid of myosin (TAUGNER et al. 1987b) are located in a small segment close to the glomerulus. More proximal, intermediate cells occur that are contractile, as judged from their morphology and from direct visualization of contraction (STEINHAUSEN et al. 1986). Further upstream, the boundary is reached where cells with the appearance of vascular smooth muscle cells, but still containing some renin granules, are continuous with "true" vascular smooth muscle cells devoid of renin.

It is most important to note that in most situations and species, intermediate cells, which all have residuals of the contractile machinary, i.e., attachment sites and myofilaments, usually abound in number over epithelioid cells. The contractile elements in intermediate cells are usually located at the periphery, where they form a dense subcortical network. Typically, renin granules in these cells are separated from the cell surface by this subcortical myofilament network, in contrast to purely secretory epithelioid cells that do not react with myosin antibodies, in which renin granules are usually located almost in direct contact with the plasma membrane (cf. Sect. 2.2.2).

Taugner and colleagues (1988a) presented direct morphological evidence that this subcortical network may hinder renin granules from gaining access to the plasma membrane. They suggest that the inverse relationship between renin secretion and intracellular calcium may be explained by assuming that an increase in intracellular calcium induces contraction of the subcortical myofilament layers (by essentially the same mechanism as in vascular smooth muscle cells), which further impairs the access of renin granules to the cell membrane and their exocytosis, whereas a low intracellular calcium concentration relaxes and loosens the network and thus facilitates the contact of granules with the cell surface and promotes renin secretion. This interpretation is consistent with the general experience that vasodilators usually stimulate, while vasoconstrictors inhibit, renin secretion.

In this view, the subcortical network of myofilaments in intermediate renin-producing cells corresponds to an analogous structure of the cytosceleton in other secretory cells. In extension of previous work of Orci et al. (1972), recent studies suggest that in secretory cells a subcortical filamental network provides a mechanical barrier for secretory vesicles to fuse with the plasma membrane. The stability of this cytosceletal network, which includes F-actin and caldesmon, appears to be regulated by calcium and thus to participate in the control of secretion. However, in contrast to renin-secreting cells, this network may not be stabilized, but rather loosened, by an increase in cytosolic calcium, perhaps by activation of calcium-dependent proteases (Burgoyne et al. 1986; Cheek and Burgoyne 1986; Hutton 1986; Lelkes et al. 1986; Lew et al. 1986; Creutz et al. 1987; Perrin et al. 1987; Mellgren 1987). This mechanistic view of the secretory event, both in the conventional secretory process and in renin secretion is, of course, not exclusive with other mechanisms participating in the control of secretion, such as phosphorylation reactions, pore formation and osmotic swelling (see Chap. 6 and below).

If the hypothetical explanation of the calcium paradox of renin secretion of Taugner et al. 1988a is correct, it implies that the process of renin exocytosis follows basically the conventional pattern of secretion, i.e., the actual exocytotic event would require calcium. But, in intermediate cells, calcium would also tend to prevent exocytosis by tightening the myofilament barrier. This speculative view could explain the divergent data with respect to the direction in which a change in intracellular calcium is transformed into renin secretion; the net effect of these two opposing mechanisms of action of calcium would depend on the actual location of epithelioid cell granules with regard to the cytoskeletal fusion barrier and on the relative proportion of intermediate cells to pure epithelioid cells. If the latter predominate, calcium is expected to serve preferentially as a stimulus-secretion coupling agent, just as in other secretory cells. If intermediate cells abound in the preparation studied, calcium would produce net inhibition by contracting the myofilament network. If, however, intermediate cells are maintained in a relaxed state for some time, such as during calcium deprivation, a sufficient quantity of renin granules may have penetrated the myofilament network without being able to fuse with the plasma membrane (because this requires calcium) and, upon reintroduction of calcium, give rise to the transient sharp increase in renin release that has been observed in several studies (see above).

In this context, it is of interest to note that the majority of studies in which an inverse relationship between renin release and calcium have been observed have been performed in the rat, in which intermediate myofilament-carrying cells clearly outnumber granulated cells under control conditions.

Still another concept of the intracellular function of calcium in the control of renin secretion has been proposed by Skøtt (1986a). Based on morphological studies by Taugner et al. (1984b) and on secretion studies in isolated superfused glomeruli, Skøtt suggests that the principal mechanism of renin release is osmotic swelling of renin granules, by which the probability of contact and fusion of these vesicles with the plasmalemma and subsequent exocytosis is increased. In this scheme, intracellular calcium is proposed to inhibit exocytosis by preventing disaggregation of intravesicular particulate matter and thus conversion of osmotically inactive into osmotically active material, whereas a decrease of intracellular free calcium accelerates the physiological process of matrix solubilization, leading to an increased rate of swelling, membrane fusion and exocytosis. Osmotic swelling has been recognized in many cell types as an essential step in exocytosis which immediately preceeds the exocytotic event. However, here again calcium is thought to trigger this event (GRINSTEIN et al. 1982; ZIMMERBERG and WHITAKER 1985; FINKELSTEIN et al. 1986; HUTTON 1986). As possible mechanisms of calcium-mediated osmotic swelling, the opening of calcium-activated potassium channels in vesicles resulting in an increased influx of potassium and anions or stimulation of Na-H^+ exchange and subsequent osmotic swelling have been proposed (STANLEY and EHRENSTEIN 1985). Interestingly, two groups of investigators have recently demonstrated by capacitance measurements that in mast cell secretion the calcium-dependent formation of a small pore (fusion pore) between the interior of the secretory vesicle and the extracellular space precedes osmotic swelling and exocytosis and concluded that osmotic swelling cannot be the driving force for membrane fusion (BRECKENRIDGE and ALMERS 1987; ZIMMERBERG et al. 1987). However, regardless of the question whether this observation can be extended to exocytosis in other cell types, calcium maintains its central role in stimulus-secretion coupling because this event of pore formation is also a calcium-dependent process. Therefore, the proposal by SKØTT (1986a, 1987) of an opposite function of calcium in osmotic swelling of renin granules needs further experimental support. It will also be of interest to extend earlier observations of HAMMERSEN et al. (1971) and Schmidt et al. (1971) on the biochemical properties of isolated renin granules, demonstrating no correlation of renin release with changes of the osmolality of the incubation medium.

7.6.9 Summary and Conclusions

Despite considerable efforts to elucidate the role of intracellular messengers in the control of renin secretion, progress in this field was slow. This can be attributed to the fact that isolated granulated cells are not readily available for the measurement of intracellular events. Collectively, the data reported in the literature permit the conclusion that intracellular calcium is an important second messenger in renin secretion, its intracellular concentration usually being inversely correlated with renin secretion. However, as described in detail, the status of this inverse relationship is not of dogmatic nature, as there is a considerable number of opposite findings. Explanations for the unusual inverse relationship have been offered by some authors, but these must still be considered as hypotheses, at best.

Activation of adenylate cyclase and formation of cAMP have clearly been established as a stimulatory pathway in the control of renin secretion. However, the interaction of this pathway with that of calcium, or the biochemical sequelae of cAMP generation in renin secreting cells, remain unknown. Activation of particulate guanylate cyclase and formation of cGMP have been described as an inhibitory pathway. Its general significance and interaction with other pathways remains to be established. Evidence for the participation of protein kinase C in the control of renin release must be considered preliminary at present.

Further developments in the evaluation of the role of intracellular messengers in renin secretion will largely depend on the availability of isolated renin-secreting cells, on the separation of different types of renin secreting cells, i.e., fully granulated secretory from intermediate cells, and on the experimental dissociation of renin synthesis from renin secretion. One of the prominent issues in defining the role of second messengers in the control of renin secretion will probably be the identification and differentiation of different types of renin release (e.g. regulated vs. constitutive), which, if existent, might be regulated by a different set of intracellular messengers.

In this context it may be permitted to add the speculative thought that perhaps we are looking in the wrong direction when studying the control mechanisms for renin secretion; the "efforts" of renin-producing cells and the mechanisms that control renin release may not be directed at actively *promoting* but rather at actively *preventing* secretion. In this case, stimulation of renin "secretion" would be a process in which the grip on the retention of renin inside the cell is weakened, albeit in a controlled manner. This speculative view, first expressed by BLENDSTRUP et al. (1975), is born out of the general experience described by many investigators that in all isolated preparations there is a rapid spontaneous loss of renin, amounting to 100- to 1000-fold the rate of spontaneous renin release in vivo. This is in disagreement with the experience in many other secretory systems and is difficult to explain solely by the removal of physiological inhibitory control mechanisms operating in vivo. Furthermore, maneuvers which impair cell energy production, such as oxygen deprivation or metabolic inhibitors, were found to stimulate renin release (BLENDSTRUP et al. 1975; PARK et al. 1981). Most striking is the observation that lowering the incubation temperature of rat isolated glomeruli or rat kidney cortex slices dramatically enhanced the rate or renin release (BLENDSTRUP et al. 1975; BAUMBACH et al. 1976; PARK et al. 1981). It may be worthwhile to look into this aspect more closely because the solution to the calcium paradox may be buried herein. In fact, the myofilament-contraction hypothesis, as described in the preceding section, would be in line with this view.

7.7 Secretion of Prorenin

The existence of an inactive (or less active) forms of renin in plasma or amniotic fluid was first described by LUMBERS (1971), who observed an increase in renin activity of human amniotic fluid upon exposure to a low pH. Subsequently, SKINNER et al. (1972, 1975) demonstrated that plasma of pregnant women and normal subjects also contains an acid-activatable form of renin. MORRIS and LUMBERS reported in 1972 that inactive renin in amniotic fluid can be activated not only by acidification, but also by incubation with proteolytic enzymes. The same was shown to occur in human plasma by OSMOND et al. (1974). Finally, inactive renin in plasma could be activated by prolonged storage (i.e., for several days) at $-5\,°C$ (OSMOND et al. 1973; SEALEY and LARAGH 1975). These early observations of the existence of an inactive form of renin in the circulation set the stage for intensive research efforts in many laboratories directed at the elucidation of the following questions: what are the biochemical mechanisms of acid activation, cryoactivation, or protease activation of inactive renin; is circulating inactive renin identical to prorenin as identified in the kidney; is the kidney the source of plasma-inactive renin; if so, is the release of prorenin controlled by the same mechanisms that also control the release of active renin; and what is the physiological function of inactive renin. In the following, these questions will be briefly discussed.

Most research on the experimental activation of inactive renin in plasma has been reviewed by SLATER and HABER (1979), SEALEY et al. (1980, 1983), LECKIE (1981), and POULSEN and JACOBSEN (1983) and will not be discussed here in detail. Taken together, these studies permit the following conclusions. Inactive renin in plasma has a higher molecular weight than active renin. The activation at low pH, e.g., by dialysis at pH 3.4, is a reversible process which does not involve proteolytic cleavage but rather leads to reversible denaturation of inactive renin by inducing conformational changes. These conformational changes expose the active site of the enzyme which is buried in the interior of the polypeptide in the native state of the proenzyme (cf. DERKX et al. 1986). In contrast, proteolytic cleavage activates inactive renin irreversibly by removing part of the polypeptide chain that probably covers the active site. Consequently, acid activation (and also cryoactivation) does not change the molecular weight, whereas protease treatment usually does. In most studies it was found that proteolytic activation of inactive renin in whole plasma is a slow process which can be accelerated by prior acid denaturation. Denaturation of proteinase inhibitors by the acidification step may also contribute to the increased effectiveness of subsequent proteinase activation, as indicated by the observation reported by INAGAMI et al. (1982) that the rate of activation by proteases was not increased by previous acidification when purified inactive renin from human plasma instead of whole plasma was studied.

Experimentally, inactive renin can be activated by proteases of all classes, e.g., kallikrein (SEALEY et al. 1979a; INAGAMI et al. 1982), plasmin (SEALEY 1979a), trypsin (SEALEY et al. 1979a; GALLAGHER et al. 1980), tonin (GUTKOWSKA et al. 1982), nerve growth factor associated protease (MORRIS and

CATANZARRO 1980), cathepsin B (LUETSCHER et al. 1982), cathepsin D, and pepsin (MORRIS 1978). INAGAMI et al. (1982) compared the potency of various proteases to activate purified human plasma inactive renin and found trypsin to be the most active, followed by plasmin, plasma kallikrein, human liver cathepsin B, nerve growth factor, and epidermal growth factor associated peptidase, in this order.

The endogenous plasma protease responsible for the activation during prolonged cold treatment (cryoactivation) or subsequent to dialysis at pH 3.4 and redialysis to neutral pH has been identified as a serine protease, most likely kallikrein (LECKIE 1978; SEALEY et al. 1979b; SEALEY 1980; MORRIS and MCGIRR 1981; DERKX et al. 1986). It should be kept in mind, however, that in all of these studies on the activation of plasma-inactive renin, artifical conditions have been used which do not occur in vivo, and it remains questionable whether activation of plasma-inactive renin by kallikrein occurs in the circulation to a significant degree (see below). Before discussing this point further, the origin of plasma-inactive renin will be considered.

There can be no doubt that active renin present in the circulation comes predominantly from the kidney. Consequently, the kidney has also been considered the major source of plasma-inactive renin. There are indeed several different observations that support this contention. The release of inactive renin into the circulation has been demonstrated in the isolated hog kidney (OHDE et al. 1982) and the isolated perfused human kidney (ATLAS et al. 1980). Release of prorenin into the lymph of the dog kidney has been described by DZAU et al. (1986). In several clinical studies, a significant arteriovenous difference of inactive renin across the renal circulation has been measured (e.g., BIRKENHÄGER et al. 1978; MILLAR et al. 1978; AOI et al. 1979; HSUEH et al. 1983), indicating the release of inactive renin (prorenin) from the kidney. Arteriovenous differences have also been observed in the dog (DZAU et al. 1986). Several studies have demonstrated the similarity of kidney prorenin and plasma-inactive renin with regard to physicochemical and enzymatic properties (ATLAS et al. 1980, 1982; HSUEH et al. 1983) or by peptide mapping analysis (MCINTYRE et al. 1984). Furthermore, plasma-inactive renin has been shown to react not only with antibodies raised against pure kidney renin (YOKOSAWA et al. 1980) but also with antibodies raised against the carboxy-terminal portion (ATLAS et al. 1985; BOUHNIK et al. 1985; KIM et al. 1985) or the amino-terminal sequence of the prosegment (DAY et al. 1986). This latter observation is in contrast to that of HIROSE et al. (1985) who studied the reaction of human plasma inactive renin with different antibodies, recognizing the amino-terminal sequence, the middle portion, and the carboxy-terminal sequence, respectively, of the prosegment of human kidney prorenin. Only the antibody directed against the carboxy terminus inhibited the activation of plasma-inactive renin by trypsin and bound to plasma-inactive renin during affinity chromatography. The authors concluded that plasma-inactive renin is a truncated version of intact prorenin, lacking most of the amino-terminus of the prosegment. A possible explanation for the apparent discrepancy of these results and conclusions to those of DAY et al. (1986) may be the difference in methodology. DAY et al. (1986) report that their antibody, which is specific for the amino terminus, recognizes plasma inactive renin only following denaturation. As HIROSE et al. (1985) did not use denaturating conditions in their experiments, it appears possible that their antibody directed against the amino-terminus and the middle portion could not recognize the respective epitopes because they were not accessible in the native state of prorenin. Thus, the existence of a truncated form of prorenin in human plasma needs to be confirmed by more rigorous criteria.

In conclusion, there is convincing evidence that kidneys of humans, dogs, and hogs secrete prorenin into the circulation. However, attempts to identify the kidney as the source of plasma-inactive renin have failed in other species, e.g., in the rat, where no inactive renin could be detected in the perfusate of the isolated perfused kidney (VANDONGEN et al. 1977; HACKENTHAL, unpublished observations). The observation of DOI et al. (1984) that plasma-inactive renin in the rat increased rather than decreased following bilateral nephrectomy is also consistent with the view that in this species the kidney does not significantly contribute to plasma-inactive renin.

Even though the human kidney secretes prorenin into the circulation, this does not account for all of the inactive renin found in plasma. For example, high concentrations of inactive renin have been detected in anephric subjects (SEALEY et al. 1977; WEINBERGER et al. 1977; KONRADS et al. 1980; TAYLOR et al. 1986) and in nephrectomized rats

(Madeddu et al. 1984), the source of which is still unknown. The biochemical identity of this form of inactive renin, the source and characteristics of big renin with a molecular mass in excess of 60 kd (Day et al. 1976; Takii et al. 1980), as well as of other forms of inactive renin or active renin with a higher molecular weight than compatible with prorenin, remains to be clarified.

Most studies on the possible control of prorenin secretion have been performed in man. Thus, Kappelgaard and colleagues (1978) infused the angiotensin II antagonist saralasin in order to acutely remove the inhibitory control of angiotensin II on renin secretion. As expected, there was a sharp rise in the plasma concentration of active renin, whereas inactive renin remained unchanged. Furthermore, no arteriovenous difference of inactive renin across the kidney was found either before or during saralasin infusion. Similary, an acute stimulation of renin release by the injection of furosemide had only negligible effects on plasma inactive renin. However, when diuretic treatment was maintained for 5 days, inactive renin in plasma gradually increased. The authors concluded that inactive renin has a time constant of secretion differing from that of active renin by orders of magnitude. Similar observations have been reported by Rumpf et al. (1978), McKenzie and Reisin (1978), and Derkx et al. (1979). A furosemide-induced increase in active, and decrease in inactive, renin have also been observed in the rabbit (Richards et al. 1981b). Interruption of the feedback control of renin secretion by angiotensin II can also be accomplished by the inhibition of converting enzyme. The acute effect of the oral or parenteral administration of converting-enzyme inhibitors, i.e., occurring within 1 or 2 h, was a consistent, several-fold increase in active plasma renin and a less consistent reciprocal fall in inactive renin (between zero and 70%) (Goto et al. 1980, 1986; Glorioso et al. 1982; Goldstone et al. 1983; Hsueh et al. 1985; Ogawa et al. 1985; Patrassi et al. 1985). Similar changes were found after the infusion of diazoxide (Derkx et al. 1979). When, however, captopril was given chronically, inactive renin also increased to reach the "normal" ratio of active to inactive renin within a few days (Ogawa et al. 1985). During long-term treatment with propranolol, active and inactive plasma renin decreased concomitantly (McKenzie and Reisin 1978). Another condition in which active renin was found to rise sharply and inactive renin to fall is that of acute stress induced by immobilization in rats (Jindra and Kvetnansky 1982; Barrett and Eggena 1986). In contrast, acute hemorrhage induced a synchronous rise in the plasma concentrations of both active and inactive renin in the rabbit (Richards et al. 1979).

Taken together, these studies seem to suggest that under basal conditions prorenin is continously secreted along with active renin by the kidney. When active renin secretion is acutely stimulated, prorenin secretion remains unchanged or decreases in most situations. However, after long-term changes of active renin secretion, the secretion pattern of inactive renin approaches that of active renin. Most studies in vitro seem to be in accordance with this view. Isoprenaline or lithium chloride induced an increase in the release of active renin from rabbit kidney slices without affecting the release of inactive renin (Ginesi et al. 1983; Richards et al. 1981b). Kallikrein has been reported to increase active renin release from hog kidney slices, which was accompanied by a decline in inactive renin release (Okamura and Inagami 1984). A dissociation between active and inactive renin was also seen in studies with calcium antagonists. The addition of verapamil or flunarizine to the incubation medium of rabbit kidney slices did not affect the secretion of active renin but stimulated that of inactive renin (Ginesi and Noble 1984). In the isolated hog kidney, prostacyclin (PGI_2) induced a stimulation of active renin release, while the concentration of inactive renin in the perfusate was reduced (Ohde et al. 1982). Since PGI_2 also stimulated kallikrein release from the kidney, the authors proposed that the increase in active renin and the concomitant decrease in inactive renin reflects intrarenal enzymatic conversion of inactive to active renin by kallikrein. An inhibitory effect of exogenous or endogenous prostaglandins on the apparent release of inactive renin from the kidney is also suggested by the observations that the decrease of inactive renin in plasma upon administration of converting-enzyme inhibitors can be prevented by inhibitors of cyclooxygenase (Lijnen et al. 1985; Hsueh et al. 1985), and that infusion of PGA_1 in man increased active and decreased inactive renin (Hsueh et al. 1985).

When interpreting the data on changes of plasma-inactive renin obtained in studies in man or the intact animal, it should be kept in mind that the kidney is not the only source of inactive renin in plasma, as indicated by the observation that plasma-inactive renin is also found in anephric patients, as described above. At present, no data on the relative contribution of the kidney to total inactive renin in plasma are available. Other sources may be the ovaries (SEALEY et al. 1983) and chorionic cells of the placenta (ACKER et al. 1982; POISNER et al. 1982). Prorenin is also secreted by vascular smooth muscle cells in culture (OHASHI et al. 1985) and the local generation of prorenin in the vascular wall has been suggested by MIZUNO et al. (1985). However, the main extrarenal sources of plasma prorenin remain unknown. Therefore, it is difficult to determine to what extent changes of plasma-inactive renin reflect changes of the secretion rate of prorenin from the kidney.

An additional problem in interpreting such data arises from the possiblity that prorenin may be converted to active renin, not only inside epithelioid cells but also postsecretionally. As pointed out above, this possibility has been discussed by OHDE et al. (1982) in view of the inverse relationship between inactive renin and kallikrein release from the isolated hog kidney. Activation of prorenin during passage of blood through the kidney has also been considered as the cause of a lower renal venous than arterial concentration of inactive renin in patients with essential hypertension treated with diazoxide (DERKX et al. 1979). Furthermore, inconsistencies in the time course of changes in plasma concentrations of inactive renin exist, which may eventually find an explanation in the peripheral activation of prorenin. From the study of DERKX et al. (1978), a half-life of inactive renin in circulation of about 150-165 min has been calculated following bilateral nephrectomy in man, much longer than the half-life of active renin of about 30-60 min. Similar results have been reported by GOTO et al. (1986). On the other hand, in some studies a very rapid decline in plasma-inactive renin has been observed in man. For example, following the administration of converting-enzyme inhibitors, plasma-inactive renin dropped by 75% within 1 h (GOTO et al. 1980). This may suggest that the kidney eliminates and/or activates inactive renin to a considerable extent. However, as pointed out before, there is as yet no direct evidence for a significant extracellular activation of prorenin either in the kidney or in other organs.

Although release of inactive renin from the kidney has undoubtedly been demonstrated in some species, the intracellular pathway of the secretory process is still debated. GALEN and colleagues (1984) studied the release of renin in a human juxtaglomerular cell tumor and found that dispersed tumor cells actively synthetize prorenin but apparently lack the machinery for packaging and condensing prorenin into protogranules for the maturation of immature granules as well as for the processing of prorenin to active renin. These cells release only prorenin into the medium. From these observations GALEN et al. (1984) proposed the existence of a dual pathway of renin secretion, one leading through the conventional maturation process and exocytosis of active renin from mature granules, the other by endoplasmic reticulum derived transport vesicles that bypass the Golgi complex and deliver their content, i.e., prorenin, directly to the extracellular space by fusion with the plasma membrane. Such a secretory pathway for prorenin without storage would be compatible with the observation that it is difficult to stimulate the acute release of prorenin. Furthermore, immunohistochemical data by LACASSE et al. (1985) have been taken to support the existence of a "constitutive" pathway of secretion. However, firm evidence for this pathway is lacking, and our own material, as discussed in Sect. 5.2 and Chapt. 6, would favor exocytosis of immature granules as the pathway for the release of prorenin, although a constitutive type of secretion cannot be excluded.

Finally, all these observations on the production and secretion of prorenin from epithelioid cells, its presence in plasma and modulation of plasma concentrations as well as its elimination, merge into the principal question of the physiological significance of prorenin in the circulation. Is prorenin release from the kidney an "intended," controlled process, or is it merely inadvertent leakage of an intermediate form during processing with no biological function. Does prorenin constitute a reservoir for active renin in blood plasma, or is it activated at specialized extravascular sites, e.g., in the interstitium of the vascular wall. For example, DZAU and colleagues (1987) identified a serine protease released from human neutrophils which is capable of activating prorenin in plasma and amniotic fluid. They speculated that this pathway may play a role in the participation of the RAS in inflamed tissue for the local control of vascular tone and permeability.

There is no ready answer available to any of these questions, and it will take considerable research effort to clarify not only these questions but also to determine which organ, in addition to the kidney, is the major source of plasma prorenin, and whether extrarenally synthesized prorenin, e.g., in the heart and the vascular system (DZAU 1987; RE 1987), has a local function.

7.8. Renin Synthesis and Secretion

As discussed extensively in the previous sections, the rate of renin secretion from epithelioid cells varies widely with the state of blood pressure control and extracellular volume homeostasis. It is obvious that depending on the intensity and duration of stimulatory or inhibitory signals an increased or decreased rate of renin release must eventually be matched by an increased or decreased rate of renin synthesis in order to maintain an appropriate supply. Although at present only few data on the control of renin synthesis are available, a discussion of this topic seems necessary in order to provide a framework into which these and forthcoming data can be integrated and along which future directions of research may be oriented.

To clarify the conditions for a functional relationship between renin synthesis and secretion, a short consideration of the quantitative aspect of this relationship appears in place. If we consider the normal laboratory rat as an example, plasma renin concentration in the conscious, unstimulated animal does not exceed the equivalent of 20 ng angiotensin I per milliliter per hour. In a 200-g rat with a plasma volume of about 10 ml this amounts to 200 ng angiotensin I/h of renin present in the circulation. Plasma half-life of renin in the circulation as reported in the literature ranges from 10 to 90 min, with an average in the order of 30 min (PETERS-HAEFELI 1971; SIEMENSEN et al. 1972; OATES et al. 1974; BIDANI and CHURCHILL 1981; KEISER et al. 1983; SESSLER et al. 1986; KIM et al. 1987). Hence, an amount of renin equivalent to 100 ng angiotensin I/h has to be replaced every 30 min, or 200 ng angiotensin I/h every hour. Assuming that extravascular (interstitial) renin does not significantly contribute to the total turnover of renin in the circulation, this figure represents the rate of renin to be secreted by the kidney in order to keep plasma renin concentration constant. Renal renin content in this species is about 2×1000 μg angiotensin I/h for both kidneys (e.g., JELINEK et al. 1986; LUDWIG et al. 1987). Consequently, under the assumed conditions, 0.01% of stored renin is released per hour, or 0.25% per day. Even if we consider a situation in which the average daily secretion is fivefold the normal rate, the fractional secretion would be only 0.05% per hour or 1.2% per day. This calculation, which may even overestimate the normal fractional secretion rate, has several consequences. (a) Under in vivo conditions, it will be virtually impossible to estimate the rate of renin synthesis by measuring changes of kidney renin content even if secretion is acutely blocked because the former is always an extremely small fraction of the latter. (b) Studies performed in vitro on renin secretion as well as on the relationship between renin secretion and renin synthesis may produce a distorted picture, as fractional secretion rates (or release rates) are usually much higher than in vivo, e.g., in the isolated perfused rat kidney between 2% and 6% per hour (NAKANE et al. 1980; VANDONGEN 1976) and in kidney slices up to 15% per hour (e.g., de VITO et al. 1970a,b; NAFTILAN and OPARIL 1981). According to the calculation shown above, a fractional secretion rate of 5% per hour is 500 times that expected in vivo. Since there is good reason to assume from morphological studies that renin stores are heterogeneous with respect to their availability for immediate secretion (Chapt. 6 and Sect. 7.7), it is conceivable that in experiments performed in vivo or in vitro different mechanisms of secretion and different levels of functional coupling between renin secretion and synthesis are unwillingly addressed. (c) In extension of this point and from a teleological point of view, there appears to be no need for a strict and immediate functional coupling between renin secretion and renin synthesis in vivo because the supply is so large in comparison to the fractional secretion rate. In contrast, under in vitro conditions with a very high fractional secretion rate, even a high threshold for synthesis, if existing, may have been reached already under "basal" conditions of release.

With these quantitative aspects in mind, the reader is referred to Fig. 7.6, where several possibilities for a functional coupling between renin synthesis and secretion are shown in a schematic and simplified fashion. Panel I of the figure illustrates the "conventional" view of this relationship, i.e., a single stimulus induces both secretion by exocytosis from mature, fusogenic granules positioned close to the plasmalemma, as well as de novo synthesis of renin, perhaps utilizing the same intracellular messengers. Newly synthesized pre-prorenin is then processed to prorenin and renin through the Golgi-proto-granule-juvenile-granule-mature-granule pathway. If this view were correct, newly synthesized renin queued up last in the waiting line would not be secreted into the plasma for several weeks. One would also predict from this scheme that inhibition of renin synthesis, e.g., inhibitors of ribosomal protein synthesis, would not affect renin secretion, since there is a sufficient supply for even a prolonged increase in secretion rate. Some experimental data seem to support this view. De Vito et al. (1970a, b) as well as Katz and Malvin (1982) found renin release from rat kidney slices not to be impaired when protein synthesis was blocked by either puromycin or cycloheximide. Similarly, Atkinson et al. (1983) reported that the increase in plasma renin activity induced by clipping the renal artery in uninephrectomized rats was not attenuated by a moderate dose of cycloheximide, which did not produce systemic hypotension. On the other hand, Bunag et al. (1970), who studied the effect of puromycin on renin release in the anesthetized dog, observed that basal renin release was not affected, but the increase following renal artery constriction was attenuated. The authors concluded from these observations that two different renin stores exist, one available for continuous release, the other for stimulated renin release, the latter depending on new synthesis. In contrast, Vandogen (1976) reported that in the isolated perfused rat kidney both basal and isoproterenol-stimulated renin release were depressed when the animal was treated with a very high dose of puromycin before removal of the kidney. Taken together, these studies do not permit any firm conclusion as to the relationship between renin secretion and synthesis. General toxic effects of the protein synthesis inhibitors, which have been used in high concentrations, may have contributed to the results. Furthermore, the high fractional release rate in the in vitro preparations may have placed the preparation into a situation where the renin pool destined for rapid release was no longer available.

Panel II of Fig. 7.6 depicts the possibility of different stimuli for secretion and synthesis. For example, one might speculate that in face of a multitude of stimuli for an acute increase in secretion, only a single stimulus of sufficient duration and intensity is the adequate signal for a given cell to increase its rate of renin synthesis. This specific signal might indicate a decreased sodium chloride load at the distal tubule for epithelioid cells in contact with the macula densa and other parts of the distale tubule or a decreased perfusion pressure for granulated cells in the afferent arteriole. However, there is as yet no evidence either in favor or against such a possibility (cf. Sect. 4.1.2 and 11.2).

Panels III and IV of Fig. 7.6 suggest that the stimulus for renin secretion acts exclusively on the secretory mechanism, and it is the depletion of a certain amount of mature renin, the increase in membrane material recycled to the interior, or an other associated event which signals the need for replenishment of the stores to the maturation process and the nucleus, either sequentially (panel IV) or in parallel (panel III). If this sequence of events controls renin synthesis, one would expect that in vitro preparations with a much higher fractional secretion rate than occurring in vivo would exhibit a much higher level of renin gene expression than the kidney in vivo. No data on this question are available.

Panel V of Fig. 7.6 illustrates a situation which, in its simple form, is very unlikely to reflect the relationship of renin secretion and synthesis. According to this scheme, the first response to a given stimulus would be an increased rate of renin gene expression, the increased rate of renin synthesis would then accelerate or augment the subsequent reaction sequence of processing and storage as well as immediate secretion. The latter sequence, which, in some secretory systems, does not include storage of the secretory product, is called "constitutive secretion." It is executed in secretory systems where continuous secretion without short-term changes of the secretion rate is required, as, for example, in the secretion of plasma proteins by hepatocytes. As mentioned before, this cannot be the principal type of control for renin secretion, if alone for the reason that renin secretion depends on stores and responds to secretory stimuli within seconds, while control of secretion by gene expression usually requires hours to adapt to a stimulus.

Nevertheless, this constitutive type of secretion is being discussed as an additional pathway to that of regulated renin secretion in epithelioid cells (cf. Morris 1986), since Galen et al. (1984) and Pinet et al. (1986) described the secretory behavior of a human juxtaglomerular cell tumor in culture and a cell line derived from the tumor by transfec-

Fig. 7.6 Possible interrelationships between renin secretion and renin synthesis. Schematic view of an epithelioid, renin-secreting cell. The regular pathway of renin secretion *(solid arrows)* includes gene expression in the nucleus *(N)*, translation (not shown), processing of prorenin to renin and condensation of protogranules and juvenile granules to mature storage granules *(STOR)*, the surface contact of secretory vesicles *(SV)*, and exocytosis. This system is stimulated by external stimuli *(St)*.

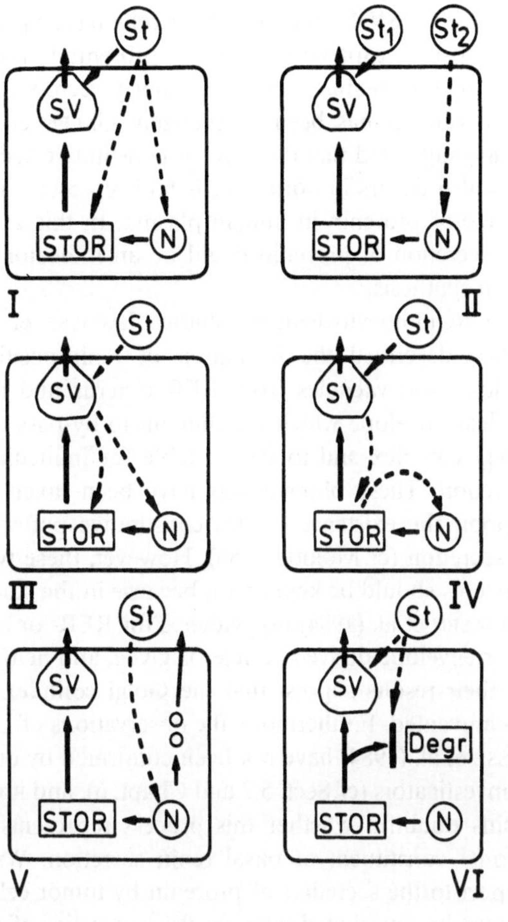

Panel I depicts the conventional view: a single stimulus activates both the secretory process and renin synthesis in parallel.
Panel II assumes that renin secretion and renin synthesis may be controlled by different mechanisms.
Panel III indicates the possibility that the external stimulus acts exclusively on the secretory process, the signal for synthesis being provided by the depletion of secretory vesicles or an associated event (e.g., membrane recycling from emptied vesicles).
Panel IV, same as in panel III, except that the signal acts sequentially "backwards" from secretory to storage vesicles to synthesis.
Panel V describes the existence of a so-called "constitutive" pathway of secretion in addition to the conventional pathway. A stimulus acts directly and perhaps exclusively on the renin gene expression. The newly synthetized renin (or prorenin) is only partly stored, and a major portion is transported directly to the cell surface. See text for discussion.
Panel VI introduces another site of control of renin stores; it assumes that part of the stored renin is continuously degraded by enzymatic cleavage. A signal for renin secretion could, at the same time, prevent further degradation of renin, thus increasing the supply available for secretion. *Degr*, lysosomal degradation. See also discussion in the text

tion with a SV 40 mutant. These cells have apparently lost the capability to convert prorenin to renin and to store renin. They continuously secrete prorenin once it has been synthesized. GALEN et al. (1984) suggested that this type of constitutive secretion also occurs in normal cells and represents the source of prorenin in human plasma. In this case, the secretion of prorenin could be an indicator of renin synthesis.

In immunocytochemical studies LACASSE et al. (1985) described the formation of renin-positive vesicles and vacuoles from RER cisterns and the nuclear envelope which are thought to by-pass the Golgi complex and to be available for immediate secretion. These observations have been taken to support the existence of this constitutive pathway of secretion (cf. MORRIS 1986). However, these two concepts should be kept apart, because in the study of GALEN et al. (1984) no evidence for RER- or nuclear envelope-derived vesicles is given, and neither do their results suggest that the Golgi complex is circumvented. Furthermore, the observations of LACASSE et al. (1984) have not been confirmed by other investigators (cf. Sect. 5.2 and Chapt. 6), and it remains doubtful whether this process represents a normal constituent of basal renin secretion. With respect to the secretion of prorenin by tumor cells, it must be questioned whether the properties of an obviously defective type of secretion in these cells are pertinent to the physiological situation in normal cells. Nevertheless, the observation by PINET and colleagues (1986) that prorenin secretion from these cells can be doubled by agents known to increase intracellular cAMP is of great interest, because it suggests that this intracellular signal, which in normal epithelioid cells is thought to induce the fusion of storage vesicles with the plasmalemma, also acts directly on the transcription and/or the translation process. It should be recalled, however, that such results may also be explained by the exocytosis of juvenile granules containing prorenin.

Finally, still another possibility is shown in panel VI of the figure. In this scheme, prorenin and renin are synthesized continously at a higher rate than needed for secretion, and the surplus renin is continuously degraded by "lysosomal" digestion (cf. TAUGNER et al. 1986a). Adjustment of the renin stores to changes in the stimulation level of secretion could then be accomplished by alteration of the rate of renin degradation. Although there is no direct evidence for this pathway, the lysosomal nature of renin granules, including the colocalization of renin and proteases in these granules, as discussed in Sect. 5.3, make this at least a feasable way of how epithelioid cells may respond to a changing demand on its secretory capacity. There is little doubt that enzymatic degradation of stored renin does occur. This is indicated by the common observation that with complete inhibition of renin release from a kidney, e.g., by constriction of the renal artery of the contralateral kidney or by aortic coarctation, its renin content declines to very low levels within a few days. Based on this observation, MOFFET et al. (1986) have suggested that this metabolic event could be important in regulating the ultimate levels of plasma renin activity. It may therefore have a significant function in addition to other regulatory pathways, such as depicted in panels I to IV.

More insight into the functional coupling of renin secretion and renin synthesis is to be expected from studies which permit the separate measurement of both events, e.g., by the estimation of the nuclear transcription rate of the renin gene, the concentration of renin mRNA, the rate of renin synthesis by pulse-chase experiments, and the posttranslational metabolism of kidney renin. Of these possibilities, only the estimation of renin mRNA in the kidney under various conditions has been described so far (MORRIS et al. 1984; DARBY et al. 1985; CATANZARRO et al. 1985; NAKAMURA et al. 1985; LUDWIG et al. 1986, 1987; MOFFET et al. 1986). In these studies performed in various species, renin secretion had been stimulated by sodium depletion, diuretic treatment, treatment with converting enzyme inhibitors, or by renal artery constriction. Consistently, a two- to fivefold increase in kidney renin mRNA was found, whereas the increase in renin secretion rate, as reflected by plasma renin activity, was usually between 5- and 20-fold. The only exception was reported by DARBY et al. (1985), who depleted sheep of sodium by a parotid duct fistula and after 3 days found a ten-fold increase in plasma renin activity, while kidney renin mRNA increased 25-fold. Although these studies clearly demonstrate that renin gene expression in the kidney is controlled by the same factors as secretion, they do not provide any information on the interrelationship between secretion and synthesis or on the nature of the signal for an increased transcription rate because the earliest time of measurement after the onset of stimulation was 3 days. More relevant information can be expected if a closer examination of the time course of the induction process becomes possible, including measurement of the transcription rate. This has been attempted by NAKAMURA et al. (1985). These authors first stimulated renin secretion in rats by sodium depletion in combination with captopril treatment for 15 days and then inhibited renin release abruptly be the infusion of angiotensin II. Complete inhibition of renin release was evident from a 84% decline of plasma renin activity within 1 h. However, there was no change of kidney renin mRNA during this 1 h period. As pointed out by the authors, this does not necessarily indicate that mRNA synthesis was not inhibited by the treatment with angiotensin II because the lack of change of renin mRNA could have also been due to the stability of mRNA already present. Recent experiments in our laboratory have shown that renin mRNA starts to increase in response to acute stimuli of secretion, e.g. forskolin in the isolated kidney, clamping of the renal artery in vivo, not before 5 hours (HACKENTHAL and MÜNTER, unpublished observations). In any case, such and similar studies will eventually provide the information necessary to unravel the delicate interactions of renin gene expression, translational control, posttranslational processing, granule maturation, intracellular translocation, and the secretory event.

Chapter 8

Electrophysiology of Epithelioid Cells

In a variety of exo- and endocrine systems, there is evidence for a control of the secretion via changes in the membrane potential of the involved secretory cells. These changes may be caused by humoral agents or by the action of neurotransmitters. As far as renin secretion by granulated cells is concerned, in addition to circulating humoral factors and the activity of sympathetic nerve terminals, two mechanisms influencing the membrane potential of these cells are conceivable: changes in wall tension of the preglomerular arteriole and the so-called macula densa signal. The identification of possible membrane effects associated with stimulation or inhibition of renin release appeared all the more important, as, in contrast to other secretory processes, there seems to be an inverse correlation between renin release and intracellular free calcium concentration. Yet, the only notion on this topic available until 1984 was derived from the experiments of FISHMAN (1976).

FISHMAN (1976) investigated membrane potential changes of epithelioid cells adhering to isolated glomeruli from mouse kidney. He found resting potentials in the order of -45 mV and reported on hyperpolarizing effects of epinephrine. Assuming that bath concentrations of $10^{-4} M$ epinephrine stimulate renin release, he concluded that hyperpolarization enhances, whereas depolarization as produced by high extracellular K^+ suppresses, renin secretion. By implication, this concept has been extended to other substances, e.g., angiotensin II and, taking into account the inhibitory action of Ca^{2+} on epithelioid cells, the suppression of renin secretion attributed to the opening of Ca^{2+} channels, increasing Ca^{2+} influx (CHURCHILL and CHURCHILL 1982a).

However, the results of FISHMAN (1976) could not be confirmed in experiments by BÜHRLE and NOBILING in our laboratory. First, clearly higher resting potentials between -60 and -75 mV were found with stable impalements. Second, epinephrine, by its α_1-mimetic component, led only to depolarizations of epithelioid cells; hyperpolarizations were never observed, not even after application of compounds with exclusive β-mimetic properties like isoproterenol. The hydronephrotic kidney preparation of rats (STEINHAUSEN et al. 1983; NOBILING et al. 1986) and mice (BÜHRLE et al. 1984, 1986a) proved to be very valuable for these electrophysiological experiments. In both species, 8-12 weeks after unilateral ureteral ligation, the tubular system of the involved kidney has atrophied, leaving only the renal vasculature intact, including glomeruli and the epithelioid cells in the walls of the afferent arterioles. The excised and superfused kidney tissue used for these experiments was a transparent sheet of about 150-220 μm in thickness particularly suitable for microelectrode impalement. After determining in separate experiments the lengths of the renin-positive portions of the preglomerular arterioles, the juxtaglomerular granulated as well as the smooth muscle cells located further upstream could be impaled under visual control and the results from both cell types compared.

Angiotensin II, AVP, and α_1-adrenergic agents reversibly depolarized epithelioid cells. On a molar basis, angiotensin II was strongest. Isoprenaline and orciprenaline, histamine, prostaglandin E_2, VIP, NPY, and atrial natriuretic peptide did not influence the membrane potential of either cell type (BÜHRLE et al. 1986b). As all substances of the first group have an inhibitory effect on renin secretion, while those of the second group, except NPY, increase the secretion of the enzyme, it was proposed that suppression of renin secretion from granulated cells is mediated or assisted by depolarization and calcium influx, whereas stimulation of renin release is triggered independently from membrane potential changes, e.g., by adenylate cyclase activation (cf. Sect. 7.6). The observation that NPY, which inhibits renin secretion, does not depolarize epithelioid cells, is not inconsistent with such a general scheme, since HACKENTHAL et al. (1987) have shown that NPY probably exerts this effect by inhibition of adenylate cyclase and not by a calcium-dependent mechanism.

Surprisingly, identical electrophysiological results were obtained in granulated cells and smooth muscle cells of the afferent arteriole. This applies not only to their resting potential, but also to their reactions upon application of the above-mentioned substances and to spontaneous depolarizations resembling excitatory junction potentials in vascular smooth muscle cells of blood vessels in different organs of various species (BÜHRLE et al. 1986c). This favors the assumption that - in spite of the fundamental morphological and functional changes during metaplastic transformation of these cells - their equipment with membrane receptors mediating events detectable with electrophysiological techniques is conserved.

Intracellular recordings of granulated cells in the hydronephrotic kidney preparation have also been used to investigate the intrarenal generation of and the tachyphylaxis induced by angiotensin II (BÜHRLE et al. 1987a, b), and the electrotonic coupling between media cells of the preglomerular arteriole (BÜHRLE, NOBILING and TAUGNER, unpublished observations; cf. Sect. 2.7 and Chapt. 9).

Chapter 9

Intrarenal Renin-Angiotensin System

The actions of angiotensin II in the kidney are often described in terms of the intrarenal RAS; a recent symposium carried the title "Physiological Role of the Intrarenal Renin-Angiotensin System" (NAVAR 1986). Obviously, the intention of using this term is to suggest that the intrarenal effects of angiotensin II are not, or not exclusively, mediated by systemically generated angiotensin II reaching the kidney via the circulation but rather by intrarenally formed angiotensin II. However, beyond the general agreement that local angiotensin II formation in the kidney occurs, there is considerable uncertainty and disagreement on the understanding of an intrarenal RAS. Therefore, we will first discuss briefly the characteristics of locally functioning RASs in general, consider the potential for the intrarenal formation of angiotensin II, then proceed to describe the intrarenal actions of exogenous and endogenous angiotensin II, and finally return to the consideration of whether an intrarenal RAS exists.

9.1 Local Renin-Angiotensin Systems

Basically, the concept of an intrarenal RAS has two major sources, one relating to the kidney itself, the other seeking parallelisms to local RASs in other organs. With regard to the kidney, the unique anatomical and functional arrangement of the JGA and adjacent structures has intrigued many investigators to speculate that renin secretion from epithelioid cells of the afferent arteriole serves the dual purpose of supplying the systemic circulation with renin and of providing the renal interstitium with the key enzyme for the extravascular generation of angiotensin II, which then controls renal functions in a "short-circuit" fashion and without systemic dilution. These intrarenal functions, thought to depend on or to be modulated by interstitially generated angiotensin II, include TGF and other vasoconstrictor effects of angiotensin II neccessary to maintain GFR, as well as tubular effects. This concept was strengthened by the demonstration of renin, angiotensin I, and angiotensin II in renal hilar lymph, which drains the renal cortical interstitium, and by the recent discovery of angiotensin II in granulated cells of a few species (CELIO and INAGAMI 1981; TAUGNER and HACKENTHAL 1981). These aspects will be discussed in more detail below.

A second major line of argumentation in favor of an intrarenal RAS is indirect, that is, the extrapolation from a local RAS in other organs. For several decades, the RAS has been described as a systemic, hormonal system, in which the effector peptide angiotensin II is generated within the circulation and reaches various target organs, such as the adrenal cortex, brain, vascular smooth muscle, liver or intestine via the blood stream, interacts with specific receptors, and induces changes collectively aimed at maintaining blood pressure and salt and water homeostasis. More recently, the local occurrence of the components of the RAS within several organs has been described. These organs seem to be capable of generating angiotensin II without requiring all, or sometimes even any, of the components of the systemic circulating RAS.

The best studied local RAS so far is that of the brain. Facilated by the very limited diffusion or uptake of macromolecules from the blood into the interstitium of the brain, the easy experimental access to cerebrospinal fluid, as well as the advent of new immunocytochemical and molecular biological techniques, it has now been established that the brain not only contains all the components necessary to produce angiotensin II, but also that angiotensin II is indeed produced locally and influences certain functions of the brain independently of the systemic RAS (see reviews by GANONG 1984; GANTEN et al. 1984; REID 1984; GANTEN et al. 1987).

The RAS of the brain not only operates largely independently of the systemic RAS, it also appears to operate differently from the latter. In contrast to the intravasal or extracellular generation of angiotensin II in the systemic RAS, angiotensin II in the brain has the characteristics of a neuromodulator, i.e., it is synthetized intraneuronally, translocated by axonal transport, and released upon neuronal stimulation to modulate the functions of other neurons. In accordance with the presumed intracellular generation of angiotensin II, renin does not seem to be secreted from neuronal tissue, and does not enter the cerebrospinal fluid. The intracellular precursor for angiotensin II has not yet been identified. Although the cerebrospinal fluid contains angiotensinogen, and angiotensinogen gene expression has been demonstrated in several areas of the brain (HEALY and PRINTZ 1984; CAMPBELL and HABENER 1986; OHKUBO et al. 1986; DZAU et al. 1987), neither uptake into nor intracellular synthesis of angiotensinogen in angiotensin II-producing neurons has been demonstrated. Altogether, the brain RAS appears to differ profoundly from the systemic RAS, even though many facets of this system are still unknown.

Many other local RASs have been postulated to exist, mainly because the local occurrence and/or synthesis of renin has been demonstrated by immunocytochemistry or by the isolation of renin mRNA. However, in none of these organs has the local synthesis of all components of the RAS been unequivocally demonstrated.

One of the reasons for this unclear situation is the fact that the separation of locally formed renin and/or angiotensinogen from plasma-derived components of the RAS is difficult to accomplish, if not impossible. For example, an arterial wall RAS has been postulated to be responsible for maintaining vascular tone independent of short-term changes of plasma renin activity (SWALES et al. 1983; OLIVER and SCIACCA 1984; DZAU 1986, 1987). There is indeed evidence for the synthesis of renin in vascular smooth muscle and endothelial cells leading to the local production of angiotensin II (RE et al. 1981; DAHLHEIM et al. 1983; LILLY et al. 1985). However, the major, if not exclusive, source of interstitial angiotensinogen in the vascular wall seems to be the circulation, and also a major part of interstitial renin is of systemic origin (LOUDON et al. 1983). Furthermore, in contrast to the situation in the brain, angiotensin II appears to be generated mainly extracellularly, although intracellular synthesis and secretion may contribute to total interstitial angiotensin II (DZAU 1984; NAKAMURA et al. 1986; KIFOR and DZAU 1987). Thus, as pointed out be CAMPBELL (1985, 1987), the implied distinction between the systemic circulating RAS and local tissue systems is by no means clear-cut, and both systems can in general be assumed to integrate in the local formation of angiotensin II. CAMPBELL (1985) carried this idea even further by suggesting that the primary function of the circulating RAS is not the systemic delivery of angiotensin II, but rather that of renin and angiotensinogen to the tissues, the angiotensins found in plasma representing spill over from the tissue sites of production.

In summary, the concept of local tissue RASs is not a uniform one. Large differences in the degree of independence from the systemic RAS as well as differences in the mode of operation seem to exist between various organs, e.g., extra- versus intracellular generation of angiotensin II. Therefore, with respect to the kidney, a comparison with RASs in other organs will be of little help in deciding whether an intrarenal RAS exists. Rather, it will be necessary to determine to what extent intrarenal angiotensin II formation occurs extravascularly and independently of systemic components, and whether this interstitial angiotensin II has functions different from those of systemically delivered angiotensin II.

9.2 Intrarenal Formation of Angiotensin II

With regard to the existence and functions of an intrarenal RAS, several sources and/or topographical pathways for the generation and the access of angiotensin II to vascular and tubular sites can be envisioned. (a) Angiotensin II is generated in the systemic circulation, carried into the kidney, and reaches its targets directly from the vascular lumen or via the peritubular capillaries. This includes angiotensin II that has been generated from angiotensin I within the renal circulation by plasma-converting enzyme or by converting enzyme of endothelial cells. Alternatively, angiotensin II may be formed extravascularly from systemic angiotensin I that has left the circulation via the peritubular capillaries and/or at other sites. (b) Angiotensin II is thought to be generated in the interstitium of the kidney by newly secreted renin acting on plasma-derived angiotensinogen that has escaped into the interstitium via the peritubular capillaries and by subsequent conversion of angiotensin I to angiotensin II by interstitial converting enzyme. (c) There are also data which indicate that angiotensin II can be generated within the kidney by nonrenin and/or nonconverting enzyme dependent pathways. (d) Angiotensin II may be formed intracellularly from angiotensinogen that has been synthetized in epithelioid and/or other cells or in the extracellular space subsequent to the release of angiotensinogen into the interstitium.

9.2.1 Systemic Origin of Angiotensin I and II

Considering the intravascular generation of angiotensin II, it must first be remembered that the blood entering the kidney contains angiotensin II in concentrations (e.g., about $5 \times 10^{-12} M$ in human and $5 \times 10^{-11} M$ in the rat) sufficient to elicit the known vascular and tubular responses and inhibition of renin release. On the other hand, angiotensin II is extensively degraded in the kidney. According to ROSIVALL et al. (1987) almost 90% of angiotensin II entering the kidney with the circulation is destroyed or extracted during passage through the kidney. Therefore, in order to make up for most of the renal venous angiotensin II (and perhaps for intrarenally active angiotensin II) the kidney must also produce angiotensin II. First, intra- and extravascular conversion of angiotensin I to angiotensin II will be considered as a source of this angiotensin II.

The rate of conversion of arterial angiotensin I to angiotensin II during passage through the kidney has been calculated in several studies from the ratio of the vasoconstrictor dose-response curve for the two peptides applied exogenously. In the normal blood-perfused dog kidney, a conversion rate of 10% was obtained (FRANKLIN et al. 1970), whereas in the isolated perfused rat kidney conversion rates of 2%-3% were found (HOFBAUER et al. 1973; SCHMIDT et al. 1986). These data are consistent with each other taking into account that in the isolated rat kidney perfused with a synthetic medium, flow rates are at least three times the normal plasma flow rate, and hence the exposure time to intrarenal converting enzyme is correspondingly shorter. Angiotensin I conversion has also been measured directly by infusion of tracer doses of radiolabeled angiotensin I into the renal artery of the dog and found to be 20% (ROSIVALL et al. 1983).

These studies did not clarify whether the conversion of angiotensin I was confined to the intravascular compartment. Converting-enzyme activity can be found over the endothelial layer along the whole vascular tree of the kidney, as revealed by immunohistochemical studies (TAUGNER and GANTEN 1982), and the glomerulus has also been shown to contain considerable concentrations of converting-enzyme activity (e.g., BURGHARDT et al. 1982; CHANSEL et al. 1987). However, converting-enzyme activity is most likely also present on cell surfaces facing the renal interstitium (cf. Sect. 4.3). In addition, converting-enzyme activity has been found in renal lymph (HORKY et al. 1971; PROUD et al. 1984), suggesting that this enzyme may be present in a soluble form in the renal interstitium. Later, ROSIVALL and colleagues (1987) attempted to calculate the portion of net intrarenal formation of angiotensin II from angiotensin I that appears in the renal vein of the dog kidney. They concluded that 80% of renal venous angiotensin II was formed in the kidney, corresponding to a generation rate of about 40 pg angiotensin II/g kidney weight per minute. Taking the data given in these studies for plasma renin activity of 10-15 ng angiotensin I/ml per hour, a renal plasma flow of 2 ml/g and an intrarenal conversion rate of 20%, exclusive intravascular generation of angiotensin II by the cleavage of angiotensinogen and subsequent conversion of angiotensin I could well account for the intrarenal angiotensin II generation rate and the concentrations found in renal vein.

Another interesting aspect that can be derived from the studies of ROSIVALL et al. (1983, 1987) as well as others is that renal venous angiotensin II concentrations are always lower than arterial concentrations by a factor of 2 or more. This means that the kidney extracts more angiotensin II than it contributes to venous outflow. Since, as explained above, a considerable portion, if not all, of the angiotensin II measured in renal venous plasma, probably originates from conversion of angiotensin I within the circulation, the contribution of interstitially generated angiotensin II to venous angiotensin, although existent, must be small.

This seems to be in contrast to many other organs which are thought to contribute actively to plasma angiotensin II by interstitial generation of both angiotensin I and II and spill-over of these peptides into the circulation (cf. CAMPBELL 1985; DZAU 1987). In any case, the role of the renal interstitium raises two questions: What are the interstitial concentrations of angiotensin I and II? Is interstitial angiotensin II derived from plasma angiotensin I and II, or is it generated from angiotensinogen within the interstitium?

9.2.2 Components of the Renin-Angiotensin System in the Renal Interstitium and Renal Lymph

Although the direct measurement ore immunohistochemical identification of components of the RAS in the interstitial space of the kidney has not yet been accomplished, it is generally assumed that the renal interstitium contains high concentrations of angiotensin II formed within the interstitium from angiotensinogen. This assumption rests mainly on the following arguments and observations.

Renin. The interstitial concentrations of renin are probably high, at least in the interstitial space surrounding renin-secreting cells. This conclusion can be derived from the consideration that renin secreted by a very limited number of renin-secreting cells first distributes in a relatively small interstitial space before it reaches the circulation into which it is diluted. Furthermore, renin has been found in renal hilar lymph at concentrations usually three to tenfold those of plasma renin in the renal vein or in the systemic cirulation (LEVER and PEART 1962; HOSIE et al. 1970; O'MORCHOE et al. 1981; PROUD et al. 1984; DZAU 1986; HACKENTHAL et al., unpublished observations; cf. Table 9.1). Since renal hilar lymph drains the renal cortical interstitium, and dilution of renin on its way from the juxtaglomerular interstitium to renal lymph is likely to occur, the true interstitial concentrations of renin at the site of secretion may be considerably higher than those found in renal lymph (cf. Sect. 3.4).

Converting Enzyme. As mentioned above and as discussed in detail in Sect. 4.3, it is likely that converting enzyme is available in the interstitium both in cell-bound and free form. The latter conclusion is derived from the demonstration of ACE activity in renal lymph (HORKY et al. 1970; PROUD et al. 1984).

Table 9.1. Components of the renin-angiotensin system in rat renal lymph

	Renal lymph	Plasma
Renin (ng AI/ml per hour)	75 ± 30	28 ± 8
Angiotensin I (pg/ml)	1133 ± 200	208 ± 35
Angiotensin II (pg/ml)	106 ± 8	69 ± 8
angiotensinogen (ng ANG I/ml)	463 ± 51	862 ± 72

Renal lymph was obtained by cannulating one of the major hilar lymph vessels while ligating the other visible lymphatics. Collected lymph was immediately cooled to prevent further enzymatic activity. Renin, angiotensin I (A I) or II, and angiotensinogen were estimated by conventional radioimmunoassay techniques (HACKENTHAL, SCHWERTSCHLAG, DUSSEL, and STEINHAUSEN, unpublished data)

Angiotensin. Angiotensin I and II have been found in renal hilar lymph (LEVER and PEART 1962; BAILIE et al. 1970, 1971; PROUD 1984; EMMERSON 1985; INAGAMI et al. 1986; HACKENTHAL et al.; cf. Table 9.1). However, in contrast to renin, reported concentrations for angiotensin II were extremely variable from one study to another, ranging from 100 pg/ml (our own study), 70-1500 pg/ml (BAILIE et al. 1970, 1971), about 3000 pg/ml (INAGAMI et al. 1980), up to 60000 pg/ml (EMMERSON and JOHNS 1985). Concentrations of angiotensin I, when measured, showed similar variability, e.g., 1100 pg/ml in our own study and 170000 pg/ml in the study of PROUD et al. (1984). These large variations in the concentrations of both angiotensin I and II probably reflect a methodological problem: we have shown that rat renal lymph contains angiotensinogen in high concentrations (see below). Thus renin, also present in lymph in high concentrations, continuously generates angiotensin I, which is partly converted to angiotensin II by converting enzyme. It therefore depends on the lymph flow rate, the length of the tubing used for sampling, and the method used to arrest further enzymatic activity in collected lymph, how much of the angiotensinogen has been metabolized to the angiotensins. In any case, it appears likely that a major portion of angiotensin I and angiotensin II measured in hilar lymph has been generated after the lymph has left the kidney. Hence, the measured concentrations of both angiotensin I and II in renal lymph are of questionable significance for the prediction of interstitial concentrations (see further discussion of this aspect below).

Before turning to angiotensinogen, another approach to the problem of intrarenal angiotensin generation should be considered, that is, the measurement of whole kidney angiotensin II under a variety of conditions as described by MENDELSOHN (1976, 1979, 1980, 1982). In these studies MENDELSOHN demonstrated that the rat kidney contains angiotensin II in concentrations exceeding those found in arterial blood of the same animal, which can therefore not be attributed to blood contamination. The origin and the exact localization of angiotensin II could not be determined, and it remained unknown whether the angiotensin II extracted from the whole kidney was localized in the interstitium, or intracellularly. An intracellular generation would be compatible with the finding of immunoreactive angiotensin II in renin-secreting cells of the rat kidney (cf. Sect. 4.2). In view of the presence of angiotensin II in renal lymph (see above), it is also conceivable that periarterial and lymph angiotensin II has contributed to these results. Since plasma renin activity, kidney renin content, and kidney angiotensin II did not always change in parallel during various experimental maneuvers, MENDELSOHN concluded that angiotensin represents an intrarenal hormone acting locally to modify renal function. Another indication for the local generation of angiotensins is the report of a continuous release of immunoreactive angiotensin I and II from the isolated perfused rat kidney (Mizuno et al. 1987). However, there is no evidence for an interstitial origin of this angiotensin II, and the occasional divergence between plasma renin activity and intrarenal angiotensin II may have other explanations.

Angiotensinogen. As described recently (TAUGNER et al. 1982a, c; RICHOUX et al. 1983; cf. Sect. 4.4), angiotensinogen could not be detected in the rat kidney by immunohistochemical techniques, except for some immunoreactivity over proximal tubular cells, reflecting uptake and intracellular degradation of filtered angiotensinogen. There are reports on the generation of angiotensin I in kidney preparations in vitro (FINKIELMAN and NAHMOD 1969; MORRIS and JOHNSTON 1976), suggesting the presence of angiotensinogen. Furthermore, angiotensinogen mRNA has been demonstrated in the kidney by several groups (CAMPBELL and HABENER 1986; FRIED and SIMPSON 1986; INGELFINGER et al. 1986). However, there is as yet no clear biochemical evidence for a substantial intrarenal synthesis and release of angiotensinogen. Therefore, the demonstration of angiotensinogen in renal lymph is of importance for the interpretation of an intrarenal RAS. Studies by HORKY et al. (1971) revealed the presence of angiotensinogen in abdominal lymph. However, actual concentrations were not measured, and the renal origin of this angiotensinogen could not be ascertained because renal lymph contributes only a small portion to total abdominal lymph flow. In a recent study in rats, angiotensinogen was measured in renal hilar lymph, and high concentrations were found, averaging about 45% of plasma concentrations (Table 9.1). An interesting observation was that during stimulated renin secretion, e.g., by ureteral occlusion, renin, angiotensin I, and angiotensin II increased, whereas angiotensinogen decreased in renal lymph. In no instance was angiotensinogen found in higher concentrations than in plasma (HACKENTHAL et al., unpublished observations).

The demonstration of high concentrations of angiotensinogen in renal lymph raises several questions: is this angiotensinogen derived from plasma angiotensinogen or is it synthetized and released intrarenally; where does it enter the interstitial space; and is it relevant for intrarenal angiotensin formation and action. None of these questions can be answered conclusively at present. The available data seem to favor a plasmatic origin of lymph angiotensinogen. Rat and dog renal lymph has been shown to contain plasma proteins at a concentration of about 20%-50% of their plasma level when measured as total proteins (KEYL et al. 1965; HENRY et al. 1969; HARGENS et al. 1977). As was to be expected, the relative proportions were not exactly the same as in plasma, as proteins with a smaller molecular weight, such as albumin, were represented in relatively higher proportion than high-molecular-weight proteins, the average albumin concentration in renal lymph being about 40% of the plasma concentration. These observations clearly demonstrate that, as in other organs, plasma proteins partially equilibrate with the interstitial fluid of the kidney, the equilibration rate (or the volume of distribution) of individual proteins depending on their molecular size. The concentration of angiotensinogen in renal lymph relative to plasma concentration predicted from its molecular weight of about 54 kd would be slightly higher than that of albumin, with a molecular weight of 67 kd, i.e., between 40% and 50%, which is exactly what has been found in rat renal lymph (Table 9.1). It would have been surprising, indeed, if there were no angiotensinogen in renal lymph.

The functional significance of the presence of angiotensinogen in renal lymph is crucially dependent on whether the composition of renal lymph reflects the composition of interstitial fluid throughout the renal cortex. Thus, an even distribution of angiotensinogen in the interstitium would permit angiotensin formation wherever renin is present. If, on the other hand, angiotensinogen enters the renal lymphatics and is drained to the renal hilus without significant distribution in the interstitium, its extravascular presence could be without any significance for renal function. It is generally held that the composition of renal lymph largely reflects the composition of the renal interstitial fluid. For example, GÄRTNER and colleagues (1973) measured extracellular distribution volumes of several molecular weight markers in comparison to their concentration in renal lymph and came to the conclusion that renal lymph is identical to renal interstitial fluid. This is probably an oversimplification, if only for the simple reason that formation of renal interstitial fluid and renal lymph is a dynamic process that is controlled by a variety of different mechanisms, such as pressure gradients directing flow, tubular water reabsorption, osmotic gradients, and anatomical structures impeding free distribution and equilibration.

There is also a theoretical argument against an equal distribution of angiotensinogen in the renal interstitium. If angiotensinogen were present in the interstitium surrounding renin-secreting cells in the same concentrations as found in renal lymph, this would result in an extremely high generation rate of angiotensin I and angiotensin II (provided converting enzyme is available), similar to those found in renal lymph. Such concentrations, far in excess of physiologically relevant concentrations, would lead to saturation of angiotensin II receptors and to complete loss of sensitivity towards vascular angiotensin II reaching the kidney in physiological concentrations. Since this is not the case, one must assume that the rate of generation of angiotensin II in the interstitium close to sensitive structures of the vessels and tubules is low, and hence the concentration of angiotensinogen is low.

Although it is not known at present where angiotensinogen enters the lymphatics, or what the local concentrations are, the most likely pathway of angiotensinogen from plasma into renal lymph is that proposed for the other plasma proteins occurring in renal lymph (cf. HENRY et al. 1969; KRIZ and NAPIWOTZKI 1979; JONES and O'MORCHOE 1983; NIRO et al. 1986; KRIZ 1987): The intrarenal arteries of the kidney are surrounded by a broad and continous periarterial sheath of loose connective tissue, which has recently been characterized in more detail by KRIZ (1987). This periarterial sheath is intimately connected with lymphatic capillaries, as well as with peritubular capillaries and veins, both equipped with a fenestrated endothelium. The morphological features of these structures as well as tracer studies (cf. KRIZ 1987) indicate that an extensive exchange of fluid and solutes, including proteins, between vascular structures and the periarterial interstitium is possible.

It may therefore be assumed that the fluid travelling in the periarterial interstitial space is continuously loaded with plasma protein including angiotensinogen from the peritubular capillaries and the fenestrated veins. The interstitial concentrations of angiotensinogen are therefore probably low at the level of the distal afferent arteriole, and increase continuously as the fluid moves along the interlobular and arcuate arteries, equilibrates with lymphatic capillaries and is eventually drained into the renal hilar lymph vessels. The capacity to generate angiotensin I and II would therefore also be low in the vicinity of the glomerulus and probably increase along this path in the periarterial interstitium until reaching renal hilar lymph. Subsequently, renal lymph is diluted by lymphatic flow from other organs, and hence angiotensin formation again decreases. This is, of course, little more than a speculative view of how angiotensinogen may reach renal lymph and interact with renin. Clearly, this situation calls for immunohistochemical studies on the distribution of angiotensinogen and renin in the renal interstitium. Unfortunately, suitable techniques are not yet available to localize and immunostain soluble substances to the extracellular space.

Recently, ultrastructural features in the region of the vascular pole of the glomerulus have been described by ROSIVALL and TAUGNER (1986) which would permit bulk fluid flow into the interstitium of the JGA. It remains to be clarified, however, whether such fluid, which was thought to compete with the reabsorbate of the macula densa, would be capable of delivering angiotensinogen to the interstitium of the JGA.

9.2.3 Alternative Pathways of Angiotensin Formation

As discussed in Sect. 4.6, there is some biochemical evidence that the kidney may form angiotensin II from angiotensinogen and from angiotensin I by pathways which are not dependent on renin and/or converting-enzyme activity, and the enzymes which may participate in these alternative pathways have been described. There are also some physiological studies that seem to support the contention of alternative pathways of angiotensin formation. For example, HOFBAUER and coworkers (1973) compared the vasoconstrictor response of angiotensin II, angiotensin I, and angiotensinogen in the isolated perfused rat kidney. The surprising finding was that the vasoconstrictor response to angiotensinogen was not affected by a converting-enzyme inhibitor (SQ 20881) even when the concentration was 15 times that needed to completely suppress the vasoconstrictor response to angiotensin I. While the study of HOFBAUER et al. (1973) suggests a non-renin-dependent formation of angiotensin from angiotensinogen in the kidney, other studies indicate the possibility that angiotensin I may be converted to angiotensin II by enzymes different from the "classical" converting enzyme. Thus, SCHMIDT et al. (1986) examined the vasoconstrictor response to angiotensin I in the isolated perfused rat kidney. Under conditions where converting-enzyme activity was completely inhibited in the kidney by captopril or ramiprilat, angiotensin I retained one-fourth of its vasoconstrictor activity. Similarly, RASSIER and coworkers (1986) found the vasoconstrictor response to angiotensin I in the dog kidney in situ attenuated only by some 30% by a dose of captopril which completely prevented conversion. Also the electrophysiological data of BÜHRLE et al. (1987b) suggest some nonrenin, nonconverting enzyme dependent generation of angiotensin II. These authors used the angiotensin II-induced depolarization of juxtaglomerular epithelioid cells of mouse kidney as an indicator for the local concentration of angiotensin II and observed that the depolarization in response to angiotensin I or renin tetradecapeptide substrate were only partly blocked by several converting-enzyme inhibitors in high concentrations or renin inhibitors whereas all responses were completely blocked by saralasin.

In summary then, some studies seem to suggest that angiotensin II can be generated intrarenally by enzymes other than renin and/or converting enzyme. However, there are also studies which fail to support this view, and the overall significance of such alternative pathways remains to be established.

9.2.4 Intracellular Generation of Angiotensin II

As discussed extensively in Sect. 4.2, the origin of angiotensin II found in renin-containing granules of epithelioid cells in the rat kidney is a matter of debate. Both intracellular synthesis and uptake from the interstitium have been suggested. Regardless of the origin, it has not been possible to verify whether angiotensin II is released together with renin, although this is to be expected in view of the costorage with renin in specific granules. It has recently been reported by Mizuno et al. (1987) that both angiotensin I and angiotensin II may be continuously released from the isolated perfused rat kidney. Yet even if released together with renin, the resulting interstitial concentrations of angiotensin II would probably be very low, as judged from the immunocytochemical staining intensity of the granules, and are unlikely to have but a very circumscribed local effect. Furthermore, angiotensin II storage in epithelioid cells is not detectable in many other species (cf. Sect. 4.2).

On the other hand, the angiotensinogen gene is expressed in the kidney, as the mRNA coding for angiotensinogen has been found in the renal cortex (CAMPBELL and HABENER 1986; INGELFINGER et al. 1986; OHKUBU et al. 1986). Although the concentration of mRNA is not a direct estimate for the rate of synthesis of the gene product, the measured concentrations, in comparison to those of the liver, suggest that the synthesis of angiotensinogen in the kidney is not negligible. There is no evidence for angiotensinogen synthesis from immunocytochemical studies, although in human kidneys of patients with Bartter's syndrome or related tubulopathies some angiotensinogen immunoreactivity was detected in hypertrophied cells of Bowman's capsule (Taugner et al., 1988). There are also no indications for angiotensin I or II generation in renal cells other than epithelioid cells. Nevertheless, local synthesis of angiotensinogen and its immediate secretion, or rapid intracellular degradation and subsequent release of angiotensin I and II into the interstitium have to be considered as possible sources of interstitial angiotensinogen or angiotensin. In view of the regulation of angiotensinogen mRNA in the kidney by changes in sodium balance (INGELFINGER et al. 1986) and by various hormones (CAMPBELL and HABENER 1986), this aspect will probably attract considerable attention in the future.

9.3. Renal Effects of Angiotensin II

9.3.1 Effects on Renal Hemodynamics

There is general agreement that the RAS plays an important role in the control of renal hemodynamics through the vasoconstrictor effects of angiotensin II. However, there is also considerable disagreement in the interpretation of differences in the segmental response of the renal vasculature to angiotensin II and their functional consequences, especially with regard to autoregulation of RBF and GFR and the TGF response. Divergent results and opinions may in part derive from differences in methodology, since, for the evaluation of the intrarenal function of the RAS, not only a large variety of experimental models, but also different pharmacological approaches have been used.

First, the infusion of angiotensin II into the renal artery or its local application to exposed vessels may, as pointed out by NAVAR and ROSIVALL (1984), not provide direct insight regarding the function of intrarenally formed angiotensin II. It nevertheless permits the characterization of the responsiveness of the renal vasculature to angiotensin II.

Second, the adminstration of the synthetic tetradecapeptide renin substrate or angiotensin I mimics the in vivo conditions somewhat more closely than exogenous angiotensin II, because the site of angiotensin II formation is confined to where the necessary enzymes, i.e., renin and angiotensin-converting enzyme are located within the kidney. However, this approach may not correctly reflect the topographical distribution of angiotensinogen and thus the local rates of endogenous angiotensin II formation. In addition, renin tetradecapeptide may serve as substrate for the generation of angiotensin II by enzymes other than renin (cf. Sect. 4.6).

Third, as another means of evaluating the role of the RAS in the control of renal hemodynamics and tubular functions, pharmacological blockade of the formation or action of endogenous angiotensin II by converting-enzyme inhibitors or competitive antagonists of angiotensin II has found widespread application. These studies have considerably advanced our understanding of the intrarenal functions of angiotensin II. The disadvantages or limitations of using inhibitors lie in the fact that they have additional biological or pharmacological effects which may interfere with the functions of the RAS. For example, ACE is identical with kininase II, which inactivates bradykinin and other kinins by enzymatic cleavage. Thus, inhibition of converting-enzyme activity would not only prevent angiotensin II formation but also protect kinins from degradation. Since kinins are thought to play an important role in the control of vascular and tubular functions in the kidney (MARIN-GREZ 1982; MARGOLIUS 1984), some of the effects observed during treatment with converting-enzyme inhibitors might be due to or modified by kinins. There are also reports that converting-enzyme inhibitors stimulate intrarenal prostaglandin synthesis by an unknown mechanism (cf. OLIVER et al. 1983), an effect which may also influence renovascular responses and renin secretion. Furthermore, there are several observations suggesting alternative pathways for the intrarenal generation of angiotensin II that are not dependent on converting-enzyme activity (see preceeding sections). Also the use of competitive angiotensin II antagonists has its methodological limitations, since these antagonists, such as saralasin, have considerable agonistic effects at higher concentrations. Furthermore, quite often significant effects of inhibitors of the RAS could only be demonstrated when the RAS was stimulated, e.g., by anesthesia and/or severe salt depletion of the experimental animals.

With these limitations in mind we will briefly discuss the renovascular effects of exogenous and endogenous angiotensin II. In many different experimental models using various experimental approaches it has been established that most segments of the renal vascular tree beyond the interlobar or arcuate artery respond to angiotensin II. This includes the interlobular artery, the afferent and efferent arteriole (BLANTZ et al. 1976; BAYLISS and BRENNER 1978; DAVALOS et al. 1978; PLOTH and NAVAR 1979; BLANTZ 1980; STEINHAUSEN et al. 1983, 1985, 1986; CARMINES et al. 1986; HALL 1986b; HALL et al. 1986; HELLER and HORAZEK 1986; OLSEN et al. 1987; TUCKER et al. 1987), glomerular capillaries (HORNYCH et al. 1972), peritubular capillaries (JENSEN and STEVEN 1977), and the medullary vessels (cf. CHOU et al. 1986). There appears to be consensus that the renal vasculature as a whole is at least as responsive to the vasoconstrictor effect of angiotensin II as other vascular beds (see KIMBROUGH et al. 1977 for direct comparison). Conflicting data and opinions exist with regard to differences in the responsiveness of individual segments of the renal vasculature and their functional interpretation.

To start with the more proximal part of the arteriolar system, HEYERAAS and AUKLAND (1987) examined the response of arterial vessels proximal to the afferent arteriole to angiotensin II in the "corticotomized" kidney. They concluded that the constriction of the distal part of the interlobular artery in response to angiotensin II reflects mainly an autoregulatory response to the increase in arterial pressure and, to a smaller extent, a direct vasoconstrictor effect of angiotensin II. In contrast, STEINHAUSEN and colleagues (1986), evaluating the segmental vasoconstrictor effect of endogenous angiotensin II in the split hydronephrotic rat kidney by the local application of saralasin, found the largest increase in inner vascular diameter in the distal part of the interlobular artery. Similarly, when angiotensin II was applied topically to the vessels of the hydronephrotic kidney, a strong vasoconstriction of the interlobular arteries and a smaller response of arcuate arteries was observed (STEINHAUSEN et al. 1987). Since blood pressure was not altered by both saralasin and angiotensin II when applied locally, autoregulatory responses can be excluded as the cause of the observed vasoconstriction.

A major controversial issue regarding the role of the RAS in the control of renal hemodynamics is the balance between the pre- and postglomerular arteriolar actions of angiotensin II. This aspect is of particular interest for the evaluation of macula densa dependent changes of segmental vascular resistances and the possible participation of the RAS in the TGF response. It is a consistent finding that exogenous and endogenous angiotensin II reduces RBF and, to a smaller and more variable extent, GFR, such that filtration fraction usually increases (cf. ROSIVALL and NAVAR 1983; NAVAR and ROSIVALL 1984). This has been taken to suggest that angiotensin II has a preferential vasoconstrictor effect on the efferent arteriole. Thus, DAVALOS et al. (1978) as well as FREGA et al. (1980), when examining the effect of antiotensin II infusion in the isolated perfused rat kidney, found an increase in GFR despite a reduction in renal perfusate flow. The same effect was also observed with the synthetic renin-substrate and could be prevented by a converting-enzyme inhibitor, indicating that this preferential efferent action is also characteristic for endogenously formed angiotensin II. HALL and colleagues (HALL et al. 1977, 1981, 1986; HALL and GRANGER 1983) similarly concluded from studies in anesthetized dogs and rats with infusions of angiotensin II, angiotensin I and/or converting-enzyme inhibitors, that angiotensin II has a preferential postglomerular vasoconstrictor effect.

This conclusion received support from the work of EDWARDS (1983), who examined the response of isolated afferent and efferent arterioles of the rabbit kidney to angiotensin II and norepinephrine. EDWARDS observed that the internal diameter of afferent arterioles was reduced by norepinephrine but not by angiotensin II over a large dose range, whereas the efferent arteriole contracted in response to both norepinephrine and angiotensin II.

In order to reconcile the presumed preferential postglomerular site of action of angiotensin II with data reported by other investigators showing a parallel increase in pre- and postglomerular resistance by angiotensin II (see below), HALL and GRANGER (1983) conducted a study in anesthetized dogs, in which the endogenous production of angiotensin II was blocked by a converting-enzyme inhibitor, and the myogenic response to changes in blood pressure minimized by keeping renal perfusion pressure constant. In this situation the infusion of angiotensin II induced a parallel increase in both pre- and postglomerular resistance. When, however, the TGF response was inhibited by ureteral occlusion, the infusion of angiotensin II resulted again in a selective increase in efferent arteriolar resistance. The authors concluded that a otherwise observed increase in preglomerular resistance in response to angiotensin II can be attributed exclusively to a combination of a myogenic response to the increase in systemic blood pressure and the TGF response to an increase in distal tubular flow rate (cf. Sect. 9.4).

KASTNER et al. (1984) studied the role of endogenous angiotensin II in the maintenance of GFR during reduction of renal perfusion pressure and/or captopril treatment. While in untreated dogs the reduction in perfusion pressure to 70 mmHg did not alter RBF and GFR, the same maneuver decreased GFR and filtration fraction in captopril-pretreated dogs without affecting RBF. Calculated efferent resistance had decreased considerably more than afferent resistance and could be restored to the level prior the captopril treatment by angiotensin II infusion. In a recent review, HALL and colleagues (1986b) summarized their data and concluded that the renal constrictor effect of angiotensin II is confined primarily to the efferent arteriole with little or no direct effect on preglomerular vessels, and that this selective action is an important factor in maintaining GFR in a variety of pathophysiological states.

In contrast to these observations on a preferential or even exclusive postglomerular vasoconstrictor effect of angiotensin II are a large number of studies by other investigators which failed to demonstrate such a preferential action. Studies in the rat kidney by BLANTZ et al. (1976) and PLOTH and NAVAR (1979) and in the dog kidney by NAVAR et al. (1979) and VIKSE et al. (1983) indicated a parallel vasoconstriction of afferent and efferent arterioles by angiotensin II. As pointed out by BLANTZ (1980), NAVAR and ROSIVALL (1984), and CARMINES (1987), an increase in filtration fraction can occur even with balanced increases in pre- and postglomerular resistances and can therefore not be taken as evidence for a selective postglomerular contraction. Also, when angiotensin I was infused into the kidney to permit generation of angiotensin II at the appropriate intrarenal site, an increase of both pre- and postglomerular resistance was observed (ROSIVALL et al. 1984). The same conclusion was reached by TUCKER et al. (1987) in a study in rats with chronic sodium depletion. Infusion of enalapril, a converting-enzyme inhibitor, increased nephron plasma flow through decreases in both afferent and efferent resistance. Interestingly, renal denervation further increased nephron plasma flow by an additional increase solely in afferent resistance.

Direct observations of segmental vascular reactivity to angiotensin II have been reported by CLICK et al. (1979), CARMINES et al. (1986), STEINHAUSEN et al. (1986, 1987), and WILSON (1986). CLICK and colleagues (1979) examined the vascular reactivity of renal cortical tissue transplanted into the hamster cheek pouch and found that topical application of angiotensin II caused constriction of both the afferent and efferent arteriole. STEINHAUSEN and coworkers (1986, 1987) reported on analogous observations in the split hydronephrotic kidney, a model in which the vascular tree can be visualized and vascular diameters measured directly by transillumination videomicroscopy. In this model, topical application of saralasin, a competitive angiotensin II antagonist, produced parallel increases in the diameter of pre- and postglomerular arterioles. Also the topical application of angiotensin II produced almost identical vasoconstriction in pre- and postglomerular arterioles. These observations provide a strong argument against the view of HALL and colleagues (HALL and GRANGER 1983; HALL et al. 1986b) that any increase in preglomerular resistance in response to angiotensin II reflects the combined effect of a myogenic response to the angiotensin II induced increase in systemic blood pressure and of TGF, since in the model of STEINHAUSEN and colleagues, TGF is absent due to lack of tubular structures, and a myogenic response can be excluded, as the topical application of angiotensin II or saralasin had no effect on systemic pressure. The same argumentation can be derived from the experiments of CLICK et al. (1979), mentioned above. An interesting observation in the studies by STEINHAUSEN et al. (1986, 1987) was that the distal part of the afferent arteriole close to the glomerulus was contracted by angiotensin II with the same sensitivity as was the proximal part of this segment, whereas the vasodilator response to saralasin was excellent in the proximal part but was very poor in the distal part of the afferent arteriole. This apparent discrepancy might reflect reduced or absent generation of endogenous angiotensin II in the distal part of the afferent arteriole due to the peculiar anatomical topography in the hydronephrotic kidney. Direct visualization of pre- and postglomerular vessels was also accomplished by CARMINES and colleagues (1986), who studied juxtamedullary nephrons in captopril-treated rats by epifluorescence videomicroscopy. In this study topical application of angiotensin II significantly reduced the diameter of both pre- and postglomerular vessels by 12%-15%. A different approach was used by WILSON (1986), who used perfusion fixation and vascular casting to visualize vasoconstrictor effects in the rat kidney. In these experiments angiotensin II as well as norepinephrine induced focal constrictions of the afferent arteriole which could not be attributed to an autoregulatory response.

In the attempt to reconcile the divergent data and conclusions regarding a differential response of pre- and postglomerular vessels, several investigators have studied the interaction of angiotensin II with other vasoactive hormones within the kidney. Among the vasodilator agents, prostaglandins have attracted most attention and seem indeed to offer an explanation for the discrepancies described above. Thus, BAYLISS and BRENNER (1978) found angiotensin II to produce an effective preglomerular vasoconstriction only when prostaglandin synthesis had been blocked by indomethacin. HELLER and HORAZEK (1986) studied this interaction in dogs, in which endogenous angiotensin II formation had been suppressed by a high-salt diet and captopril treatment. The intrarenal infusion of angiotensin II decreased RBF and increased efferent resistance by 50% with little or no change in afferent resistance. Accordingly, filtration fraction was increased by 20%. When prostaglandin synthesis was inhibited acutely, a comparable increase in afferent and efferent resistance could be induced by angiotensin II. The authors concluded that prostaglandins have a "protective" effect, at least in superficial glomeruli, in which these measurements had been made. Similar observations have been reported by KASISKE et al. (1986). Also OLSON and coworkers (1987) concluded from their studies on the intrarenal effects of angiotensin II in anesthetized dogs that concomitant increases in prostaglandin synthesis minimize angiotensin II induced reductions in GFR by preventing preglomerular vasoconstriction without significantly affecting postglomerular vessels. This view is in accordance with observations that angiotensin II increases the renovascular synthesis and release of PGE_2 and PGI_2 (SCHLONDORFF et al. 1980; SCHARSCHMIDT et al. 1983; CURRIE and NEEDLEMAN 1984; COOPER et al. 1985; PODGARNY et al. 1986; DUNN and SCHARSCHMIDT 1987).

In only a few studies has the effect of angiotensin II on medullary blood flow been examined. Vasoconstriction of the efferent arteriole in juxtamedullary nephrons would tend to reduce medullary blood flow and thereby decrease the washout of the medullary osmotic gradient, increase sodium reabsorption from the ascending limb of Henle's loop, and increase water reabsorption from the collecting tubule. Recent studies have shown that angiotensin II has indeed a direct vasoconstrictor effect on vasa recta and peritubular capillaries that may contribute to the sodium-retaining and antidiuretic tubular effects of angiotensin II, especially in chronic salt-retaining states (JENSEN and STEVEN 1977; CHOU et al. 1986; FAUBERT et al. 1987). These observation are corroborated by the demonstration of a dense representation of angiotensin II receptors along the vasa recta bundles (MENDELSOHN 1983).

As summarized by NAVAR and ROSIVALL (1984), there is no consensus regarding the role of the RAS in the regional distribution of renal blood flow, probably due to differences in the flow measurements. In general, a tendency for a proportionally greater decrease in inner cortical blood flow by endogenous angiotensin II relative to its effect on outer cortical blood is apparent from these studies (e.g., HORNYCH et al. 1972; ITSKOVITZ et al. 1973; BRITTON 1981; GÖRANSSON et al. 1986).

9.3.2 Angiotensin II and the Glomerulus

During the past decade the glomerulus itself has come into focus as a possible target site for the intrarenal actions of angiotensin II. The interest was spurred by reports on contractile properties of components of the glomerulus, the recognition of the significance of mesangial cells for glomerular function, and the demonstration of angiotensin II receptors on the glomerulus and on mesangial cells.

However, despite numerous studies on the role of angiotensin II in the control of glomerular function, the exact site and nature of the response of glomeruli to angiotensin II and the functional sigificance of this interaction are still unclear and intensely debated. Because direct access to glomeruli in situ is difficult, and glomerular functional characteristics are also dependent on the effects of angiotensin II on pre- and postglomerular vessels, most studies have been performed with isolated glomeruli and isolated mesengial cells.

Studies with isolated human or rat glomeruli indeed suggest that angiotensin II may have an important function in the regulation of glomerular hemodynamics and filtration by a direct action on glomerular structures. Several investigators have observed a decrease in the diameter of glomeruli exposed to angiotensin II with or without accompanying changes in filtration surface or filtration characteristics (BERNIK 1968; SRAER et al. 1974; CALDICOTT et al. 1981; SCHARSCHMIDT et al. 1983, 1986; SAVIN 1986; HALEY 1987).

BLANTZ and PELAYO (1983), discussing the possible site of action of angiotensin II, suggested that angiotensin may reduce the ultrafiltration coefficient K_f by contracting mesangial cells and/or inducing uniform contraction of glomerular capillaries. However, apparent vasoconstriction of glomerular capillaries, i.e., reduction in size of capillary loops, has been reported only by HORNYCH et al. (1982), whereas more recent studies failed to demonstrate such an effect. For example, STEINHAUSEN and colleagues (STEINHAUSEN et al. 1983, 1986; ZIMMERHACKL et al. 1985) examined the vascular effects of angiotensin II in the split hydronephrotic kidney, which permits direct visualization of the renal vascular tree and the glomerular capillary network. These authors found no indication for an angiotensin II induced capillary constriction, while other segments of the arterial vasculature were responsive (see above). Also in the studies of WEINBERGER et al. (1986) using vascular casting and of HALEY et al. (1987) with scanning electron microscopy, no evidence for a general capillary contraction was obtained. Rather, these authors identified a reduction in the size of the glomerular tuft as the preferential site of action of angiotensin II in the glomerulus and possibly as the major mechanism by which angiotensin II reduces the glomerular diameter. These observations and conclusions are in accordance with several reports on the contractile response of isolated or cultured glomerular mesangial cells to angiotensin II (AUSIELLO et al. 1980; FOIDART et al. 1980; SCHOR et al. 1981; VENKATACHALAM and KREISBERG 1985; KREISBERG et al. 1985; ARDAILLOU et al. 1987; SCHLONDORFF 1987), and the recent studies of Kriz and colleagues on the ultrastructure of the mesangial matrix (Sakai and Kriz 1987; Mundel et al. 1988; Kriz and Sakai 1988).

Taken together, the available evidence strongly suggests that angiotensin II has a direct action on the glomerulus; however, the exact site of action(s) as well as the functional consequences remain to be elucidated (see also reviews by MICHAEL 1980; BLANTZ and PELAYO 1983; DWORKIN et al. 1983; KON and ICHIKAWA 1985; KREISBERG et al. 1985; SCHLONDORFF 1987).

As pointed out by KON and ICHIKAWA (1983), the glomerular effects of angiotensin II should always be considered as part of an integrated reaction of the glomerulus to various hormonal agents and to renal sympathetic nerve stimulation. This view is emphasized by the observation that angiotensin II stimulates glomerular and mesangial synthesis of prostaglandins, which may offset its constrictor effect (SCHLONDORFF et al. 1980, 1985; FOIDART and MALBIEU 1986; PODJARNA et al. 1986; DUNN and SCHARSCHMIDT 1987; SCHLONDORFF 1987).

Perhaps the most convincing argument for a direct physiological role of angiotensin II in the control of glomerular function is the striking abundance of high-affinity receptors for angiotensin II on the glomerulus (SRAER et al. 1974; BROWN et al. 1980; CALDICOTT et al. 1981; CHANSEL et al. 1982; SKORECKI et al. 1983; MENDELSOHN et al. 1983, 1986) which can be localized preferentially to mesangial cells (OSBORN et al. 1975; FOIDART et al. 1980; DOUGLAS 1987b). Glomerular angiotensin II receptors are down-regulated during salt depletion (SKORECKI et al. 1983; MENDELSOHN 1986), an effect which is probably not directly dependent on the state of sodium balance but rather on the elevated concentration of circulating angiotensin II (BELUCCI and WILKES 1984; FOIDART and MATHIEU 1986; KITAMURA et al. 1986; WILKES 1987). Downregulation of glomerular angiotensin II receptors by glucocorticoids has also been described (DOUGLAS 1987a). Similar to vascular angiotensin II receptors, glomerular and mesangial angiotensin II receptors are coupled to phospholipase C activation, initiation of phosphatidylinositol breakdown and intracellular calcium activation (PFEILSCHIFTER et al. 1984, 1986; DOUGLAS 1987b; OCHI et al. 1987). Accordingly, mesangial cells are depolarized by angiotensin II (NOBILING and BÜHRLE 1987).

In summarizing the available data on the vascular effects of exogenous or endogenous angiotensin II in the kidney, there is consensus regarding the general responsiveness of most segments of the intrarenal vascular tree starting with the arcuate artery and descending to the peritubular capillaries. However, the relative contribution of these segmental effects to total renovascular resistance, as well as the consequences for the functional role of the systemic and the intrarenal RAS in the control of renal hemodynamics, is still debated. This is especially true for the controversy regarding the balance between pre- and postglomerular effects of angiotensin II and the direct actions on glomerular hemodynamics and function. Incomplete evidence suggests that the observed discrepancies may reflect differences in the activity of the RAS, variable interactions with other intrarenal hormones, especially prostaglandins, and interactions with renal nerve activity, which has a profound effect on arteriolar resistance (e.g., BELL and JOHNS 1982; HANDA and JOHNS 1985). The possible role of the RAS in the autoregulation of RBF and GFR as well as in the TGF response will be discussed below.

9.3.3 Control of Sodium Excretion by Angiotensin II

It is well established that both exogenous and endogenous angiotensin II induces antinatriuresis and antidiuresis in the mammalian kidney, and it has repeatedly been suggested that in addition to indirect extra- and intrarenal mechanisms a direct action of angiotensin II on proximal tubular sodium transport is, in part, responsible for this effect (for survey of the literature see PLOTH 1983; HARRIS and NAVAR 1985; HALL 1986a,b; NAVAR et al. 1987b).

Angiotensin II can influence sodium excretion by several extrarenal mechanisms, such as enhancement of renal sympathetic nerve activity, an increased release of norepinephrine from the adrenal medulla and stimulation of aldosterone secretion (see HALL 1986a for discussion). Furthermore, even under conditions where these extrarenal mechanisms can be excluded or controlled, e.g., in denervated or isolated perfused kidneys, or by intrarenal infusion of small, systemically ineffective doses, angiotensin II may still influence sodium excretion indirectly by changing intrarenal hemodynamics. In many studies focusing on the antinatriuretic and antidiuretic effect of angiotensin II, either by its infusion or by preventing its formation with converting-enzyme inhibitors, profound changes of RBF, GFR, and filtration fraction occurred, which could account for the observed changes in proximal tubular sodium reabsorption (cf. HALL 1986a).

However, there are also studies which convincingly demonstrate a direct effect of angiotensin II on proximal tubular sodium and fluid reabsorption. For example, when angiotensin II was infused directly into the peritubular capillaries, sodium and fluid reabsorption from microperfused or micropunctured proximal tubules was increased (STEVEN 1977; HARRIS and YOUNG 1977; HARRIS and NAVAR 1985; MITCHEL and NAVAR 1987). In accordance with these observations HUANG et al. (1982) as well as HARRIS et al. (1984) reported that in micropuncture studies in normotensive or Goldblatt hypertensive rats, the inhibition of endogenous angiotensin II formation by a converting-enzyme inhibitor decreased proximal sodium reabsorption. In a different approach, OLSEN et al. (1985) and ZHUO et al. (1986) used lithium clearance as a marker for proximal tubular sodium handling. They found that the response to the infusion of low concentrations of angiotensin II or to treatment with a converting-enzyme inhibitor was best explained by a direct stimulation of proximal tubular sodium reabsorption by angiotensin II. An interesting experiment has been reported by SIRAGY and colleagues (1986), who analyzed the intrarenal effects of frog skin angiotensin. This peptide carries a three amino acid extension at the aminoterminus of angiotensin II (Ala-Pro-Gly-angiotensin II). When infused into conscious uninephrectomized dogs, this peptide, at low concentrations, had no effect on GFR, RBF, plasma renin activity, or aldosterone secretion, but produced a significant antinatriuresis and antidiuresis. Conflicting results have been obtained with preparations of isolated tubules. HEALY et al. (1969) found no alteration of tubular transport by angiotensin II, whereas DOMINGUEZ et al. (1987) observed inhibition of sodium transport (rather than stimulation) by angiotensin II at concentrations of $10^{-8} M$ and higher in isolated rabbit proximal tubules. However, experiments by SCHUSTER et al. (1984) and SCHUSTER (1986) seem to provide a convincing explanation for this unexpected result. These authors observed an biphasic response of sodium transport in isolated proximal tubules from the rabbit kidney; low concentrations, i.e., 10^{-13}-$10^{-10} M$, of angiotensin II clearly stimulated sodium transport, whereas at concentrations of $10^{-7} M$ and above sodium transport was inhibited.

A direct tubular site of action of angiotensin II is also suggested by the presence of angiotensin II receptors on tubular cells. High-affinity binding of angiotensin II to both basolateral membranes and brush border fragments obtained from rat renal cortex has been reported by BROWN and DOUGLAS

(1982, 1983) and Cox et al. (1983a, b). SIMPSON and GOODFRIEND (1984) described the presence of both low- and high-affinity angiotensin II receptors on cultured renal tubules from fetal calf. From autoradiographic studies of MENDELSOHN (1985) it is apparent that cortical tubular binding is much weaker than ligand binding to glomeruli or the renal medulla. This is in accordance with data of DOUGLAS (1987a), who characterized tubular angiotensin II receptors in the kidney cortex as having lower affinities to angiotensin II than mesangial receptors. These tubular receptors appear to be up-regulated by high levels of angiotensin II and are coupled to adenylate cyclase via the inhibitory guanylnucleotide binding protein (G_i), whereas the high-affinity mesangial receptors are down-regulated by angiotensin II and mediate the activation of phospholipase C and intracellular calcium. However, a sparse representation of high-affinity receptors on tubular cells cannot be excluded (DOUGLAS 1987a).

In conclusion, several lines of evidence strongly suggest that angiotensin II has a direct action on the proximal tubule by stimulating sodium transport and fluid reabsorption. In contrast, an effect on Henle's loop and the distal tubule has not been convincingly demonstrated (see HALL 1986a for discussion), and it can be assumed that the intrarenal RAS, by way of interstitial angiotensin II generation, significantly contributes to the tubular effects.

9.4 Renin-Angiotensin System in the Control of Renal Hemodynamics

The prominent intrarenal mechanisms controlling renal hemodynamics are autoregulation of RBF, autoregulation of GFR, and TGF. Autoregulation of RBF and GFR describe the phenomenon that RBF and GFR are maintained at a stable level when renal perfusion pressure is varied over a wide (autoregulatory) pressure range. TGF, first described and evaluated in its functional significance by THURAU and SCHNERMANN (1965), is a process that operates at the single-nephron level where changes in the distal tubule flow rate are balanced by opposite changes in GFR of that nephron. The sensor for changes of the tubular flow rate is thought to be the macula densa of the distal tubule. The signal perceived is probably not the flow rate itself, but rather the delivery of sodium chloride to the distal tubule and thus the absorptive transport of sodium and/or chloride by the macula densa cells (cf. Sect. 7.3). As the effector mechanism of TGF by which GFR is adjusted to stabilize distal tubular flow, changes in resistance of pre-, post-, and intraglomerular vessels, as well as changes in glomerular filtration coefficient (not necessarily through vascular effects) have been proposed (see reviews by THURAU and MASON 1974; THURAU 1981; DWORKIN et al. 1983; HÄBERLE and v. BAEYER 1983; BLANTZ and PELAYO 1984; WRIGHT 1984; KON and ICHIKAWA 1985; HALL 1986b; SCHNERMANN and BRIGGS 1986; ARENDSHORST 1987; BELL et al. 1987; BRIGGS and SCHNERMANN 1987).

Obviously, changes or adjustments in RBF and or GFR will affect the TGF response and vice versa, and models describing the intrarenal control of renal hemodynamics and filtration must take the interrelationship of these mechanisms into account. The ongoing debate on the relative contribution of these mechanisms to the regulation of kidney function is best illustrated by describing the extreme positions in a simplified way.

Autoregulation of RBF is often described as a myogenic response. In this model, which has been quantitatively described by OIEN and AUKLAND (1983) as well as by LUSH and FRAY (1984), an increased perfusion pressure induces an increase in circumferential tension of the vessel wall. This "stretch" provides a signal that induces the vessel to contract until the circumferential wall tension is the same as before the pressure increase. Conversely, when perfusion pressure falls, the vessel relaxes until circumferential wall tension is again the same as before. This mechanism permits to maintain RBF constant over a wide range of perfusion pressures. The myogenic response may be modulated by, but is not dependent on, renal nerve activity or hormonal signals. It is rather a cellular response to an alteration in tension, the expression of which may be a change in permeability of the cell membrane of vascular smooth muscle cells to calcium. Thus, an increase in tension would promote calcium entry, resulting in an increase in the contractile state and vice versa (for review see COHEN and FRAY 1982; LOUTZENHISER and EPSTEIN 1985; JOHNSON 1986; NAVAR et al. 1986; OGAWA and ONO 1987). As the sensing device for changes in vascular stretch, club-like myoendothelial junctions have been proposed by JOHNSON (1980). Interestingly, such myoendothelial junctions have also been described in the interlobular artery and the afferent arteriole of the dog, rat, and *Tupaia* kidney (TAUGNER et al. 1984c). In the kidney, the myogenic theory implies that autoregulation of GFR is mainly the consequence of RBF autoregulation, as during graded reductions of renal perfusion pressure the myogenic dilatation descends along the preglomerular vessels, and thus automatically maintains not only RBF but also GFR until the total autoregulatory capacity is exhausted, i.e., all preglomerular vessel segments are dilated. In this theory, TGF is not denied to exist - which would hardly be possible in view of the experimental data. However, the TGF mechanism is variably considered as a "back-up system" or as a "second-line defense" coming into play only under extreme pathophysiological conditions, such as reduction of perfusion pressure below the autoregulatory range, or as a mechanism of "fine tuning" of GFR with no major contribution to the overall regulation of GFR under most conditions.

Among the direct evidences for the myogenic theory of autoregulation is the demonstration of an autoregulatory response of interlobular arteries in the rat kidney (KALLSKOG et al. 1976; HEYERAAS-TONDER and AUKLAND 1979/1980). These vessels have no contact with the macula densa and can probably not be reached by humoral signals from this tubular site. Furthermore, a myogenic response of afferent arterioles transplanted into a hamster cheek pouch chamber (GILMORE et al. 1980) and of isolated perfused afferent arterioles (EDWARDS 1983) has been described.

The opposite position holds that autoregulation of both RBF and GFR is predominantly, if not exclusively, mediated by TGF, represented by the sequence: increased arterial pressure, increased filtration rate, increased distal flow rate, TGF, preglomerular vasoconstriction, normalized RBF, normalized GFR, or vice versa (cf. NAVAR 1978; NAVAR et al. 1981, 1982b). Also postglomerular and intraglomerular vascular changes, as well as changes of ultrafiltration coefficient have been implied as the efferent limb of TGF (see below).

This is not the place to discuss in detail the arguments for or against one or the other hypothesis. It should suffice to say that there is compelling evidence for the participation of both mechanisms in the control of GFR, and that opinions are now beginning to converge around the view that both mechanisms operate in concert. This is best illustrated by a recent publication of AUKLAND and OIEN (1987), who analyzed several mathematical models of renal autoregulation and came to the conclusion that neither the myogenic response nor TGF alone can fully account for the experimental data. In a model combining a myogenic response of preglomerular vessels with TGF regulation of preglomerular resistance, good autoregulation of RBF and GFR as well as consistence with experimental data on TGF was obtained. An alternative model with a good fit to experimental data was the combination of preglomerular myogenic response with both pre- and postglomerular TGF operating in the same direction.

In the following, the possible participation of the RAS in the regulation of renal hemodynamics will be discussed.

In view of the profound reduction of RBF by endogenous or exogenous angiotensin II, as described in the preceeding sections, the role of angiotensin II in the autoregulatory adjustment of renal blood flow has repeatedly been examined. Earlier theoretical considerations (THURAU 1964, 1967) and experimental observations of an impaired autoregulation of RBF in dogs with the RAS suppressed by DOCA-salt treatment (KALOYANIDES et al. 1974) have led to the view that an intrarenal RAS may participate in the expression of RBF autoregulation. However, numerous subsequent studies have failed to show such a relationship. For example, HALL and coworkers (1977a) and MURRAY and MALVIN (1979) observed a fully maintained autoregulatory response to graded pressure reductions in DOCA-treated dogs on a high-salt diet which had almost undectable levels of plasma renin. Interruption of the RAS by saralasin or converting-enzyme inhibitors did not impair the autoregulatory capacity in rats (AHRENDSHORST and FINN 1977; JOHNSTON et al. 1977a, b) and dogs (ABE et al. 1976; KASTNER et al. 1984; OGAWA and ONO 1985) even when the RAS was stimulated by salt depletion (ROSIVALL et al. 1986).

The data related to autoregulation of GFR are more controversial. In the experiments of POTKAY and GILMORE (1973) and those of MURRAY and MALVIN (1979) in renin-depleted dogs, neither RBF nor GFR autoregulation were impaired. Other investigators observed a dissociation between RBF and GFR autoregulation, i.e., impairment of the GFR autoregulatory response in the face of a maintained RBF autoregulation in animals with a suppressed RAS or during inhibition of converting enzyme (HALL et al. 1977a, b; CLAPPISON et al. 1981; KASTNER et al. 1984; GÖRANSSON et al. 1986b; ROSIVALL et al. 1986). The general conclusion from these studies was that the RAS modulates the GFR autoregulatory response under certain conditions but is not a mediating system. This is illustrated by the observation of KASTNER et al. (1984) that the GFR autoregulatory response was attenuated by converting-enzyme inhibition but could be restored by the infusion of a fixed concentration of angiotensin II. It should be noted, however, that most studies demonstrating an impaired GFR autoregulation have been performed in salt-depleted, anesthetized animals with acute surgical trauma, in which the systemic blood pressure was elevated, the RAS was drastically stimulated and, most likely, the renal sympathetic nerve activity was high. In contrast, in the conscious, chronically instrumented dog, intrarenal infusion of a converting-enzyme inhibitor or of saralasin did not significantly impair or modify the autoregulation of RBF or GFR (HACKENTHAL et al. 1983; PERSSON et al. 1988).

HALL (1986) reviewed the available evidence on the role of the RAS in the control of GFR and concluded that angiotensin II assists mainly in maintaining GFR in states of an elevated RAS, particularly at reduced arterial pressure. As pointed out above, HALL and colleagues consider preferential postglomerular vasoconstriction as the principal mechanism by which angiotensin II helps to maintain GFR in these conditions. Human pathophysiology provides an impressive illustration of this view. In the vast majority of patients receiving converting-enzyme inhibitors for the treatment of hypertension or congestive heart failure, no impairment of renal function is observed. However, in patients with renal artery stenosis, converting-enzyme inhibition significantly reduces GFR in the affected kidney, where renin secretion rate is increased. Usually, this effect is compensated by the intact contralateral kidney. If, however, both kidneys are affected, or the patient has renal artery stenosis of a solitary kidney, renal excretory function can be dramatically reduced, and acute renal failure may ensue (e.g., CANZANELLO et al. 1987; LEVENSON and DZAU 1987; WENTING et al. 1987).

One of the most controversial issues in the discussion of intrarenal functions of the RAS is its role in the TGF mechanism. It was initially suggested by THURAU and colleagues (cf. THURAU 1964; THURAU and MASON 1974; THURAU 1981; THURAU et al. 1982) that renin secretion from juxtaglomerular cells increases in response to an increased distal tubule flow rate (equivalent to an elevated sodium chloride concentration reaching the macula densa cells), and that the subsequent interstitial formation of angiotensin II induces contraction of the afferent arteriole. Basically, this hypothesis has three components, first, the implication of a positive correlation between distal tubule salt delivery and renin secretion, second, the assumption that the intrarenal RAS in the principal effector system mediating, third, a preglomerular vasoconstrictor response.

The concept of a direct correlation between distal flow rate and renin secretion was mainly based on the interesting observations of THURAU and coworkers (DAHLHEIM et al. 1970; GRANGER et al. 1972; THURAU et al. 1982) and those of GILLIES and MORGAN (1978) that the renin activity of individual glomeruli increased when the associated distal nephron had been perfused with a higher than normal concentration of sodium chloride. As discussed in Sect. 7.3, an increased renin activity of isolated glomeruli, which many result from the intracellular activation of prorenin, does not necessarily imply an increased secretion of active renin. Furthermore, the concept of a positive correlation between the distal salt load and renin release is counter to the general observation that, in the whole kidney, renin secretion is suppressed by a high salt load to the kidney (cf. Sect. 7.3). In order to reconcile these discrepancies, WRIGHT (1984) has advanced the view that although an increase in salt delivery to the macula densa cells would tend to increase renin secretion, this effect may not always be detectable because it may be overridden by a stronger inhibition of renin secretion from arteriolar baroreceptor activation.

However, evidence has now accumulated favoring the notion originally proposed by VANDER and MILLER (1964) and VANDER (1967) of an inverse relationship between distal salt delivery and renin secretion. In addition to the evidence discussed in Sect. 7.3 and the micropuncture studies of LEYSSAC (1984, 1986), the most clearcut demonstration of an inverse relationship is that by SKØTT and BRIGGS (1987). The authors perfused the macula densa segment of microdissected JGAs of the rabbit kidney, and found a prompt increase in renin release from the associated glomerulus when the sodium chloride concentration of the tubule perfusion fluid was lowered.

The concept of a mediator function of an intrarenal RAS in the TGF response, originally based on theoretical considerations (THURAU 1964), was later supported by several studies demonstrating that the TGF response was significantly attenuated when the RAS was suppressed, e.g., by DOCA and high-salt treatment (DEV et al. 1974; PLOTH et al. 1977), by converting-enzyme inhibitors or saralasin (PLOTH et al. 1979; STOWE et al. 1979; PLOTH and ROY 1982a).

It should be emphasized, however, that in these studies the feedback response was never completely eliminated, but only attenuated to about half the normal response. Furthermore, when kidney renin content was drastically reduced by DOCA and high-salt treatment, a normal feedback response could be restored by acute volume depletion with furosemide (MOORE et al. 1980), indicating that a normal renin content and probably secretion rate is not essential for the TGF response. PLOTH and ROY (1982b) also contributed convincing evidence that angiotensin II is not the mediator of the TGF response by demonstrating that the magnitude of the TGF response, which had been attenuated by converting-enzyme inhibition, could partly be restored by infusion of a fixed concentration of angiotensin II.

In addition, the majority of data on the effector site of the TGF response suggest a vasoconstrictor effect on preglomerular vessels (cf. NAVAR and ROSIVALL 1984; NAVAR 1986; BRIGGS and SCHNERMANN 1987; AHRENDSHORST 1987). This would not be compatible with an inverse relationship between sodium chloride delivery to the macula densa and renin secretion. For these and other reasons there appears to be increasing acceptance of the view that macula densa dependent changes in renin secretion and the TGF response are parallel events initiated by the same tubular signal (BRIGGS and SCHNERMANN 1986; SCHNERMANN and BRIGGS 1986; WILCOX and PEART 1987). Consequently, the RAS cannot be considered as the mediator of the TGF response but rather as an important modulator, acting in concert with other hormonal factors such as prostaglandins (SCHOR et al. 1981; SCHNERMANN et al. 1984a), bradykinin (SCHNERMANN et al. 1984b), and ANF (BRIGGS et al. 1982; HUANG et al. 1985; see also reviews by KON and ICHIKAWA 1985; AHRENDSHORST 1987; BRIGGS and SCHNERMANN 1987). Since angiotensin II augments the TGF response, and renin release is inversely related to distal salt delivery, a selective preglomerular vasoconstriction by angiotensin II is unlikely to play a role in modulating TGF. As discussed above, a combined pre- and postglomerular vasoconstriction (BLANTZ and PELAYO 1983; NAVAR et al. 1984), preferential postglomerular vasoconstriction (HALL 1986), and/or intraglomerular reduction of the ultrafiltration coefficient (ICHIKAWA 1982; DWORKIN et al. 1983) have to be considered as possible mechanisms by which angiotensin II modulates the TGF response.

9.5 Summary and Conclusion

Returning to the question of an intrarenal RAS, the available data suggest that renin is secreted from epithelioid cells almost exclusively into the renal interstitium next to the glomerulus, from where it travels with the interstitial fluid, including the periarterial interstitial network, to reach the systemic circulation at the level of accompanying veins and the peritubular capillaries. A small portion, i.e., less than 5% of secreted renin, is drained by the renal lymphatics. Somewhere along its intrarenal path, renin meets angiotensinogen that diffuses from the blood stream into the interstitial space, most likely at the level of the peritubular capillaries and the fenestrated veins. Consequently, is appears unlikely that the highest interstitial angiotensin II generation occurs in the immediate vicinity of the juxtaglomerular apparatus. Furthermore, it should be considered that the distal part of the afferent arteriole is the least contractile segment of this vessel and that the concentration of angiotensinases is high at juxtaglomerular sites (cf. Sect. 4.5).

It is likely, but not proven, that this local angiotensin formation in the renal interstitium results in both vascular and tubular effects. There is no experimental evidence to assume that this interstitial angiotensin II has intrarenal effects and functions different from those of systemically delivered angiotensin II, and neither is there reason to believe that renin secretion is controlled locally with the only purpose of providing intrarenal angiotensin II formation.

Renin secretion, and hence local angiotensin II formation, is probably always associated with the delivery of renin into the renal vein, so that, usually, changes in renin secretion are directly mirrored by corresponding changes in plasma renin activity. The only known and predictable exception is the situation in which the two kidneys are in different functional states, best examplified by the one-clip, two-kidney Goldblatt hypertension model, in which plasma renin activity dissociates from intrarenal angiotensin II in the untouched contralateral kidney (Mendelsohn 1982; Taugner et al. 1984d).

With regard to renal functions, it appears that intrarenal angiotensin II is not the mediator either of the autoregulation of RBF, the autoregulation of GFR, or of the TGF response. However, in addition to tubule effects, angiotensin has an important modulating function in the control of intrarenal vascular reactivity, which becomes most apparent when systemic circulatory homeostasis and control of GFR are severely compromised. It is likely that angiotensin II generated from angiotensinogen in the renal interstitium contributes to this function.

Without precise definition of a local RAS, it remains a matter of convention whether we call the pathways of formation and intrarenal actions of angiotensin II, as described so far, an intrarenal RAS. When thinking of an intrarenal RAS in the strict sense, i.e., comparable to that in the brain, we are largely confined to speculations. The elucidation of the functional significance of the expression of the angiotensinogen gene in the kidney and the intracellular occurrence of angiotensin II in epithelioid cells will perhaps be helpful in clarifying this question.

Chapter 10

Metaplastic Transformation Between Smooth Muscle Cells and Epithelioid Cells

The number of epithelioid cells increases upon stimulation of the RAS and returns to control values after removal of the respective stimulus. Conversely, the number of epithelioid cells reversibly decreases when the level of stimulation is reduced below control values. Immunohistochemical experiments show that these changes take place mainly at the transition between the renin-negative/myosin-positive (proximal) and the renin-positive/myosin-negative (distal) portion of the afferent arteriole (Fig. 10.1). Apparently depending on the level of stimulation, the media cells of a certain vessel segment show either the immunoreactivity and ultrastructure of epithelioid cells, or that of plain smooth muscle cells (GOORMAGHTIGH 1942, 1945a; LATTA and MAUNSBACH 1962a; BARAJAS and LATTA 1963a; DUNIHUE and BOLDOSSER 1963; BOHLE and SITTE 1966; CAIN and KRAUS 1971; CANTIN et al. 1977a).

If cell migration is excluded, principally two different ways for such a - reversible - transformation of secretory cells into a contractile cell population exist. Firstly, one type of cells would multiply at the expense of the other. In this case, the hyperplasia of one cell type would be reflected in an increased mitotic activity, while the hypoplasia of the other would show up in an increased rate of cell disintegration. Here, intermediate forms between both cell types would not be expected. Secondly, cells of one type would transform into cells of the other, e.g., smooth muscle cells into epithelioid cells or vice versa. In this case, an increase in mitotic activity would not be anticipated, whereas intermediate cells would be expected to occur.

Binucleated cells and mitotic figures have long been reported to occur in the region of the JGA (GOORMAGHTIGH 1939, 1944; GOORMAGHTIGH and GRIMSON 1939; DUNIHUE and BOLDOSSER 1963), and the example depicted by TSUDA et al. (1971) demonstrates that granulated cells may in principle have the ability to divide. Nevertheless, the number of mitoses published is far too small to permit the assumption that mitosis may be a significant mode of increasing the number of granulated cells (cf. DUNIHUE 1941; BARAJAS and LATTA 1963a; ROSENBAUER 1965; JOHNSTON et al. 1967). In our studies of several hundred of renal corpuscles from various species, no mitosis was observed in the region of the JGA, although mostly specimens were investigated from animals in which the renin-positive portion of the afferent arteriole and therewith the number of granulated cells could be expected to change markedly within a few days upon stimulation or inhibition of the RAS (partial occlusion of the renal artery, subchronic application of furosemide or captopril, thirst in diabetes insipidus rats, adrenalectomy; salt and DOCA substitution in adrenalectomized animals). Above all, the existence of cells intermediate between those of the smooth muscle and those of the secretory type, with myofilaments and attachment sites adjacent to secretory granules prompted several authors to suspect that epithelioid cells originate from smooth muscle cells (GOORMAGHTIGH 1940, 1942, 1945a; LATTA and MAUNSBACH 1962a; BARAJAS and LATTA 1963a; BOHLE and SITTE 1966; CAIN and KRAUS 1971; ROJO-ORTEGA et al. 1973a; cf. Sects. 2.2.2, 4.1.2).

CANTIN et al. (1977a) finally succeeded to dissipate any doubts as to the existence of metaplastic transformation, in particular of smooth muscle cells into epithelioid cells. The authors investigated mitotic activity in the vasculature of the ischemic (endocrine) kidney by means of light and electron microscopic autoradiography with labeled thymidine. This experimental model was thought to be ideally suited for such experiments because of the extraordinarily rapid increase in the number of granulated cells not only in the JGA region, but also in the wall of the arterial tree far upstream from the glomerulus. CANTIN et al. (1977a) found that neither epithelioid nor intermediate cells of the vascular wall incorporate labeled thymidine injected during a 10-day period. In contrast to tubular cells, [^{3}H] thymidine was practically never incorporated

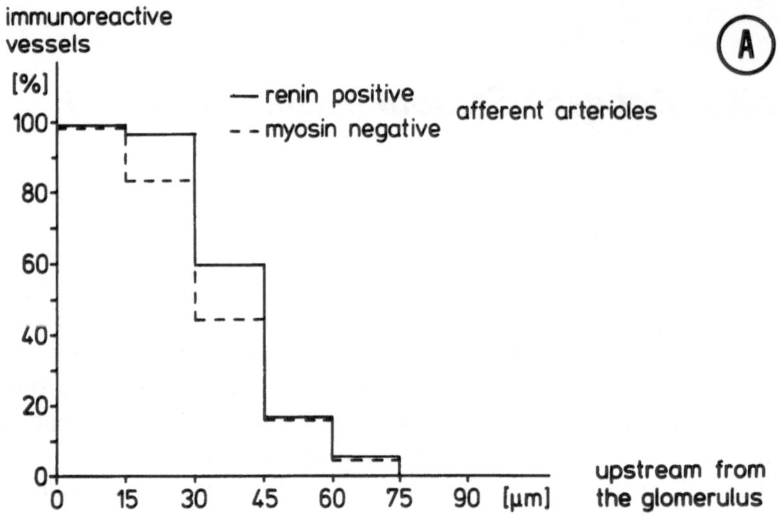

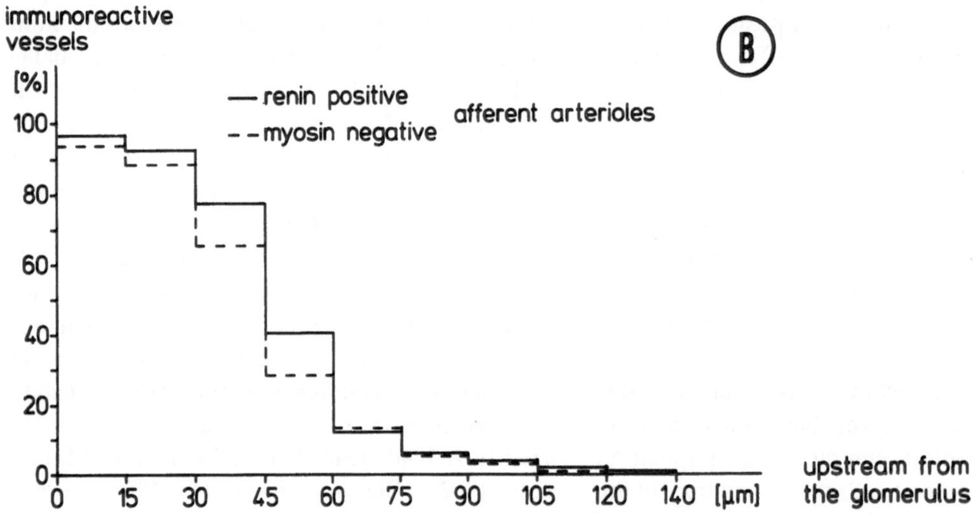

Fig. 10.1. Length of the renin-positive and myosin-negative segment of the afferent arteriole in control rats **A** and in diabetes insipidus rats after dehydration **B.** *Ordinate*, percentage of immunoreactive vessels; *abscissa*, length of the immunoreactive vessel segment upstream from the parent glomerulus. *n* in **A**, 89 for renin and 90 for myosin; in **B**, 241 for renin and 202 for myosin. (From TAUGNER et al. 1987b).

into the nuclei of plain smooth muscle cells during the experimental period. As the mitotic activity in the region of the JGA and in the media of the renal arteries was small even upon maximal stimulation of the RAS, the authors concluded that intermediate and epithelioid cells in the ischemic and probably also in the normal kidney develop by metaplasia of preexistent smooth muscle cells (for the likewise low mitotic activity in glomeruli see PABST and STERZEL 1983). Thus, CANTIN et al. (1977a) definitely confirmed the observations of GOORMAGHTIGH (1945b) on the slightly ischemic rabbit kidney that some of the fibrillar smooth muscle cells become afibrillar and granular.[1]

The ultrastructural aspects of the transformation of smooth muscle cells into epithelioid cells have been described in detail, among others, by CAIN and

[1] Changes of the JGA induced by stimulation of the RAS, particularly in the area of the incoming afferent arteriole, are commonly included under the terms "hypertrophia" or "proliferation" of the JGA or, rather, of the epithelioid cells. This could be misinterpreted to suggest that the number of these cells - and not only their individual volumes - increase upon stimulation. Unfortunately, also the expression "hyperplasia" would not be correct, as the media cells involved are going through fundamental changes of their structure and function. Although somewhat imprecise semantically, we prefer the term "metaplastic transformation", which is apt to accentuate the special aspects of this process (cf. CANTIN et al. 1977a).

KRAUS (1971) and CANTIN et al. (1977a; cf. also Sect. 2.2.2). According to these authors, swelling of the nucleus and nucleolus is followed by an increase in glycogen, hypertrophy of the RER and the Golgi area, and, finally, development of the first mature secretory granules from protogranules. Concomitantly, the number of myofilaments and attachment sites decreases. In fully transformed cells, myofilaments may be virtually absent, so that the resulting epithelioid cells can no longer be regarded as contractile (CAIN and KRAUS 1971; cf. Sects. 2.2.2, 2.2.3). There are no reports on the participation of autophagosomes in the metaplastic transformation of smooth muscle cells into epithelioid cells. This may indicate that mainly cytoplasmic enzymes are involved in the catabolic processes accompanying this drastical remodeling of media cells.

Comparatively little reasearch has been dedicated to the ultrastructural alterations of the JGA upon acute and drastic inhibition of the RAS. There is agreement that retransformation of epithelioid cells into smooth muscle cells takes place after a decrease in the level of stimulation, but, as far as we know, there have been no reports on the associated ultrastructural changes. HATT (1967) reported on hypoactive epithelioid cells of rats on a sodium-rich diet treated with a high dose of DOCA. He also gave examples of degranulated cells from the nonischemic kidney of animals with unilateral renal ischemia, showing only a few voided membrane-bound sacs instead of secretory granules. As the NaCl plus DOCA treatment and the renal ischemia had lasted for 21 days and 3 months, respectively, these examples should rather represent epithelioid cells in the state of chronic suppression of renin synthesis, referred to in Chap. 11, than in that of acute retransformation.

As discussed earlier, the process of retransformation from epithelioid cells to smooth muscle cells should be characterized by the absence of mitoses and, secondly, by the occurrence of intermediate cell forms. In our studies with adrenalectomized animals substituted with NaCl and DOCA, mitoses were not encountered. Ultrastructurally, during retransformation of epithelioid cells into smooth muscle cells, the disappearance or elimination by autodigestion, exocytosis, or sequestration of the numerous large mature secretory granules is imperative. Consequently, the process of retransformation cannot simply be the reversal of the transformation process and, thus, other forms of intermediate cells have to be anticipated during retransformation.

In order to increase the probability of finding intermediate cells in the process of retransformation, we first stimulated the RAS of mice by adrenalectomy and reduced the stimulation level of the RAS a few days later by the administration of NaCl and DOCA. The adrenalectomized animals were given daily doses of 0.5 ml isotonic NaCl and 3 mg DOCA (Percorten) per kilogram, injected subcutaneously for 1-4 days prior to perfusion fixation (METZ et al. 1988). In addition, the drinking water of the mice contained 1% NaCl during this period. In other experiments, the kidneys of mice from the spontaneous recovery period after adrenalectomy were investigated, 5-10 days after the operation (cf. Chap. 11).

Figure 10.2 shows an example where part of the cytoplasm including secretory granules is separated by a double membrane from the rest of the cell, with the sequestered organelles apparently in the course of autolytic digestion. As such changes were seldomly seen in control mice and practically absent in stimulated but unsubstituted animals, the hypothesis is implied that the observed demarcation of part of the cytoplasm including renin granules represents the first step in the retransformation of epithelioid cells into smooth muscle cells after resetting of the stimulation level. After sequestration, the assembly of a contractile machinery, starting in the perikaryon of the cell, would be essential to accomplish the retransformation. Crinophagy, i.e., the uptake of secretory granules by autophagosomes, as known for other secretory systems, was not observed in our material (TAUGNER et al. 1985b). However, upon resetting of the stimulation level, the macroautophagic activity of the renin granules seemed to increase (TAUGNER et al. 1988b). This observation would be in accordance with our assumption that the secretory granules of epithelioid cells may take over some functions of lysosomes, with their acidic hydrolases involved not only in the cleavage of surplus secretory product (cf. Sect. 5.3), but also in the autodigestion of cell organelles. Perhaps epithelioid cells possess two mechanisms for the reduction of surplus secretion product: a scavenger process, i.e., the sequestration and degradation of parts of their cytoplasm including secretory granules upon a drastic, long-lasting inhibition of the RAS as outlined above and, sec-

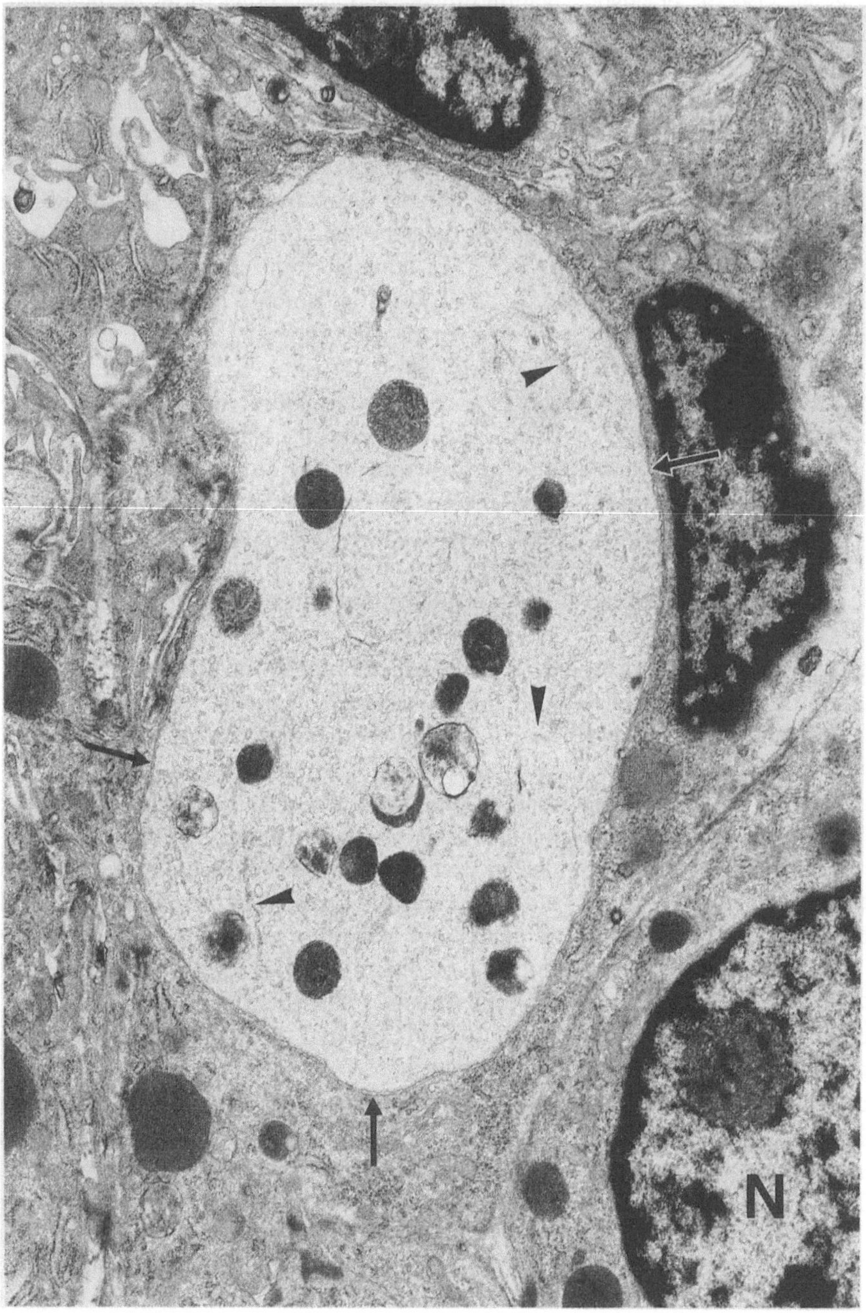

Fig. 10.2. Epithelioid cells from the juxtaglomerular portion of the afferent arteriole in mouse kidney. In one cell, a large portion of the cytoplasm, recognizable by its low electron density, is sequestered by a double membrane (*arrows*). The contents of the sequester: small vesicular structures, secretory granules, and remnants of Golgi or RER cisterns (*arrowheads*) show signs of deterioriation. Note that the sequester is surrounded by a small rim of unaltered cytoplasm, and the tangentially cut nucleus of the involved cell shows no signs of pyknosis. The animal was in the spontaneous recovery period from adrenalectomy, 7 days after the operation; × 14000. (Courtesy F. TAUGNER)

ond, the capacity to diminish the granular store by partial intragranular hydrolysis during phases of the ephemeral short-lasting fluctuations of the stimulation level (cf. Sect. 7.8).

The authors are well aware of the largely hypothetical character of such an interpretation of the retransformation process. Nevertheless, they take the risk of presenting it here, as there are indications that sequestration of large cytoplasmic portions is an ability of vascular media cells in general. Extreme vasoconstriction by angiotensin II or norepinephrine in arteries of several organs is reported to induce processes which, up to now, were considered to be herniations, i.e., protrusions of one smooth muscle cell into another (for reference see LEE et al. 1984). However, by serial sectioning of afferent arterioles and interlobular arteries, we observed that some of these "herniae" are in fact sequestered portions of the cytoplasm of the involved vascular smooth muscle cells (METZ et al. 1988). Thus the metaplastic retransformation of epithelioid cells into smooth muscle cells by sequestration processes could represent an "atavistic" scavenger mechanism characteristic for media cells in general.

In Sects. 2.4 and 4.1.3 it was reported that during long-term stimulation of the RAS in man, renin-negative cells of the lacis may be transformed into granulated renin-positive cells (CHRISTENSEN et al. 1989a). This transformation was not accompanied by an increase of the mitotic rate (cf. CANTIN et al. 1977a). These findings indicate the possibility of a transformation of plain Goormaghtigh cells into epithelioid cells upon drastic stimulation and vice versa also for a retransformation after cessation of the stimulation. However, experience shows that it is extremely difficult to reproduce these processes in animal experiments. At present, therefore, little is known of the exact prerequisites for the transformation of lacis cells and the ultrastructural changes accompanying such a process.

Chapter 11

Long-term Modulation of Renin Synthesis and Secretion

Subchronic or chronic stimulation of renin secretion, e.g., by sodium deprivation or by narrowing of the renal artery, must obviously be accompanied by an increase in renin synthesis. At present very few details are known as to how such "stimuli" are converted into an increase in renin synthesis. This deficit in knowledge - especially as compared with our understanding of renin secretion - of the conditions and mechanisms in the control of renin synthesis has several reasons.

(a) In contrast to the acute secretory event, the relatively slow process of renin synthesis (DESORMEAUX et al. 1982) has not yet been studied under the well-controlled conditions of in vitro experiments. (b) Changes in the rate of renin synthesis could be estimated until now only by the determination of kidney renin. However, the renin depot of the kidney is large in comparison to the rate of synthesis and secretion, and only dramatic changes in the rate of synthesis can be expected to result in measurable differences of kidney renin content. The functional evaluation of kidney renin content is further complicated by the fact that it depends also on the rate of secretion. We will show later that this aspect is especially important under conditions of exaggerated release of renin, such as adrenalectomy. (c) In most secretory systems, the activation of synthesis can be assessed by examining the ultrastructural state of the protein synthesizing, processing, and secreting machinery. However, this ap-

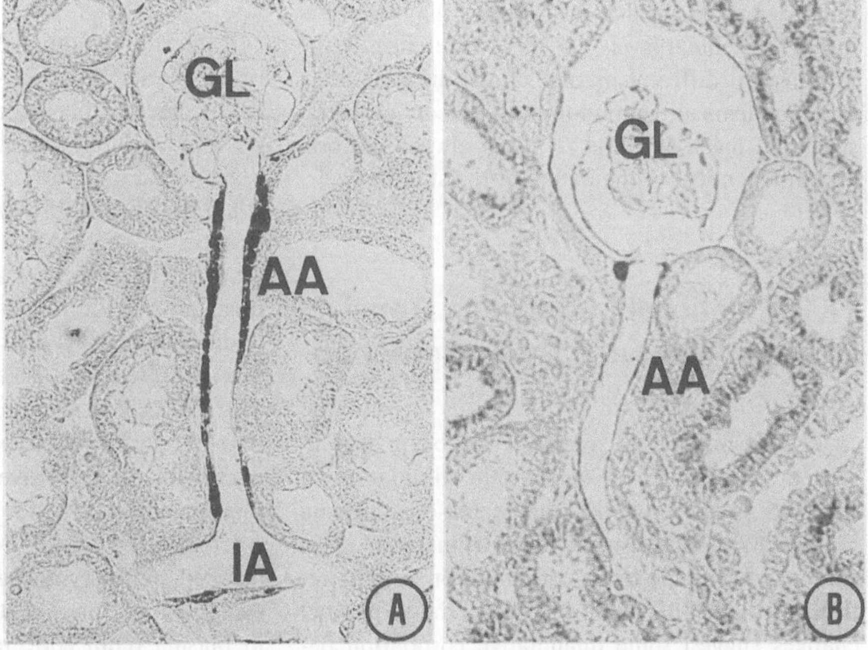

Fig. 11.1. Immunohistochemical staining for renin of paraffin sections from mouse kidney after sodium depletion (**A**) and sodium loading (**B**). Note the great difference in the length of the renin-positive portion of the afferent arteriole (AA). *IA*, interlobular artery; *GL*, glomerulus. Sodium content of the diet was 4 and 23 g/kg, respectively, given for 2 weeks; perfusion fixation with Bouin's fluid containing 0.1% glutaraldehyde; antiserum dilution 1:1000; ×275 and 335, respectively

proach requires a homogeneous behavior of all competent cells - which does not apply for granulated renin-producing cells. Apart from internephron heterogeneity discussed in Sect. 4.1.1, in nearly every afferent arteriole cells of very different secretory activity are found, so that a survey and quantitative evaluation of the general state of synthesizing activity could only be obtained by exhaustive morphometric analysis. (d) Since during long-term stimulation of renin secretion the length of the renin-positive portion of the afferent arteriole increases by metaplastic transformation (see preceding section), estimation of this change by immunohistochemical techniques could also serve as an indicator of an increased rate of renin synthesis (Fig. 11.1). Still, the question remains whether this recruitment of renin-producing cells responds to the same stimuli as does renin synthesis in already transformed cells.

The assessement of renin synthesis independently of secretion and storage has recently become possible by the determination of renin mRNA. Since information on the control of renin synthesis obtained with this technique is still scarce (cf. Sect. 7.8), we will restrict the present discussion to three questions: Is there a relation between the size of the renal renin depot and the secretion rate? Which manipulations lead to an enlargement of the renin depot, and which mechanisms may be effective at the cellular level? Are there differences in the response to different manipulations discernable at the biochemical, immunohistochemical, or ultrastructural level, or is the response always uniform and independent of the nature of the stimulus?

11.1 Relationship Between Renal Renin Content and Renin Secretion

In a great number of studies it has been demonstrated that chronic or subchronic stimulation of the RAS induces not only an increase of plasma renin activity, but also an increase in renal renin content. As plasma renin concentration usually reflects the rate of renin secretion, these parallel changes suggest some quantitative correlation between renal renin content and renin secretion. This parallel shift has been observed during chronic or subchronic sodium depletion (Gross 1964, in Gross et al. 1965; Oda and Ziegler 1965; Sparks and Suzik 1977; Fray 1978a; Park et al. 1978; Morimoto et al. 1979; Nakane et al. 1980b; Nakamura et al. 1985), following renal artery constriction in the Goldblatt hypertension model (Gross 1964, in Gross et al. 1965; Park et al. 1978), in captopril-treated animals (Nakamura et al. 1985), and in some studies with adrenalectomized rats (Schaechtelin et al. 1963). A correlation between renin concentration and release has also been described with respect to zonal heterogeneities in the renal cortical renin distribution (Jones et al. 1979). Similar results have been obtained by comparing the histochemical JGI or the so-called granularity of epithelioid cells with plasma renin concentration. This is not surprising in view of the parallelism of the morphometric and the biochemical data on renal renin content, as discussed in Sect. 4.1.4.

Manipulations in the opposite direction, i. e., subchronic or chronic suppression of the RAS, e. g., by sodium loading with or without addition of DOCA, lead to a parallel decrease of plasma renin concentration and kidney renin content.

However, the parallel alteration of kidney renin content and secretion rate is neither a proportional change, nor is it a consistent finding. Usually the fractional increase in kidney renin content is much smaller than the concomitant increase in secretion rate.

If the stimulation of the RAS is less intense and/or of shorter duration (i. e., for 2-8 days), significantly elevated plasma renin concentrations can be observed without a concomitant increase in kidney renin content (Gross et al. 1965; Ludwig et al. 1987). If, on the other hand, the secretory stimulus is excessive, e. g., when induced by narrowing the renal artery or the aorta between the origin of the renal arteries or following adrenalectomy, kidney renin content may even decrease (Dauda and Dévényi 1971; Peter et al. 1974a; Atkinson et al. 1983; F. Taugner, unpublished observations). Apparently, renin synthesis is not capable of keeping pace with the excessive rate of renin secretion in this situation.

Conversely, when renin secretion is acutely blocked, renin synthesis seems to continue at an elevated level for some time, resulting in a decrease in plasma renin concentration and an increase in kidney renin content (Silverman and Barajas 1974; De Senarclens et al. 1977).

Another interesting aspect of the relationship between renin synthesis, kidney renin content, and renin secretion rate is the mechanism by which an elevated level of renin secretion is maintained during chronic stimulation. Several investigators have observed that isolated preparations - such as the isolated perfused kidney, kidney cortex slices, or isolated glomeruli - when taken from animals in which renin secretion had been stimulated, e.g., by salt depletion plus diuretic treatment, continue to release renin in vitro at an elevated rate (DE JONG 1969; FRAY 1978a; JONES et al. 1979; NAKANE et al. 1980a; BRAVERMAN et al. 1971; SCHRYVER 1984; LUDWIG et al. 1986, 1987). This increased rate of renin release could not be accounted for by the increased renin content of the preparation because the fractional increase of renin content was usually much smaller than the fractional increase of renin release, or the renin content was not different from controls at all. In contrast, when renin secretion rate in vivo had been stimulated by treatment with converting-enzyme inhibitors for 8 days to the same level as by salt depletion, the isolated perfused kidney taken from these rats may have lower than normal secretion rates (LUDWIG et al. 1986, 1987). As an explanation for this latter finding, the authors proposed that during chronic captopril treatment epithelioid cells do not change their basic secretory ability but maintain an elevated secretion rate in vivo because the inhibitory signal, angiotensin II, is not present; other functional stimuli, such as increased renal nerve activity or decreased renal perfusion pressure may contribute to the chronic stimulation. When these kidneys are isolated, they relapse to their basic state. On the other hand, in those situations where the isolated preparation continues to secrete renin at a higher level, it must be assumed that epithelioid cells had undergone an "intrinsic" change during chronic stimulation to the end that an increased activity of the secretory mechanism is encoded in the cellular organization which persists when the chronic stimulus is acutely withdrawn by removing the preparation from the organism.

An illustration of this phenomenon, which has been called "stable transformation" (ROSS 1978), is given in Fig. 11.2. In the experiment shown, rats had been kept on a low-, normal- or high-salt diet for 10 days before the kidneys were isolated for per-

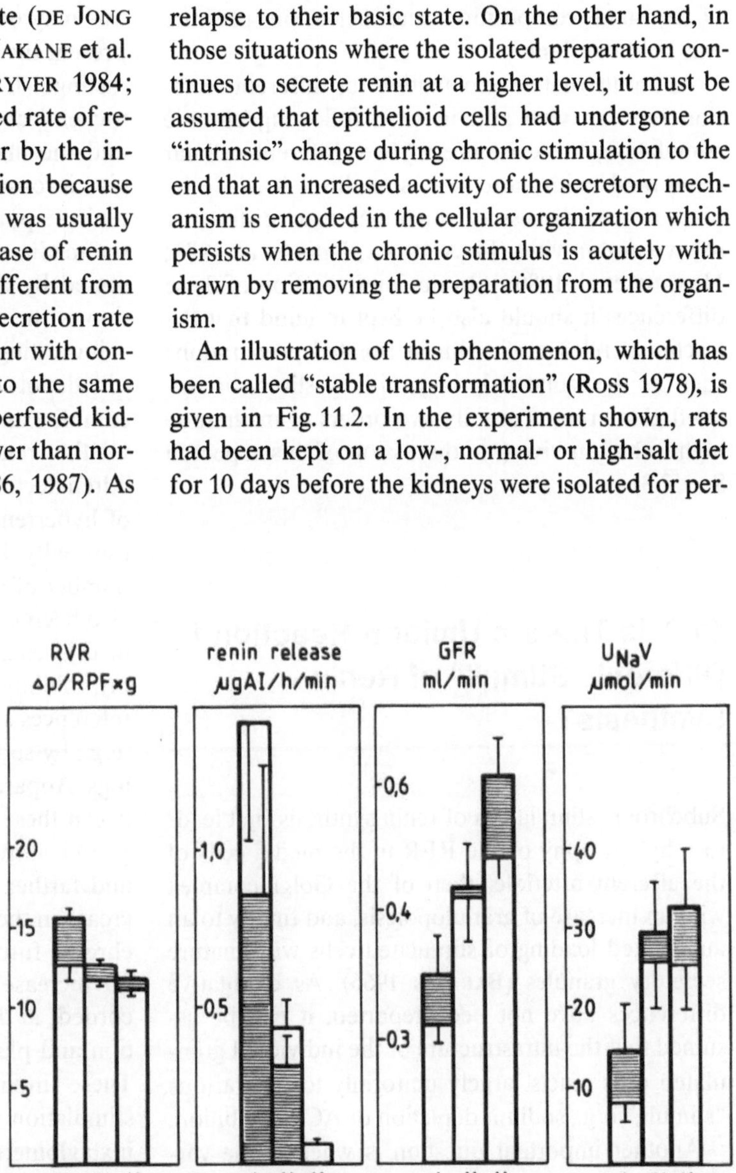

Fig. 11.2. History of salt intake and functional properties of the isolated perfused rat kidney. Male Sprague-Dawley rats were subjected for 10 days to different salt regimens as indicated in the figure. Kidneys were taken for isolated perfusion as described by SCHWERTSCHLAG and HACKENTHAL (1982) with a synthetic medium at 105 mm Hg. *Shaded parts of bars* indicate the increase or, in renin release, the decrease induced by 10 nM angiotensin II. *RVR*, renovascular resistance; *Δp*, pressure difference; *GFR*, glomerular filtration rate (ml/min); $U_{Na}V$, urinary sodium excretion (μmol/min); *renin release* (μg angiotensin/h per minute); *L*, low salt (0.12 mEq NaCl/kg diet, 10 days); *N*, normal salt (ssniff, 85 mEq/kg diet); *H*, high salt (1000 mEq/kg diet, 150 mEq/l H_2O). (SCHWERTSCHLAG and HACKENTHAL, unpublished observations)

fusion. Interestingly, not only renin secretion rate, but also GFR and sodium excretion differed dramatically between the three groups, although the renovascular resistance exhibited little change. Renin content of the kidney differed by 40% between the low- and high-sodium diet group. It should be noted that the renal vasculature in this preparation is no longer under the control of renal nerves or humoral signals, the perfusion medium is of constant composition and perfusion pressure was held constant.

As an alternative explanation for different secretion rates in vivo and in vitro following chronic stimulation, the existence of two different cellular renin pools with more or less readily releasable renin has been discussed which may react differently in vivo and in vitro (FRAY 1978a; PARK et al. 1978; NAKANE et al. 1980a). In the interpretation of these differences it should also be kept in mind that the fractional release of renin in isolated preparations such as the isolated kidney, kidney slices, or isolated glomeruli, is usually two orders of magnitude higher than in vivo (see discussion of this aspect in Sect. 7.8).

11.2 Is There a Uniform Reaction to Different "Stimuli" of Renin Synthesis?

Subchronic stimulation of renin synthesis first leads to a hypertrophy of the RER in the media cells of the afferent arteriole, then of the Golgi complex with an increase of granulopoiesis, and finally to an augmented loading of stimulated cells with mature secretory granules (BARAJAS 1966). As qualitative differences have not been reported, it can be assumed that the ultrastructure of the individual granulated cells reacts largely uniformly to the various "stimuli," e.g., sodium depletion or ACE inhibition.

Another important question is whether the various stimuli of renin synthesis differ with respect to their microtopographical realization. According to Sect. 2.2 and 4.1, upon stimulation of the RAS the renin concentration of the kidney can be augmented not only by an increase in the renin depot of cells already renin-positive but also by the recruitment of new granulated cells from the pool of smooth muscle cells in the media of the afferent arteriole. The former process occurs, above all, in the juxtaglomerular area, the latter farther upstream near the boundary between the renin-positive and renin-negative portions of the afferent arteriole. However, it is conceivable that with one stimulus - perhaps under the influence of the macula densa - renin synthesis is predominantly increased in the juxtaglomerular media cells, while in other cases this may occur in media cells farther upstream. At present there are only indirect methods available to distinguish between these possibilities. Both the histochemical JGI of HARTROFT and HARTROFT (1953) and the immunohistochemical JGI (cf. Sect. 4.1.4) give information on the size of the juxtaglomerular renin depot. The stimulation of renin synthesis in the media cells located farther upstream can be assessed by determination of the renin-positive portion of the afferent arteriole. An increase of the JGI - indicating an increase of the juxtaglomerular renin depot - was observed under the effect of all stimuli examined so far, i.e., reduction of pressure in the renal artery, sodium deprivation, thirst, adrenalectomy, inhibition of ACE, and in a number of hyperreninemic conditions in man, among these especially Bartter's syndrome. An increase in the number of renin-positive cells farther upstream has also been documented for the reduction of pressure in the renal artery, sodium depletion, adrenalectomy, inhibition of ACE, and Barrter's syndrome (for references see Sect. 4.1.4). An inhibition of the RAS (e.g., by sodium loading) lead to the opposite findings. Apparently there is no principal difference between these "stimuli" in regard to their microtopographical realization in the juxtaglomerular area and farther upstream. In contrast, although an increase in the immunohistochemical JGI during subchronic furosemide application could be observed, no increase in the renin-positive portion was recorded; at the same time the renal renin concentration and plasma renin were significantly increased. These findings appear to indicate that subchronic stimulation with furosemide may be confined to the juxtaglomerular portion of the afferent arteriole and does not lead to recruitment of additional renin-producing cells farther upstream (F. TAUGNER, unpublished observations).

Chapter 12

Interspecies Variations of the Juxtaglomerular Apparatus in Mammals

12.1 Qualitative Differences

Fundamental differences in the morphology of the JGA and the glomerular hilus in mammals are largely unknown. Minor interspecies heterogeneities have been pointed out in the respective sections. These concern, among others, the dimensions of the hilar arterioles, the Goormaghtigh cell field and the macula densa, the incidence of podocyte processes in the parietal layer of Bowman's capsule, and the number of peripolar cells. Also, the localization and ultrastructure of epithelioid cells appears to be basically similar in all mammals studied so far, just as does the phenomenon of metaplastic transformation of smooth muscle cells into epithelioid cells. Larger differences seem to exist in the incidence of paracrystalline structures that are found especially in juvenile granules.

BARAJAS (1966) has shown that protogranules pinching off from the innermost Golgi cistern contain paracrystalline structures, particularly after stimulation of the RAS (cf. Sect. 5.1). In humans and other primates, these paracrystals occur more frequently and occasionally even in intermediate and mature secretory granules (BARAJAS 1966; ROSEN and TISHER 1968). In Sects. 5.2 and 5.3 arguments are given in favor of the assumption that the paracrystalline contents of protogranules in humans may consist of prorenin. In contrast to all other mammals examined, in *Tupaia belangeri*, a primitive primate, paracrystalline matrix structures are common not only for small nascent, but also for large storage granules of epithelioid cells (FORSSMANN and TAUGNER 1977; cf. Fig. 12.1).

It may therefore be speculated that these paracrystalline structures consist of prorenin, the epithelioid cells of *Tupaia* thus containing an outstandingly large supply of inactive renin that would first have to be activated upon stimulation of the RAS by cleavage of the prosegment. Indeed, it has already been shown that the characteristic paracrystalline structures in the secretory granules of *Tupaia* disappear upon electric stimulation of the renal sympathetic nerves, the matrix of mature granules becoming homogeneously electron dense (TAUGNER et al. 1978a). If the assumption is correct that this change corresponds with the cleavage of the prosegment, then epithelioid cells of *Tupaia* would be a suitable model for the study of renin activation using morphological methods.

A conspicuous species heterogeneity which is not well understood, both in view of its origin and its functional implications, refers to the coexistence of renin and angiotensin in the secretory granules of epithelioid cells typical for rats (cf. Sect. 4.2); it may be related to species differences in the activity of angiotensinases (cf. Sect. 4.5).

Finally, another conspicuous finding should be mentioned, of which it is unclear whether it may be traced back to an interspecies heterogeneity. According to the observations of BIAVA and WEST (1966b) and BIAVA (1967), there are two different types of organelles within human epithelioid cells: specific and nonspecific cytoplasmic granules, i.e., renin granules and lysosomes. However, atypic granules and also renin-negative organelles resembling secondary lysosomes - except MVBs - are quite rare in the laboratory animals studied most extensively, i.e., in rats and mice (TAUGNER and METZ 1986), suggesting that, in these species, there is only one type of granular organelles which may, altogether, be classified as lysosomes (TAUGNER et al. 1985b; TAUGNER and HACKENTHAL 1987; cf. Sect. 5.3). Here, it should be remembered that the epithelioid cells of laboratory animals are usually studied following perfusion fixation of the kidneys, generally resulting in an optimal tissue preservation. The tissue specimens obtained from human kidneys, on the other hand, can only be embedded after immersion fixation, with inferior structural

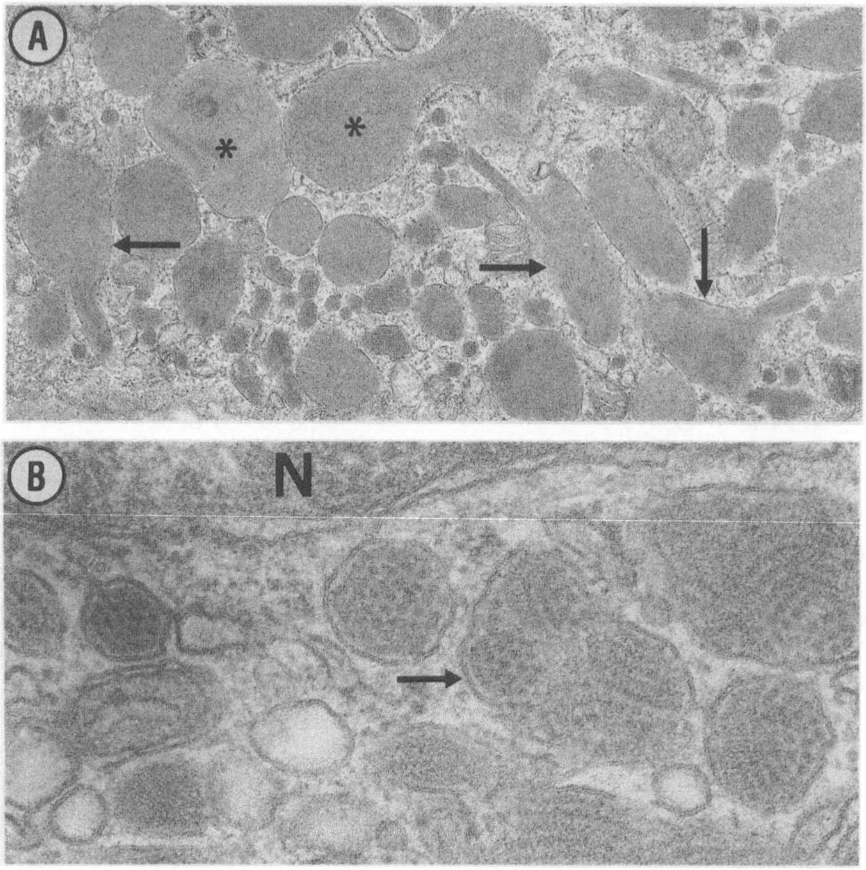

Fig. 12.1 A, B. Epithelioid cell secretory granules of *Tupaia belangeri.* **A** The cytoplasma of the epithelioid cell depicted contains large ovoid granules with a homogeneous internum (*asterisks*), large polymorphous granules showing in their matrix accumulations of electron-dense filamentous material (*arrows*), and small granules exhibiting only paracrystalline filaments; ×21000. **B** Epithelioid cell secretory granules of *Tupaia* at higher magnification. *N*, nucleus. Note the paracrystalline filamentous structures, which in some of the larger granules show different orientations (*arrow*); ×92000

preservation. Since epithelioid cells are particularly sensitive, the question arises whether the nonspecific granules in the literature may not be altered renin granules. Presently, there are no immunocytochemical experiments answering this question (cf. Chap. 14).

12.2 Quantitative Differences

Unlike the relatively small qualitative differences in the morphology of the JGA, considerable quantitative differences seem to exist in the equipment of the JGAs and afferent arterioles of the individual species with granulated cells, which may reflect respective differences in the renal renin content, and also in the plasma renin activity.

Unfortunately, the available data are remarkably inconsistent. Some authors simply state that the number of granulated or renin-positive cells is large or small in a particular species, without defining a standard for this statement or giving numerical data. In other studies, two species are compared, but paired experiments combining the determination of renal and/or plasma renin with morphometric data on the equipment of the JGA with granulated cells are an exception. This becomes even more important in view of the intra- and interzonal heterogeneities discussed in section 4.1.1 and the well-known variations in the renin status of the kidney, depending on several parameters, first of all the sodium balance. It is evident that the problem of standardization is greatest with human material. Nevertheless, some clearly outstanding findings show that the number of granulated or epithelioid

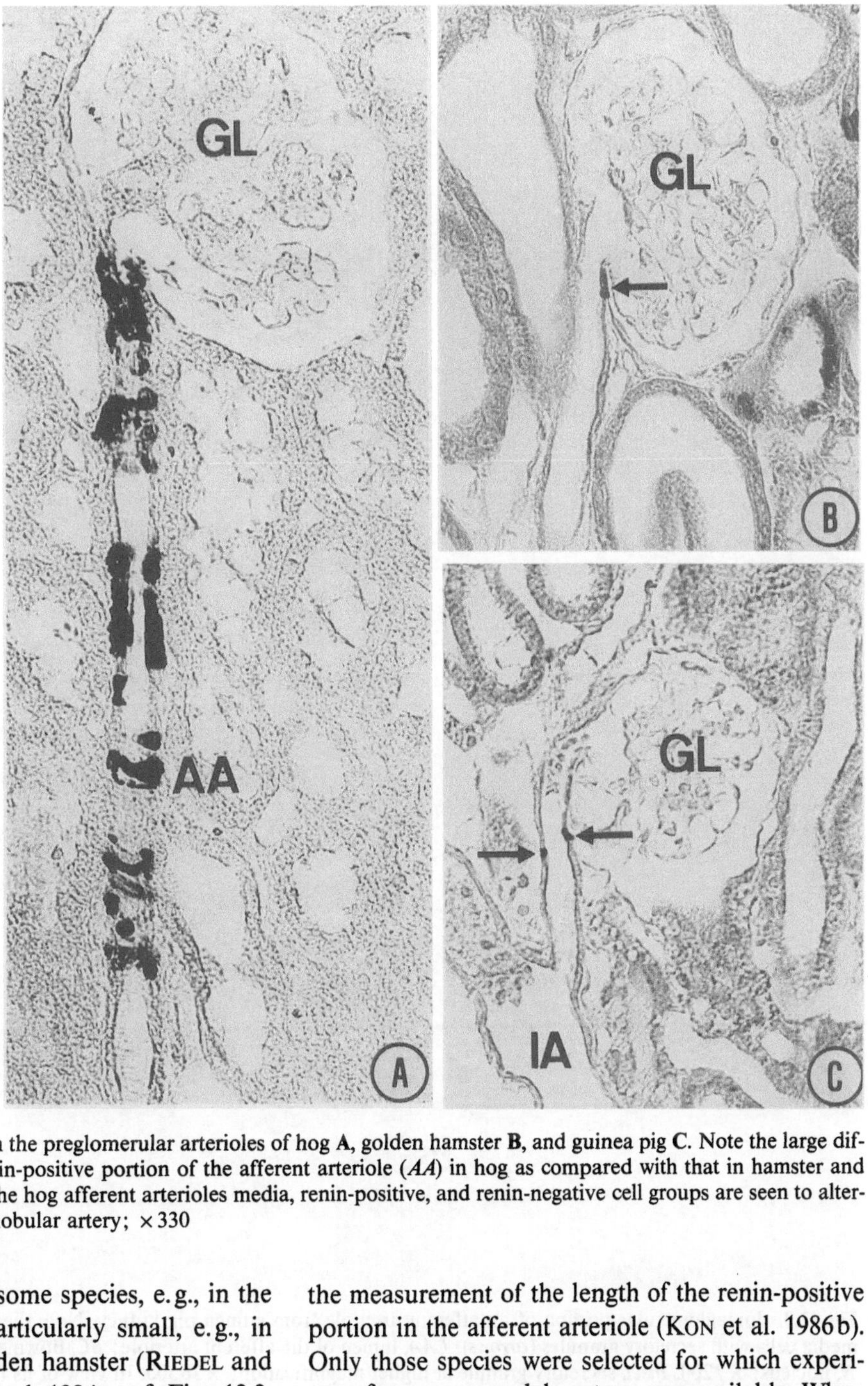

Fig. 12.2. Renin-positive cells in the preglomerular arterioles of hog **A**, golden hamster **B**, and guinea pig **C**. Note the large difference in the length of the renin-positive portion of the afferent arteriole (*AA*) in hog as compared with that in hamster and guinea pig kidney (*arrows*). In the hog afferent arterioles media, renin-positive, and renin-negative cell groups are seen to alternate. *GL*, glomerulus; *IA*, interlobular artery; ×330

cells is especially large in some species, e.g., in the mouse and the hog, or particularly small, e.g., in the guinea pig and the golden hamster (RIEDEL and BUCHER 1967; TAUGNER et al. 1984a; cf. Figs. 12.2, 12.3).

Table 12.1 may illustrate these problems. Selected and partially converted data from three studies are listed addressing the comparison of the kidney renin status of several species. Three methods have been used: the determination of kidney renin (SCHAFFENBURG et al. 1960), the assessment of the histochemical JGI (MATSUHASHI et al. 1977a), and the measurement of the length of the renin-positive portion in the afferent arteriole (KON et al. 1986b). Only those species were selected for which experience from our own laboratory was available. When the individual species are compared as to the renin status found in the different studies, agreements but also great discrepancies are apparent. Conspicuously, the hamster, ranging last in the reports of MATSUHASHI et al. (1977a) and HACKENTHAL et al. (1987), is rated by KON et al. (1986b) even above the mouse although the mouse has been considered high ranking from the beginning (RUYTER 1925; cf.

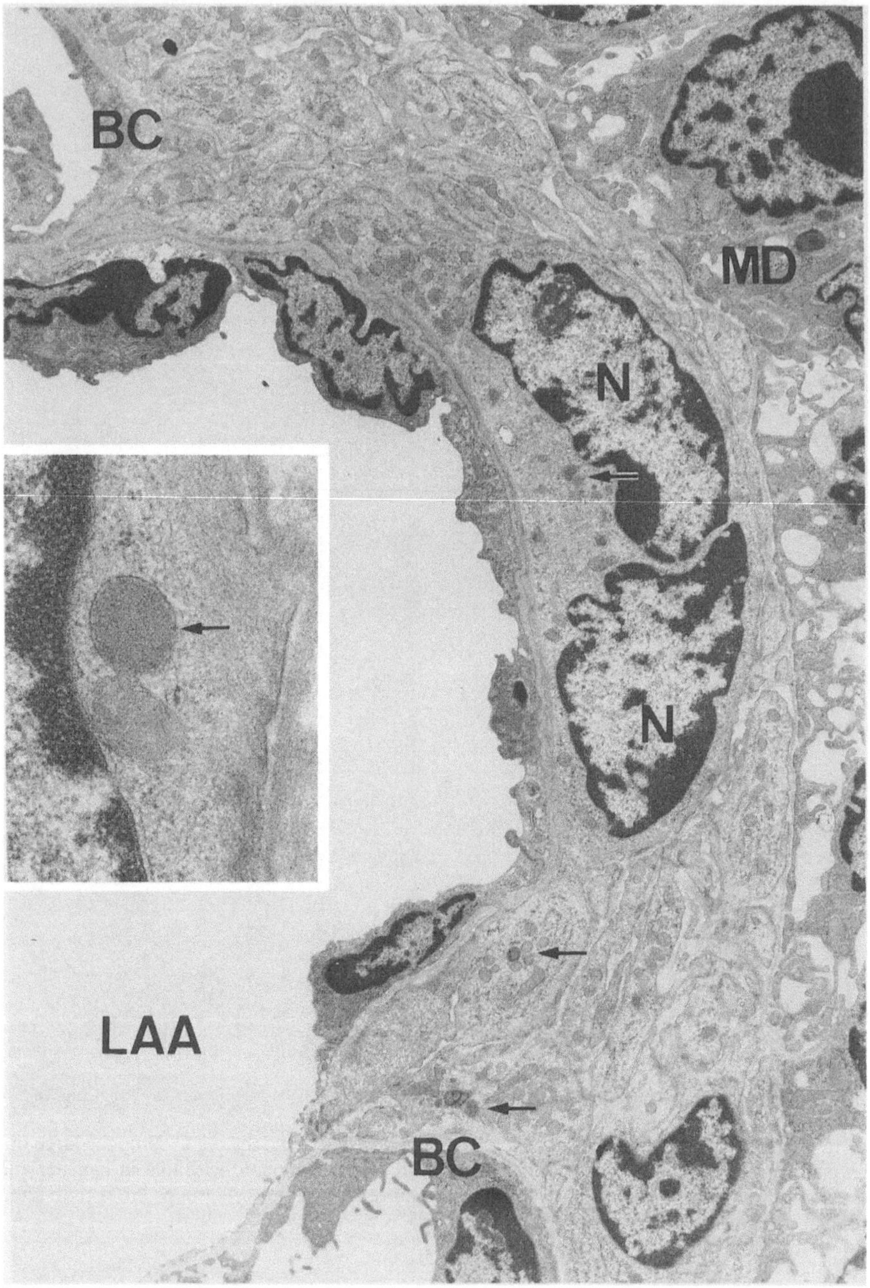

Fig. 12.3. Juxtaglomerular portion of an afferent arteriole from guinea pig kidney. Note the modest equipment of the vessel media cells with secretory granules (*arrows*). *LAA*, lumen of the afferent arteriole; *BC*, Bowman's capsule; *MD*, macula densa; *N*, nucleus; ×7200. *Inset*, secretory granule at higher magnification; ×36500. In view of its lengthy shape and modest equipment with elements of the secretory machinery, the cell depicted may be classified as an intermediate cell

also HARTROFT 1963). Discrepancies are also evident with regard to the human kidney, whose renin is the very lowest in the study of SCHAFFENBURG et al. (1960). In order to improve interspecies comparison, technical standard criteria and the consideration of variables such as age and sex as well as that of the current stimulation level of the RAS would be required (cf. Fig. 12.4).

This is particularly important for the biochemical estimation of kidney renin content, which, at first glance, appears to be the most reliable and the best standardized parameter, since data are expressed in

Table 12.1. Renin status in six selected species assessed with three different methods

	Juxtaglomerular index[a] (MATSUHASHI et al. 1977a)	Renin-positive portion of the afferent arteriole		Kidney renin[b] (SCHAFFENBURG et al. 1960)
		KON et al. (1986b)	HACKENTHAL et al. (1987)	
Mouse	49.6 (%)	22.3 (μm)[a]	45.4 (μm)	-
Rat	11.9	43.8	36.8	53
Hog	10.8	44.9	68.5	36
Monkey	10.8	-	39.9	-
Hamster	3.7	38.9	24.0	-
Human	-	-	43.0	1.2

[a] The numerical data were derived, at the best approximation, from the figures in the cited papers
[b] Kidney renin content is expressed as the generation rate of angiotensin in kidney extracts incubated with homologous angiotensinogen preparations, and subsequent estimation of angiotensin by rat blood pressure bioassay

quantitative terms suitable for interspecies comparability, e.g., in nanograms of angiotensin I generated from angiotensinogen per gram of tissue per unit of time. Although this view is in principle correct, the pertinent data reported in the literature are by and large unsuitable for such comparisons. Not only have incubations of kidney extracts with angiotensinogen been performed at different pH values, using different extraction procedures, and different concentrations of angiotensinogen, but also, and this is of major importance, angiotensinogen preparations from different species, mostly heterologous, have been used. The reaction rate of a given renin preparation may vary as much as 30-fold depending on the species from which the angiotensinogen preparation has been obtained, even if all other variables are held constant (e.g., POULSEN and JACOBSEN 1986), making a direct quantitative interspecies comparison of such data meaningless. It therefore appears more appropiate to select only those data for interspecies comparison which have been obtained by incubation of kidney extracts with homologous angiotensinogen preparations under standardized conditions, thus giving a better estimate of the "true" biological angiotensin generation capacity of kidney renin. It is for this reason that for the computation of Table 12.1 the kidney renin data of SCHAFFENBURG et al. (1960) have been selected, although in these studies complete inhibition of angiotensinase activity during incubation has not been ascertained, which may possibly explain the unexpectedly low values for human kidney renin.

Interestingly enough, the reaction rate of renin from a given species with homologous angiotensinogen is usually lower than with heterologous angiotensinogens. This observation, in addition to the low affinity and low reaction rate of renin and angiotensinogen (in comparison to other hydrolytic enzymes and their natural substrates), has led POULSEN and JACOBSEN (1986) to hypothesize that angiotensinogen may in fact represent an inhibitor rather than a substrate of renin, the "real" substrate still being unknown.

Interspecies comparisons of the renin status of the kidney are, of course, of interest for the question of whether they reflect major interspecies differences in the relative contribution of the renin-angiotensin system to the overall regulation of blood pressure and salt and water homeostasis. It must be emphasized that this question cannot be answered at present; not only because of the poor comparability of kidney renin data, but also because extrapolation from the kidney renin content to its systemic importance is not valid. Other factors, such as spontaneous and stimulated secretion rates of renin, its plasma half-life, plasma concentrations of angiotensin II, its clearance rate, the distribution, density, and reactivity of angiotensin II receptors, and many other related factors, may vary considerably between species and determine the overall activity of the RAS to a larger extent than the kidney renin content. This point is illustrated by the comparison of plasma renin activity to kidney renin content: Interspecies differences in plasma renin activity estimated under identical conditions in plasma taken from conscious rats, dogs, or humans were much smaller (HACKENTHAL, unpublished observations) than those of kidney renin content, as

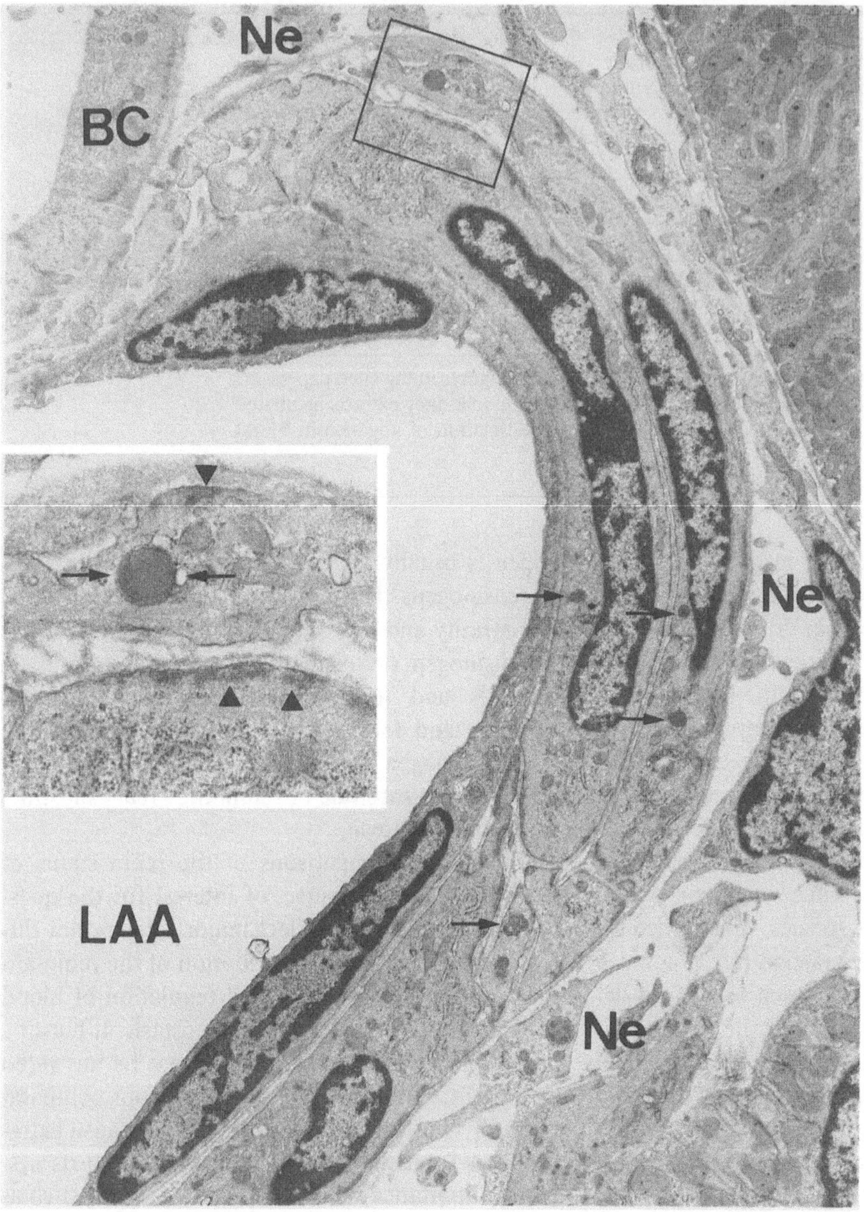

Fig. 12.4. Juxtaglomerular portion of an afferent arteriole from mouse kidney after 2 weeks of sodium loading (1 500 mg Na/kg synthetic diet). Note the modest equipment of the vessel media cells with secretory granules (*arrows*). *LAA*, lumen of the afferent arteriole; *Ne*, nerve axon bundles; *BC*, Bowman's capsule; ×7600. *Inset*, detail marked in the overview at higher magnification; *arrows*, renin granule; *arrowheads*, attachment sites, ×23300. Note that the media of the vessel consists of intermediate cells which have probably been retransformed from epithelioid cells upon sodium loading. As a consequence, the juxtaglomerular portion of the afferent arteriole deviates from the picture characteristic for control mice (cf. Fig. 2.16), thus rather resembling that of guinea pigs (cf. Fig. 12.3) or golden hamsters

reported by SCHAFFENBURG et al. (1960) (mean values of plasma renin activity, 7, 1.5, and 1.5 ng angiotensin I/ml per hour, values for kidney renin 53, 28, and 1.2 units/g, respectively).

Thus, in order to identify species in which the RAS has a more prominent regulatory function than in others, an integrated approach would be necessary, taking into account these and many additional variables. To the best of our knowledge, no major attempt has yet been made in this direction.

Chapter 13

Phylogeny and Ontogeny of the Juxtaglomerular Apparatus

13.1 Evolution of the Juxtaglomerular Apparatus in Nonmammalian Vertebrates

The evolution of the RAS in conjunction with the JGA has been summarized by SOKABE and OGAWA (1974), NISHIMURA (1978, 1980), and WILSON (1984a). In accordance with the objective of this monograph, the conditions in nonmammalian species - referring to these reviews - will only be dealt with summarily. For a comparison between the various mammalian species see Chap. 12. For the renal actions of angiotensin in lower vertebrates, the reader is referred to the comprehensive review of NISHIMURA (1980), and for the coevolution of the RAS and the nervous control of blood circulation to that of WILSON (1984b).

For a long time it was controversial as to whether a RAS with its focal point in the kidney exists in primitive vertebrates (BEAN 1942; WEICHERT 1965). In the meantime, there is convincing, although mostly indirect, evidence for the presence of a RAS comparable to that of mammals in all classes of lower vertebrates with the exception of *cyclostome* and *elasmobranch fishes*. Thus, elements of the JGA have not been found in the lamprey and the hagfish (SOKABE et al. 1969; NISHIMURA et al. 1970). However, angiotensin II, as in many other nonmammalian vertebrates, is known to elevate blood pressure through catecholamine release also in the hagfish (CARROLL and OPDYKE 1982), and there is evidence compatible with the assumption that the kidneys of agnathian fish may contain an enzyme homologous with mammalian renin (cf. WILSON 1984a, with comments on the reliability of renin assay in nonmammalian species).

The situation is similar in elasmobranch fish, one of the surviving members of the class of *Chondrichthyes*. BOHLE and WALVIG (1964) first reported on the absence of granulated juxtaglomerular cells in *Acanthias vulgaris*. Since then, several other elasmobranch species have been investigated. Although the distal tubule is in contact with the vascular pole of its parent renal corpuscle, neither granulated cells nor Goormaghtigh cells or a macula densa could be identified in elasmobranchs (for references and functional aspects see OGAWA and OGURI 1978; NISHIMURA 1980; WILSON 1984a). In holocephalian fish, the other surviving member of the class of Chondrichthyes, in contrast, granulated cells have been found in a juxtaglomerular position as well as farther upstream in the renal arterial tree, but no macula densa or extraglomerular mesangium (NISHIMURA et al. 1973; OGURI 1978). Holocephalians are reported to resemble teleosts in some endocrinological aspects, although morphlogically they do so with elasmobranchs (NISHIMURA 1980).

Except for some primitive bony fish *(Chondrostei* and, partly, *Holostei*), granulated cells and upon adequate incubation renin-like pressor activity have been found in all other *Osteichthyes*, including more than 100 teleost species (MEYER et al. 1966, 1967; CAPRÉOL and SUTHERLAND 1968; WILSON 1984a). The ultrastructure of the granulated cells in the kidney of a freshwater and a saltwater teleost (goldfish and English sole) has been studied in detail by BULGER and TRUMP (1969). Besides homogeneous epithelioid cell granules, granules containing hexagonally packed paracrystalline tubular structures were found similar to those encountered in *Pleuronectes* by CHRISTENSEN et al. (1987). In sarcopterygian fish, granulated cells resembling those in a juxtaglomerular position are also located in large renal arteries (OGAWA et al. 1972; LAGIOS 1974). As in glomerular teleosts, no obvious relation between renal tubules and myoepithelioid cells associated with renal arteries seemed to exist in teleosts of the aglomerular variety.

CHRISTENSEN et al. (1987) reinvestigated this relationship and the ultrastructure of the granulated cells in the aglomerular lemon sole (*Pleuronectes microcephalus* Donovani), originally described by BOHLE and WALVIG (1964). Serial sectioning revealed anastomosing arteriolar networks in the caudal half of the *Pleuronectes* kidney connected with peritubular capillaries and also with veins. The walls of these arterioles were composed of epithelioid cells the granules of which had a paracrystalline substructure, thus closely resembling those of the English sole (BULGER and TRUMP 1969). The epithelioid cell granules of *Pleuronectes* cross-reacted with antibodies directed against murine and human renin (Fig. 13.1). These results were taken to suggest that the RAS in the lemon sole may be important not only for blood pressure homeostasis, but also for the blood supply to the peritubular capillaries. Furthermore it was speculated that the arteriolar networks, situated between arterial branches and peritubular capillaries, may represent retia mirabilia or modified glomeruli (Fig. 13.2).

In fish inhabiting hypertonic environments, the RAS may regulate drinking; in teleosts acclimatized to freshwater, it may contribute to osmoregulation. Adaptation phenomena to osmotically different habitats of the renin-angiotensin system have been studied in the Japanese eel, in *Tilapia mossambica*, and in the Atlantic salmon by SOKABE et al. (1973), KRISHNAMURTY and BERN (1973), and CHRISTENSEN et al. (1982a, 1989b), respectively. Renin activity increases with bleeding or hypotensive treatment in the toadfish, an aglomerular teleost (NISHIMURA et al. 1979), indicating the involvement of the RAS, possibly mediated by the renal baroreceptor, in cardiovascular homeostasis (NISHIMURA 1980). GANNON (1972) showed that in some teleosts blood vessels are contacted by adrenergic nerves, but the anatomical relationship between granulated cells and the nerve endings has not been studied.

Angiotensin II is thought to exert its pressor action in fish as in some phylogenetically younger nonmammalian species mainly by mediation of the sympathetic system (KHOSLA 1985; for references see WILSON 1984a). As in other primitive vertebrates, the structure of angiotensin I deviates also in fish at positions 1, 5, and 9 from that in mammals (HAYASHI et al. 1978; TAKEMOTO et al. 1983). However, conversion into the corresponding octapeptides is required for pressor activity also in these animals (KHOSLA et al. 1985).

Except for BEAN (1942) and WEICHERT (1965), all authors agree on the occurrence of granulated cells, and on that of renin-like enzymatic activity in the *amphibian kidney* (OKKELS 1929; OKKELS and PÉTERFI 1929; for further references see LAMERS et al. 1973). In some contrast to mammals and similar to teleosts, the granular cells are dispersed over a rather large part of the afferent arteriole (VAN DONGEN and VAN DER HEIJDEN 1969; LAMERS et al. 1973). In the frog and the toad, UNSICKER et al. (1975) found nerve endings close to both smooth muscle and epithelioid cells of the afferent arteriole. The occurrence of converting enzyme in the kidney of *Rana catesbeiana* and *Bufo arenarum* was demonstrated by YAMAGUCHI et al. (1986) and FERNANDEZ-PARDAL et al. (1986).

Whereas granulated cells seem to have emerged during the early evolution of bony fishes, the Goormaghtigh cell field and the macula densa evolved later in the phylogeny of vertebrates (MCKELVEY 1963; MEYER et al. 1966; CAPRÉOL and SUTHERLAND 1968). Thus, Goormaghtigh cells have never been found in the amphibian kidney (for references see LAMERS et al. 1973). As far as the macula densa

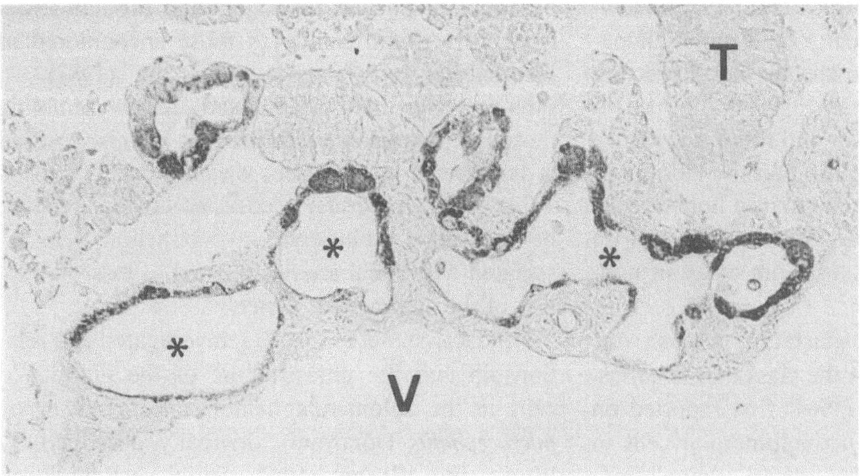

Fig. 13.1. Arteriolar network (*asterisks*) in the kidney of *Pleuronectes microcephalus* (lemon sole) with epithelioid cells showing cross-reactivity against human renin antibody. *V*, vein; *T*, tubules; ×1000

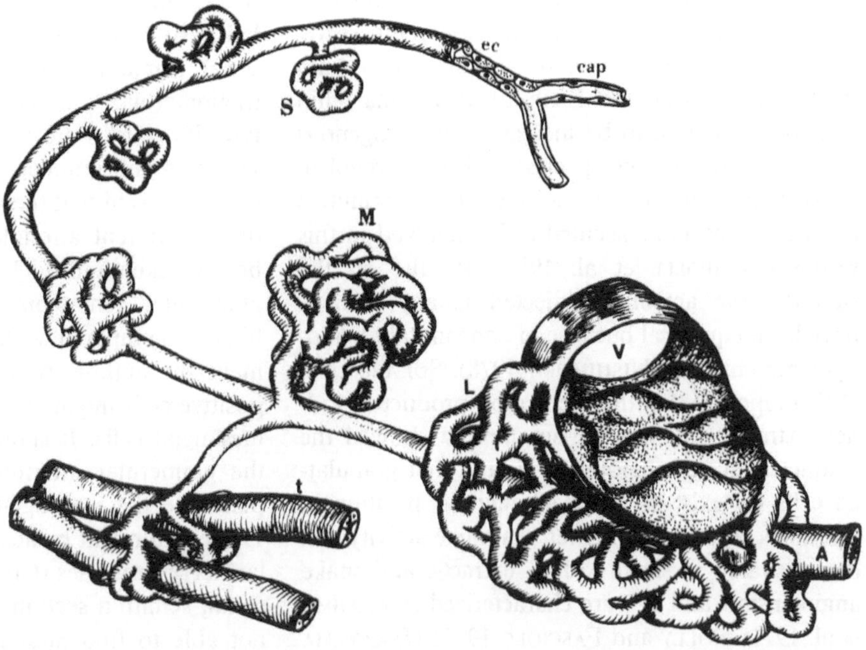

Fig. 13.2. Schematic drawing of an intrarenal artery from *Pleuronectes* kidney with many arteriolar networks. The terminal end of the artery (*A*) with epithelioid cells (*ec*) continues into the peritubular capillaries (*cap*). *L*, large network; *M*, medium-sized network; *S*, small network, *V*, large vein which shows how the arteriolar networks make impressions into the lumen; *t*, renal tubules. (From CHRISTENSEN et al. 1987)

is concerned, there has been some dispute about its presence in amphibia (EDWARDS 1940; CAPELLI et al. 1970; LAMERS et al. 1973, 1974, 1977; UNSICKER et al. 1975; but see SOKABE et al. 1969). These discrepancies might be a matter of definition: While in mammals, the macula densa lies against the hilus of the renal corpuscle, in the toad it may be turned away from it, covering a larger part of the distal tubule (LAMERS et al. 1973).

In a series of papers LAMERS and coworkers reported on several aspects of the granulated cells in toad kidney. The existence of a specialized macula densa-like portion of the distal tubule in intimate contact with the afferent arteriole was documented by graphic reconstruction (Lamers et al. 1973) and glucose-6-phosphate dehydrogenase histochemistry (LAMERS et al. 1977). The granulated cells were shown to be innervated (LAMERS et al. 1974) and renin positive (LAMERS et al. 1985). Besides the distal (juxtaglomerular) afferent arteriole, also media cells of larger arterial vessels were immunoreactive. Acid phosphatase activity of the granules encountered in these cells (LAMERS and STADHOUDERS 1985) and some of their ultrastructural features (LAMERS et al. 1974) seemed to indicate that renin granules in *Bufo* like those in mammals may be modified lysosomes. As to the complex problem of whether and how the RAS of amphibians is involved in blood pressure homeostasis, steroidogenesis, and osmoregulation, we refer to the reviews mentioned above.

The distal tubule does not contact its parent glomerulus in *reptiles*, so that both a macula densa and a Goormaghtigh cell field are missing. However, granulated cells are found in a juxtaglomerular position, and renin activity can be extracted from kidney tissue (for references see WILSON 1984a). The increase of plasma renin activity by trypsinization or acidification suggests the existence of an inactive form of the enzyme in the freshwater turtle (*Pseudemys scripta*) (CIPOLLE and ZEHR 1984). Upon incubation of blood plasma, an angiotensin I-like peptide ([Asp^1, Val^5, His^9] Ang I) was generated by an enzyme with renin-like characteristics in *P. scripta elegans* (HASEGAWA et al. 1986). ACE activity has been demonstrated by CIPOLLE and ZEHR (1984) in *P. scripta*. According to UVA and VALLARINO (1982), salt loss increases and salt loading decreases kidney renin activity in the terrestrial turtle (*Testudo hermanni* Gmelin). In *Pseudemys scripta*, severe hemorrhagic as well as nonhemorrhagic hypotension failed to increase renin secretion, although the animals did respond to sodium depletion (CIPOLLE and ZEHR 1985). Plasma renin activity is reported to be increased by endogenous catecholamines in this species; as isoproterenol in contrast to epinephrine was ineffective, primitive adrenergic receptors seemed to be involved in this response (CIPOLLE et al. 1986). Results on the steroidogenic action of injected kidney extracts have been equivocal because of concomitant hemodynamic changes (NISHIMURA 1978). SOKABE et al (1969) reported that the incubation product of kidney extracts and homologous plasma showed the characteristics of angiotensin II, and that granulated cells do exist in a juxtaglomerular position in three species of snakes. Later, renin-like activity was demonstrated in snake kidney extracts, and snake angiotensin I and II were characterized (NAKAJIMA et al. 1971; NOLLY and FASCIOLO 1972; NAKAYAMA et al. 1977; HO et al. 1984; GERVITZ et al. 1987). In addition, the presence of ACE (LAVRAS et al. 1977) and angiotensinogen (GERVITZ et al. 1987) in the plasma of *Bothrops jararaca* has been reported, suggesting the existence of a complete renin-angiotensin system comparable to that found in mammals. Although angiotensin II has a pressor effect which depends in part on catecholamine release, the RAS is ascribed only a minor role in the blood pressure control of reptiles (cf. WILSON 1984a).

In *birds*, there are two types of nephrons: a reptilian type without, and a mammalian type with, a Henle's loop; the latter is only found in the deeper parts of the cortex (HUBER 1917; SPERBER 1960; SHOEMAKER 1972; cf MORILD et al. 1985a). The existence of a JGA in the avian kidney has been claimed by some authors (EDWARDS 1940; MILLER 1967; SOKABE et al. 1969; OGAWA and SOKABE 1971; WIDEMAN et al. 1981), but denied by others (for references see MORILD et al. 1985b). In the kidneys of domestic fowl fixed by perfusion, MORILD et al. (1985b) found all components of the JGA including Goormaghtigh cells and the macula densa in both reptilian- and mammalian-type nephrons. Granulated cells were occasionally found in the afferent arteriole of mammalian type nephrons, but not in those of the reptilian type, or in any efferent arterioles. When present, granulated cells were also seen in the intraglomerular part of the afferent arteriole, ramifying within the prominent mesangial cell mass typical for this species. Mesangial cells transformed into granulated cells after subchronic stimulation by furosemide injections were not found on serial, semithin sections (MORILD et al. 1987). Intraglomerular granulated cells termed "mesangial-type JG cells" were observed by KON et al. (1984). Using an antiserum against mouse renin, KON et al. (1986) showed that these cells, like those in the wall of the afferent arteriole, were immunoreactive in both chicken and ducks. As the incidence of intraglomerular renin-containing cells was extremely high as compared with their occurrence in mammals, the authors concluded that, in birds, renin-positive cells might also be derived from peripheral mesangial cells. It should be noted, however, that the glomerular arterioles of domestic fowl often penetrate the mesangial cell mass (MORILD et al. 1986), and might be accompanied by intraglomerular nerve processes (CHRISTENSEN et al. 1982b). On serial, semithin sections MORILD et al. (1987) were not able to find mesangial cells transformed into granulated cells after subchronic stimulation by furosemide injections. Altogether, with avian glomeruli, the distinction between arteriolar media and mesangial cells may meet with special difficulties. Related questions pertain to the homology of avian and mammalian mesangial cells and to the functional significance of the peculiar centroglomerular cell mass suggested to be of special relevance for the regulation of the GFR (cf. MORILD et al. 1985a, b).

Granulated cells and renin activity have been found in the kidneys of several avian species (for references see WILSON 1984a). Plasma renin activity is reported to increase after blood loss, upon sodium depletion and sustained hypotension (CHAN and HOLMES 1971; NISHIMURA and BAILEY 1982). Subchronic hypovolemia and hypotension induced by repeated furosemide injections is followed by hyperplasia and hypergranulation of epithelioid cells in the JGAs of both mammalian- and reptilian-type nephrons (MORILD et al. 1987). Angiotensin II increases blood pressure, in chicken after a preceding depressor response, and in other species in a monophasic reaction. However, this rise in blood pressure is probably not due to a direct action of the octapeptide on vascular smooth muscle cells, but rather to the release or potentiation of noradrenaline and adrenaline (for references see WILSON 1984a). The biphasic response to [Asp^1, Val^5] angiotensin II in domestic fowl was analyzed by NAKAMURA et al. (1982) and NISHIMURA et al.

(1982), suggesting that the depressor effect may be due to a direct vasodilator action of the octapepetide or to the release of an unknown depressor substance.

13.2 Ontogeny of the Juxtaglomerular Apparatus in Mammals

The development of the JGA is closely related to the morphogenesis of the nephron. The glomerulus originates from capillary invagination into the lower cleft of the S-shaped body (OSATHANONDH and POTTER 1966). Primitive mesenchymal cells of the nephrogenic zone, not utilized in the formation of the nephron, follow the first capillary on its way into the glomerular cavity. As the renal corpuscle differentiates, they become mesangial cells and also form the media of the glomerular arterioles. A few of these mesenchymal cells invading the early glomerular cavity remain in a hilar position and initiate the formation of the Goormaghtigh cell field continuous with the glomerular stalk and the peripheral mesangium (KAZIMIERCZAK 1971). The macula densa differentiates as the lips of the glomerular cup forming the edges of Bowman's capsule constrict, thus drawing together the prospective glomerular arterioles (MITCHELL et al. 1982). In stage IV of nephrogenesis, the afferent arteriole, having early and close contact with the macula densa, has a more adult appearance than the efferent vessel: while the latter may still be regarded as a capillary devoid of a media, the afferent arteriole is equipped with a continuous layer of media cells some of which are already granulated (KAZIMIERCZAK 1971).

The prenatal appearance of granules in media cells of the preglomerular arteriole in the *metanephric kidney* has been observed in several species (LJUNGQVIST and WÅGERMARK 1966; ERTL 1967; KAYLOR and CARTER 1967; SUTHERLAND and HARTROFT 1968; KAZIMIERCZAK 1971; ZIMMERMANN 1971; BRÜHL et al. 1974; SMITH et al 1974; MOLTENI et al.1974; EGUCHI et al. 1975; MITCHELL et al. 1982; EGERER et al. 1984).

As expected, the granulated cells in the fetal kidney proved to be renin-positive in man (PHAT et al. 1981), the mouse (MINUTH et al. 1981), the rat (RICHOUX et al. 1987), the rabbit (AMSAGUINE 1983/1984), and the pig (EGERER et al. 1984). According to MINUTH et al. (1981), renin is found in the fetal mouse kidney prior to all other organs. At 3-4 days before delivery, immunoreactivity was preferentially located in media cells of interlobular arteries. Subsequently, the formation of new nephrons and the maturation of their glomeruli was accompanied by a shift in renin localization from the interlobular arteries to the afferent arterioles (Fig. 13.3) and, in these, to a juxtaglomerular position. At the same time, kidney renin content and concentration increased rapidly (Fig. 13.4). Similar observations have been made in the fetal metanephric kidney of humans (MOLTENI et al. 1974; LINDOP and GARDINER 1986), rabbits (DRUKKER et al. 1983), pigs (EGERER et al. 1984), and rats (JELÍNEK et al. 1986). An especially impressive shift in the localization of renin-positive cells during nephrogenesis from large to small arteries and arterioles was observed by RICHOUX (personal communication) in the fetal mouse, by EGERER et al. (1984) in the fetal pig, and by RICHOUX et al. (1987) in the fetal rat. According to RICHOUX, in mice on the 13th day of gestation, renin-positive cells can even be found at the origin of various arteries from the abdominal aorta. In the rat, immunoreactive cells were detectable as early as the 15th day of gestation outside the nephrogenic territories within the walls of mesonephric gonadic and renal arteries (RICHOUX et al. 1987). This is in some contrast to the observations by PHAT et al. (1981) in humans. Although renin-positive cells were found in 5-week-old fetuses, most of these cells were located in the vicinity of the prospective vascular pole. Only rarely were immunolabeled cells encountered in the wall of the major branches of renal arteries (PHAT et al. 1981). This may indicate species differences in the distribution of renin-producing cells in the arterial tree of the developing metanephros.

The possible functional importance of the intrarenal distribution of renin up to the time of birth was discussed by MINUTH et al. (1981) in relation to their experiments in mice. In newborn mice, the localization of renin closely resembles that of adult animals in that most of the immunoreactivity occurs in the epithelioid cells of the afferent arteriole. In addition, at this early stage of kidney development the macula densa of the more mature nephrons is already differentiated and sympathetic innervation is present. It may thus be assumed that the major

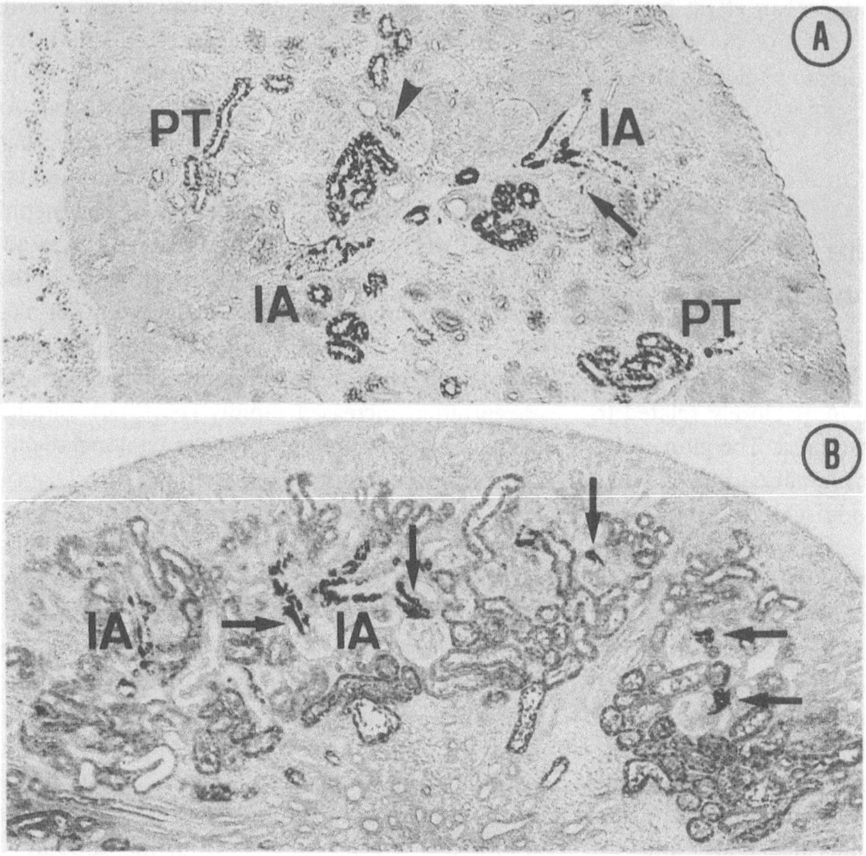

Fig. 13.3. Immunostaining for renin in the differentiating mouse kidney 2 days before birth (**A**) and in the newborn (**B**). **A** Two days before birth, most of the reaction product occurs in the wall of interlobular arteries (*IA*). The vascular pole of the renal corpuscle is only beginning to show renin reactivity, corresponding to this early stage of kidney differentiation (*arrow*). In some cases, also cells of the glomerular stalk exhibit specific staining (*arrowhead*). In addition, many of the more mature proximal convolutions (*PT*) are renin positive; ×75. **B** As compared with the late fetal kidney in **A**, the number of filtering glomeruli and proximal convolutions with pinocytotic activity has significantly increased in the newborn kidney. Consequently, more renin-positive afferent arterioles (*arrows*) and proximal tubules are seen at this time. *IA*, immunoreactive interlobular arteries; ×100. Paraffin sections (7 μm) after immersion fixation in a mixture of 10% formaldehyde and saturated picric acid (1:1, v/v). Antiserum dilution, 1:1000

stimuli of renin secretion, i.e., the intrarenal baroreceptor mechanism, sympathetic control, and the macula densa mechanism are already in operation at or soon after birth.

This view is supported by biochemical studies of the perinatal RAS. As has been shown by Minuth et al. (1981), the renin content of the mouse kidney rises sharply at term. Broughton Pipkin et al. (1974a) and Jelínek et al. (1986) have demonstrated that renin is released into the circulation and the other components of the RAS are active at this stage of development. Broughton Pipkin et al. (1974b) observed low plasma angiotensin II concentrations in fetal animals of various species, with a drastic increase during and shortly after birth, which parallels the rapid increase of plasma renin concentration during birth observed by others (Granger et al. 1971; Pohlová and Jelínek 1974; Osborn et al. 1980). Postnatally, the further increase of renin and angiotensin II depends on the state of differentiation of the kidney. Thus in lambs, born with mature, fully developed kidneys, highest plasma angiotensin II concentrations were recorded in newborns less than 8 h old, whereas in rabbits, in which kidneys are less mature at term, plasma angiotensin II continues to rise up to 2 weeks after birth (Broughton Pipkin et al. 1974b; for man see Kotchen et al. 1972). In rats, even 5 weeks may elapse before maximum plasma angiotensin II concentrations are reached (Wallace et al. 1980).

As to the target tissues and organs, the vascular smooth muscle is sensitive to the vasoconstrictor ac-

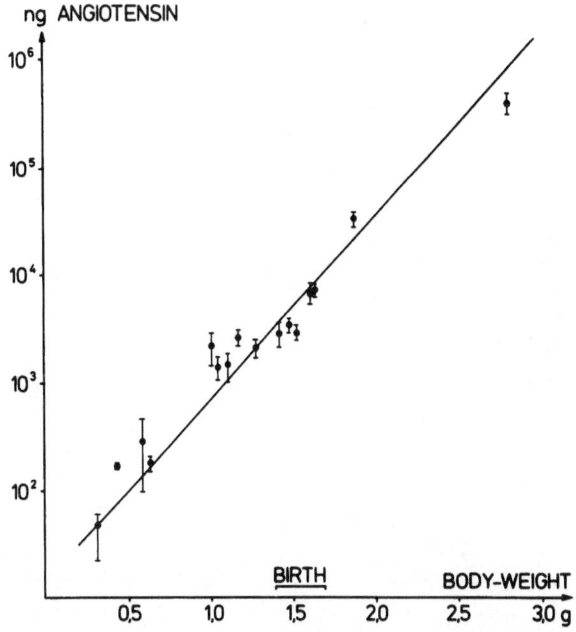

Fig. 13.4. Steep rise of mouse kidney renin during the perinatal period, paralleling the increased immunohistochemical reactivity. Renal renin content (*ordinate*) is a logarithmic function of body weight (*abscissa*) and kidney weight, respectively (not shown). (From MINUTH et al. 1981)

tion of angiotensin II very early in fetal development, and aldosterone synthesis and secretion, which is very low in the prenatal period, appears to respond to angiotenins II at birth, as plasma aldosterone concentration closely follows angiotensin II concentration, both exhibiting a sharp rise during birth (DALLE et al. 1978).

Thus, as judged from the summarized morphological and biochemical findings, all components and stimuli of the RAS are present or functioning at birth, the main target organs are responsive, and the major regulatory mechanisms are probably in operation. It may therefore be assumed that, in principle, the RAS serves the same functions in newborn mice as it does in the mature animal.

In fetal animals, e.g., in fetal pigs (EGERER et al. 1984) and in the mouse, 3-4 days before term (MINUTH et al. 1981), the situation is different. At this time, renin is predominantly located in the wall of the interlobular arteries with little or no specific immunoreactivity detectable at the "classical" location, the juxtaglomerular afferent arterioles. Furthermore, the macula densa is not differentiated, peripheral angiotensin II concentrations are very low (BROUGHTON PIPKIN et al. 1974a, b), and the adrenal cortex is still at a low level of differentiation, as reflected in very low plasma aldosterone concentrations (DALLE et al. 1978). Taken together, these data suggested to MINUTH et al. (1981) that at this stage of development two of the three major functions of the RAS, namely, the control of the tubuloglomerular balance through the macula densa and the regulation of the aldosterone secretion from the zona glomerulosa, are not yet in operation or still at a modest level of responsiveness and activity.

It is unknown to what extent renin contributes to the third major RAS function, the systemic blood pressure regulation at this early stage of development. The low peripheral concentrations of angiotensin II (BROUGHTON PIPKIN et al. 1974b) would rather suggest that its role is a minor one. The distinct localization of renin in the interlobular arteries indicates the possibility that the major function of the RAS in fetal animals is to help maintain renal perfusion pressure.

In conclusion, the vasoconstrictor function, local or systemic, of the RAS appears to be expressed ontogenetically well before the maturation of the putative macula densa associated functions of renin. This ontogenetic sequence would have its correspondence in the phylogenetic development of the RAS, as the evolution of juxtaglomerular cells precedes that of the macula densa on the phylogenetic scala. Furthermore, granulated cells are widely distributed along the renal arteries also in primitive animals and tend to be located close to the glomeruli only in the more advanced organism (cf. Sect. 13.1). One might, therefore, postulate that the RAS has evolved as a humoral regulator of blood pressure homeostasis and kidney perfusion and that later in phylogenesis

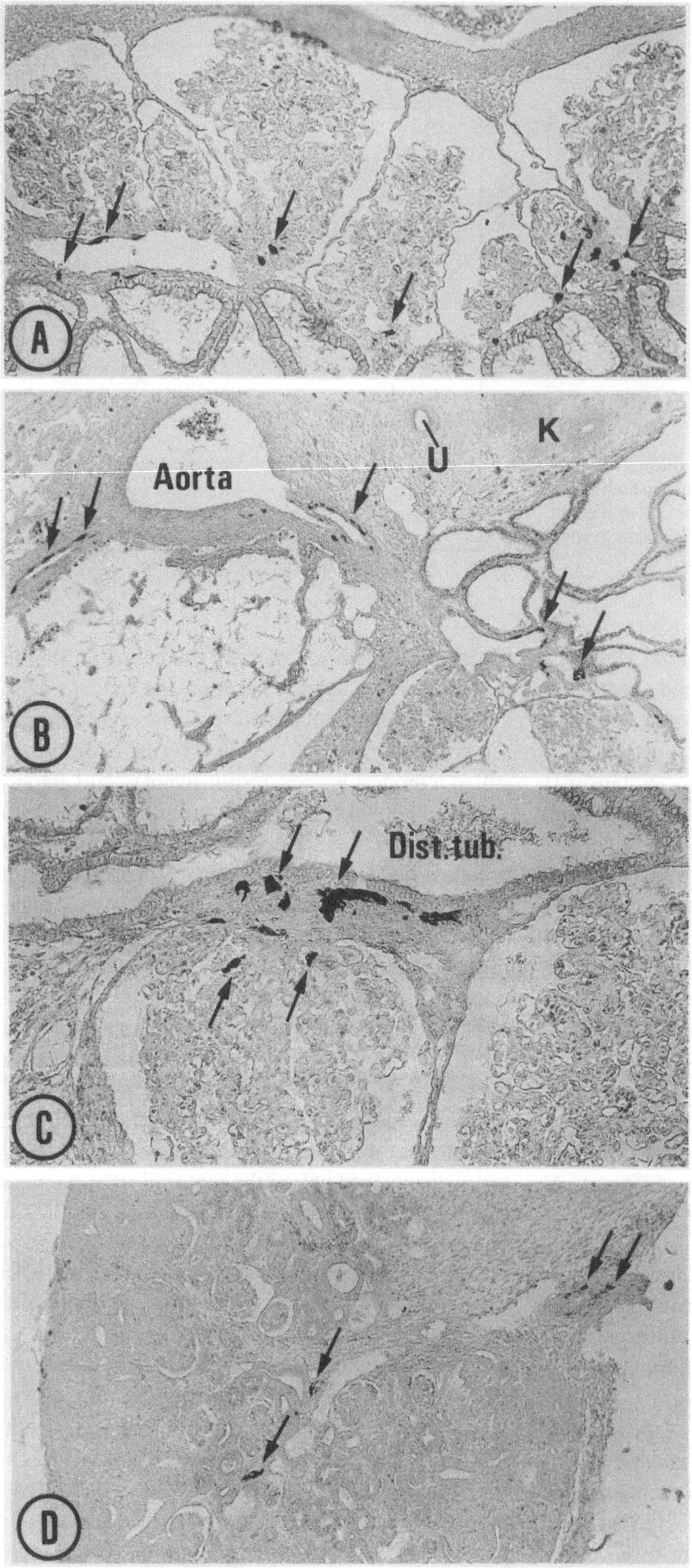
A
Aorta
U
K
B
Dist.tub.
C
D

a shift from a systemic endocrine function to a local regulator of glomerular function has occurred (cf. NISHIMURA 1978). For reactions and function of the RAS during the perinatal period see BROUGHTON PIPKIN (1984a, b) and JELÍNEK et al. (1986).

There is evidence that, at least in some mammals, the prenatal RAS is at first expressed in the mesonephric kidney. Like the pronephros, the *mesonephros* is a transitory organ, the nephrons of which, however, largely differentiate in some species, as in pigs, sheep, and rabbits, before falling to involution (STANIER 1960). The pig embryo has an exceptionally large and long-lasting mesonephros (BREMER 1916) that lends itself well for comparative studies with the development of the definitive kidney. A biologically active protein with renin-like properties was extracted from mesonephric tissue of hog fetus as early as 1942 (KAPLAN and FRIEDMAN). Granulated cells in the JGA of the mesonephros have been shown both in light and electron micrographs (MATSUHASHI et al. 1975; TIEDEMANN and EGERER 1984). Renin-positive cells have been demonstrated in the media of the afferent arterioles in the mesonephric kidney of rabbits (AMSAGUINE 1983/84; EGERER et al. 1984) and of a human embryo, 4 weeks old (AMSAGUINE 1983/84). EGERER et al. (1984) found immunoreactive cells in the mesonephros of a 12-mm pig embryo, corresponding to the 21st day of gestation. Most of these cells were encountered in the media of the juxtaglomerular afferent arteriole. Besides, scattered immunoreactive cells were also found in the media of the ventral aorta, of the mesonephric arteries, and of the efferent arterioles (Fig. 13.5A–C). The disappearance of mesonephric renin-positive cells at later stages of development was not investigated by EGERER et al. (1984). Granulated cells were missing in 76- and 90- to 100-mm pig embryos (MATSUHASHI et al. 1975; SUTHERLAND and HARTROFT 1968). In the definitive pig kidney, renin-positive cells did not appear until approximately the 31st day of gestation, establishing the second generation of glomeruli (Fig. 13.5D). Again, most of the cells occurred in the juxtaglomerular portions of the metanephric afferent arterioles. Additionally, immunoreactive cells were detected in the media of interlobular and arcuate arteries and also in the renal artery within the hilar region of the metanephric kidney (EGERER et al. 1984). The observed distribution of renin-positive cells all along the renal arterial tree in both the developing meso- and metanephric kidneys was interpreted to confirm the proposal that the major function of the RAS in the fetal animal is to participate in blood pressure homeostasis. Serving this function, a transitory RAS seems to be installed in the mesonephric organ, apparently comparable with that of the metanephric kidney.

Fig. 13.5. Metanephric **A–C** and definitive **D** kidneys from pig embryos, immunoreacted for renin. Paraffin sections, 7 µm; no counterstain. **A** Longitudinal section of a mesonephros from pig embryo (approximately 21st day of gestation, CR length, 12 mm). Renin-positive cells in the wall of large mesonephric artery branches both on the glomerular side and close to the macula densa attachement (*arrows*). Some of the immunoreactive cells lie within the glomerular stalk; ×200. **B** Cross-section of a pig embryo (approximately 29th day, CR length, 21 mm). Strongest immunoreaction in mesonephric arteries close to subjacent glomeruli (*arrows*). Somewhat weaker reaction in the media of mesonephric arteries branching from the aorta. *U*, ureter; *K*, kidney; ×60. **C** Metanephric renal corpuscle from larger pig embryo (approximately 38th day, CR length, 45 mm) with renin-positive cells in glomerular stalk and tangentially sectioned mesonephric arteries; ×170. **D** Definitive kidney from pig embryo (approximately 31st day) with renin-positive cells in the renal artery and, more strongly reactive, at juxtaglomerular sites (*arrows*). At this stage, the glomeruli of the third generation begin to form below the organ capsule; ×80

Chapter 14

Pathology of the Human Juxtaglomerular Apparatus

R. WALDHERR

To our current knowledge, there are no fundamental differences between humans and other mammalian species either in the histology of the JGA or in the ultrastructure of the granulated renin-producing cells. Common features are discussed in the preceding chapters. In Chap. 12, for example, the normal human JGA is compared with the JGA of other mammals; the extensive studies on the synthesis and maturation of renin in humans are dealt with in Sect. 5.2; and investigations on the secretion of inactive renin in Sect. 7.7. It would have been difficult, however, to integrate the pathological reactions of the human JGA in various disorders within the long-term experimental modulations of renin synthesis and secretion discussed in Chap. 11, for the following reasons. Most relevant disorders are accompanied by chronic stimulation of the JGA over months and years, not feasible for reproduction in animal experiments. In contrast to experimental models, most conditions associated with alterations of the human JGA are heterogenous, with their precise pathogenesis frequently unknown; in consequence, the stimuli of renin synthesis and secretion are obscure in many instances, thus limiting the basis of comparison with the various experimental models (cf. Sect. 7.1 and Chap. 11). Finally, the relevance of the human JGA in these disorders, emphasized by the number of pertinent publications, called for a separate discussion easier to survey for the reader interested most of all in the pathology of the human JGA.

The following review summarizes our current knowledge of the response of the JGA in various human renal and nonrenal disorders. Most of these conditions are characterized by a marked stimulation of the JGA, whereas inhibition of renin synthesis and secretion is exceptional (Table 14.1). Particular reference is made to Bartter's syndrome, including a recently described variant, the so-called hyperprostaglandin E syndrome (SEYBERTH et al. 1985; TAUGNER et al. 1988), and pseudo-Bartter's syndrome since these disorders are associated with a rather "pure" form of chronic stimulation of the JGA and only minor additional parenchymal alterations.

Table 14.1. Renal and nonrenal disorders with pathological reactions of the JGA

Stimulation of the Juxtaglomerular Apparatus
Ischemic kidney
Renal artery stenosis
Renal infarction
Thrombotic microangiopathy/malignant nephrosclerosis
End-stage kidneys
Reflux nephropathy/segmental hypoplasia
Primary and secondary glomerulopathies (various forms)
Bartter's syndrome and related disorders
Addison's disease
Renin-secreting tumors
Miscellaneous conditions
Inhibition of the Juxtaglomerular Apparatus
Primary hyperaldosteronism
Feedback inhibition in normal renal cortex as a consequence of renin hypersecretion from affected tissue

14.1 Stimulation of the JGA

14.1.1 Ischemic Kidney

Examination of renal tissue obtained by biopsy, by surgery, or at autopsy has confirmed findings in classical animal studies that the JGA is markedly activated in renovascular hypertension (cf. Chap. 11). In diffuse or localized ischemia of the kidney due to renal artery stenosis or renal infarction, respectively, the number of granulated cells is substantially increased (TURGEON and SOMMERS 1961; CROCKER 1962; ITSKOVITZ et al. 1963; BARAJAS et al. 1967; HELBER et al. 1970; MEYER 1972, 1978; MALTINTI et al. 1977; CHRISTENSEN et al. 1978; MCLAREN and MACDONALD 1982). On immunohistochemistry (MENARD et al. 1979; CAMIL-

LERI et al. 1980; FARAGGIANA et al. 1982; CAMILLERI et al. 1983; MAST et al. 1983; NOCHY et al. 1983), renin-positive JGAs are far more numerous, and individual JGAs contain more immunoreactive cells as compared with controls (Fig. 14.1). In serial sections from "hyperplastic" JGAs, groups of renin-producing cells encircling the glomerular tuft at the visceroparietal junction of Bowman's capsule have been demonstrated (CAMILLERI et al. 1983). Renin-positive cells are even detected near completely obsolescent glomeruli. In addition, immunoreactive cells are found in the arteriolar walls upstream from the renal corpuscle, not only in the proximal portion of the afferent arteriole but also in the interlobular artery. With in situ hybridization, renin mRNA has been demonstrated in these renin-positive cells (BRUNEVAL et al. 1988). Goormaghtigh cells are consistently renin-negative in this condition (CAMILLERI et al. 1983). In renal tissue outside the ischemic areas of partially infarcted kidneys, the JGAs are normal or even atrophic (MENARD et al. 1979; CAMILLERI et al. 1980; MAST et al. 1983) comparable with the contralateral untouched kidney in animals with unilateral renal artery constriction (TAUGNER et al. 1982d). The apical portions of the proximal tubules, however, disclose positive immunostaining, in particular in the noninfarcted cortical areas, representing pinocytosed renin.

Likewise, in thrombotic microangiopathy (hemolytic-uremic syndrome, endotheliotropic microangiopathy), in systemic sclerosis, and in primary and secondary malignant nephrosclerosis, an enlargement of the JGAs (MEYER et al. 1972, 1978; CAIN and KRAUS 1976; BOHLE et al. 1977, 1982), an increased number of renin-positive JGAs, and an elevated number of renin-positive cells in the individual JGA have been observed (CAMILLERI et al. 1983; NOCHY et al. 1983).

In all these disorders, RAS stimulation is compatible with an increased renin synthesis via the renal vascular receptor (cf. Chap. 11). "JGA hypoplasia" of the nonaffected kidney in unilateral renal

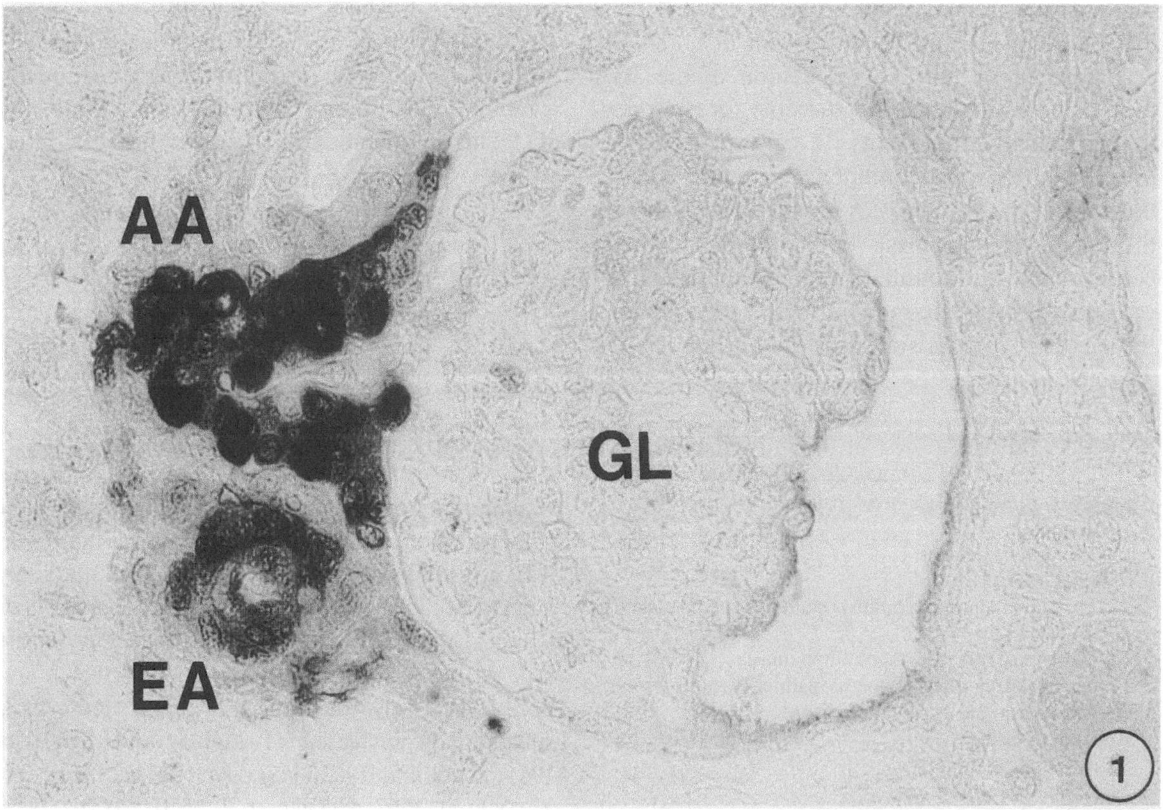

Fig. 14.1 Renal artery stenosis. Ischemic retraction of the glomerular tuft *(GL)* with dilatation of Bowman's capsule. Marked enlargement of the JGA with an increase in renin-positive cells in the media of the afferent *(AA)* and efferent *(EA)* glomerular arterioles; 5 μm paraplast section; PAP-method. Antiserum dilution, 1:5000; × 800

artery stenosis and of noninfarcted cortical areas in partial renal infarction is presumably caused by feedback inhibition.

Other conditions resulting in decreased RBF and activation of the JGA include contracted kidneys of different etiology, kidneys in protracted shock, liver cirrhosis, eclampsia, radiation nephritis, classic panarteritis, chronic transplant rejection, and Addison's disease (ALEXANDER 1968; HELBER et al. 1970; MEYER 1972, 1978; CHRISTENSEN et al. 1976; HELMCHEN 1987). Here, in addition to a decrease of the intravascular pressure affecting the renal vascular receptor, the stimulation of the JGA may in some instances be due to an increase in sympathetic outflow.

14.1.2 Reflux Nephropathy

Reflux nephropathy is considered as a form of chronic atrophic pyelonephritis caused by and/or associated with vesicoureteral reflux. There is now general agreement that segmental renal scarring and marked parenchymal atrophy in undersized kidneys previously attributed to primary hypoplasia (so-called Ask-Upmark kidney), represent an acquired abnormality which is strongly related to vesicoureteral reflux occuring in early life or even in utero. Hypertension is found in 60%-70% of the patients associated with elevated plasma renin activity and increased renin activity in renal veins. In children, reflux nephropathy is the most frequent cause of severe hypertension.

The typical lesion appears as strongly demarcated shrunken kidney lobes with calyceal dilatation and loss of nephrons. There is a marked increase in the number of granulated renin-containing cells within the scarred atrophic areas (GODARD 1973; BARAJAS et al. 1977b; MEYER 1978; AMAT et al. 1981; CAMILLERI et al. 1983), in particular in the subcapsular region, whereas normal or even negative immunostaining is observed in the adjacent nonatrophic spared tissue except for a positive reaction in proximal tubules suggestive of filtered renin reabsorbed by pinocytosis (cf. Sect. 14.1.1). Renin-containing cells are numerous at the vascular pole of severely altered or small obsolescent glomeruli as well as in the walls of tortuous afferent arterioles and interlobular arteries (Fig. 14.2). In addition, renin-positive cells are present in the mesangial stalk of hyalinized glomeruli and in nests scattered throughout the cortical tissue. In agreement with AMAT et al. (1981), we found these scattered renin-positve cells always adjacent to capillary-type vascular channels but not in the interstitium proper. Whereas in pre- and postglomerular arterioles as well as in interlobular arteries metaplastic transformation of smooth muscle cells is well known to account for the increase in renin-positive cells (CANTIN et al. 1977a; cf. Chap. 10), the spindle-shaped appearance of renin-positive cells adjoining peritubular capillaries rather suggests the transformation of pericytes. On electron microscopy, the granulated media cells in atrophic scars show evidence of a markedly enhanced synthetic activity with large amounts of crystalline protogranules in Golgi cysterns, aggregates of protogranules, and large mature secretory granules (BARAJAS et al. 1977b).

Activation of the RAS in reflux nephropathy seems to be mediated by locally decreased perfusion pressure due to scar formation and arteriolar occlusion. Feedback inhibition might be responsible for the weaker immunostaining in the adjacent nonatrophic areas.

14.1.3 Glomerular Disorders

Early planimetric studies demonstrated a significant increase in the mean size of the "juxtaglomerular cell complex" in various forms of glomerulonephritis, including mesangial-proliferative glomerulonephritis, membranoproliferative glomerulonephritis, and membranous glomerulonephritis, particularly in advanced stages and when associated with an increased serum creatinine level (TURGEON and SOMMERS 1961; MEYER 1972, 1978; SKAANE et al. 1975). In minimal-change nephrotic syndrome, the juxtaglomerular cell complexes are strikingly enlarged in the presence of a marked nephrotic syndrome (HARA and MEYER 1975; MEYER 1978). Recent immunohistochemical examinations of renal biopsies from patients with various glomerular disorders, such as minimal-change nephrotic syndrome, mesangial IgA glomerulonephritis (Fig. 14.3) and membranous glomerulonephritis disclosed an increase in the percentage of renin-positive JGAs and an increment in the number of immunoreactive cells per JGA when compared with controls (CAMILLERI et al. 1983; NOCHY et al. 1983).

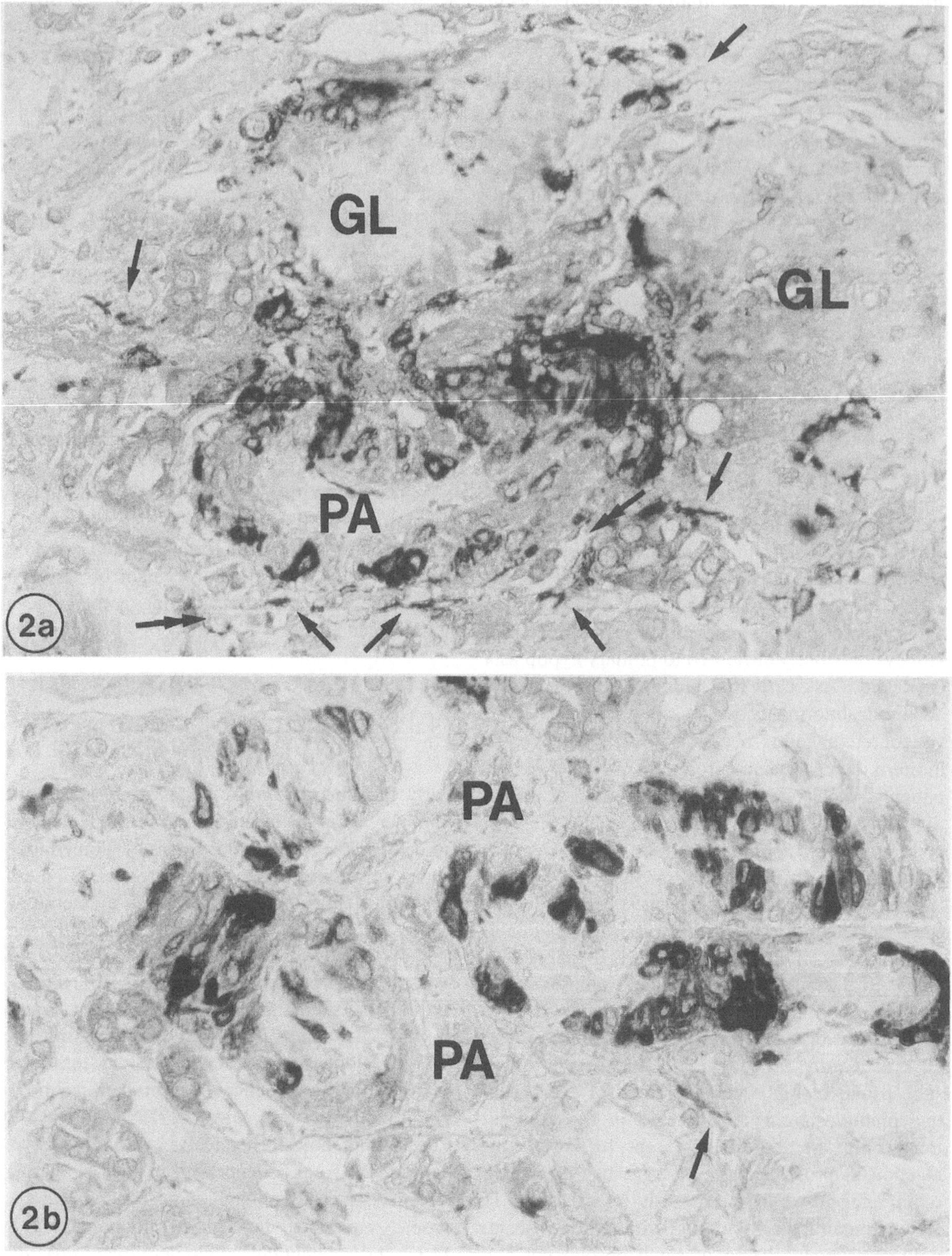

Fig. 14.2 a, b. Reflux nephropathy. **a** Scarred atrophic renal tissue with numerous renin-positive cells at the vascular poles of obsolescent glomeruli *(GL)* and in the media of preglomerular arterioles *(PA)*. In addition, spindle-shaped renin-positive cells *(arrows)* are seen related to capillary-type vascular channels. The presumable course of these channels is marked by *double arrows*. Antiserum dilution, 1 : 5000; × 500. **b** Scattered renin-positive cells in the media of tortuous preglomerular arterioles *(PA)* distant from the vascular pole. Spindle-shaped immunoreactive cell *(arrow)* adjacent to a capillary vessel. Antiserum dilution, 1 : 5000; × 500

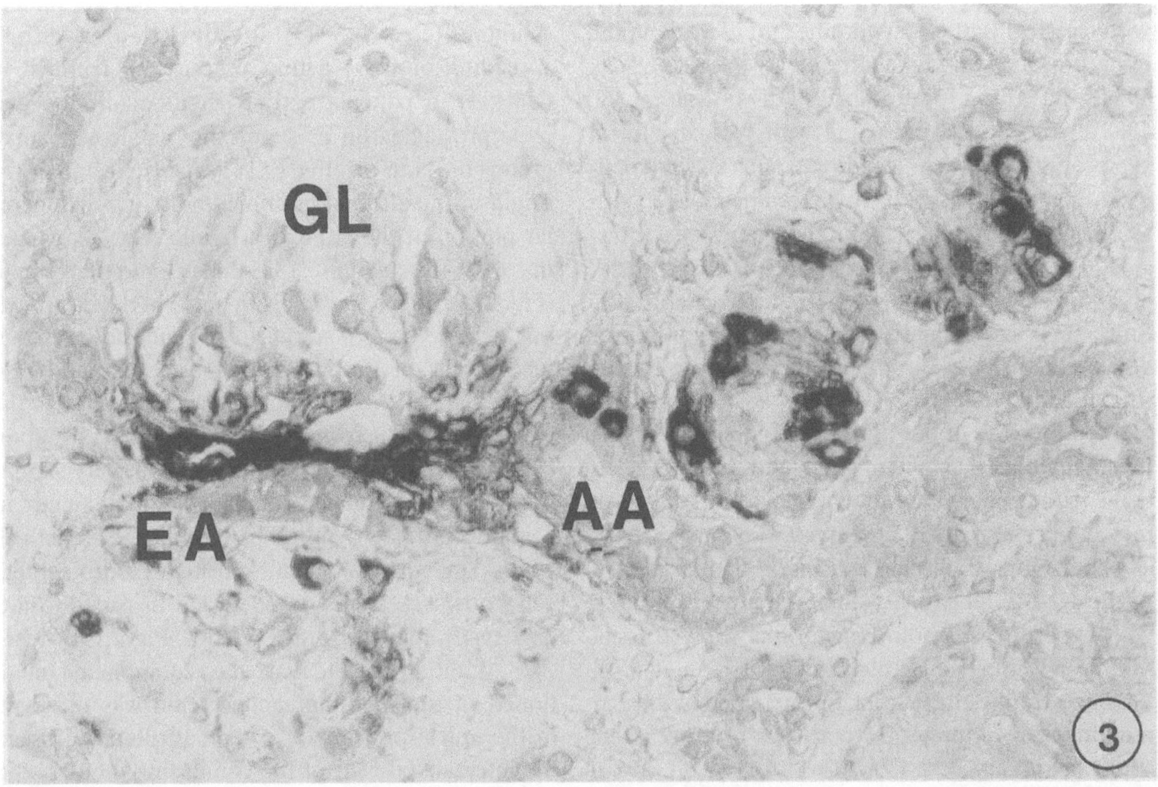

Fig. 14.3. Chronic glomerulonephritis (mesangial IgA glomerulonephritis). The number of renin-containing cells is markedly increased in the media of the afferent *(AA)* and the efferent glomerular arteriole *(EA)*. Scattered immunoreactive cells are present in the arteriolar media at some distance upstream from the glomerulus *(GL)*. Anti-renin dilution, 1:5000; ×500

There is no correlation between blood pressure and the JGA status.

JGA enlargement in glomerulonephritis is certainly induced by a combination of different stimuli affecting several or all relevant receptors (cf. Chaps. 7, 11). When associated with a nephrotic syndrome, hypovolemia probably represents the primary cause.

14.1.4 Bartter's Syndrome and Related Disorders

In 1962, BARTTER et al. described a syndrome characterized by hypokalemia, metabolic alkalosis, hyperreninemia, secondary hyperaldosteronism, and normal blood pressure. The initial observation by BARTTER et al. that the JGA is enlarged in this disorder has been repeatedly confirmed (BRYAN et al. 1966; BRACKETT et al. 1968; GOODMAN et al. 1969; WEGMANN 1970; SUTHERLAND et al. 1970; SCHMIDT et al. 1973; RAMANATHAN et al. 1973; CHRISTENSEN et al. 1976; VERBERCKMOES et al. 1977; MEYER 1978; FARAGGIANA et al. 1980; NAKADA et al. 1980; BOHLE et al. 1982; MCLAREN and MACDONALD 1982; CAMILLERI et al. 1983). In subsequent years it has become apparent that Bartter's syndrome is not a uniform disease but rather comprises a spectrum of conditions. Even if these disorders share the above-mentioned basic features, there are some fundamental clinical and pathophysiological differences which justify the separation of variants (STEIN 1985), or even distinct entities (SEYBERTH et al. 1985). Morphologically, all these conditions are characterized by a marked enlargement of the JGA on renal biopsy, and it was postulated that this finding constitutes an essential requisite of the syndrome (BARTTER and RODRIGUEZ 1982). Very similar if not identical metabolic and morphological changes are also observed in the so-called pseudo-Bartter's syndrome, due to diuretic abuse, chronic diarrhea, repeated vomiting, anorexia nervosa, etc. (WOLFF et al. 1968; HELBER et al. 1970; MEYER 1972, 1978; WAGNER et al. 1979; BARTTER and RODRIGUEZ, 1982; MIZUIRI et al. 1987).

A detailed immunohistochemical and electron

microscopic study on renal biopsies from children with classical Bartter's syndrome and its variant, the hyperprostaglandin E syndrome (SEYBERTH et al. 1985), has been performed by TAUGNER et al. 1988c. All patients presented with typical clinical symptoms of Bartter's syndrome such as hypokalemia, metabolic alkalosis, hyperreninemia, secondary hyperaldosteronism, and normal blood pressure. Furthermore, the patients with hyperprostaglandin E syndrome showed some remarkable features, such as prenatal onset with polyhydramnios and prematurity at birth, episodic diarrhea, hypercalciuria and nephrocalcinosis, associated with a marked increase of not only renal but also systemic prostaglandin E_2 activity. Kidney biopsies from patients with isolated intermittent hematuria and completely normal renal tissue served as controls.

Renal biopsies disclosed a marked enlargement of the JGA in all patients. In addition, the children with hyperprostaglandin E syndrome showed focal tubular and interstitial calcification associated with tubular atrophy, interstitial fibrosis, and scattered mononuclear infiltrates. Hyperplasia of renal interstitial medullary cells (VERBERCKMOES et al. 1976) was absent in biopsies containing medullary tissue. Features indicating a recently described familial proximal tubulopathy (GÜLLNER et al. 1983) with an increased staining of proximal tubular cells and thickening of tubular basement membranes resembling Bartter's syndrome on clinical grounds were not detected on light or electron microscopy (TAUGNER et al. 1988c).

Immunohistochemical investigations showed no difference between the JGAs from kidneys of patients with Bartter's syndrome and those with hyperprostaglandin E syndrome. The bulk of renin-positive cells was constantly found to be localized in the media of the distal - juxtaglomerular - portion of the afferent arterioles in control kidneys as well as in those of Bartter's syndrome and hyperprostaglandin E syndrome (Fig. 14.4a-c). However, the length of the renin-positive segment of the preglomerular arteriole was significantly elevated in the diseased kidneys (100 ± 32 µm; $n=15$; Fig. 14.4a, b) when compared with controls (49 ± 17 µm; $n=57$; Fig. 14.4c). Characteristically, there was an increase not only in the number of renin-positive cells but also in their immunoreactivity. In control kidneys, the antiserum could not be diluted more than 1:2500 to produce reliable immunostaining (Fig. 14.4c); in the diseased kidneys, a dilution of 1:10000 was still effective (Fig. 14.4b); smaller dilutions (e.g., 1:5000) already led to a diffuse spreading of the reaction product. In contrast to controls, in patients with Barrter's syndrome and hyperprostaglandin E syndrome single or small groups of renin-positive cells were also found at the origin of the afferent arteriole, in the juxtaglomerular portion of the efferent arteriole (Fig. 14.4d) and in the mesangium of the glomerular stalk (Fig. 14.4e). This finding is in agreement with previous reports (BARAJAS et al. 1976) and indicates that media cells of the glomerular arterioles and also mesangial cells of the glomerular stalk possess the capacity to transform into epithelioid cells upon chronic stimulation (cf. Chapts. 10, 11). Very similar immunohistochemical results were obtained in pseudo-Bartter's syndrome (CHRISTENSEN et al. 1988). Unequivocally renin-positive Goormaghtigh cells, however, were only detected in pseudo-Bartter's syndrome (cf. Sects. 2.4, 4.1.3; Fig. 2.15).

In addition to the vascular components mentioned so far, immunoreactive renin was observed in the apical portions of proximal tubular cells, in the intercalated cells of the connecting tubules, and in cortical as well as outer medullary collecting ducts. This reaction - like in other conditions with elevated plasma renin - probably represents filtered renin reabsorbed by pinocytosis (cf. Sect. 4.1.1).

In contrast to control kidneys, hypertrophied cells of the parietal layer of Bowman's capsule in Bartter's syndrome and hyperprostaglandin E syndrome reacted intensely with an antibody against human angiotensinogen. Converting-enzyme reactivity was encountered in both unchanged and diseased kidneys, particularly in the area of the brush border of proximal tubules. Angiotensin II could be detected in epithelioid cells neither of control kidneys nor of kidneys from patients with Bartter's syndrome or hyperprostaglandin E syndrome. This observation, which is at variance with positive results reported by CELIO (1981), as well as the negative findings with angiotensinogen, angiotensin I, and converting-enzyme may be considered as a strong argument against any hypothesis of hormone-like actions of angiotensin II based on the presence of the octapeptide in epithelioid cells of the rat (HACKENTHAL et al. 1987c; cf. Sect. 4.2).

On electron microscopy, the media of the juxtaglomerular afferent arteriole in Bartter's syndrome and hyperprostaglandin E syndrome was composed of a broadened collar of epithelioid cells

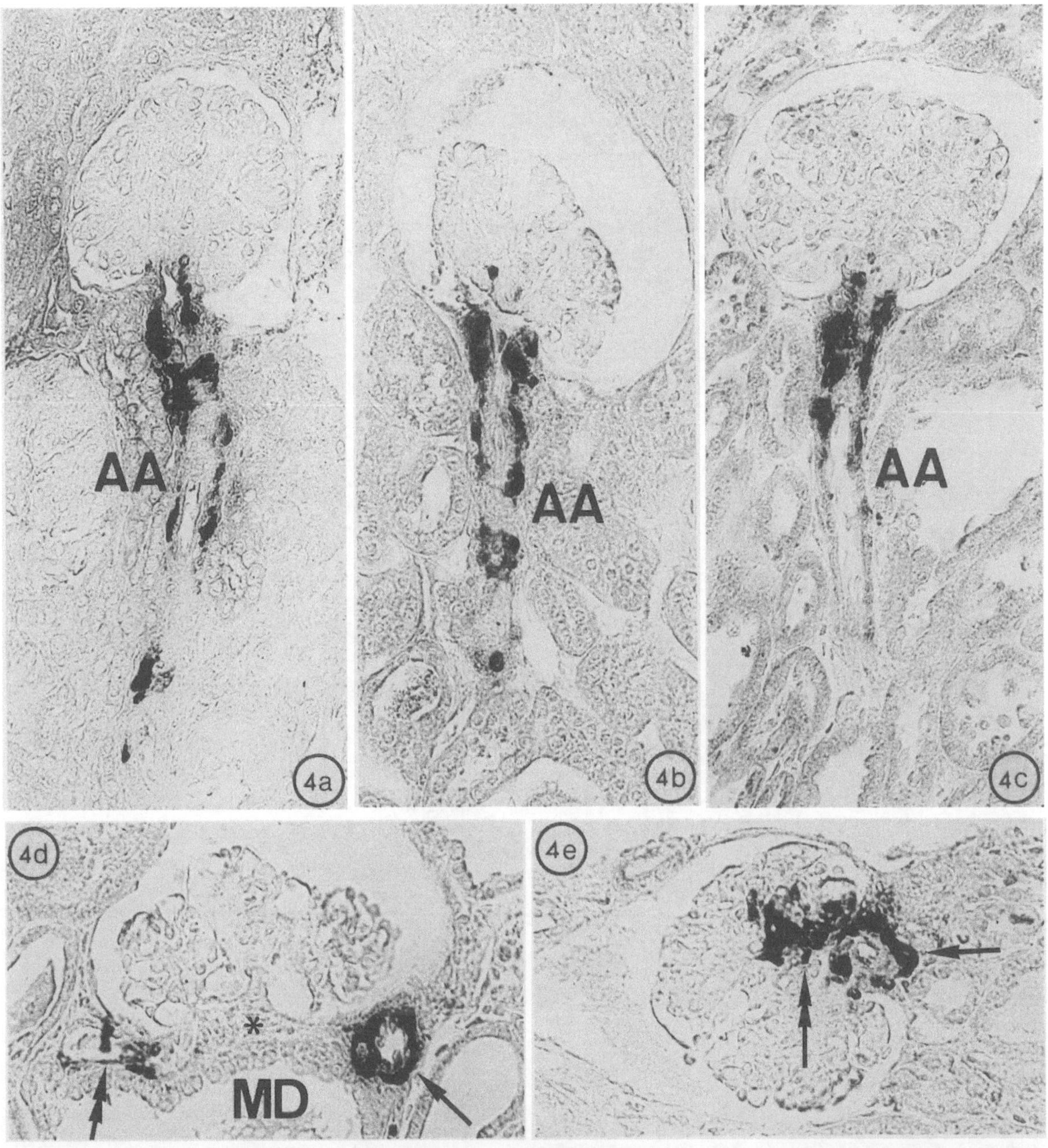

Fig. 14.4a–e. Hyperprostaglandin E syndrome. **a, b** Renin-positive reactions in the media of afferent arterioles *(AA)*. Note the alternation between renin-positive and renin-negative media cells in the proximal part of the vessel. Antiserum dilution, 1:5000 and 1:10000, respectively; ×320. **c** Renin-positive reaction in the media of an afferent arteriole *(AA)* from control kidney. Antiserum dilution; 1:2500; ×320. **d** Renin-positive reactions in the afferent *(arrow)* and the efferent glomerular arteriole *(double arrow)*. MD, macula densa; ×380. **e** Renin-positive cells in the the wall of the afferent arteriole *(arrow)* and in the glomerular stalk *(double arrow)*; ×380

(Fig. 14.5 a). These cells generally had distinct and rarely extremely widened RER cisterns (Fig. 14.5 e). Their Golgi complex was hypertrophied and contained, or was associated with, some paracrystalline protogranules (Fig. 14.5 c) or conglomerate granules (Fig. 14.5 d). In addition, irregularly shaped paracrystalline deposits were found in dilated RER cisterns of some epithelioid cells (Fig. 14.5 a, b). Using the protein A-gold method, renin could be identified in the mature secretory granules as well as in

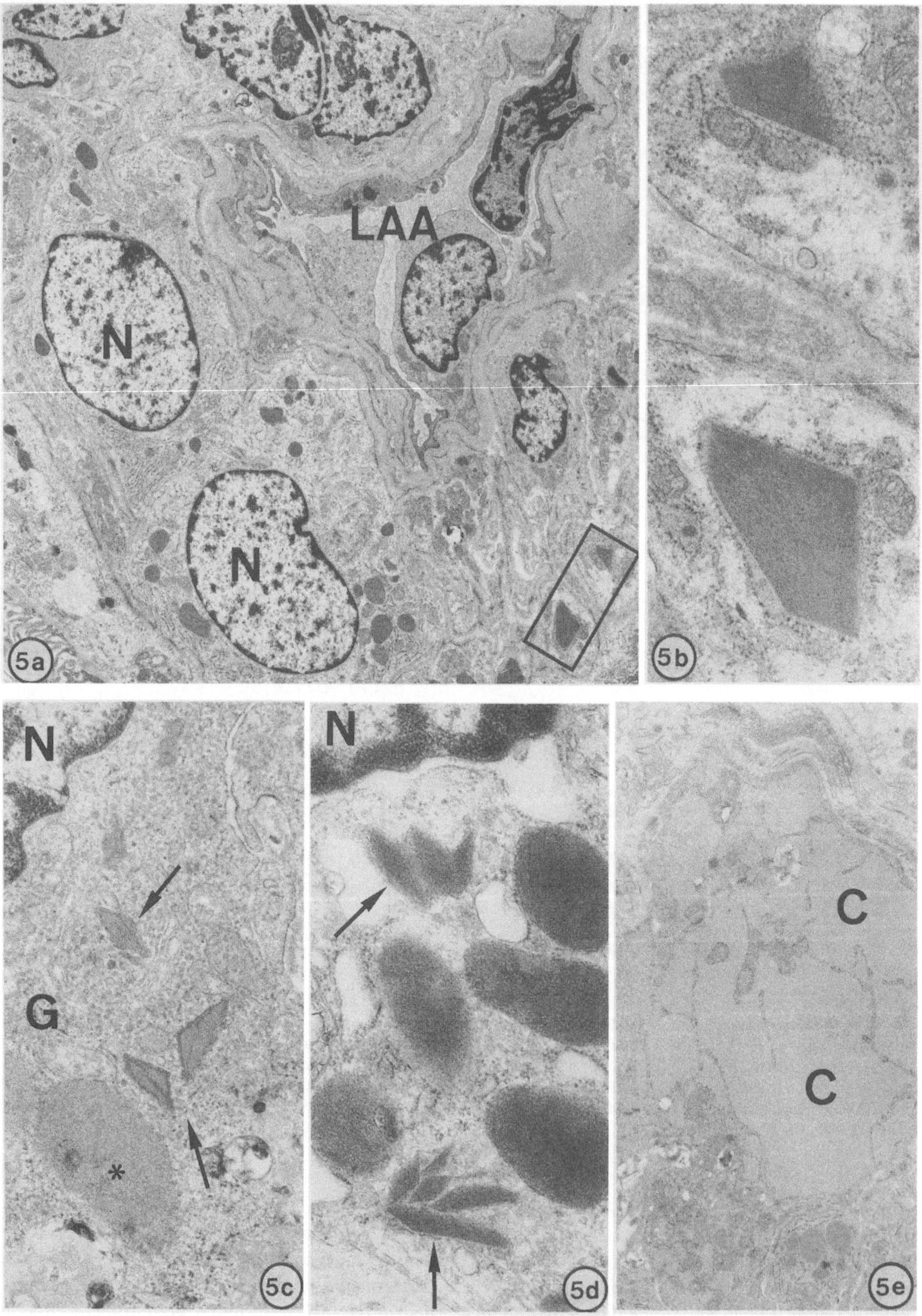
LAA
N
N
5a
5b
N
G
*
5c
N
5d
C
C
5e

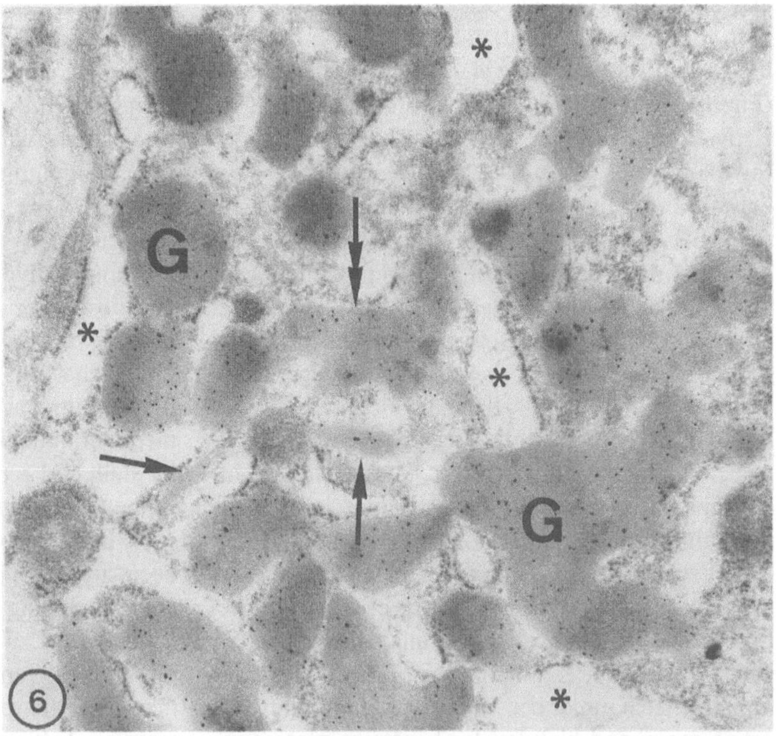

Fig. 14.6. Hyperprostaglandin E syndrome. Detail of an epithelioid cell from the afferent arteriole showing renin-positive mature granules *(G)*, protogranules *(arrows)*, protogranule conglomerates *(double arrow)* and dilated RER cisterns *(asterisks)*. Protein A-gold method; antiserum dilution, 1:100; ×27200

the protogranules of these cells (Fig. 14.6). Renin-negative electron-dense organelles with the ultrastructural appearance of secondary lysosomes could not be identified in the granulated cells in Bartter's syndrome and hyperprostaglandin E syndrome. This negative result, which corresponds to similar observations in rats and mice, is hardly compatible with the assumption that two classes of relatively large electron-dense organelles exist in epithelioid cells, namely, specific and nonspecific granules (BIAVA and WEST 1965, 1966b; BIAVA 1967; TAUGNER and METZ 1986; cf. Sect. 5.3).

According to BARAJAS (1966), granulopoiesis in epithelioid cells proceeds via protogranules, shown to pinch off from the dilated rims of the innermost Golgi cisterns. Upon stimulation of the JGA, e.g., by reduction of the pressure in the renal arterial tree, the number of rhomboid protogranules with paracrystalline contents increases. Subsequently, several protogranules may merge into conglomerate granules, in which the individual structure of the original protogranules may still be seen, until finally the large mature granules with amorphous contents emerge (BARAJAS 1966; LINDOP and DOWNIE 1984; TAUGNER and METZ 1986; cf. Sect. 5.1). Juvenile granules with paracrystalline contents were regularly seen in the epithelioid cells of patients with Bartter's syndrome or hyperprostaglandin E syndrome. They have been described as "renin bodies," with sharply angulate shapes in a case of Bartter's syndrome by ZAVAGLI et al. (1983), and have been shown to contain prorenin (TAUGNER et al. 1986b, 1987a). However, paracrystalline deposits were also found in the dilated cisterns of the RER in Bartter's syndrome as well as in hyperprostaglandin E syndrome kidneys. These RER deposits differed from the rhomboid protogranules, which, as post-Golgi deposits, are enclosed by a smooth membranous sack, particularly by their membranous envelope studded with ribosomes. In

Fig. 14.5a–e. Bartter's syndrome (BS) and hyperprostaglandin E syndrome (HES). Epithelioid cells in the media of the afferent glomerular arteriole. **a–c** From a patient with BS, **d–e** from a patient with HES. **a** Overview. *LAA*, lumen of the afferent glomerular arteriole; *N*, nucleus; ×6000. **b** Area marked in **a** at higher magnification with paracrystalline inclusions in cisterns of the RER; ×32000. **c** Golgi complex *(G)* with adjacent mature granule *(asterisk)* and several juvenile granules *(arrows)* showing the lines of previous protogranule fusion. *N*, nucleus. ×31000. **d** Detail of an epithelioid cell with mature granules and conglomerates of protogranules *(arrows)*; ×37700. **e** Epithelioid cell process with bizarre dilatation of RER cisterns *(C)*; ×27000 addition, the RER deposits were usually bigger than protogranules and, instead of being rhomboid, had a sharp-edged, asymmetrical profile (Fig. 14.5a, b). Although paracrystalline deposits in RER cisterns of other cell types have been known for many years, there have been no reports on the existence of polyhedral paracrystalline structures in the RER of human epithelioid cells. In mice, the corresponding RER inclusions of epithelioid cells are globular (KANETA et al. 1981; HACKENTHAL et al. 1987c). Since both protogranules and paracrystalline RER inclusions prevail upon severe stimulation of renin synthesis, they may be interpreted as a sign of overcharge in the processing and/or packaging of the secretory product. Paracrystalline inclusions have repeatedly been observed in the cells of renin-producing tumors (MIMRAN et al. 1978; BALDET et al. 1983; CAMILLERI et al. 1984; GALEN et al. 1984; SQUIRES et al. 1984; TETU et al. 1984; LINDOP et al. 1986), indicating a similar disturbance of activation and packaging of renin. It is of particular note that such tumor cells secrete predominantly inactive renin, and high levels of inactive plasma renin have also been reported in Bartter's syndrome (NAGAI et al. 1984; MCKENZIE et al. 1988).

Unlike controls, in which the media of the postglomerular arteriole consisted of a uniform type of pericyte-like cells, two types of granulated cells were observed in the efferent vessels of diseased kidneys: oval cells with an electron-lucent cytoplasm, largely corresponding to the epithelioid cells of the preglomerular arteriole, and osmiophilic cells with numerous ramifications resembling Goormaghtigh cells. At higher magnification, all characteristics of an increased synthetic activity were observed in both types of postglomerular epithelioid cells: dilated RER cisterns containing material of moderate electron density, hypertrophied Golgi complexes, protogranules, and mature secretory granules.

Morphological indications of stimulation were not only found in the media of the hilar arterioles but also in the area of the Goormaghtigh cell field and in the glomerular mesangium. In Bartter's syndrome and hyperprostaglandin E syndrome, several profiles of otherwise typical Goormaghtigh cells and also those of mesangial cells presented round homogeneous electron dense organelles suggestive of secretory granules. However, because of their scarcity, these organelles could not be identified with certainty in the immunocytochemical experiments (TAUGNER et al. 1988c). This holds also true for the mesangial cells in pseudo-Bartter's syndrome. In contrast, Goormaghtigh cells in pseudo-Bartter's syndrome were not only renin-positive but also loaded with secretory granules, thus strongly resembling epithelioid cells (CHRISTENSEN et al. 1988; cf. Sect. 2.4).

In summary, kidneys from patients with Bartter's syndrome, hyperprostaglandin E syndrome, and pseudo-Bartter's syndrome disclose all signs of marked chronic stimulation of renin synthesis, implying that this stimulation has been followed by the recruitment of further granulated cells in the media of both glomerular arterioles, in the glomerular stalk and - in pseudo-Bartter's syndrome - also in the Goormaghtigh cell field.

In Bartter's syndrome, several mechanisms have been suggested as the primary cause of hyperreninemia, including effects of the sympathetic nervous system (MCLAREN and MACDONALD 1982), hyperplasia of the renomedullary interstitial cells with an inappropriate release of prostaglandins (VERBERCKMOES et al. 1976), and various functional defects of the renal tubule (STEIN 1985), in particular a deficient chloride reabsorption in the ascending limb of Henle's loop (BARTTER et al. 1962; GILL and BARTTER 1978; BAEHLER et al. 1980; BARTTER and RODRIGUEZ 1982). The latter suggestion recalls recent experimental studies suggesting that a decrease in NaCl reabsorption in the juxtaglomerular portion of the distal tubule is followed by an increase in renin secretion (LEYSSAC 1986; SKØTT and BRIGGS 1987) and, thus, by inference also in renin synthesis (cf. Chap. 11). In pseudo-Bartter's syndrome, inhibition of the reabsorption by furosemide in this part of the tubule may be decisive, assisted by the hypovolemic condition (CHRISTENSEN et al. 1988). In hyperprostaglandin E syndrome, primary variations of the prostaglandin synthesis could play a role (SEYBERTH et al. 1985).

14.1.5 Renin-Secreting Tumors

14.1.5.1 Nonrenal Tumors

Hypertension associated with an increased plasma renin activity has been observed in rare cases of nonrenal neoplasms, i.e., in small cell carcinoma and adenocarcinoma of the lung (HAUGER-KLEVENE 1970; GENEST et al. 1975), pancreatic carcinoma (RUDDY et al. 1982; ATLAS et al. 1984), epithelial liver hamartoma (COX et al. 1975), retroperitoneal leiomyosarcoma (FROMME et al. 1985), orbital hemangiopericytoma (YOKOYAMA et al. 1979) as well as in carcinoma (AURELL et al. 1979; ATLAS et al. 1984) and SERTOLI cell tumor of the ovary (KORZETS et al. 1986).

In these nonrenal tumors with ectopic renin production high plasma renin levels are usually characterized by a disproportional increase in inactive renin as compared to active renin (RUDDY et al. 1982; SOUBRIER et al. 1982; ATLAS et al. 1984).

The immunocytochemical localization of renin in alveolar soft part sarcoma reported by DESCHRYVER-KECSKEMETI et al. (1982) could not be confirmed by other investigators (MUKAI et al. 1983; TOMITA et al. 1987). Recently, scattered renin-positive cells have been detected in an epithelioid sarcoma. Synthesis of renin in these cells has been confirmed by in situ hybridization using a radiolabeled mRNA probe (BRUNEVAL et al. 1988). An adrenal tumor associated with hypertension and the adrenogenital syndrome has been described in a 27-year-old woman (IIMURA et al. 1986). The tumor contained dehydroepiandrosterone, dehydroepiandrosterone sulfate, testosterone, and renin. Immunohistochemical studies revealed renin reactivity in some of the tumor cells. Renal biopsy showed some atrophy of juxtaglomerular granulated cells, presumably due to negative feedback, and pronounced arteriosclerosis as a consequence of systemic hypertension.

14.1.5.2 Nephroblastoma and Renal Cell Carcinoma

In renal tumors, renin secretion by the tumor itself should be differentiated from secretion by the nontumorous kidney tissue as a consequence of mechanical compression and ischemia.

Nephroblastoma (Wilms' tumor) is a usually malignant renal tumor occurring in infancy and childhood which derives from pluripotent nephrogenic blastema. In some cases of nephroblastoma elevated preoperative plasma renin levels were detected which returned to normal after nephrectomy (MITCHELL et al. 1970; SPAHR et al. 1981), and some of these tumors contained higher concentrations of renin than the adjacent nontumorous renal tissue (MITCHELL et al. 1970; GANGULY et al. 1973; SHETH et al. 1978; SPAHR et al. 1981). The presence of renin-containing cells in nephroblastomas has recently been confirmed by immunohistochemical studies. LINDOP et al. (1984) observed renin-positive cells in 10 out of 19 of surgically removed nephroblastomas. Immunoreactive cells were also identified in distant metastases from 3 out of 12 autopsy cases. TOMITA et al. (1987) detected scattered renin-positive cells in one of five tumors examined. There was no relationship between the presence of renin-positive cells and the histological subtype of nephroblastoma. Predominantly spindle-shaped renin-containing cells are found close to capillary vessels. The presence of such cells not only in the primary tumor but also in metastases argues against the assumption according to which in the kidney they may grow into the tumor from the adjacent normal renal tissue. In Lindop's series (1984) all patients had normal blood pressure and serum potassium concentrations, suggesting that the secreted renin was biologically inactive. Likewise, DAY and LUETSCHER (1974) reported a nephroblastoma which secreted inactive renin with higher molecular weight. Quantitative determination of the tumoral renin content in one case showed a concentration of about 2% that of the normal renal cortex; approximately half was active and half was inactive renin (TOMITA et al. 1987). In a recent prospective study of eight children with nephroblastoma and abnormally high plasma renin levels, it has been demonstrated that the increase in total renin was predominantly due to the secretion of inactive renin (CARACHI et al. 1987). The elevated levels returned to normal after nephrectomy. Increased plasma levels of inactive renin might, therefore, even be useful as a "tumor marker" in some patients with nephroblastoma.

As in nephroblastomas, several reports indicate that renin can be produced in and secreted by *renal cell carcinomas,* in most cases associated with marked hypertension (HOLLIFIELD et al. 1975; LEBEL et al. 1977; LINDOP et al. 1983a, 1986; BERNARD et al. 1986). Immunoreactive renin was found in 7 of 19 primary renal cell carcinomas by LINDOP and FLEMING (1984) and in 3 of 10 carcinomas examined by TOMITA et al. (1987). Renin-positive cells were also detected in distant metastases in other organs but not in secondary (metastatic) tumors within the kidney, suggesting that the immunoreactive cells did not derive from the adjacent tissue (LINDOP and FLEMING 1984). In the majority of cases, renin-containing cells were scant in number and usually located close to capillary blood vessels. No patient in Lindop's series demonstrated clinical, biochemical or histological evidence of excess renin secretion (LINDOP and FLEMING 1984). In cases of renal cell carcinoma with marked hypertension, preoperative measurements indicated that plasma levels of inactive renin were disproportionally more elevated than those of active renin (LECKIE et al. 1978; LINDOP et al. 1983a, 1986). Likewise, quantitative determination of renin in tumor tissue revealed that more than half of this renin was inactive (LECKIE et al. 1978; TOMITA et al. 1987).

14.1.5.3 Various Renal Tumors

Except for nephroblastomas and renal cell carcinomas, renin-containing cells could be identified in oncocytoma, congenital mesoblastic nephroma, and angiomyolipoma but were absent in renal adenomas (YUM et al. 1984; LINDOP and LEVER 1986; TOMITA et al. 1987).

The presence of renin-positive cells in primary renal tumors and even in distant metastases from these tumors suggests that renin-containing cells do not derive from the adjacent nontumoral tissue but are a characteristic endocrine component of these tumors. A primary origin from vascular smooth muscle cells and/or pericytes by the way of transformation into renin-producing cells may be considered. It is questionable, however, whether stimuli leading to such transformation are really tumor-specific since similar immunoreactive cells occur in various other pathological conditions (cf. Sect. 14.1.2).

14.1.5.4 Juxtaglomerular Cell Tumor

A rare benign renin-secreting tumor, the so-called JGA cell tumor, originally described by ROBERTSON et al. (1967), is composed of renin-producing granulated cells. More than 30 cases have been reported in the literature (ROBERTSON et al. 1967; KIHARA et al. 1968; EDDY et al. 1971; LEE et al. 1971; BONNIN et al. 1972; CONN et al. 1972; PHILIPPS and MUKHERJEE 1972; BROWN et al. 1973; MACCALLUM et al. 1973; SCHAMBELAN et al. 1973; DAVIDSON et al. 1974; GHERARDI et al. 1974; HIROSE et al. 1974; MORE et al. 1974; ORJAVIK et al. 1975; TAKAHASHI et al. 1976; BARAJAS et al. 1977; CONNOR et al. 1978; MIMRAM et al. 1978; HANNA et al. 1979; WARSHAW et al. 1979; ZHANG 1979; EL MATRI et al. 1980; VALDÉS et al. 1980; LAM et al. 1982; MOSS et al. 1982; SANFILIPPO et al. 1982; BALDET et al. 1983; CAMILLERI et al. 1983, 1984; DUNNICK et al. 1983; FURUSATO et al. 1983; LINDOP et al. 1983b; BARUCH et al. 1984; GALEN et al. 1984; TETU et al. 1984; SQUIRES et al. 1984; DENNIS et al. 1985; JORDAN and GUNNELLS 1985; HANDA et al. 1986; TOMITA et al. 1987). JGA cell tumors occur mainly in children and young adults; females are more frequently affected than males. Patients usually present with severe hypertension associated with elevated peripheral plasma levels of renin and angiotensin II, secondary hyperaldosteronism and consecutive hypokalemia. Angiography reveals a circumscribed hypovascular solid renal mass (DUNNICK et al. 1983). The tumor is usually solitary and measures up to 5 cm in diameter. Partial or total nephrectomy is commonly followed by disappearance of hypertension and biochemical abnormalities. In rare cases, however, mild hypertension may persist because of established hypertensive angiopathy (SQUIRES et al. 1984). JGA cell tumors are invariably benign. No patient has developed tumor recurrence or metastases.

The juxtaglomerular cell tumor is rich in extractable renin, and immunohistochemical studies have demonstrated a diffuse staining for renin within the tumor cells (PHILIPPS and MUKHERJEE 1972; CONN et al. 1972; LAM et al. 1982; CAMILLERI et al. 1983, 1984; GALEN et al. 1984; DENNIS et al. 1985; TAMITA et al. 1987). In situ hybridization studies have shown the presence of renin mRNA in all the tumor cells (BRUNEVAL et al. 1988). On electron microscopy, renin-positive cells contain typical juvenile and mature secretory granules of different types and, in addition, atypical giant crystalloid protogranules (MACCALLUM et al. 1973; MIMRAN et al. 1978; BALDET et al. 1983; CAMILLERI et al. 1984; GALEN et al. 1984; SQUIRES et al. 1984; TETU et al. 1984; LINDOP et al. 1986). In contrast to other renin-positive renal tumors, which secrete lesser amounts of renin, most of which is inactive, JGA tumors produce enormous amounts of the active enzyme.

On light microscopy, the tumor is composed of interlacing cords and occasional nodules of relatively uniform cells without any mitotic activity. The predominant cell type is a polygonal or short spindle cell. These cells are aggregated around

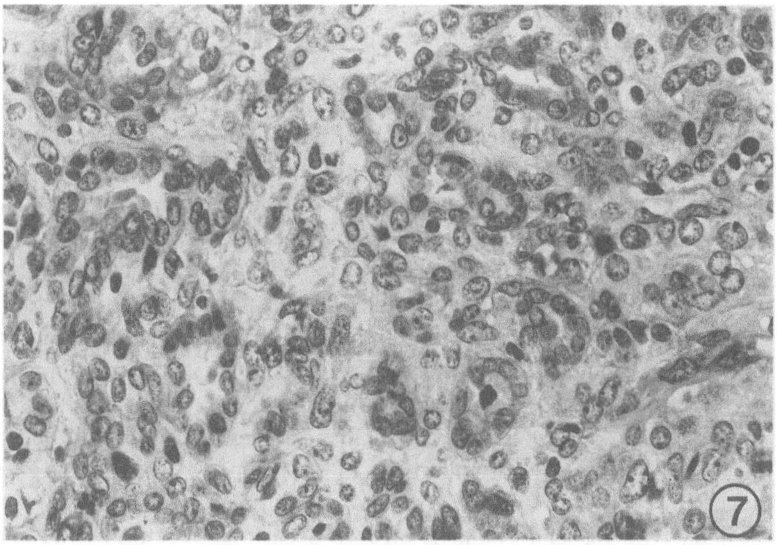

Fig. 14.7. Juxtaglomerular cell tumor with a haemangiopericytoma-like pattern. Masson trichrome stain, ×400. (Photograph courtesy Prof. Dr. V. TOTOVIC, Institute of Pathology, University of Bonn, FRG; cf. TETU et al. 1984).

blood vessels resembling the pattern of a hemangiopericytoma (Fig. 14.7). In some cases, an occasional tubular component has been identified (ROBERTSON et al. 1967; SCHAMBELAN et al. 1973; GHERARDI et al. 1974; ORJAVIK et al. 1975; TAKAHASHI et al. 1976; BALDET et al. 1983; FURUSATO et al. 1983; TETU et al. 1984; CAMILLERI et al. 1985). Some authors hold that this biphasic pattern may support the view that JGA cell tumors represent a primary hamartomatous lesion. An additional rare feature is the presence of a dense sympathetic innervation (BARAJAS et al. 1977a; BALDET et al. 1983; CAMILLERI et al. 1984).

On the basis of morphological and cell culture studies, GALEN et al. (1984) have postulated the existence of two pathways for the processing, packaging, and secretion of renin in tumor cells: in native juxtaglomerular cell tumors, renin is synthesized as pre-prorenin and rapidly converted into prorenin which, in turn, is packaged in secretory granules, processed into active renin, and secreted. In contrast, in cultured juxtaglomerular tumor cells, renin is synthesized as pre-prorenin, converted into prorenin, and neither stored nor processed into active renin but released in its inactive form (cf. Sect. 7.7). The factors responsible for the secretion of active renin in juxtaglomerular cell tumors and the dedifferentiation of tumor cells in cell cultures only capable of secreting inactive renin are unknown.

14.2 Inhibition of the JGA

14.2.1 Primary Hyperaldosteronism

A significant reduction in the number of juxtaglomerular granulated cells and in the juxtaglomerular granulation index has repeatedly been reported in patients with primary hyperaldosteronism (COHEN et al. 1965; CONN et al. 1965; HELBER et al. 1970; WEGMANN 1970; MEYER 1972, 1978; CHRISTENSEN et al. 1976).

14.2.2 Miscellaneous Conditions

"Hypoplasia" of the JGA in the contralateral kidney of patients with unilateral processes associated with an increased renin secretion or in normal renal tissue adjacent to circumscribed hypersecretory areas is discussed in the preceding sections. In addition, a reduction in renin-positive cells has been demonstrated in renal biopsies from hypertensive pregnant women who had either preeclampsia, isolated elevation of blood pressure or previous renal disease, in comparison with nonpregnant normotensive women (NOCHY et al. 1984).

14.3 Conclusions

Nonneoplastic stimulation of the RAS in humans is followed by morphological reactions of the JGA closely resembling those of experimental investigations. These changes include an increase in the renin depot of the individual granulated cell and the recruitment of additional renin-positive cells, in particular in the preglomerular but also in the postglomerular arteriole, and the mesangial stalk. The morphological alterations of the JGA are accompanied by an elevated secretion of predominantly active renin. In contrast to experimental models, renin-positive cells are also observed in the Goormaghtigh cell field and adjacent to peritubular capillary-type vascular channels on rare occasions. On electron microscopy, excessive stimulation of renin synthesis and epithelioid cell granulopoiesis in humans is associated not only with an increase in the number of rhomboid protogranules but also with the appearance of asymmetric paracrystalline deposits in RER cisterns. Nonspecific granules, i.e., renin-negative electron-dense organelles with the ultrastructural characteristics of secondary lysosomes, could not be detected in our material.

Renin-producing tumors can hardly be compared with experimental models and other human disorders accompanied by a chronic stimulation of the JGA. On electron microscopy, certain pecularities of granulopoiesis are noted which are probably related to the predominant secretion - except for the so-called juxtaglomerular cell tumor - of inactive renin by the tumor cells.

References

ABDEL-LATIF AA (1986) Calcium-mobilizing receptors, polyphosphoinositides, and the generation of second messengers. Pharmacol Rev 38: 227-272

ABE Y, OKAHARA T, KISHIMOTO T, YAMAMOTO K, UEDA J (1973) Relationship between intrarenal distribution of blood flow and renin secretion. Am J Physiol 225: 319-323

ABE Y, IWAO H, OKAHARA T, YAMAMOTO K (1977) Control of renin secretion. Jpn Circ J 41: 251-257

ABE Y, YUKIMURA T, IWAO H, MORI N, OKAHARA T, YAMAMOTO K (1983) Effects of EDTA and Verapamil on renin release in dogs. Jpn J Pharmacol 33: 627-633

ABRAHAM GE (1969) Solid-phase radioimmunoassay of estradiol-*17β*. J Clin Endocrinol Metab 29: 866-870

ACKER GM, GALEN FX, DEVAUX C, FOOTE S, PAPERNIK E, PESTY A, MENARD J, CORVOL P (1982) Human chorionic cells in primary culture: A model for renin biosynthesis. J Clin Endocrinol Metab 55: 902-909

ACKERMANN U (1986) Cardiovascular effects of atrial natriuretic extract in the whole animal. Fed Proc 45: 2111-2114

ADAM WR, ADAMS BA (1985) Production and excretion of dopamine by the isolated perfused rat kidney. Renal Physiol 8: 150-158

ADEBAHR G (1962) Beitrag zur Morphologie der vasa afferentia und efferentia der juxtamedullären Glomeruli der menschlichen Niere. Z Mikrosk Anat Forsch 68: 48-60

AEIKENS B, HILDEBRAND U (1981) Morphometrische Untersuchungen am juxtaglomerulären Apparat und Glomerulum der Rattenniere. Micros Acta 84: 185-193

AIKEN JW, VANE JR (1972) Inhibition of converting enzyme of the renin-angiotensin system in kidneys and hindlegs of dogs. Circ Res 30: 263-273

AKABAS MH, COHEN FS, FINKELSTEIN A (1984) Separation of the osmotically driven fusion event from vesicle-planar membrane attachment in a model system for exocytosis. J Cell Biol 98: 1063-1071

ALBERTINE KH, O'MORCHOE CCC (1980) Renal lymphatic ultrastructure and translymphatic transport. Mircovasc Res 19: 338-351

ALCORN D, EMSLIE KR, ROSS BD, RYAN GB, TANGE JD (1981) Selective distal nephron damage during isolated kidney perfusion. Kidney Int 19: 638-647

ALCORN D, CHESHIRE GR, COGHLAN JP, RYAN GB (1984) Peripolar cell hypertrophy in the renal juxtaglomerular region of newborn sheep. Cell Tissue Res 236: 197-202

ALCORN D, ANDERSON WP, RYAN GB (1986) Morphological changes in the renal macula densa during natriuresis and diuresis. Renal Physiol 9: 335-347

ALEXANDER F (1968) The juxtaglomerular apparatus in Addison's disease. J Pathol 96: 27-32

ALLEN JM, BIRCHAM PMM, EDWARDS AV, TATEMOTO K, BLOOM SR (1983) Neuropeptide Y (NPY) reduces myocardial perfusion and inhibits the force of contraction of the isolated perfused rabbit heart. Regul Pept 6: 247-253

AMARA SG, JONAS V, ROSENFELD MG, ONG ES, EVANS RM (1982) Alternative RNA-processing in calcitonin gene expression generates mRNAs encoding different polypeptide products. Nature 298: 240-244

AMAT D, CAMILLERI JP, PHAT VN, BARIETY J, CORVOL P, MENARD J (1981) Renin localization in segmental renal hypoplasia. Virchows Arch [A] 390: 193-204

AMSAGUINE S (1983/1984) Le système rénine-angiotensine rénal: Caractérisation immunocytochimique des antisérums antirénine et application à l'étude du mésonéphros chez le lapin, le porc et l'homme. Thesis, Université de Nancy I

ANAND-SRIVASTAVA MB, CANTIN M (1986) Atrial natriuretic factor receptors are negatively coupled to adenylate cyclase in cultured atrial and ventricular cardiocytes. Biochem Biophys Res Commun 138: 427-436

ANAND-SRIVASTAVA MB, FRANKS DJ, CANTIN M, GENEST J (1984) Atrial natriuretic factor inhibits adenylate cyclase activity. Biochem Biophys Res Commun 121: 855-862

ANAND-SRIVASTAVA MB, CANTIN M, GENEST J (1985) Inhibition of pituitary adenylate cyclase by atrial natriuretic factor. Life Sci 36: 1873-1879

ANDERSON JV, DONCKIER J, PAYNE NN, BEACHAM J, SLATER JDH, BLOOM SR (1987) Atrial natriuretic peptide: evidence of action as a natriuretic hormone at physiological plasma concentrations in man. Clin Sci 72: 305-312

ANDERSON WP (1982) Prostaglandins do not mediate renin release during severe reduction of renal blood flow in conscious dogs. Clin Exp Pharmacol Physiol 9: 259-263

ANDERSON WP, BARTLEY PJ, CASLEY DJ, SELIG SE (1983) Comparison of aspirin and indomethacin pretreatments on the responses to reduced renal artery pressure in conscious dogs. J Physiol (Lond) 336: 101-112

ANDREWS PM, COFFEY AK (1983) Cytoplasmic contractile elements in glomerular cells. Fed Proc 42: 3046-3052

ANGUS JA, KORNER PI (1977) Regional vascular resistance and heart rate responses mediated through H1- und H2-histamine receptors in the unanaesthetized rabbit. Eur J Pharmacol 45: 45-53

ANTONIPILLAI I, HORTON R (1985) Role of extra- und intracellular calcium and calmodulin in renin release from rat kidney. Endocrinology 117: 601-606

ANTONIPILLAI I, VOGELSANG J, HORTON R (1986) Role of atrial natriuretic factor in renin release. Endocrinology 119: 318-322

ARAKAWA K, YUKI M, IKEDA M (1980) Chemical identity of tryptensin with angiotensin. Biochem J 187: 647-653

ARAUJO-NASCIMENTO DMF, DÉSORMEAUX Y, CANTIN M (1976) Ultrastructural cytochemistry of the ischemic (endocrine) kidney. Am J Pathol 82: 527-548

ARDAILLOU R, SRAER J, CHANSEL D, ARDAILLOU N, SRAER JD (1987) The effects of angiotensin II on isolated glomeruli and cultured glomerular cells. Kidney Int 31 [Suppl 20]: 74-80

AREND LJ, HARAMATI A, THOMPSON CI, SPIELMAN WS (1984) Adenosine-induced decrease in renin release: dissociation from hemodynamic effects. Am J Physiol 247: F447-F452

ARENDSHORST WJ (1987) Altered reactivity of tubuloglomerular feedback. Ann Rev Physiol 49: 295-317

ARENDSHORST WJ, FINN WF (1977) Renal hemodynamics in the rat before and during inhibition of angiotensin II. Am J Physiol 233: F290-F297

ARREGUI A, IVERSEN LL (1978) Angiotensin-converting enzyme: presence of high activity in chorioid plexus of mammalian brain. Eur J Pharmacol 52: 147-150

ASHLEY PL, MACDONALD RJ (1985) Tissue-specific expression of kallikrein-related genes in the rat. Biochemistry 24: 4520-4527

ASHLEY PL, MACDONALD RJ (1985) Kallikrein-related mRNAs of the rat submaxillary gland: nucleotide sequences of four distinct types including tonin. Biochemistry 24: 4512-4520

ATKINSON J, LUTHI P, BOILLAT N (1983) Cyclohexamide and renin release following renal artery constriction. J Pharmacol (Paris) 14: 161-169

ATLAS SA, LARAGH JH (1986) Atrial natriuretic peptide: a new factor in hormonal control of blood pressure and electrolyte homeostasis. Ann Rev Med 37: 397-414

ATLAS SA, SEALEY JE, DHARMGRONGARTAMA B, HESSON TE, LARAGH JH (1981) Detection and isolation of inactive, large molecular weight renin in human kidney and plasma. Hypertension 3 [Suppl 1]: 30-40

ATLAS SA, HESSON TE, SEALEY JE, LARAGH JH (1982) Reversible acid-activation of inactive renin: evidence favouring a unimolecular reaction. Clin Sci 63: 167-170

ATLAS SA, HESSON TE, SEALEY JE, DHARMGRONGARTAMA B, LARAGH JH, RUDDY MC, AURELL M (1984) Characterization of inactive renin ("prorenin") from renin-secreting tumors of nonrenal origin. J Clin Invest 73: 437-447

ATLAS SA, CHRISTOFALO P, HESSON T, SEALEY JE, FRITZ JC (1985) Immunological evidence that inactive renin is prorenin. Biochem Biophys Res Commun 132: 1038-1045

ATLAS SA, VOLPE M, SOSA RE, LARAGH JH, CAMARGO MJF, MAACK T (1986) Effects of atrial natriuretic factor on blood pressure and the renin-angiotensin-aldosterone system. Fed Proc 45: 2115-2121

AUKLAND K, OIEN AH (1987) Renal autoregulation: models combining tubuloglomerular feedback and myogenic response. Am J Physiol 252: F768-F783

AURELL M, RUDIN A, TISELL LE, KINDBLOM LG, SANDBERG G (1979) Captopril effect on hypertension in patient with renin-producing tumour. Lancet 2: 149-150

AUSIELLO DA, KREISBERG JI, ROY C, KARNOVSKY MJ (1980) Contraction of cultured rat glomerular cells of apparent mesangial origin after stimulation with angiotensin II and arginine vasopressin. J Clin Invest 65: 764-760

AYERS CR, HARRIS RHJ, LEFER LG (1969) Control of renin release in experimental hypertension. Circ Res 24/25 [Suppl 1]: 103-113

AYERS CR, VAUGHAN EDJ, YANCEY MR, BING KTJCC, MORTON C (1974) Effect of 1-sarcosine-8-alanine angiotensin II and converting enzyme inhibitor on renin release in dog acute renovascular hypertension. Circ Res 34 [Suppl 1]: 27-33

BACHMANN S, KRIZ W, KUHN C, FRANKE WW (1983) Differentiation of cell types in the mammalian kidney by immunofluorescence microscopy using antibodies to intermediate filament proteins and desmoplakins. Histochemistry 77: 365-394

BADER H, KIRCHERTZ EJ, KNEISSLER U, TAUGNER R, HELMCHEN U (1982) Zur Struktur und Funktion juxtaglomerulärer Apparate im heterotop autotransplantierten Nierengewebe. Verh Dtsch Ges Path 66: 528

BAEHLER RW, WORK J, KOTCHEN TA, MCMORROW G, GUTHRIE G (1980) Studies on the pathogenesis of Bartter's syndrome. Am J Med 69: 933-938

BAILIE MD, OPARIL S (1977) Relation of renal hemodynamics to metabolism of angiotensin II by the canine kidney. Circ Res 41: 283-287

BAILIE MD, RECTOR FC, SELDIN DW (1971) Angiotensin II in arterial and renal venous plasma and renal lymph in the dog. J Clin Invest 50: 119-126

BAILIE MD, LOUTZENHISER R, MOYER S (1972) Relation of renal hemodynamics to angiotensin II in renal hilar lymph of the dog. Am J Physiol 222: 1075-1078

BAKER PF, KNIGHT DE (1986) Exocytosis: control by calcium and other factors. Br Med Bull 42: 399-404

BAKHLE YS, REYNARD AM, VANE JR (1969) Metabolism of the angiotensins in isolated perfused tissues. Nature 222: 956-959

BALDET P, MIMRAN A, GRANIER M, DUPONT M (1983) Forms histologiques et ultrastructurales des tumeurs bénignes du rein avec sécrétion de rénine. Ann Pathol 3: 225-234

BALL S, JOHNS EJ (1982) Influence of the renin-angiotensin system in the renal haemodynamic responses to modest renal nerve stimulation in the rat. J Endocrinol 93: 65-70

BALLESTA J, POLAK JM, HERNANDEZ FJ, ALLEN JM, BLOOM SR (1983) NPY localization and distribution in the juxtaglomerular apparatus. Regul Pept 7: 413 (Abstract)

BALLESTA J, POLAK JM, ALLEN JM, BLOOM SR (1984) The nerves of the juxtaglomerular apparatus of man and other mammals contain the potent peptide NPY. Histochemistry 80: 483-485

BANKIR L, FARMAN N (1973) Hétérogénéité des glomérules chez le lapin. Arch Anat Microsc Morphol Exp 62: 281-291

BANKS RO, FONDACARO JD, SCHWAIGER MM, JACOBSON ED (1978) Renal histamine H1 and H2 receptors: characterization and functional significance. Am J Physiol 235: F570-F575

BARAJAS L (1964) The innervation of the juxtaglomerular apparatus. An electron microscopic study of the innervation of the glomerular arterioles. Lab Invest 13: 916-929

BARAJAS L (1966) The development and ultrastructure of the juxtaglomerular cell granule. J Ultrastruct Res 15: 400-413

BARAJAS L (1970) The ultrastructure of the juxtaglomerular apparatus as disclosed by three-dimensional reconstructions from serial sections. The anatomical relationship between the tubular and vascular components. J Ultrastruct Res 33: 116-147

BARAJAS L (1971) Renin secretion: an anatomical basis for tubular control. Science 172: 485-487

BARAJAS L (1972) Anatomical considerations in the control of renin secretion. In: ASSAYKEEN TA (ed) Control of renin secretion. Plenum, New York, pp 1-16

BARAJAS L (1978) Innervation of the renal cortex. Fed Proc 37: 1192-1201

BARAJAS L (1979) Anatomy of the juxtaglomerular apparatus. Am J Physiol 237: F333-F343

BARAJAS L (1981) The juxtaglomerular apparatus: anatomical considerations in feedback control of glomerular filtration rate. Fed Proc 40: 78-86

BARAJAS L, LATTA H (1963a) A three dimensional study of

the juxtaglomerular apparatus in the rat. Lab Invest 12: 257-269

BARAJAS L, LATTA H (1963b) The juxtaglomerular apparatus in adrenalectomized rats. Light and electron microscopic observations. Lab Invest 12: 1046-1059

BARAJAS L, LATTA H (1965) The development of the juxtaglomerular cell granule. Anat Rec 151: 321

BARAJAS L, LATTA H (1967) Structure of the juxtaglomerular apparatus. Circ Res 20/21 [Suppl 2]: 15-28

BARAJAS L, MÜLLER J (1973) The innervation of the juxtaglomerular apparatus and surrounding tubules: a quantitative analysis by serial section electron microscopy. J Ultrastruct Res 43: 107-132

BARAJAS L, POWERS K (1984) The structure of the juxtaglomerular apparatus and the control of renin secretion. An update. J Hypertens 2 [Suppl 1]: 3-12

BARAJAS L, WANG P (1975) Demonstration of acetylcholinesterase in the adrenergic nerves of the renal glomerular arterioles. J Ultrastruct Res 53: 244-253

BARAJAS L, WANG P (1978) Myelinated nerves of the rat kidney. A light and electron microscopic autoradiographic study. J Ultrastruct Res 65: 148-162

BARAJAS L, WANG P (1979) Localization of tritiated norepinephrine in the renal arteriolar nerves. Anat Rec 195: 525-534

BARAJAS L, LUPU AN, KAUFMANN JJ, LATTA H, MAXWELL MH (1967) The value of renal biopsy in unilateral renovascular hypertension. Nephron 4: 231-247

BARAJAS L, WANG P, BENNETT CM, WILBURN RL (1976) The renal sympathetic system and juxtaglomerular cells in experimental renovascular hypertension. Lab Invest 35: 574-587

BARAJAS L, BENNETT CM, CONNOR G, LINDSTROM RR (1977a) Structure of a juxtaglomerular cell tumor: the presence of a neural component. Lab Invest 37: 357-368

BARAJAS L, MARKS LS, TRYGSTAD CW (1977b) Unilateral renal hypoplasia with associated venous anomaly and hypertension. A study of juxtaglomerular cells. Virchows Arch [A] 374: 169-182

BARAJAS LB, SOKOLSKI KN, LECHAGO J (1983) Vasoactive intestinal polypeptide-immunoreactive nerves in the kidney. Neurosci Lett 43: 263-269

BARAJAS L, POWERS K, WANG P (1984) Innervation of the renal cortical tubules: a quantitative study. Am J Physiol 247: F50-F60

BARAJAS L, POWERS K, WANG P (1985) Innvervation of the late distal nephron: an autoradiographic and ultrastructural study. J Ultrastruct Res 92: 146-157

BARAJAS L, POWERS K, CARRETERO O, SCICLI AG, INAGAMI T (1986) Immunocytochemical localization of renin and kallikrein in the rat renal cortex. Kidney Int 29: 965-970

BARAJAS L, SALIDO EC, POWERS KV (1988) Anatomical basis of the tubuloglomerular feedback mechanism: the juxtaglomerular apparatus. In: PERSSON AEG, BOBERG U (eds) The juxtaglomerular apparatus, 11th Fernström Symposium. Elsevier, Amsterdam, pp 7-26

BARGMANN W (1978) Niere und ableitende Harnwege. Springer; Berlin Heidelberg New York

BARKA T (1980) Biologically active polypeptides in submandibular glands. J Histochem Cytochem 28: 836-859

BARNETT R, SINGHAL PC, SCHARSCHMIDT LA, SCHLONDORFF D (1986) Dopamine attenuates the contractile response to angiotensin II in isolated rat glomeruli and cultured mesangial cells. Circ Res 59: 529-533

BARRETT AJ (1985) The cystatins: small protein inhibitors of cysteine proteinases. In: KHAIRALLAH EA, BOND JS, BIRD JWC (eds) Intracellular protein catabolism. Liss, New York, pp 105-116

BARRETT JD, EGGENA P (1986) The influence of acute stress on active and inactive renin in the rat. Clin Exp Hypertens [A], 8 (7): 1171-1178

BARRETT JD, EGGENA P, SAMBHI MP (1977) Enzymatic and partial physical characterization of a new high molecular weight renin from normal human kidney. Circ Res 41 [Suppl 2]: 7-11

BARTTER FC, RODRIGUEZ A (1982) Bartter's syndrome. Adv Intern Med Pediatr 50: 79-103

BARTTER FC, PRONOVE P, GILL JR Jr, MACCARDLE RC (1962) Hyperplasia of the juxtaglomerular complex with hyperaldosteronism and hypokalemic alkalosis. A new syndrome. Am J Med 33: 811-828

BARUCH D, DUFLOUX M, GUYENNE TT, GAUX JC, RAYNAUD A, BRISSET JM, DUCLOS JM, CORVOL P, MENARD J (1983) Trois cas de tumeur à rénine d'origine rénale. Méthodes diagnostiques et traitment. Arch Mal Coeur 76 [Suppl]: 81-86

BARUCH D, CORVOL P, ALHENC-GELAS F, DUFLOUX MA, GUYENNE TT, GAUX JC, RAYNAUD A, BRISSET JM, DUCLOS JM, MENARD J (1984) Diagnosis and treatment of renin-secreting tumors. Report of three cases. Hypertension 6: 760-766

BAUMBACH L, LEYSSAC PP (1977) Studies on the mechanism of renin release from isolated superfused rat glomeruli: effects of calcium, calcium ionophore and lanthanum. J Physiol (Lond) 273: 745-764

BAUMBACH L, SKØTT O (1981) Renin release from isolated rat glomeruli: seasonal variations and effects of D600 on the response to calcium deprivation. J Physiol 310: 285-292

BAUMBACH L, SKØTT O (1982) Isolated glomeruli in vitro: An approach to the macula-densa-mediated renin release. Kidney Int 22 [Suppl 12]: 73-77

BAUMBACH L, SKØTT O (1986) Renin release from different parts of rat afferent arterioles in vitro. Am J Physiol 251: F12-F16

BAUMBACH L, LEYSSAC PP, SKINNER SL (1976) Studies on renin release from isolated superfused glomeruli: effects of temperature, urea, ouabain and ethacrynic acid. J Physiol (Lond) 258: 243-256

BAXTER CR, LAZZARO V, DUGGIN GG, HORVATH JS, TILLER DJ (1985) Stimulation of renin secretion by 8-(N,N-diethylamino) acetyl-3,4,5-trimethoxybenzoate (TMB-8). Biochem Pharmacol 34: 1523-1527

BAYLIS C, BRENNER BM (1978) Modulation by prostaglandin synthesis inhibitors of the action of exogenous angiotensin II on glomerular ultrafiltration in the rat. Circ Res 43: 889-898

BEAN JW (1942) Specificity in the renin-hypertensinogen reaction. Am J Physiol 136: 731-742

BECHER H (1936) Über besondere Zellengruppen und das Polkissen am Vas afferens in der Niere des Menschen. Z Wiss Mikrosk 53: 205-214

BECHER H (1949) Die gestaltlichen Grundlagen der Strombahnsteuerung am Gefäßpol der Malpighischen Körperchen in der menschlichen Niere. Ärztl Forsch 3: 351-367

BECKER CG (1972) Demonstration of actomyosin in mesangial cells of the renal glomerulus. Am J Pathol 66: 97-107

BEEUWKES R (1980) The vascular organization of the kidney. Annu Rev Physiol 42: 531-542

BEEUWKES R, SHAHOOD J, ROSEN S (1975) Macula densa: absence of transport ATPase. Kidney Int 8: 467

BEIERWALTES WH, SCHRYVER S, OLSON PS, ROMERO JC (1980) Interaction of the prostaglandin and renin-angiotensin systems in isolated rat glomeruli. Am J Physiol 239: F602-F608

BEIERWALTES WH, SCHRYVER S, OLSON PS, ROMERO JC (1981) Control of renin release in isolated rat glomeruli. Hypertension 3 [Suppl 2]: 30-34

BEIERWALTES WH, PRADA J, CARRETERO OA (1985a) Effect of glandular kallikrein on renin release in isolated rat glomeruli. Hypertension 7: 27-31

BEIERWALTES WH, PRADA J, CARRETERO OA (1985b) Kinin stimulation of renin release in isolated rat glomeruli. Am J Physiol 248: F757-F761

BELL C (1982) Dopamine as a postganglionic autonomic neurotransmitter. Neuroscience 7: 1-8

BELL C, LANG WJ, LASKA J (1978) Dopamine-containing vasomotor nerves in the dog kidney. J Neurochem 31: 77-83

BELL PD, FRANCO M, NAVAR LG (1987) Calcium as a mediator of tubuloglomerular feedback. Annu Rev Physiol 49: 275-293

BELLUCCI A, WILKES BM (1984) Mechanism of sodium modulation of glomerular angiotensin receptors in the rat. J Clin Invest 74: 1593-1600

BENCOSME SA, STONE RS, LATTA H (1959) Acute alterations produced by uranyl nitrate in glomeruli of rat kidneys: light and electronmicroscopic studies. Am J Pathol 35: 670-682

BENCOSME SA, STONE RS, LATTA H, MADDEN SC (1960) Acute reaction with collagen production in renal glomeruli of rats as studied electron microscopically. J Ultrastruct Res 3: 171-185

BERL T, HENRICH WL, ERICKSON AL, SCHRIER RW (1979) Prostaglandins in the beta-adrenergic and baroreceptor-mediated secretion of renin. Am J Physiol 236: F472-F477

BERNARD C, HOEFFEL JC, SCHMITT M, ANDRE JL, BRASSE F, FREYD S (1986) Tumeur de Grawitz avec hypertension artérielle chez l'enfant. J Radiol (Paris) 67: 921-925

BERNE RM, KNABB3RM, ELY SW, RUBIO R (1983) Adenosine in the local regulation of blood flow: a brief overview. Fed Proc 42: 3136-3142

BERNIK MB (1969) Contractile activity of human glomeruli in culture. Nephron 6: 1-10

BERRIDGE MJ (1984) Inositol trisphosphate and diacylglycerol as second messengers. Biochem J 220: 345-360

BERRIDGE MJ (1986) Intracellular signalling through inositol trisphosphate and diacylglycerol. Biol Chem Hoppe Seyler 367: 447-456

BERRIDGE MJ (1987) Inositol trisphosphate and diacylglycerol: two interacting second messengers. Annu Rev Biochem 56: 159-193

BERRIDGE MJ, IRVINE RF (1984) Inositol trisphosphate, a novel second messenger in cellular signal transduction. Nature 312: 315-321

BIANCHI C, GUTKOWSKA J, THIBAULT G, GARCIA R, GENEST J, CANTIN M (1986) Distinct localization of atrial natriuretic factor and angiotensin II binding sites in the glomerulus. Am J Physiol 251: F594-F602

BIAVA CG (1967) Ultrastructural observations on the morphogenesis of nonspecific granules in human juxtaglomerular and renal vascular cells. Circ Res 20/21 [Suppl 2]: 47-67

BIAVA C, WEST M (1965) Lipofuscin-like granules in vascular smooth muscle and juxtaglomerular cells of human kidneys. Am J Pathol 47: 287-313

BIAVA CG, WEST M (1966a) Fine structure of normal human juxtaglomerular cells. I. General structure and intercellular relationships. Am J Pathol 49: 679-721

BIAVA CG, WEST M (1966b) Fine structure of normal human juxtaglomerular cells. II. Specific and non-specific cytoplasmic granules. Am J Pathol 49: 955-979

BIDANI AK, CHURCHILL PC (1981) Kinetics of the disappearance of endogenous plasma renin following nephrectomy in pregnant and non-pregnant rats. J Physiol (Lond) 315: 461-467

BING J (1973) Rapid marked increase in plasma renin in rats treated with inhibitors of the renin system. Acta Pathol Microbiol Immunol Scand [A] 81: 376-378

BING J, KAZIMIERCZAK J (1962) Renin content of different parts of the juxtaglomerular apparatus. 4. Localization of renin in the kidney. Acta Pathol Microbiol Immunol Scand 54: 80-84

BING J, NIELSEN K (1973) Cause of the prolonged pressor action of renin in nephrectomized rats. Acta Pathol Microbiol Immunol Scand [A] 81: 247-253

BING J, POULSEN K (1975) Time course of changes in plasma renin after blockade of the renin-system. Acta Pathol Microbiol Immunol Scand [A] 83: 454-466

BLACKSHEAR JL, SPIELMAN WS, KNOX FG, ROMERO JC (1979) Dissociation of renin release and renal vasodilation by prostaglandin synthesis inhibitors. Am J Physiol 237: F20-F24

BLAINE EH, DAVIS JO (1971) Evidence for a renal vascular mechanism in renin release: New observations with graded stimulation by aortic constriction. Circ Res 28 [Suppl 2]: 118-126

BLAINE EH, DAVIS JO, WITTY RT (1970) Renin release after hemorrhage and after suprarenal aortic constriction in dogs without sodium delivery to the macula densa. Circ Res 27: 1081-1089

BLAINE EH, DAVIS JO, PREWITT RL (1971) Evidence for a renal vascular receptor in control of renin secretion. Am J Physiol 220: 1593-1597

BLAIR ML (1983) Stimulation of renin secretion by α-adrenoceptor agonists. Am J Physiol 244: E37-E44

BLAIR ML, CHEN YH, HISA H (1986) Elevation of plasma renin activity by α-adrenoceptor agonists in conscious dogs. Am J Physiol 251: E695-E702

BLAIR-WEST JR, COGHLAN JP, DENTON DA, FUNDER JW, SCOGGINS BA, WRIGHT RD (1971) Inhibition of renin secretion by systemic and intrarenal angiotensin infusion. Am J Physiol 220: 1309-1315

BLANTZ RC (1980) Segmental renal vascular resistance: single nephron. Annu Rev Physiol 42: 573-588

BLANTZ RC, PELAYO JC (1983) In vivo actions of angiotensin II on glomerular function. Fed Proc 42: 3071-3074

BLANTZ RC, PELAYO JC (1984) A functional role for the tubuloglomerular feedback mechanism. Kidney Int 25: 739-746

BLANTZ RC, KONNEN KS, TUCKER BJ (1976) Angiotensin II effects upon the glomerular microcirculation and ultrafiltration coefficient of the rat. J Clin Invest 57: 419-434

BLENDSTRUP K, LEYSSAC PP, POULSEN K, SKINNER SL (1975) Characteristics of renin release from isolated superfused glomeruli in vitro. J Phyiol (Lond) 246: 653-672

BOHLE A (1959) Elektronenmikroskopische Untersuchungen über die Struktur des Gefäßpols der Niere. Verh Dtsch Ges Pathol 43: 219-225

BOHLE A, HERFARTH C (1958) Zur Frage eines intercapillären Bindegewebes im Gomerulum der Niere des Menschen. Virchows Arch [A] 331: 573-590

Bohle A, Sitte H (1966) Der juxtaglomeruläre Apparat der Niere. In: Aktuelle Probleme der Nephrologie. IV. Symposium der Gesellschaft für Nephrologie, pp 3-17 and 765-775. Krück F (Ed) Springer, Berlin, Heidelberg, New York.

Bohle A, Walvig F (1964) Beitrag zur vergleichenden Morphologie der epitheloiden Zellen der Nierenarteriolen unter besonderer Berücksichtigung der epitheloiden Zellen in den Nieren von Seewasserfischen. Klin Wochenschr 42: 415-421

Bohle A, Helmchen U, Meyer D (1970) In: Kaufmann W et al. (eds). Renin-Angiotensin-Aldosteron-System und Pharmaka. Med Welt 21: 1631-1648

Bohle A, Helmchen U, Grund KE, Gärtner HV, Meyer DS, Bock KD, Bulla M, Bünger P, Diekmann L, Frotscher U, Hayduk K, Kösters W, Strauch M, Scheler F, Christ H (1977) Malignant nephrosclerosis in patients with hemolytic uremic syndrome (primary malignant nephrosclerosis). Curr Top Pathol 65: 81-113

Bohle A, Christensen J, Meyer DS, Laberke HG, Strauch M (1982) Juxtaglomerular apparatus of the human kidney: Correlation between structure and function. Kidney Int 22 [Suppl 12]: 18-23

Boknam L, Ericson AC, Åberg B, Ulfendahl HR (1981) Flow resistance of the interlobular artery in the rat kidney. Acta Physiol Scand 111: 159-163

Bolender RP (1970) Morphometric analysis of the juxtaglomerular cells of the mouse kidney. Thesis, Harvard University, Cambridge

Boll HU, Forssmann WG, Taugner R (1975) Studies on the juxtaglomerular apparatus. IV. Freeze-fracturing of membrane surfaces. Cell Tissue Res 161: 459-469

Bondar N, Cadnapaphornchai P, McDonald FD, Taher S (1984) Mechanism of effect of dibutyryl cyclic adenosine 3',5'-monophosphate on canine renal renin release. J Physiol (Lond) 355: 33-41

Bonnin JM, Hodge RL, Lumbers ER (1972) A renin-secreting renal tumor associated with hypertension. Aust NZ J Med 2: 178-181

Borges DR, Webster ME, Guimaraes JA, Prado JL (1981) Synthesis of prekallikrein and metabolism of plasma kallikrein by perfused rat liver. Biochem Pharmacol 30: 1065-1069

Boucher R, Saidi M, Genest J (1972) A new "angiotensin converting enzyme" system. In: Genest J, Koiw E (eds) Hypertension, vol 72. Springer, Berlin Heidelberg New York, pp 512-523

Boucher R, Asselin J, Genest J (1974) A new enzyme leading to the direct formation of angiotensin II. Circ Res 34/35 [Suppl 1]: 203-209

Boucher R, Demassieux S, Garcia R, Genest J (1977) Tonin-angiotensin II system: a review. Circ Res 41 [Suppl 2]: 26-29

Bouhnik J, Fehrentz JA, Galen FX, Seyer R, Evin G, Castro B, Menard J, Corvol P (1985) Immunologic identification of both plasma and human renal inactive renin as prorenin. J Clin Endocrinol Metab 60: 399-401

Bouskela E, Wiederhielm CA (1979) Microvascular myogenic reaction in the wing of the intact unanesthetized bat. Am J Physiol 237: H59-H65

Boyd GW (1974) A protein-bound form of porcine renal renin. Circ Res 35: 426-438

Boyd GW (1979) Prolonged pressor response to renin in the nephrectomized rat. Circ Res 45: 396-404

Brackett NC Jr, Koppel M, Randall RE Jr, Nixon WP (1968) Hyperplasia of the juxtaglomerular complex with secondary aldosteronism without hypertension (Bartter's syndrome). Am J Med 44: 803-819

Brain SD, William TJ, Tippins JR, Morris HR, MacIntyre J (1985) Calcitonin gene-related peptide is a potent vasodilator. Nature 313: 54-56

Brand MD, Felber SM (1984) The intracellular calcium antagonist TMB-8 (NN-diethylamino) octyl-3,4,5-trimethoxybenzoate) inhibits mitochondrial ATP production in rat thymocytes. Biochem J 224: 1027-1030

Braun-Menendez E, Fasciolo JC, Leloir LF, Muñoz JM (1939) La substancia hipertensora de la sangre del rinon isquemiado. Rev Soc Arg Biol 15: 420-430

Braverman B, Freeman RH, Rostorfer HH (1971) The influence of dietary sodium chloride on in vitro renin release from rat kidney slices. Proc Soc Exp Biol Med 138: 81-88

Breckenridge LJ, Almers W (1987) Final steps in exocytosis observed in a cell with giant secretory granules. Proc Natl Acad Sci USA 84: 1945-1949

Bremer JL (1916) The interrelations of the mesonephros, kidney and placenta in different classes of animals. Am J Anat 19: 179-210

Brenner BM, Dworkin LD, Ichikawa I (1986) Glomerular ultrafiltration. In: Brenner BM, Rector FC (eds) The kidney. Saunders, Berlin, pp 124-144

Briggs JP, Schnermann J (1986) Macula densa control of renin secretion and glomerular vascular tone: evidence for common cellular mechanisms. Renal Physiol 9: 193-203

Briggs JP, Schnermann J (1987) The tubuloglomerular feedback mechanism: functional and biochemical aspects. Annu Rev Physiol 49: 251-273

Briggs JP, Steipe B, Schubert G, Schnermann J (1982) Micropuncture studies of the renal effects of atrial natriuretic substance. Pflügers Arch 395: 271-276

Briggs JP, Marin-Grez M, Steipe B, Schubert G, Schnermann J (1984) Inactivation of atrial natriuretic substance by kallikrein. Am J Physiol 247: F480-F484

Britton SL (1981) Intrarenal vascular effects of angiotensin I and angiotensin II. Am J Physiol 240: H914-H919

Broadwell RD, Oliver C (1983) An enzyme cytochemical study of the endocytic pathways in anterior pituitary cells of the mouse in vivo. J Histochem Cytochem 31: 325-335

Brooks VL, Brownfield MS, Reid IA (1982) Measurement and localization of angiotensin-like immunoreactivity in juxtaglomerular cells of the rat. Regul Pept 4: 317-324

Broughton Pipkin F (1984) The renin-angiotensin system in mother and foetus. Equine Vet J 16: 253-255

Broughton Pipkin F, Kirkpatrick SML, Lumbers ER, Mott JC (1974a) Renin and angiotensin-like levels in foetal, newborn and adult sheep. J Physiol (Lond) 241: 575-588

Broughton Pipkin F, Lumbers ER, Mott JC (1974b) Factors influencing plasma renin and angiotensin II in the conscious pregnant ewe and its foetus. J Physiol (Lond) 243: 619-636

Brouhard RH, Gill JR, Yun JCH, Kelly GD, Bartter FC (1979) Prostaglandin dependent and independent effects of bradykinin on renal function in the dog. Renal Physiol 80: 44-53

Brown GP, Douglas JG (1982) Angiotensin II binding sites on isolated rat renal brush border membranes. Endocrinology 111: 1830-1836

Brown GP, Douglas JG (1983) Angiotensin II-binding sites in rat and primate isolated renal tubular basolateral membranes. Endocrinology 112: 2007-2014

BROWN GP, DOUGLAS JG, KRONTIRIS-LITOWITZ J (1980) Properties of angiotensin II receptors of isolated rat glomeruli: factors influencing binding affinity and comparative binding of angiotensin analogs. Endocrinology 106: 1923-1929

BROWN JJ, DAVIES DL, LEVER AF, PARKER RA, ROBERTSON JIS (1965) The assay of renin in single glomeruli in the normal rabbit and the appearance of the juxtaglomerular apparatus. J Physiol (Lond) 176: 418-428

BROWN JJ, DAVIES DL, DOAK PB, LEVER AF, ROBERTSON JIS (1966) Serial estimation of plasma renin concentration during pregnancy and after parturition. J Endocrinol 35: 373

BROWN JJ, FRASER R, LEVER AF, MORTON JJ, ROBERTSON JIS, TREE M, BELL PRF, DAVIDSON JK, RUTHVEN IS (1973) Hypertension and secondary hyperaldosteronism associated with a renin-secreting renal juxtaglomerular-cell tumor. Lancet 2: 1228-1232

BROWN WJ, GOODHOUSE J, FARQUHAR MG (1986) Mannose-6-phosphate receptors for lysosomal enzymes cycle between the Golgi complex and endosomes. J Cell Biol 103: 1235-1247

BRÜHL U, TAUGNER R, FORSSMANN WG (1974) Studies on the juxtaglomerular apparatus. I. Perinatal development in the rat. Cell Tissue Res 151: 433-456

BRUNEVAL P, HINGLAIS N, ALHENC-GELAS F, TRICOTTET V, CORVOL P, MENARD J, CAMILLERI JP, BARIETY J (1986) Angiotensin I converting enzyme in human intestine and kidney. Ultrastructural immunohistochemical localization. Histochemistry 85: 73-80

BRUNEVAL P, Fournier JG, SOUBRIER F, BELAIR MF, DA SILVA JL, GUETTIER C, PINET F, TARDIVEL I, CORVOL P, BARIETY J, CAMILLERI JP (1988) Detection and localization of renin messenger RNA in human pathologic tissues using in situ hybridization. Am J Pathol 131: 320-330

BRYAN GT, MACCARDLE RC, BARTTER FC (1966) Hyperaldosteronism, hyperplasia of the juxtaglomerular complex, normal blood pressure, and dwarfism: report of a case. Pediatrics 37: 43-50

BUCHER O (1973) Cytologie, Histologie und mikroskopische Anatomie des Menschen. Huber, Bern

BUCHER O, KAISSLING B (1973) Morphologie des juxtaglomerulären Apparates. Verh Anat Ges 67: 109-136

BUCHER O, REALE E (1961 a) Zur elektronenmikroskopischen Untersuchung der juxtaglomerulären Spezialeinrichtungen der Niere. I. Problemstellung und erste Beobachtungen. Z Zellforsch 54: 167-181

BUCHER O, REALE E (1961 b) Zur elektronenmikroskopischen Untersuchung der juxtaglomerulären Spezialeinrichtungen der Niere. II. Über die Macula densa des Mittelstückes. Z Mikrosk Anat Forsch 67: 514-528

BUCHER O, REALE E (1962 a) Zur elektronenmikroskopischen Untersuchung der juxtaglomerulären Spezialeinrichtungen der Niere. III. Die epitheloiden Zellen der Arteriola afferens. Z Zellforsch 56: 344-358

BUCHER O, REALE E (1962 b) Zur elektronenmikroskopischen Untersuchung der juxtaglomerulären Spezialeinrichtungen der Niere. IV. Die Goormaghtighschen Zellen. Z Anat Entw Gesch 123: 206-220

BUCHER O, RIEDEL B (1965) Der juxtaglomeruläre Apparat der Niere. Hippokrates 36: 857-865

BUCHER O, ZIMMERMANN E (1960) A propos de la macula densa du rein. Acta Anat 42: 352-371

BUCHER O, RIEDEL B, BAECHTOLD-FOWLER N, SCHAECHTELIN G, PETRIK P, PETERS G (1974) Juxtaglomerular cell granulation and renal cortical renin activity in normal and adrenalectomized guinea pigs. Nephron 12: 178-187

BÜHRLE CP, NOBILING R, MANNEK E, SCHNEIDER D, HAKKENTHAL E, TAUGNER R (1984) The afferent glomerular arteriole: Immunocytochemical and electrophysiological investigations. J Cardiovasc Pharmacol 6: S383-S393

BÜHRLE CP, NOBILING R, TAUGNER R (1985) Intracellular recordings from renin-positive cells of the glomerular arteriole. Am J Physiol 249: F272-F281

BÜHRLE CP, HACKENTHAL E, HELMCHEN U, LACKNER K, NOBILING R, STEINHAUSEN M, TAUGNER R (1986 a) The hydronephrotic kidney of the mouse as a tool for intravital microscopy and in vitro electrophysiological studies of renin-containing cells. Lab. Invest. 54: 462-472

BÜHRLE CP, SCHOLZ H, HACKENTHAL E, NOBILING R, TAUGNER R (1986 b) Epithelioid cells: membrane potential changes induced by substances influencing renin secretion. Mol Cell Endocrinol 45: 37-47

BÜHRLE CP, SCHOLZ H, NOBILING R, TAUGNER R (1986 c) Junctional transmission in renin-containing and smooth muscle cells of the afferent arteriole. Pflügers Arch 406: 578-586

BÜHRLE CP, HACKENTHAL E, NOBILING R, SKØTT O, BAUMBACH L, TAUGNER R (1987 a) Tachyphylaxis of juxtaglomerular epithelioid cells to angiotensin II. Differences between the electrical membrane response and renin secretion. Pflügers Arch. 410: 55-62

BÜHRLE CP, ROSIVALL L, TAUGNER R (1987 b) Intrarenal generation of angiotensin II evaluated by an electrophysiological technique. Am J Physiol 252: F635-F644

BULGER RE, NAGLE RB (1973) Ultrastructure of the interstitium in the rabbit kidney. Am J Anat 136: 183-204

BULGER RE, TRUMP BF (1969) Ultrastructure of granulated arteriolar cells (juxtaglomerular cells) in kidney of fresh and salt water teleost. Am J Anat 124: 77-88

BUMPUS FM, SCHWARZ H, PAGE IH (1957) Synthesis and pharmacology of the octapeptide angiotonin. Science 125: 886-887

BUNAG RD, PAGE IH, MCCUBBIN JW (1966) Neural stimulation of renin release. Circ Res 19: 851-858

BUNAG RD, PAGE IH, MCCUBBIN JW (1967) Inhibition of renin release by vasopressin and angiotensin. Cardiovasc Res 1: 67-73

BUNAG RD, PAGE IH, MCCUBBIN JW (1970) Reduction of renin release by puromycin. Cardiovasc Res 4: 213-216

BURGHARDT W, SCHWEISFURTH H, DAHLHEIM H (1982) Juxtaglomerular angiotensin II formation. Kidney Int 22 [Suppl 12]: 49-54

BURGOYNE RD, CHEEK TR, NORMAN KM (1986) Identification of a secretory granule-binding protein as caldesmon. Nature 319: 68-70

BURNETT JCJ, GRANGER JP, OPGENORTH TJ (1984) Effects of synthetic atrial natriuretic factor on renal function and renin release. Am J Physiol 247: F863-F866

BURNIER M, BIOLLAZ J, BRUNNER DB, BRUNNER HR (1983) Blood pressure maintenance in awake dehydrated rats: renin, vasopressin, and sympathetic activity. Am J Physiol 245: H203-H209

BURNSTOCK G, KENNEDY C (1985) Is there a basis for distinguishing two types of P2-purinoceptor? Gen Pharmacol 16: 433-440

BUSSE R, TROGISCH G, BASSENGE R (1985) The role of endothelium in the control of vascular tone. Basic Res Cardiol 80: 475-490

BUTLEN D, MISTAOUI MH, MOREL F (1987) Atrial natriuretic

peptide receptors along the rat and rabbit nephrons: [125 I] α-rat atrial natriuretic peptide binding in microdissected glomeruli and tubules. Pflügers Arch 408: 356-365

CADNAPAPHORNCHAI P, KELLER D, MCDONALD FD (1987) Role of intracellular calcium in renal nerve-mediated renin release. Proc Soc Exp Biol Med 185: 24-30

CAIN H, KRAUS B (1969) Struktur und Funktion des juxtaglomerulären Apparates der Niere unter geordneten Bedingungen. Dtsch Med Wochenschr 94: 2173-2180

CAIN H, KRAUS B (1970) Der juxtaglomeruläre Apparat der Niere bei verschiedenen pathologischen Prozessen. Dtsch Med Wochenschr 95: 282-288

CAIN H, KRAUS B (1971) Funktions- und Formwandel der Gefäßwandzellen des juxtaglomerulären Apparates der Niere beim Kaninchen. Virchows Arch [B] 7: 160-173

CAIN H, KRAUS B (1976) The juxtaglomerular apparatus in malignant hypertension of man. Virchows Arch [A] 372: 11-28

CALAM J, DIMALINE R, PEART WS, SINGH J, UNWIN R (1983a) Effects of vasoactive intestinal polypeptide on renal function in man. J Physiol (Lond) 345: 469-475

CALAM J, DIMALINE R, PEART WS, UNWIN R (1983b) Studies on the renin response to vasoactive intestinal polypeptide (VIP) in the conscious rabbit. Br J Pharmacol 80: 13-15

CALDICOTT WJH, TAUB KJ, MARGULIES SS, HOLLENBERG NK (1981) Angiotensin receptors in glomeruli differ from those in renal arterioles. Kidney Int 19: 687-693

CALDWELL PRB, SEEGAL BC, HSU KC, DAS M, SOFFER RL (1976) Angiotensin-converting enzyme: vascular endothelial localization. Science 191: 1050-1051

CAMILLERI JP, PHAT VN, BARIETY J, CORVOL P, MENARD J (1980) Use of a specific antiserum for renin detection in human kidney. J Histochem Cytochem 28: 1343-1346

CAMILLERI JP, HINGLAIS N, NOCHY C, PHAT VN, BARIETY J (1983) Immunohistochemistry of renin in human diseased kidney. Clin Exp Hypertens [A] 5: 1179-1190

CAMILLIERI JP, HINGLAIS N, BRUNEVAL P, BARIETY J, TRICOTTET V, ROUCHON M, MANICLLA-JIMENEZ R, CORVOL P, MENARD J (1984) Renin storage and cell differentiation in juxtaglomerular cell tumors: an immunohistochemical and ultrastructural study of three cases. Hum Pathol 15: 1069-1079

CAMPBELL D (1985) The site of angiotensin production. J Hypertens 3: 199-207

CAMPBELL DJ (1987) Circulating and tissue angiotensin systems. J Clin Invest 79: 1-6

CAMPBELL DJ, HABENER JF (1986) Angiotensinogen gene is expressed and differentially regulated in multiple tissues of the rat. J Clin Invest 78: 31-39

CAMPBELL DJ, BOUHNIK J, COEZY E, PINET F, CLAUSER E, MENARD J, CORVOL P (1984a) Characterization of precursor and secreted forms of rat angiotensinogen. Endocrinology 114: 776-785

CAMPBELL DJ, BOUHNIK J, MENARD J, CORVOL P (1984b) Identitiy of angiotensinogen precursors of rat brain and liver. Nature 308: 206-208

CAMPBELL WB, ITSKOVITZ HD (1976) Effect of histamine and antihistamines on renal hemodynamics and functions in the isolated perfused canine kidney. J Pharmacol Exp Ther 198: 661-667

CAMPBELL WB, GRAHAM RM, JACKSON EK (1979) Role of renal prostaglandins in sympathetically mediated renin release in the rat. J Clin Invest 64: 448-456

CANTIN M, GENEST J (1985) The heart and the atrial natriuretic factor. Endocr Rev 6: 107-127

CANTIN M, ARAUJO-NASCIMENTO MF, BENCHIMOL S, DESORMEAUX Y (1977a) Metaplasia of smooth muscle cells into juxtaglomerular cells in the juxtaglomerular apparatus, arteries, and arterioles of the ischemic (endocrine) kidney. Am J Pathol 87: 581-602

CANTIN M, DESORMEAUX Y, BENCHIMOL S (1977b) On the lysosomal function of juxtaglomerular granules. Beitr Pathol 161: 310-327

CANTIN M, GUTKOWSKA J, LACASSE J, BALLAK M, LEDOUX S, INAGAMI T, BEUZERON J, GENEST J (1984) Ultrastructural immunocytochemical localization of renin and angiotensin II in the juxtaglomerular cells of the ischemic kidney in experimental renal hypertension. Am J Pathol 115: 212-224

CANZANELLO VJ, MADAIO MP, MADIAS NE (1987) Enalapril in the management of hypertension associated with renal artery stenosis. J Clin Pharmacol 27: 32-40

CAPELLI JP, WESSON LG, APONTE GE (1970) A phylogenetic study of the renin-angiotensin system. Am J Physiol 218: 1171-1178

CAPPONI AM, VALLOTTON MB (1976) Renin release by rat kidney slices incubated in vitro. Role of sodium and of α- and β-adrenergic receptors, and effect of vincristine. Circ Res 39: 200-203

CAPPONI AM, GOURJON M, VALLOTTON MB (1977) Effect of β-blocking agents and angiotensin II on isoproterenol-stimulated renin release from rat kidney slices. Circ Res 40 [Suppl 1]: 89-93

CAPRÉOL SV, SUTHERLAND LE (1968) Comparative morphology of juxtaglomerular cells. I. Juxtaglomerular cells in fish. Can J Zool 46: 249-256

CARACHI R, LINDOP GBM, LECKIE BJ (1987) Inactive renin: a tumor marker in nephroblastoma. J Pediatr Surg 22: 278-280

CARMINES PK (1986) The blood-perfused juxtamedullary nephron preparation: videometric study of microvascular function. Fed Proc 45: 2856-2858

CARMINES PK, MORRISON TK, NAVAR LG (1986) Angiotensin II effects on microvascular diameters of in vitro blood-perfused juxtamedullary nephrons. Am J Physiol 251: F610-F618

CARMINES PK, PERRY MD, HAZELRIG JB, NAVAR LG (1987) Effects of preglomerular and postglomerular vascular resistance alterations on filtration fraction. Kidney Int 31 [Suppl 20]: 229-232

CARONE FA, PETERSON DR, OPARIL S, PULLMAN TN (1980) Renal tubular transport and catabolism of small peptides. In: MAUNSBACH AB, OLSEN TS, CHRISTENSEN EI (eds) Functional ultrastructure of the kidney. Academic, London, pp 327-340

CARRETERO OA, BEIERWALTES WH (1984) Effect of glandular kallikrein, kinins and aprotinin (a serin protease inhibitor) on renin release. J Hypertens 2 [Suppl 1]: 125-130

CARRETERO O, GROSS F (1967) Renin substrate in plasma under various experimental conditions in the rat. Am J Physiol 213: 695-700

CARROLL RG, OPDYKE DF (1982) Evolution of angiotensin II-induced catecholamine release. Am J Physiol 243: R65-R69

CARVALHO JS (1982) The effect of water deprivation on the osmotic release of renin. Kidney Int 22: 344-347

CASELLAS D (1986a) A method for scanning electron microscopic observation of glomerular vascular poles in rat kidneys. J Electron Microsc Techn 4: 63-64

CASELLAS D (1986b) The blood-perfused juxtamedullary

nephron preparation: anatomical and micropuncture studies. Fed Proc 45: 2855-2856

Casellas D, Navar LG (1984) In vitro perfusion of juxtamedullary nephrons in rats. Am J Physiol 246: F349-F358

Casellas D, Taugner R (1986) Renin status of the afferent arteriole and ultrastucture of the juxtaglomerular apparatus in 'superficial' juxtamedullary nephrons from rats. Renal Physiol 9: 348-356

Casellas D, Carmines PK, Navar LG (1985) Microvascular reactivity of in vitro blood perfused juxtamedullary nephrons from rats. Kidney Int 28: 752-759

Casellas D, Dupont M, Jover B, Mimran A (1982) Scanning electron microscopic study of arterial cushions in rats: a novel application of the corrosion-replication technique. Anat Rec 203: 419-428

Catanzaro DF, Mesterovic N, Morris BJ (1985) Studies of the regulation of mouse renin genes by measurement of renin messenger ribonucleic acid. Endocrinology 117: 872-878

Celio MR (1982) Angiotensin II immunoreactivity coexisting with renin in the human juxtaglomerular epithelioid cells. Kidney Int 22 [Suppl 12]: 30-32

Celio MR, Inagami T (1981) Angiotensin II immunoreactivity coexists with renin in the juxtaglomerular granular cells of the kidney. Proc Natl Acad Sci USA 78: 3897-3900

Chan MY, Holmes WN (1971) Studies on a «renin-angiotensin» system in the normal and hypophysectomized pigeon *(Columba livia)*. Gen Comp Endocrinol 16: 304-311

Chandra S, Skelton FR, Bernardis LL (1964) Separation of renal pressor activity by ultracentrifugation. Lab Invest 13: 1192-1197

Chandra S, Hubbard JC, Skelton FR, Bernardis LL, Kamura S (1965) Genesis of juxtaglomerular cell granules. A physiologic, light and electron microscopic study concerning experimental renal hypertension. Lab Invest 14: 1834-1847

Chang JJ, Kisaragi M, Ikamoto H, Inagami T (1981) Isolation and activation of inactive renin from human kidney and plasma. Plasma and renal inactive renins have different molecular weights. Hypertension 3: 509-515

Changaris DG, Miller JJ, Levy RS (1986) Angiotensin II generated by a human renal carboxypeptidase. Biochem Biophys Res Commun 138: 573-579

Chansel D, Ardaillou N, Nivez MP, Ardaillou R (1982a) Angiotensin II receptors in human isolated renal glomeruli. J Clin Endocrinol Metab 55: 961-966

Chansel D, Oudinet JP, Nivez MP, Ardaillou R (1982b) Histamine H2 receptors in rat renal glomeruli. Biochem Pharmacol 31: 367-375

Chansel D, Morin JP, Borghi H, Ardaillou N, Ardaillou R (1987) Angiotensin I-converting enzyme in isolated human glomeruli. FEBS Lett 220: 247-252

Cheek TR, Burgoyne RD (1986) Nicotine-evoked dissassembly of cortical actin filaments in adrenal chromaffine cells. FEBS Lett 207: 110-114

Chen DS, Poisner AM (1976) Direct stimulation of renin release by calcium. Proc Soc Exp Biol Med 152: 565-567

Chen GL, Sutrina SL, Frayer KL, Chen WW (1986) Effects of lysosomotropic agents on lipogenesis. Arch Biochem Biophys 245: 66-75

Chiou CY, Malagody CM (1975) Studies on the mechanism of action of a new calcium antagonist, 8(N,N-diethylamino)octyl 3,4,5-trimethoxybenzoate hydrochloride in smooth and sceletal muscles. Br J Pharmacol 53: 279-285

Chiu PJS, Sybertz EJ (1986) The effects of atriopeptin II on calcium fluxes in rabbit aorta. Eur J Pharmacol 124: 277-284

Chiu PJS, Brown AD, Barnett A (1984) Inhibitory effect of captopril on renal responses to frusemide in sodium-restricted rats. J Pharm Pharmacol 36: 31-35

Cho KW, Malvin RL (1979) Renin inactivation during in vitro experiments. Am J Physiol 236: F501-F504

Chokshi DS, Yeh BK, Sambhi P (1972) Effects of dopamine and isoproterenol on renin secretion in the dog. Proc Soc Exp Biol Med 140: 54-57

Chou SY, Faubert PF, Porush JG (1986) Contribution of angiotensin to the control of medullary hemodynamics. Fed Proc 45: 1438-1443

Christensen EI, Carone FA, Rennke HG (1981) Effect of molecular charge on endocytic uptake of ferritin in renal proximal tubule cells. Lab Invest 44: 351-358

Christensen JA, Bohle A (1978) The juxtaglomerular apparatus in the normal rat kidney. Virchows Arch [A] 379: 143-150

Christensen JA, Meyer DS, Bohle A (1975) The structure of the human juxtaglomerular apparatus. A morphometric, lightmicroscopic study on serial sections. Virchows Arch [A] 367: 83-92

Christensen JA, Bader H, Bohle A, Meyer DS (1976) The structure of the juxtaglomerular apparatus in Addison's disease, Bartter's syndrome, and Conn's syndrome. Virchows Arch [A] 370: 103-112

Christensen JA, Meyer DS, Jakubowski HD, Neunhöfer J, Bohle A (1978) The juxtaglomerular apparatus in a human kidney with polar artery stenosis. Acta Pathol Microbiol Immunol Scand [A] 86: 375-381

Christensen JA, Bjaerke HA, Meyer DS, Bohle A (1979) The normal juxtaglomerular apparatus in the human kidney. A morphological study. Acta Anat 103: 374-383

Christensen JA, Knutsson S, Bohle A (1982a) Granular epithelioid cells in the kidney of the Atlantic salmon (salmo salar). Kidney Int 22 [Suppl 12]: 179-184

Christensen JA, Morild I, Mikeler E, Bohle A (1982b) Juxtaglomerular apparatus in the domestic fowl *(Gallus domesticus)*. Kidney Int 22 [Suppl 12]: 24-29

Christensen JA, Taugner R, Meyer DS, Bohle A (1987) The granular epithelioid cells in the kidney of the lemon sole (*Pleuronectes microcephalus* Donovani). Cell Tissue Res 249: 137-143

Christensen JA, Bohle A, Mikeler E, Taugner R (1989a) Renin-positive granulated Goormaghtigh cells. Immunohistochemical and electron microscopical studies on biopsies from patients with pseudo-Bartter's syndrome. Cell Tissue Res 255: 149-153

Christensen JA, Mikeler E, Bohle A (1989b) Granular epithelioid cells of the kidneys in salmon adapted to fresh- and seawater. Anat Rec 223: 21-26

Churchill MC, Churchill PC (1980) Separate and combined effects of ouabain and extracellular potassium on renin secretion from rat renal cortical slices. J Physiol (Lond) 300: 105-114

Churchill MC, Churchill PC, McDonald FD (1983) Comparison of the effects of rubidium and potassium on renin secretion from rat kidney slices. Endocrinology 112: 777-781

Churchill PC (1979) Possible mechanism of the inhibitory effect of ouabain on renin secrection from rat renal cortical slices. J Physiol (Lond) 294: 123-134

Churchill PC (1980) Effect of D-600 on inhibition of in vi-

tro renin release in the rat by high extracellular potassium and angiotensin II. J Physiol (Lond) 304: 449-458
Churchill PC (1985) Second messengers in renin secretion. Am J Physiol 249: F175-F184
Churchill PC, Bidani A (1987) Renal effects of selective adenosine receptor agonists in anesthetized rats. Am J Physiol 256: F299-F303
Churchill PC, Churchill MC (1982a) Ca-dependence of the inhibitory effect of K-depolarization on renin secretion from rat kidney slices. Arch Int Pharmacodyn Ther 258: 300-312
Churchill PC, Churchill MC (1982b) Isoproterenol-stimulated renin secretion in the rat: second messenger roles of Ca and cyclic AMP. Life Sci 30: 1313-1319
Churchill PC, Churchill MC (1983) Effects of trifluoperazine on renin secretion of rat kidney slices. J Pharmacol Exp Ther 224: 68-72
Churchill PC, Churchill MC (1984) 12-0-tetradecanoylphorbol 13-acetate (TPA) inhibits renin secretion of rat renal cortical slices. J Hypertens 2 [Suppl 1]: 25-28
Churchill PC, Churchill MC (1985) A1 and A2 adenosine receptor activation inhibits and stimulates renin secretion of rat renal cortical slices. J Pharmacol Exp Ther 232: 589-594
Churchill PC, Churchill MC (1987) Bay K 8644, a calcium channel agonist inhibits renin secretion in vitro. Arch Int Pharmacodyn Ther 285: 87-97
Churchill PC, Lyons HJ (1976) Effect of intrarenal arterial infusion of magnesium on renin release in dogs. Proc Soc Exp Biol Med 152: 6-10
Churchill PC, McDonald FD, Churchill MC (1981) Effect of diltiazem, a calcium antagonist, on renin secretion from rat kidney slices. Life Sci 29: 383-389
Churchill PC, Churchill MC, McDonald FD (1983a) Evidence that 1-adrenoceptor activation mediates isoproterenol-stimulated renin secretion in the rat. Endocrinology 113: 687-692
Churchill PC, Savoy-Moore RT, Churchill MC (1983b) Lack of relationship between prostaglandin E2 release and renin secretion in rat renal cortical slices. J Pharmacol Exp Ther 226: 46-51
Cipolle MD, Zehr JE (1984) Charcterization of the renin-angiotensin system in the turtle *Pseudemys scripta.* Am J Physiol 247: R15-R23
Cipolle MD, Zehr JE (1985) Renin release in turtles: Effects of volume depletion and furosemide administration. Am J Physiol 249: R100-R105
Cipolle MD, Zehr JE, Reinhart GA (1986) Effects of autonomic agents on renin release in the turtle, *Pseudemys scripta.* Am J Physiol 251: R1103-R1108
Clappison BH, Anderson WP, Johnston CI (1981) Renal hemodynamics and renal kinins after angiotensin-converting enzyme inhibition. Kidney Int 20: 615-620
Clauser E, Bouhnik J, Coezy E, Corvol P, Menard J (1983) Synthesis and release of immunoreactive angiotensinogen by rat liver slices. Endocrinology 112: 1188-1193
Clauser E, Bouhnik J, Jaramillo HN, Auzan C, Corvol P, Menard J (1985) Angiotensinogen production and consumption in the adrenalectomized rat. Endocrinology 116: 274-280
Click RL, Joyner WL, Gilmore JP (1979) Reactivity of glomerular afferent and efferent arterioles in renal hypertension. Kidney Int 15: 109-115
Cogan MG (1986) Atrial natriuretic factor can increase renal solute excretion primarily by raising glomerular filtration. Am J Physiol 250: F710-F714
Coghlan JP, Penschow JD, Hudson PJ, Niall HD (1984) Hybridization histochemistry: use of recombination DNA for tissue localisation of specific mRNA populations. Clin Exp Hypertension (A) 6: 63-78
Cohen AJ, Fray JCS (1982) Calcium dependence of myogenic renal plasma flow autoregulation: evidence from the isolated perfused rat kidney. J Physiol (Lond) 330: 449-460
Cohen AJ, Laurens P, Fray JCS (1983) Suppression of renin secretion by insulin: dependence on extracellular calcium. Am J Physiol 245: E531-E534
Cohen FS, Akabas MH, Finkelstein A (1982) Osmotic swelling of phospholipid vesicles causes them to fuse with a planar phospholipid bilayer membrane. Science 217: 458-460
Cohen FS, Akabas MH, Zimmerberg J, Finkelstein A (1984) Parameters affecting the fusion of unilamellar phospholipid vesicles with planar bilayer membranes. J Gen Physiol 98: 1054-1062
Cohen RJ, Severance RC, Whiting EG, Lundberg GD (1965) A study of the juxtaglomerular body in primary aldosteronism. Ann Intern Med 62: 569-575
Cohen S, Taylor JM, Murakami K, Michelakis AM, Inagami T (1972) Isolation and characterization of renin-like enzymes from mouse submaxillary glands. Biochemistry 11: 4286-4293
Conn JW, Rovner DR, Cohen EL (1965) Normal and altered function of the renin-angiotensin-aldosterone system in man. Applications in clinical and research medicine. Ann Intern Med 63: 266-284
Conn JW, Cohen EL, Lucas CP, McDonald WJ, Mayor GH, Blough WM Jr, Eveland WC, Bookstein JJ, Lapides J (1972) Primary reninism. Hypertension, hyperreninemia, and secondary hyperaldosteronism due to renin-producing juxtaglomerular cell tumors. Arch Intern Med 130: 682-696
Connor G, Bennet CM, Lindstrom RR, Brosman SA, Barajas L, Edelbaum D (1978) Juxtaglomerular cell tumor. Nephron 21: 325-333
Cook WF (1971) Cellular localization of renin. In: Fisher JW (ed) Kidney hormones. Academic, New York, pp 117-128
Cook WF, Pickering GW (1958) The location of renin within the kidney. J Physiol (Lond) 143: 78P-79P
Cook WF, Pickering GW (1959) The location of renin in the rabbit kidney. J Physiol (Lond) 149: 526-536
Cooper CL, Shaffer JE, Malik KU (1985) Mechanism of action of angiotensin II and bradykinin on prostaglandin synthesis and vascular tone in isolated rat kidney. Effect of Ca++ antagonists and calmodulin inhibitors. Circ Res 56: 97-108
Cordes EH (1984) Structure and function of carboxyl proteases. In: Doyle AE, Bearn AG (eds) Hypertension and the angiotensin system: therapeutic approaches. Raven, New York, pp 77-91
Corman B, Michel JB (1986) Renin-angiotensin system, converting-enzyme inhibition, and kidney function in aging female rats. Am J Physiol 251: R450-R455
Corvol P, Devaux C, Ito T, Sicard P, Ducloux J, Menard J (1977) Large scale purification of hog renin. Physicochemical characterization. Circ Res 41: 616-622
Cowley AWJ, Guyton AC (1972) Quantification of intermediate steps in the renin-angiotensin-vasoconstrictor feedback loop in the dog. Circ Res 30: 557-566

Cox JN, Paunier L, Vallotten MB, Humbert JR, Rohner A (1975) Epithelial liver hamartoma, systemic arterial hypertension and renin hypersecretion. Virchows Arch [A] 366: 15-26

Cox HM, Munday KA, Poat JA (1983a) The binding of (125I)-angiotensin to rat renal epithelial cell membranes. Br J Pharmacol 79: 63-70

Cox HM, Poat JA, Munday KA (1983b) The effect of guanine nucleotides on (125I)-angiotensin binding in rat kidney cortex epithelial membranes. Biochem Pharmacol 32: 3601-3604

Cramb G (1986) Selective lysosomal uptake and accumulation of the beta-adrenergic antagonist propranolol in cultured and isolated cell systems. Biochem Pharmacol 35: 1365-1372

Craven PA, DeRubertis FR (1985) Ca2+-dependent modulation of renin release from isolated glomeruli: apparent independence from alterations in cGMP. Metabolism 34: 651-657

Creutz CE, Zaks WJ, Hamman HC, Martin WH (1987) The roles of Ca2+-dependent membrane-binding proteins in the regulation and mechanism of exocytosis. In: Sowers AE (ed) Cell fusion. Plenum, New York, pp 45-68

Crocker DW, Newton RA, Mahoney EM, Harrison JH (1962) Hypertension due to primary renal ischemia. A correlation of juxtaglomerular cell counts with clinicopathological findings in twenty-five cases. N Engl J Med 267: 794-800 (1962)

Cruz-Soto MA, Benabe JE, Lopez-Novoa JM, Martinez-Maldonado M (1982) Renal Na+-K+-ATPase in renin release. Am J Physiol 243: F598-F603

Cruz-Soto M, Benabe JE, Lopez-Novoa JM, Martinez-Maldonado M (1984) Na+-K+-ATPase inhibitors and renin release: relationship to calcium. Am J Physiol 247: F650-F655

Currie MG, Needleman P (1984) Renal arachidonic acid metabolism. Ann Rev Physiol 46: 327-341

Dahlheim H, Granger P, Thurau K (1970) A sensitive method for determination of renin activity in the single juxtaglomerular apparatus of the rat kidney. Pflügers Arch 321: 303-315

Dahlheim H, Jacob ICM, Pschorr J, Rosenthal J (1983) The renin angiotensin system in the extrarenal vascular walls: an approach to studies in humans. In: Fotherby K, Pal SB (eds) Hormones in normal and abnormal human tissue. Walter de Gruyter, Berlin, pp 251-284

Dalle M, Giry M, Gay M, Delost P (1978) Perinatal changes in plasma and adrenal corticosterone and aldosterone concentrations in the mouse. J Endocrinol 76: 303-309

Darby IA, Aldred GP, Coghlan JP, Fernley RT, Penschow JD, Ryan GB (1985a) Use of synthetic oligonucleotide and recombinant DNA probes to study renin gene expression. Clin Exp Pharmacol Physiol 12: 199-203

Darby IA, Aldred P, Crawford RJ, Fernley RT, Niall HD, Penschow JD, Ryan GB, Coghlan JP (1985b) Renin gene expression in vessels of the ovine renal cortex. J Hypertens 3: 9-11

Data JL, Gerber JG, Crump WJ, Frohlich JC, Hollifield JW, Nies AS (1978) The prostaglandin system: a role in canine baroreceptor control of renin release. Circ Res 42: 454-458

Datar S, McCauley FA, Wilson TW (1987) Effect of cyclooxygenase and thromboxane synthetase inhibition on furosemide-stimulated plasma renin activity. Can J Physiol Pharmacol 65: 80-83

Dauda G, Dévényi I (1971) Bilateral adrenalectomy and the renin-angiotensin system. Acta Physiol Hung 39: 335-341

Dauda G, Csapó Z, Kovács M (1976) Frequency of juxtaglomerular granulated cells in the mouse kidney. Acta Morphol Hung 24: 23-27

Davalos M, Frega NS, Saker B, Leaf A (1978) Effect of exogenous angiotensin II in the isolated perfused rat kidney. Am J Physiol 235: F605-F610

Dave JR, Witorsch RJ (1983) Indomethacin decreases both luteinizing hormone binding and fluidity of testicular microsomal membranes in rat. Prostaglandins Leukotrienes Med 12: 371-380

Davidson JK, Clark DC (1974) Renin-secreting juxtaglomerular-cell tumour. Br J Radiol 47: 594-597

Davis JO, Freeman RH (1976) Mechanisms regulating renin release. Physiol Rev 56: 1-56

Day RP, Luetscher JA (1974) Big renin: a possible prohormone in kidney and plasma of a patient with Wilms' tumour. J Clin Endocrinol Metab 38: 923-926

Day RP, Reid IA (1976) Renin activity in dog brain: enzymological similarity to cathepsin D. Endocrinology 99: 93-100

Day RP, Luetscher JA, Gonzales CM (1975) Occurrence of big renin in human plasma, amniotic fluid and kidney extracts. J Clin Endocrinol Metab 40: 1078-1084

Day RP, Luetscher JA, Zager PG (1976) Big renin: Identification, chemical properties and clinical implications. Am J Cardiol 37: 667-674

Day RP, Hui KY, Gure M, Carlson WD, Dzau VJ, Haber E (1986) A monoclonal antibody specific for the aminoterminal sequence of human prorenin identifies a common epitope on renal and amniotic fluid inactive renins. J Hypertens 4: 375-381

Dean RT, Jessup W, Roberts CR (1984) Effects of exogenous amines on mammalian cells, with particular reference to membrane flow. Biochem J 217: 27-40

De Bold AJ, Borenstein HB, Veress ATSH (1981) A rapid and potent natriuretic response to intravenous injection of atrial myocardial extracts in rats. Life Sci 28: 89-94

De Bruyn PPH, Cho Y (1986) In vivo exocytosis of lysosomes by the endothelium of the venous sinuses of bone marrow and liver: visualization at normal and low body temperature. Am J Anat 177: 35-41

De Duve C (1983) Lysosomes revisited. Eur J Biochem 137: 391-397

De Duve C, Wattiaux R (1966) Functions of lysosomes. Ann Rev Physiol 28: 435-492

Defendini R, Zimmerman EA, Weare JA, Alhenc-Gelas F, Erdös EG (1983) Angiotensin-converting enzyme in epithelial and neuroepithelial cells. Nueroendocrinology 37: 32-40

De Jong W (1969) Release of renin by rat kidney slices; relationship to plasma renin after desoxycorticosterone and renal hypertension. Proc Soc Exp Biol Med 130: 85-88

Delange RJ, Smith EL (1971) Leucine aminopeptidase and other N-terminal exopeptidases. In: Boyer PD (ed) The enzymes, Vol. III. Academic, New York, pp 81-118

De Lean A, Racz K, Gutkowska J, Nguyen TT, Cantin M, Genest J (1984) Specific receptor-mediated inhibition by synthetic atrial natriuretic factor of hormone stimulated steroidogenesis in culture bovine adrenal cells. Endocrinology 115: 1636-1638

Demopoulos H, Kaley G, Zweifach BW (1960) The histologic distribution of renin in the kidneys of the rat and rabbit. Am J Pathol 37: 443-455

De Muylder CG (1952) The "neurility" of the kidney. A monograph on nerve supply to the kidney. Blackwell; Oxford

Dennis RL, McDougal WS, Glick AD, MacDonell RC (1985) Juxtaglomerular cell tumor of the kidney. J Urol 134: 334-338

Deray G, Branch RA, Herzer WA, Ohnishi A, Jackson EK (1987) Effects of atrial natriuretic factor on hormone-induced renin release. Hypertension 9: 513-517

Deray G, Branch RA, Herzer WA, Ohnishi A, Jackson EK (1987) Adenosine inhibits β-adrenoceptor but not DBcAMP-induced renin release. Am J Physiol 252: F46-F52

Derkx FHM, Wenting GJ, Man in't Veld AJ, Verhoeven RP, Schalekamp MADH (1978) Control of enzymatically inactive renin in man under various pathological conditions: implications for the interpretation of renin measurements in peripheral and renal venous plasma. Clin Sci Mol Med 54: 529-538

Derkx FHM, Wenting GJ, Man in't Veld AJ, Verhoeven RP, Schalekamp MADH (1979) Evidence for activation of circulating inactive renin by the human kidney. Clin Sci 56: 115-120

Derkx FHM, Schalekamp PA, Schalekamp ADH (1987) Two-step prorenin-renin conversion. J Biol Chem 262: 2472-2477

de Rouffignac C, Bonvalet JP, Menard J (1974) Renin content in superficial and deep glomeruli of normal and salt-loaded rats. Am J Physiol 226: 150-154

Deschepper CF, Mellon SH, Cumin F, Baxter JD, Ganong WF (1986) Analysis by immunocytochemistry and in situ hybridization of renin and its mRNA in kidney, testis, adrenal, and pituitary of the rat. Proc Natl Acad Sci USA 83: 7552-7556

DeSchryver-Kecskemeti K, Kraus FT, Engleman W, Lacy PE (1982) Alveolar soft part sarcoma - a malignant angioreninoma. Am J Surg Pathol 6: 5-18

de Senarclens CF, Pricam CE, Banichahi FD, Vallotton MB (1977) Renin synthesis, storage, and release in the rat: A morphological and biochemical study. Kidney Int 11: 161-169

Desjardins-Giasson S, Gutkowska J, Garcia R, Genest J (1981) Renin substrate in rat mesenteric artery. Can J Physiol Pharmacol 59: 528-532

Désormeaux Y, Ballak M, Benchimol S, Lacasse J, Cantin M, Genest J (1982) Synthesis and migration of proteins and glycoproteins in juxtaglomerular cells of sodium-deficient rats. Cell Tissue Res 222: 53-67

Dev B, Drescher C, Schnermann J (1974) Resetting of tubuloglomerular feedback sensitivity by dietary salt intake. Pflügers Arch 346: 263-277

Dévényi I, Dauda G, Szabó J (1971) Juxtaglomerular index and activity of the renin-angiotensin system. Pathol Eur 6: 19-33

De Vito E, Gordon SB, Cabrera RR, Fasciolo JC (1970a) Release of renin by rat kidney slices. Am J Physiol 219: 1036-1041

De Vito E, Cabrera RR, Fasciolo JC (1970b) Renin production and release by rat kidney slices. Am J Physiol 219: 1042-1045

Dhiab al Naimi AAH, Bearn JG (1981) The juxtaglomerular apparatus and glomerular granular cells in fetuses and neonates of the rat. J Anat 133: 607-623

D'Humieres S, Russo-Marie F, Vargaftig BB (1986) PAF-acether-induced synthesis of prostacyclin by human endothelial cells. Eur J Pharmacol 131: 13-19

DiBona GF (1982) The functions of the renal nerves. Rev Physiol Biochem Pharmacol 94: 76-181

DiBona GF (1985) Neural regulation of renal tubular sodium reabsorption and renin secretion. Fed Proc 44: 2816-2822

Dieterich HJ (1973) Interstitium und venöse Strombahn in der Rattenniere. Verh Anat Ges 67: 37-45

Dieterich HJ (1978) Die Struktur der Blutgefäße in der Rattenniere. Norm Pathol Anat (Stuttg) 35: 1-127

Dimaline R, Peart WS, Unwin RJ (1983) Effects of vasoactive intestinal polypeptide (VIP) on renal function and plasma renin activity in the conscious rabbit. J Physiol (Lond) 344: 379-388

Dinerstein RJ, Vannice J, Henderson RC, Roth LJ, Goldberg LI, Hoffmann PC (1979) Histofluorescence techniques provide evidence for dopamine containing neuronal elements in canine kidney. Science 205: 497-499

Dinerstein RJ, Jones RT, Goldberg LI (1983) Evidence for dopamine-containing renal nerves. Fed Proc 42: 3005-3008

Di Nicolantonio R, Mendelsohn FAO (1986) Plasma renin and angiotensin in dehydrated and rehydrated rats. Am J Physiol 250: R898-R901

Do YS, Shinagawa T, Tam H, Inagami T, Hsueh WA (1987) Characterization of pure human renal renin. Evidence for a subunit structure. J Biol Chem 262: 1037-1043

Doi Y, Hinko A, Franco-Saenz R, Mulrow PJ (1983) Reexamination of the effect of urinary kallikrein on renin release: Evidence that kallikrein does not release renin but protects renin from destruction. Endocrinology 113: 114-118

Doležel S (1966) Monoaminergic innervation of the arteries and veins of the kidney observed using fluorescence reaction. Folia Morphol (Warsz) 14: 168-174

Doležel S, Edvinsson L, Owman C, Owman T (1976) Fluorescence histochemistry and autoradiography of adrenergic nerves in the renal juxtaglomerular complex of mammals and man, with special regard to the efferent arteriole. Cell Tissue Res 169: 211-220

Dominguez JH, Snowdowne KW, Freudenrich CC, Brown T, Borle AB (1987) Intracellular messenger for action of angiotensin II on fluid transport in rabbit proximal tubule. Am J Physiol 252: F423-F428

D'Orleans-Juste P, Dion S, Mizrahi J, Regoli D (1985) Effects of peptides and non-peptides on isolated arterial smooth muscles: role of endothelium. Eur J Pharmacol 114: 9-21

Douglas JG (1987a) Corticosteroids decrease glomerular angiotensin receptors. Am J Physiol 252: F453-F457

Douglas JG (1987b) Angiotensin receptor subtypes of the kidney cortex. Am J Physiol 253: F1-F7

Drukker A, Donoso VS, Linshaw MA, Bailie MD (1983) Intrarenal distribution of renin in the developing rabbit. Pediatr Res 17: 762-765

Drury PL, Williams BC, Edwards CRW, Oddie CJ, Horne B (1986) Development and application of a superfusion technique for the study of renin secretion in rat renal cortical cells. Clin Sci 71: 581-587

Dunihue FW (1941) Effect of cellophane perinephritis on the granular cells of the juxtaglomerular apparatus. Arch Pathol 32: 211-216

DUNIHUE FW (1946) The effect of bilateral adrenalectomy on the juxtaglomerular apparatus. Anat Rec 96: 536
DUNIHUE FW (1947) The juxtaglomerular apparatus in experimental hypotension. Am J Pathol 23: 906-907
DUNIHUE FW (1949) The effect of adrenal insufficiency and of desoxycorticosterone acetate on the juxtaglomerular apparatus. Anat Rec 103: 442-443
DUNIHUE FW, BOLDOSSER WG (1963) Observations on the similarity of mesangial to juxtaglomerular cells. Lab Invest 12: 1228-1240
DUNIHUE FW, CANDON BH (1940) Histologic changes in the renal arteriole of hypertensive rabbits. Arch Pathol Lab Med 29: 777
DUNIHUE FW, ROBERTSON WVB (1957) Effect of desoxycorticosterone acetate and of sodium on juxtaglomerular apparatus. Endocrinology 61: 293-299
DUNN BR, ICHIKAWA I, PFEFFER JM, TROY JL, BRENNER BM (1986) Renal and systemic hemodynamic effects of synthetic atrial natriuretic peptide in the anesthetized rat. Circ Res 59: 237-246
DUNN MJ, SCHARSCHMIDT LA (1987) Prostaglandins modulate the glomerular actions of angiotensin II. Kidney Int 31 [Suppl 20]: 95-101
DUNNICK NR, HARTMANN DS, FORD KK, DAVIS CJ Jr, AMIS ES Jr (1983) The radiology of juxtaglomerular tumors. Radiology 147: 321-326
DÜSING R, BARTTER FC, GILL JR Jr, KRÜCK F, KRAMER HJ (1983) Das Bartter Syndrom. Klin Wochenschr 61: 311-319
DVORAK AM, DVORAK HF, KARNOVSKY MJ (1972) Uptake of horseradish peroxidase by guinea pig basophilic leukocytes. Lab Invest 26: 27-39
DWORKIN LD, ICHIKAWA I, BRENNER BM (1983) Hormonal modulation of glomerular function. Am J Physiol 244: F95-F104
DYKES CW, BHAT K, TAYLOR JM, INAGAMI T (1986) Mouse kidney renin is synthetized in precursor form in the wheat germ cell-free protein synthesis system. Biomed Res 1: 565-568
DZAU VJ (1984) Vascular renin-angiotensin: a possible autocrine or paracrine system in control of vascular function. J Cardiovasc Pharmacol 6 [Suppl 2]: 377-382
DZAU VJ (1986) Significance of the vascular renin-angiotensin pathway. Hypertension 8: 553-559
DZAU VJ (1987) Implications of local angiotensinogen production in cardiovascular physiology and pharmacology. Am J Cardiol 59: 59-65
DZAU VJ, WILCOX CS, SANDS K, DUNCKEL P (1986) Dog inactive renin: biochemical characterization and secretion into renal plasma and lymph. Am J Physiol 250: E55-E61
DZAU VJ, ELLISON KE, BRODY T, INGELFINGER J, PRATT RE (1987a) A comparative study of the distributions of renin and angiotensinogen messenger ribonucleic acids in rat and mouse tissues. Endocrinology 120: 2334-2338
DZAU VJ, GONZALEZ D, KAEMPFER C, DUBIN D, WINTROUB BU (1987b) Human neutrophils release serine proteases capable of activating prorenin. Circ Res 60: 595-601
ECKERT H, KUNDE D (1974) Semiquantitative histochemische Untersuchungen am juxtaglomerulären Apparat der Niere im Tierexperiment. Acta Histochem (Jena) 48: 1-10
EDELMAN R, HARTROFT PM (1961) Localization of renin in juxtaglomerular cells of rabbit and dog through the use of the fluorescent-antibody technique. Circ Res 9: 1069-1077
EDDY RL, SANCHEZ SA (1971) Renin-secreting renal neoplasm and hypertension with hypokalemia. Ann Intern Med 75: 725-729
EDWARDS BR, HARMANCI MC (1983) Intramembranous particle clusters in collecting duct cells of rats. Influence of water balance. Renal Physiol 6: 275-280
EDWARDS BR, LAROCHELLE FT (1984) Antidiuretic effect of endogenous oxytocin in dehydrated Brattleboro homozygous rats. Am J Physiol 247: F453-F465
EDWARDS JG (1940) The vascular pole of the glomerulus in the kidney of vertebrates. Anat Rec 76: 381-389
EDWARDS RM (1983) Segmental effects of norepinephrine and angiotensin II on isolated renal microvessels. Am J Physiol 244: F526-F534
EDWARDS RM, WEIDLEY EF (1987) Lack of effect of atriopeptin II on rabbit glomerular arterioles in vitro. Am J Physiol 252: F317-F321
EGERER G, TAUGNER R, TIEDEMANN K (1984) Renin immunohistochemistry in the mesonephros and metanephros of the pig embryo. Histochemistry 81: 385-390
EGUCHI Y, YAMAKAWA M, MORIKAWA Y, HASHIMOTO Y (1975) Granular cells in the juxtaglomerular apparatus in perinatal rats. Anat Rec 181: 627-634
EHMKE H, PERSSON P, KIRCHHEIM H (1987a) Pressure-dependent renin release: the kidney factor in long-term control of arterial blood pressure in conscious dogs. Clin Exp Hypertens [A] 9 [Suppl 1]: 181-195
EHMKE H, PERSSON P, KIRCHHEIM H (1987b) A physiological role for pressure-dependent renin release in long-term blood pressure control. Pflügers Arch 410: 450-456
EIDE I, LOYNING E, KIIL F (1973) Evidence for hemodynamic autoregulation of renin release. Circ Res 32: 237-245
EKAS RDJ, STEENBERG ML, LOKHANDWALA MF (1983) Increased norepinephrine release during sympathetic nerve stimulation and its inhibition by adenosine in the isolated perfused kidney of spontaneously hypertensive rats. Clin Exp Hypertens [A] 5: 41-48
EKBLAD E, EDVINSSON I, WAHLESTEDT C, UDDMAN R, HAKANON R, SUNDLER F (1984) Neuropeptide Y co-exists and co-operates with noradrenaline in perivascular nerve fibers. Regul Pept 8: 225-235
ELAUT L (1934) La structure de l'artère afférente du glomérule rénal chez le chien hypertendu. C R Soc Biol (Paris) 115: 1416-1418
ELLIOT DF, PEART WS (1957) Amino acid sequence of hypertensin. Biochem J 65: 246-254
ELLIOTT ME, GOODFRIEND TL (1986) Atrial natriuretic peptide inhibits protein phosphorylation stimulated by angiotensin II in bovine adrenal glomerulosa cells. Biochem Biophys Res Commun 140: 814-820
ELLIS B, ITSKOVITZ HD (1980) Angiotensin II and norepinephrine after indomethacin in isolated perfused canine kidneys. Tachyphylaxis vs modulator effects of prostaglandins. Prostaglandins 20: 981-991
EL MATRI A, BEN AYED H, SLIM R, BEN MAIZ H, ZMERLI S, CAMILLERI JP, MILLIEZ JP (1980) Un autre cas d'hypertension artérielle due à une tumeur de l'appareil juxta-glomérulaire. J Urol (Paris) 86: 33-36
EMILSSON A, SUNDLER R (1986) Evidence for a catalytic role of phospholipase A in phorbol diester- and zymosan-induced mobilization of arachidonic acid in mouse peritoneal macrophages. Biochim Biophys Acta 876: 533-542
EMMERSON JCS, JOHNS EJ (1985) Intrarenal angiotensin II and the control of renal function in the rat. J Physiol (Lond) 365: 113
ENDES P, GOMBA S, DÉVÉNYI I (1969) Specific staining and exact quantitative evaluation of the granulation in the juxtaglomerular cells. Acta Morphol Hung 17: 47-53

EPSTEIN M, FLAMENBAUM W, LOUTZENHISER R (1979/80) Characterization of the renin-angiotensin system in the isolated perfused rat kidney. Renal Physiol 2: 244-256

EPSTEIN S, SAGEL J, BRODOVCKY H, TUFF S, EALES L (1976) Absence of an acute effect of calcium or parathyroid hormone administration on plasma renin activity in man. Clin Sci Mol Med 50: 79-81

ERCAN ZS, TURK RK (1975) Potentiation of the effect of histamine by PGE in the isolated perfused rabbit kidney and guinea pig lung. Experientia 31: 333-335

ERDÖS EG, YANG HYT (1967) An enzyme in microsomal fraction of kidney that inactivates bradykinin. Life Sci 6: 569-574

ERDÖS EG (1975) Angiotensin I converting enzyme. Circ Res 36: 247-255

ERICSSON JLE (1964) Absorption and decomposition of homologous hemoglobin in renal proximal tubular cells. An experimental light and electron microscopic study. Acta Pathol Microbiol Immunol Scand [Suppl] 168: 1-121

ERTL N (1967) Zur Entwicklung des juxtaglomerulären Apparates in Nieren von Mäuseembryonen. Z Anat Entwicklungsgesch 126: 132-137

ETTIENNE EM, FRAY JCS (1979) Influence of potassium, sodium, calcium, perfusion pressure, and isoprenaline on renin release induced by high concentrations of magnesium. J Physiol (Lond) 292: 373-380

FAARUP P (1965) On the morphology of the juxtaglomerular apparatus. Acta Anat (Basel) 60: 20-38

FAARUP P (1967) Renin location in the different parts of the juxtaglomerular apparatus in the cat kidney. 1. The afferent arteriole and the macula densa. Acta Pathol Microbiol Immunol Scand [A] 71: 509-521

FAARUP P (1968) Renin location in the different parts of the juxtaglomerular apparatus in the cat kidney. 2. Fractions of the afferent arteriole, the cell group of Goormaghtigh, the efferent arteriole and the glomerulus. Acta Pathol Microbiol Immunol Scand [A] 72: 109-117

FAARUP P (1971) Morphological aspects of the renin-angiotensin system. Acta Pathol Microbiol Immunol Scand [Suppl] 222: 1-96

FARAGGIANA T, GRESIK E, TANAKA T, INAGAMI T, LUPO A (1982) Immunohistochemical localization of renin in the human kidney. J Histochem Cytochem 30: 459-465

FARHI ER, CANT JR, BARGER AC (1982) Interactions between intrarenal epinephrine receptors and the renal baroreceptor in the control of PRA in conscious dogs. Circ Res 50: 477-485

FARHI ER, CANT JR, BARGER AC (1983) Alteration of renal baroreceptor by salt intake in control of plasma renin activity in conscious dogs. Am J Physiol 245: F119-F122

FARQUHAR MG (1977) Secretion and crinophagy in prolactin cells. Adv Exp Med Biol 80: 37-94

FARQUHAR MG (1981) Membrane recycling in secretory cells: Implications for traffic products and specialized membranes within the Golgi complex. Methods Cell Biol 23: 399-427

FARQUHAR MG (1982) Membrane recycling in secretory cells: pathway to the Golgi complex. Ciba Found Symp 92: 157-183

FARQUHAR MG (1985) Progress in unraveling pathways of Golgi traffic. Annu Rev Cell Biol 1: 447-488

FARQUHAR MG, PALADE GE (1962) Functional evidence for the existence of a third cell type in renal glomerulus. J Cell Biol 13: 55-87

FARQUHAR MG, PALADE GE (1981) The Golgi apparatus (complex) - (1954-1981) - from artifact to center stage. J Cell Biol 91: 77s-103s

FARQUHAR MG, WISSIG SL, PALADE GE (1961) Glomerular permeability: I. Ferritin transfer across the normal glomerular capillary wall. J Exp Med 112: 47-65

FARQUHAR MG, SKUTELSKY EH, HOPKINS CR (1975) Structure and function of the anterior pituitary and dispersed pituitary cells. In vitro studies. In: TIXIER-VIDAL A, FARQUHAR MG (eds) The anterior pituitary. Academic, New York, pp 81-135

FAUBERT PF, CHOU SY, PORUSH JG (1987) Regulation of papillary plasma flow by angiotensin II. Kidney Int 32: 472-478

FAUST PL, CHIRGWIN JM, KORNFELD S (1987) Renin, a secretory glycoprotein, acquires phosphomannosyl residues. J Cell Biol 105: 1947-1955

FELDER RA, BLECHER M, EISNER GM, JOSE PA (1984) Cortical tubular and glomerular dopamine receptors in the rat kidney. Am J Physiol 246: F557-F568

FERGUSON M, BELL C (1985) Substance P-immunoreactive nerves in the rat kidney. Neurosci Lett 60: 183-188

FERNANDEZ-CRUZ AJ, NOTH RH, HELDLER RG, MULROW PJ (1975) Glucagon stimulation of plasma renin activity in humans. J Clin Endocrinol Metab 41: 183-185

FERNANDEZ-PARDAL J, PERSEGUINO JC, BERMAN DM, COVIELLO A (1986) Angiotensin converting enzyme in the toad *bufo arenarum*. Comp Biochem Physiol 83A: 331-332

FEUERSTEIN G, FEUERSTEIN N (1980) The effect of indomethacin on isoprenaline-induced renin secretion in the cat. Eur J Pharmacol 61: 85-88

FIELD LJ, MCGOWAN RA, DICKINSON DP, GROSS KW (1984) Tissue and gene specificity of mouse renin expression. Hypertension 6: 597-603

FINKE R, GROSS R, HACKENTHAL E, HUBER I, KIRCHHEIM H (1983) Threshold pressure for the pressure-dependent renin release in the autoregulating kidney of conscious dogs. Pflügers Arch 399: 102-110

FINKELSTEIN A, ZIMMERBERG J, COHEN FS (1986) Osmotic swelling of vesicles: its role in the fusion of vesicles with planar phospholipid bilayer membranes and its possible role in exocytosis. Annu Rev Physiol 48: 163-174

FINKIELMAN S, NAHMOD VE (1969) In vitro production of angiotensin I by renal glomeruli. Nature 222: 1186-1188

FISHER ER (1961) Correlation of juxtaglomerular granulation, pressor activity, and enzymes of macula densa in experimental hypertension. Lab Invest 10: 707-718

FISHER ER (1966) Lysosomal nature of juxtaglomerular granules. Science 152: 1752-1753

FISHER ER, PEREZ-STABLE E, PARDO V (1966) Ultrastructural studies in hypertension. I. Comparison of renal vascular and juxtaglomerular cell alterations in essential and renal hypertension in man. Lab Invest 15: 1409-1433

FISHMAN MC (1976) Membrane potential of juxtaglomerular cells. Nature 260: 542-544

FLAMENBAUM W, HAMBURGER RJ (1974) Superficial and deep juxtaglomerular apparatus renin activity of the rat kidney. Effect of surgical preparation and NaCl intake. J Clin Invest 54: 1373-1381

FLAMENBAUM W, GAGNON J, RAMWELL P (1979) Bradykinin-induced renal hemodynamic alterations: renin and prostaglandin relationships. Am J Physiol 237: F433-F440

FLUCKIGER J, WAEBER B, NUSSBERGER J, MATSUEDA G, BRUNNER HR (1987) Effect of indomethacin and propranolol on the blood pressure and renin response to atriopeptin III in conscious rats. Regul Pept 17: 277-284

FOIDART JB, MAHIEU P (1986) Glomerular mesangial cell contractility in vitro is controlled by an angiotensin-prostaglandin balance. Mol Cell Endocrinol 47: 163-173

FOIDART J, SRAER J, DELARUE F, MAHIEU P, ARDAILLOU R (1980) Evidence for mesangial glomerular receptors for angiotensin II linked to mesangial cell contractility. FEBS Lett 121: 333-339

FOLKERT VW, SCHLONDORFF D (1977) Prostaglandin synthesis in isolated glomeruli. Prostaglandins 13: 873-892

FORSSMANN WG, REINECKE M (1984) Organ-specific innervation by autonomic nerve fibers as revealed by electron microscopy and immunohistochemistry. Front Horm Res 12: 59-73

FORSSMANN WG, TAUGNER R (1977) Studies on the juxtaglomerular apparatus. V. The juxtaglomerular apparatus in *Tupaia* with special reference to intercellular contacts. Cell Tissue Res 177: 291-305

FOURMAN J, MOFFAT DB (1964) Observations on the fine blood vessels of the kidney. In: HARRISON RG, HILL KR (eds) Cardiovascular anatomy and pathology. Academic, London, pp 57-71

FOURMAN J, MOFFAT DB (1971) The blood vessels of the kidney. Blackwell, Oxford

FRANCISCO LL, OSBORN JL, DIBONA GF (1982) Prostaglandins in renin release during sodium deprivation. Am J Physiol 243: F537-F542

FRANCO-SAENZ R, SUZUKI S, TAN SY (1980a) Prostaglandins and renin production: a review. Prostaglandins 20: 1131-1143

FRANCO-SAENZ R, SUZUKI S, TAN SY, MULROW PJ (1980b) Prostaglandin stimulation of renin release: independence of adrenergic receptor activity and possible mechanism of action. Endocrinology 106: 1400-1404

FRANK M, KRIZ W (1982) Scanning electron microscopy studies of the vascular pole of the rat glomerulus. Anat Rec 204: 149-152

FRANKLIN WG, PEACH MJ, GILMORE JP (1970) Evidence for the renal conversion of angiotensin I in the dog. Circ Res 27: 321-324

FRANSEN JAM (1987) Lysosomal characteristics of human renin-containing granules. Lab Invest 56: 124

FRAY JCS (1976) Stretch receptor model for renin release with evidence from perfused rat kidney. Am J Physiol 231: 936-944

FRAY JCS (1977) Stimulation of renin release in perfused kidney by low calcium and high magnesium. Am J Physiol 232: F377-F382

FRAY JCS (1978a) Mechanism of increased renin release during sodium deprivation. Am J Physiol 234: F376-F380

FRAY JCS (1978b) Stretch receptor control of renin release in perfused rat kidney: effect of high perfusate potassium. J Physiol (Lond) 282: 207-217

FRAY JCS (1980) Mechanism by which renin secretion from perfused rat kidneys is stimulated by isoprenaline and inhibited by high perfusion pressure. J Physiol (Lond) 308: 1-13

FRAY JCS, LAURENS NJ (1981) Mechanisms by which albumin stimulates renin secretion in isolated kidneys and juxtaglomerular cells. J Physiol (Lond) 320: 31-39

FRAY JCS, LUSH DJ (1984) Stretch receptor hypothesis for renin secretion: the role of calcium. J Hypertens 2 [Suppl 1]: 19-23

FRAY JCS, PARK CS (1986) Forskolin and calcium: Interactions in the control of renin secretion and perfusate flow in the isolated rat kidney. J Physiol (Lond) 375: 361-375

FRAY JCS, JOHNSON MD, BARGER AC (1977) Renin release and pressure response to renal arterial hypotension: effect of dietary sodium. Am J Physiol 233: H191-H195

FRAY JCS, LUSH DJ, SHARE DS, VALENTINE AND (1983a) Possible role of calmodulin in renin secretion from isolated rat kidneys and renal cells: Studies with trifluoperazine. J Physiol (Lond) 343: 447-454

FRAY JCS, LUSH DJ, VALENTINE AND (1983b) Cellular mechanisms of renin secretion. Fed Proc 42: 3150-3154

FRAY JCS, PARK CS, VALENTINE AND (1987) Calcium and the control of renin secretion. Endocrine Rev 8: 53-93

FREDERIKSEN O, LEYSSAC PP, SKINNER SL (1975) Sensitive osmometer function of juxtaglomerular cells in vitro. J Physiol (Lond) 252: 669-679

FREDHOLM BB, JANSEN I, EDVINSSON L (1985) Neuropeptide Y is a potent inhibitor of cyclic AMP accumulation in feline cerebral blood vessels. Acta Physiol Scand 124: 467-469

FREEMAN RH, ROSTORFER HH (1972) Hepatic changes in renin substrate biosynthesis and alkaline phosphatase activity in the rat. Am J Physiol 223: 364-370

FREEMAN RH, DAVIS JO, LOHMEIER TE (1975) Des-1-Asp-Angiotensin II. Possible intrarenal role in homeostasis in the dog. Circ Res 37: 30-34

FREEMAN RH, DAVIS JO, DIETZ JR, VILLARREAL D, SEYMOUR AA, ECHTENKAMP SF (1982) Renal prostaglandins and the control of renin release. Hypertension 4 [Suppl 2]: 106-112

FREEMAN RH, DAVIS JO, VILLARREAL D (1984) Role of renal prostaglandins in the control of renin release. Circ Res 54: 1-9

FREGA NS, DAVALOS M, LEAF A (1980) Effect of endogenous angiotensin on the efferent glomerular arteriole of rat kidney. Kidney Int 18: 323-327

FREISSMUTH M, HAUSLEITHNER V, TUISL E, NANOFF C, SCHÜTZ W (1987) Glomeruli and microvessels of the rabbit kidney contain both A1- and A2-adenosine receptors. Naunyn-Schmiedebergs Arch Pharmacol 335: 438-444

FRIED TA, SIMPSON EA (1986) Intrarenal localization of angiotensinogen mRNA by RNA-DNA dot-blot hybridization. Am J Physiol 250: F374-F377

FRIEDBERG EC (1964) The distribution of the juxtaglomerular granules and the macula densa in the renal cortex of the mouse. Lab Invest 13: 1003-1013

FRIEDBERG EC (1965a) Zonal variations in the juxtaglomerular granules in mice of different ages. Nephron 2: 230-238

FRIEDBERG EC (1965b) Distribution of juxtaglomerular granules. A study of mice on a salt-restricted diet. Arch Pathol Lab Med 80: 621-632

FRIEND DS, FARQUHAR MG (1967) Functions of coated vesicles during protein absorption in the rat was deferens. J Cell Biol 35: 357-376

FROMME M, STREICHER E, KRAUS B, KRUSE-JARRES J (1985) Arterielle Hypertonie bei reninsezernierendem retroperitonealem Leiomyosarkom. Klin Wochenschr 63: 158-163

FRY GN, DEVINE CE, BURNSTOCK G (1977) Freeze-fracture studies of nexuses between smooth muscle cells. Close relationship to sarcoplasmatic reticulum. J Cell Biol 72: 26-34

FUNAKAWA S, IKEMOTO F, KAWAMURA M, YAMAMOTO K (1979) Relationship between molecular weight conversion and renin activity in dog renal renin. Jpn Circ J 43: 824-826

FURCHGOTT RF (1983) Role of endothelium in responses of vascular smooth muscle. Circ Res 53: 557-573

FURUSATO M, HAYASHI H, KAWAGUCHI N, YOKATA K, SAITO K, AIZAWA S, ISHIKAWA E (1983) Juxtaglomerular cell tumour with special reference to its tubular component in regard to its histogenesis. Acta Pathol Jpn 33: 609-618

FYHRQUIST F, ROSENLÖF K, GRÖNHAGEN-RISKA C, HORTLING L, TIKKANEN I (1984) Is renin substrate an erythropoietin precursor? Nature 308: 649-652

FYNN M, ONOMAKPOME N, PEART WS (1977) The effects of ionophores (A23187 and RO2-2985) on renin secretion and renal vasoconstriction. Proc R Soc Lond [Biol] 199: 199-212

GAAL K, FORGACS I (1975) Effects of cyclic adenosine monophosphate on renal function and renin secretion. Acta Physiol Hung 46: 9-18

GAAL K, FORGACS I (1977) Interaction of cAMP and adrenoceptors on regulation of renal function and renin secretion. Acta Physiol Hung 49: 111-118

GAAL K, FORGACS I, BACSALMASY Z (1976) Effect of adenosine compounds (ATP, cAMP) on renin release in vitro. Acta Physiol Hung 47: 49-54

GALEN FX, DEVAUX C, GUYENNE T, MENARD J, CORVOL P (1979) Multiple forms of human renin. Purification and characterization. J Biol Chem 254: 4848-4855

GALEN FX, GUYENNE T, DEVAUX C, MENARD J, CORVOL P, BARIETY J, CAMILLIERI JP, N'GOC PW (1980) Purification of human kidney renin and production of highly specific antibodies. Application to direct radioimmunoassay of tissue and plasma renin. In: GROSS F, VOGEL G (eds) Enzymatic release of vasoactive peptides. Raven, New York, pp 51-65

GALEN FX, DEVAUX C, HOUOT AM, MENARD J, CORVOL P, CORVOL MT, GUBLER MC, MOUNIER F, CAMILLERI JP (1984) Renin biosynthesis by human tumoral juxtaglomerular cells. Evidences for a renin precursor. J Clin Invest 73: 1144-1155

GALL JAM, ALCORN D, COGHLAN JP, JOHNSTON CI, RYAN GB (1984) Immunohistochemical detection of urinary kallikrein-like material in juxtaglomerular peripolar cells. Proc IXth Int Cong Nephrol 451A (abstract)

GALL JAM, ALCORN D, BUTKUS A, COGHLAN JP, RYAN GB (1986) Distribution of glomerular peripolar cells in different mammalian species. Cell Tissue Res 244: 203-208

GANGULY A, GRIBBLE J, TUNE B, KEMSON RL, LUETSCHER JA (1973) Renin-secreting Wilms' tumour with severe hypertension. Ann Intern Med 79: 835-837

GANNON BJ (1972) Comparative and developmental studies of autonomic nerves in visceral and cardiovascular systems. PhD Thesis, Department of Zoology, University of Melbourne, Australia

GANONG WF (1984) The brain renin-angiotensin system. Annu Rev Physiol 46: 17-31

GANTEN D, GANTEN U, SCHELLING P, BOUCHER R, GENEST J (1975) The renin and iso-renin angiotensin systems in rats with experimental pituitary tumors. Proc Soc Exp Biol Med 148: 568-572

GANTEN D, HUTCHINSON JS, SCHELLING P, GANTEN U, FISCHER H (1976) The iso-renin angiotensin systems in extrarenal tissue. Clin Exp Pharmacol Physiol 3: 103-126

GANTEN D, PRINTZ M, UNGER T, LANG RE (1984) The brain renin-angiotensin system: problems and answers: In: VILLAREAL H, SAMBHI MP (eds) Topics in pathophysiology of hypertension. Martinus Nijhoff, Boston, pp 440-454

GANTEN D, LIND RW, UNGER T, LANG RE (1987) The Renin-Angiotensin System in the Brain. In: MCCANN, WEINER (eds) Integrative neuroendocrinology: molecular and clinical aspects. Karger, Basel, pp 92-101

GARCIA R, SCHIFFRIN EL, THIBAULT G, GENEST J (1981) Effects of tonin, an angiotensin II-forming enzyme, on vascular smooth muscle in the normal rabbit. Hypertension 3 [Suppl 1]: 101-106

GARCIA R, THIBAULT G, GUTKOWSKA J, HAMET P, CANTIN M, GENEST J (1985) Effect of chronic infusion of synthetic atrial natriuretic factor (ANF 8-33) in conscious two-kidney, one-clip hypertensive rats. Proc Soc Exp Biol Med 178: 155-159

GARDINER DS, LINDOP GBM (1985) The granular peripolar cell of the human glomerulus: a new component of the juxtaglomerular apparatus? Histopathology 9: 675-685

GARDINER DS, MORE IAR, LINDOP GBM (1986) The granular peripolar cell of the human glomerulus: an ultrastructural study. J Anat 146: 31-43

GÄRTNER K, BANNING E, VOGEL G, ULBRICH M (1973) The identity of interstitial fluid and hilar lymph in the kidney. Pflügers Arch 343: 331-340

GATTONE VH, EVAN AP, WILLIS LR, LUFT FC (1983) Renal afferent arteriole in the spontaneously hypertensive rat. Hypertension 5: 8-16

GATTONE VH, LUFT FC, EVAN AP (1984) Renal afferent and efferent arterioles of the rabbit. Am J Physiol 247: F219-F228

GAUNT R, EVERSOLE WJ (1949) Notes on the history of the adrenal cortical problem. Ann NY Acad Sci 50: 511-521

GAUNT R, GAUNT JH, TOBIN CE (1934/35) Colony differences in survival of adrenalectomized rats. Proc Soc Exp Biol Med 32: 888-892

GAVRAS H, BROWN JJ, LEVER AF, ROBERTSON JIS (1970) Changes of renin in individual glomeruli in response to variations of sodium intake in the rabbit. Clin Sci 38: 409-414

GENEST J, ROJO-ORTEGA JM, KUCHEL O, BOUCHER R, NOWACZYNSKI W, LEFEBVRE R, CHRETIEN M, CANTIN J, GRANGER P (1975) Malignant hypertension with hypokalemia in a patient with renin-producing pulmonary carcinoma. Trans Assoc Am Physicians 88: 192-201

GERBER JG, NIES AS (1983) The role of histamine receptors in the release of renin. Br J Pharmacol 79: 57-61

GERBER JG, NIES AS (1986) Renal vasoconstrictor response to hypertonic saline in the dog: effects of prostaglandins, indomethacin and theophylline. J Physiol (Lond) 380: 35-43

GERBER JG, BRANCH RA, NIES AS, HOLLIFIELD JW, GERKENS JF (1979a) Influence of hypertonic saline on canine renal blood flow and renin release. Am J Physiol 237: F441-F446

GERBER JG, KELLER RT, NIES AS (1979b) Prostaglandins and renin release. The effect of PGI2, PGE2, and 13,14-dihydro PGE2 on the baroreceptor mechanism of renin release in the dog. Circ Res 44: 796-799

GERBER JG, NIES AS, OLSEN RD (1981a) Control of canine renin release: Macula densa requires prostaglandin synthesis. J Physiol (Lond) 319: 419-429

GERBER JG, OLSON RD, NIES AS (1981b) Interrelationship between prostaglandins and renin release. Kidney Int 19: 816-821

GERSTHEIMER FP, SIMON T, KÖLB J, HÖPKER W, METZ J (1987) Computer-assisted morphometric study of the innervation of the guinea pig heart. Histochemistry 88: 543-549

GERVITZ RK, HIRAICHI E, FICHMAN M, LAVRAS AAC (1987) The renin-angiotensin system in the snake *Bothrops jarara-*

ca (Serpentes, Crotalinae). Comp Biochem Physiol [A] 86 (3): 503-507

Gherardi GJ, Arya S, Hickler RB (1974) Juxtaglomerular body tumor: a rare occult but curable cuase of lethal hypertension. Hum Pathol 5: 236-240

Giacomelli F, Wiener J (1976) Specialized junction in the distal convoluted tubule of rat kidney. Anat Rec 185: 197-208

Gibbons GH, Dzau VJ, Farhi ER, Barger AC (1984) Interaction of signals influencing renin release. Annu Rev Physiol 46: 291-308

Gill JR, Bartter FC (1978) Evidence for a prostaglandin-independent defect in chloride reabsorption in the loop of Henle as a proximal cause of Bartter's syndrome. Am J Med 65: 766-772

Gillies A, Morgan T (1978) Renin content of individual juxtaglomerular apparatuses and the effect of diet, changes in nephron flow rate and in vitro acidification on the renin content. Pflügers Arch 375: 105-110

Gillies A, Morgan T (1982) Activity of renin in the juxtaglomerular apparatus. Kidney Int 22 [Suppl 12]: 67-72

Gillies A, Morgan T, Fitzgibbon W (1982) Changes in "active" and "inactive" renin in the juxtaglomerular apparatuses of rat nephrons and plasma induced by different salt intake. Pflügers Arch 393: 308-312

Gilman AG (1984) Guanine nucleotide-binding regulatory proteins and dual control of adenylate cyclase. J Clin Invest 73: 1-4

Gilmore JP, Cornish KG, Rogers SD, Joyner WL (1980) Direct evidence for myogenic autoregulation of the renal microcirculation in the hamster. Circ Res 47: 226-230

Ginesi LM, Noble AR (1984) Secretion of active and inactive renin by rabbit kidney cortex slices: effects of verapamil, flunarizine and A23187. Clin Exp Hypertens 6: 1331-1343

Ginesi LM, Munday KA, Noble AR (1981) Active and inactive renin release by rabbit kidney cortex slices: effect of calcium and potassium. J Physiol (Lond) 315: 41P-42P

Ginesi LM, Munday KA, Noble AR (1983) Direct effect of lithium on active and inactive renin secretion. Br J Pharmacol 78: 3-4

Glaumann H, Ericsson JLE, Marzella L (1981) Mechanism of intralysosomal degradation with special reference to autophagocytosis and heterophagocytosis of cell organelles. Int Rev Cytol 73: 149-182

Glenner GG, Folk JE (1961) Glutamyl peptidases in rat and guinea-pig kidney slices. Nature 192: 338-340

Glenner GG, McMillan PJ, Folk JE (1962) A mammalian peptidase specific for the hydrolysis of N-terminal α-L-glutamyl and aspartyl residues. Nature 194: 867

Glorioso N, Dessi-Fulgheri P, Madeddu P, Gois G, Palermo M, Cocco F, Dettori S, Rapelli A (1982) Active and inactive renin after a single dose of captopril in hypertensive subjects. Am J Cardiol 49: 1552-1557

Godard C, Vallotton MB, Broyer M (1973) Plasma renin activity in segmental hypoplasia of the kidneys with hypertension. Nephron 11: 308-317

Goldberg LI (1972) Cardiovascular and renal actions of dopamine: potential clinical applications. Pharmacol Rev 24: 1-29

Goldberg LI, Volkman PH, Kohli JD (1978) A comparison of the vascular dopamine receptor with other dopamine receptors. Annu Rev Pharmacol Toxicol 18: 57-79

Goldblatt H, Lynch J, Hanzal RF, Summerville WW (1934) Studies on experimental hypertention. I. The production of persistent elevation of systolic blood pressure by means of renal ischemia. J Exp Med 59: 347-380

Goldstone R, Horton R, Carlson EJ, Hsueh WA (1983) Reciprocal changes in active and inactive renin after converting enzyme inhibition in normal men. J Clin Endocrinol Metab 56: 264-268

Golgi C (1889) Annotazioni intorno all'Istologia dei reni dell'uomo e di altri mammiferi e sull'Istogenesi dei canalicoli oriniferi. Atti R Acad Naz Lincei Rendiconti 5: 334-342

Gomba SZ, Soltesz MB (1969) Histochemistry of lysosomal enzymes in juxtaglomerular cells. Experientia 25: 513

Gomba SZ, Soltesz MB, Szokoly V (1967) Studies of the histochemistry of phosphatase enzymes in the juxtaglomerular complex. Histochemie 8: 264-274

Gomba SZ, Soltesz MB, Szokoly V, Endes P (1970) Histochemical characterization of the juxtaglomerular apparatus. Proc 4th Int Congr Nephrol, Stockholm (1969) 2: 7-11, Karger, Basel

Gomperts BD, Barrowman MM, Cockcroft S (1986) Dual role for guanine nucleotides in stimulus-secretion coupling. Fed Proc 45: 2156-2161

Gonzalez E, Salomonsson M, Müller-Suhr C, Persson AEG (1988) NaCl transport and osmotic water permeability of macula densa cells contained in isolated and perfused rabbit kidney tubules. In: Persson AEG, Boberg U (eds) The juxtaglomerular apparatus, 11th Fernström Symposium. Elsevier, Amsterdam, pp 97-119

Goodman D, Vagnucci AH, Hartroft PM (1969) Pathogenesis of Bartter's syndrome. N Engl J Med 281: 1435-1439

Goodwin FJ, Ledingham JGG, Laragh JH (1970) The effects of prolonged administration of vasopressin and oxytocin on renin, aldosterone and sodium balance in normal man. Clin Sci 39: 641-651

Goormaghtigh N (1932) Les segments neuro-myo-artériels juxta-glomérulaires du rein. Arch Biol 43: 575-591

Goormaghtigh N (1937) L'appareil neuro-myo-artériel juxta-glomérulaire du rein; ses réactions en pathologie et ses rapports avec le tube urinifère. CR Soc Biol (Paris) 124: 293-296

Goormaghtigh N (1939) Existence of an endocrine gland in the media of the renal arterioles. Proc Soc Exp Biol Med 42: 688-689

Goormaghtigh N (1940) Les cellules afibrillaires artériolaires dans l'ischémie rénale chez le chien. Rev Belg Sci Med 12: 85-107

Goormaghtigh N (1942) "La fonction endocrine des artérioles rénales. Son rôle dans la pathogénie de l'hypertension artérielle". Librairie Fonteyn, Louvain

Goormaghtigh N (1945a) La fonction endocrine des artérioles rénales. Rev Belg Sci Med 16: 1-155

Goormaghtigh N (1945b) Facts in favour of an endocrine function of the renal arterioles. J Pathol Bacteriol 57: 392-393

Goormaghtigh N, Grimson KS (1939) Vascular changes in renal ischemia: cell mitosis in the media of arteries. Proc Soc Exp Biol Med 42: 227-228

Goormaghtigh N, Handovsky H (1938) Effect of vitamin D_2 (calciferol) on the dog. Arch Pathol 26: 1144-1182

Göransson A, Isaksson B, Sjöquist M (1986a) Whole kidney response to reduced arterial pressure during converting enzyme inhibition in the rat. Renal Physiol 9: 287-301

Göransson A, Sjöquist M, Ulfendahl HR (1986b) Superficial and juxtamedullary nephron function during converting enzyme inhibition. Am J Physiol 251: F25-F33

GORDON MB, MOORE TJ, DLUHY RG, WILLIAMS GH (1983) Dopaminergic blockade of the renin-angiotensin-aldosterone system: effect of high and low sodium intakes. Clin Endocrinol 19: 415-425

GORGAS K (1978a) Structure and innervation of the juxtaglomerular apparatus of the rat. Adv Anat Embryol Cell Biol 54 (2): Springer, Berlin, Heidelberg, New York

GORGAS K (1978b) The renal juxtaglomerular apparatus. In: COUPLAND ME, FORSSMANN WG (eds) Peripheral neuroendocrine interaction. Springer, Berlin, Heidelberg, New York, pp 144-152

GOTO T, ABE K, OTSUKA Y, ITOH T, IMAI Y, SATOH M, OMATA K, YOSHINAGA K (1980) Active and inactive renin after SQ 14225 (Captopril) administration. Tohoku J Exp Med 132: 363-364

GOTO T, ABE K, TSUNODA K, SEINO M, YASUJIMA M, IMAI Y, CHIBA S, SATO M, HARUYAMA T, OMATA K, et al. (1986) The changes in active and inactive renin induced by various maneuvres in hypertensive patients. Tohoku J Exp Med 149: 169-181

GOULD AB, GOODMAN S, DEWOLF R, ONESTI G, SWARTZ C (1980) Interrelation of the renin system and erythropoietin in rats. J Lab Clin Med 96: 523-534

GRANGER P, ROJO-ORTEGA JM, PEREZ SC, BOUCHER R, GENEST J (1971) The renin-angiotensin system in newborn dogs. Can J Physiol Pharmacol 49: 134-138

GRANGER P, DAHLHEIM H, THURAU K (1972) Enzyme activities of the single juxtaglomerular apparatus in the rat kidney. Kidney Int 1: 78-88

GREGORY LC, REID IA (1984) Effect of renal denervation on the suppression of renin secretion by vasopressin in conscious dogs. Am J Physiol 247: F881-F887

GRENIER FC, ROLLINS TE, SMITH WL (1981) Kinin-induced prostaglandin synthesis by renal papillary collecting tubule cells in culture. Am J Physiol 241: F94-F104

GRINSTEIN S, MEULEN JV, FURUYA W (1982) Possible role of H+-alkali cation countertransport in secretory granule swelling during exocytosis. FEBS Lett 148: 1-4

GRISE C, BOUCHER R, THIBAULT G, GENEST J (1981) Formation of angiotensin II by tonin from partially purified human angiotensinogen. Can J Biochem 59: 250-255

GRONOW GHJ, COHEN JJ (1984) Substrate support for renal functions during hypoxia in the perfused rat kidney. Am J Physiol 247: F618-F631

GROSS DM, BARAJAS L (1978) Studies on the characterization of isolated renin-containing granules: the storage form of renin. J Med 9: 53-66

GROSS F (1958) Renin und Hypertensin, physiologische oder pathologische Wirkstoffe? Klin Wochenschr 36: 693-706

GROSS F (1964) In: BAULIEU EE, ROBEL P (eds) Aldosterone. Blackwell, Oxford, p 307

GROSS F, BRUNNER H, ZIEGLER M (1965) Renin-angiotensin system, aldosterone, and sodium balance. Recent Prog Horm Res 21: 119-177

GROSS R, HACKENBERG HM, HACKENTHAL E, KIRCHHEIM H (1981) Interaction between perfusion pressure and sympathetic nerves in renin release by carotid baroreflex in conscious dogs. J Physiol (Lond) 313: 237-250

GROSS R, KAYSER M, SCHRAMM M, TANIEL R, THOMAS G (1985) Cardiovascular effects of the calcium-agonistic dihydropyridine Bay K 8644 in conscious dogs. Arch Int Pharmacodyn Ther 277: 203-216

GÜLLNER HG, BARTTER FC (1979) Participation of substance P in the control of renin release. Life Sci 24: 2449-2454

GÜLLNER HG, BARTTER FC, GILL JR, DICKMAN PS, WILSON CB, TIWARI JL (1983) A sibship with hypokalemic alkalosis and renal proximal tubulopathy. Arch Intern Med 143: 1534-1540

GUTKOWSKA J, CORVOL P, THIBAULT G, GENEST J (1982) Tonin as activator of renin. Can J Biochem 60: 843-846

GUTKOWSKA J, CORVOL P, FIGUEIREDO AFS, INAGAMI T, BOUHNIK J, GENEST J (1984) Kinetic studies of rat renin and tonin on purified rat angiotensinogen. Can J Biochem Cell Biol 62: 137-142

GUYTON AC, LANGSTON JB, NAVAR LG (1964) Theory of renal autoregulation by feedback of the juxtaglomerular apparatus. Circ Res 15: I-187-I-197

HÄBERLE D, BAEYER VH (1983) Characteristics of glomerulotubular balance. Am J Physiol 244: F355-F366

HACKENTHAL E (1988) ANF und Reninsekretion. Untersuchungen an der isoliert perfundierten Rattenniere. In: KREYE VAW, BUSSMANN WD (eds) Atriales natriuretisches Peptid und das kardiovaskuläre System. Dietrich Steinkopff; Darmstadt, pp 41-49

HACKENTHAL E, MÜNTER E (1987) Effects of adenosine receptor agonists in the isolated perfused rat kidney. J Hypertens 5 [Suppl 5]: S 393-S 395

HACKENTHAL E, TAUGNER R (1983) The specificity of angiotensin-antisera. A cautionary note. Histochemistry 77: 201-207

HACKENTHAL E, TAUGNER R (1986) Hormonal signals and intracellular messengers for renin secretion. Mol Cell Endocrinol 47: 1-12

HACKENTHAL E, KOCH C, BERGEMANN T, GROSS F (1972) Partial purification and characterization of a renin-like enzyme from rat submandibular gland. Biochem Pharmacol 21: 2779-2792

HACKENTHAL E, HACKENTHAL E, HOFBAUER KG (1977) No evidence for product inhibition of the renin-angiotensinogen reaction in the rat. Circ Res 41 [Suppl 2]: 49-54

HACKENTHAL E, HACKENTHAL R, HILGENFELDT U (1978) Isorenin, pseudorenin, cathepsin D and renin. A comparative enzymatic study of angiotensin-forming enzymes. Biochim Biophys Acta 522: 574-588

HACKENTHAL E, POULSEN P, TAUGNER C, TAUGNER R (1980a) Localization of renin in mouse kidney by the peroxidase-antiperoxidase technique. In: GROSS F, VOGEL G (eds) Enzymatic release of vasoactive peptides. Raven, New York, pp 67-72

HACKENTHAL E, SCHWERTSCHLAG U, SEYBERTH HW (1980b) Prostaglandins and renin release. Studies in the isolated perfused rat kidney. Prog Biochem Pharmacol 17: 98-107

HACKENTHAL E, KIRCHHEIM H, TAUGNER R (1983a) Renal blood flow and the renin-angiotensin system. Proc Int Union Physiol Sci 15: 436

HACKENTHAL E, SCHWERTSCHLAG U, TAUGNER R (1983b) Cellular mechanisms of renin release. Clin Exp Hypertens [A] 5: 975-993

HACKENTHAL E, AKTORIES K, JAKOBS KH (1985a) Pertussis toxin attenuates the inhibition of renin release by angiotensin II. Mol Cell Endocrinol 42: 113-117

HACKENTHAL E, AKTORIES K, JAKOBS KH (1985b) Mode of inhibition of renin release by angiotensin II. J Hypertens 3 [Suppl 3]: 263-265

HACKENTHAL E, LANG RE, BÜHRLE CP (1985c) Atrial natriuretic factor stimulates renin release from the isolated rat kidney. J Hypertens 3 [Suppl 3]: 323-325

HACKENTHAL E, AKTORIES K, JAKOBS KH, LANG RE (1987a) Neuropeptide Y inhibits renin release by a pertussis toxin sensitive mechanism. Am J Physiol 252: F543-F550

HACKENTHAL E, LANG RE, BÜHRLE CP (1987b) Atrial natriuretic peptides and renin secretion. In: Diuretics: basic, pharmacological, and clinical aspects. ANDREUCCI VE, DAL CANTON (eds) Martinus Nijhoff Publ Boston, pp 137-139

HACKENTHAL E, METZ R, BÜHRLE CP, TAUGNER R (1987c) Intrarenal and intracellular distribution of renin and angiotensin. Kidney Int 31 [Suppl 20]: 4-17

HALEY DP, SARAFIAN M, BULGER RE, DOBYAN DC, EKNOYAN G (1987) Structural and functional correlates of effects of angiotensin-induced changes in rat glomerulus. Am J Physiol 253: F111-F119

HALL ER, KATO J, ERDÖS EG, ROBINSON CJG, OSHIMA G (1976) Angiotensin I-converting enzyme in the nephron. Life Sci 18: 1299-1303

HALL JE (1982) Regulation of renal hemodynamics. In: GUYTON AC, HALL JE (eds) Cardiovascular physiology, vol 4. University Park Press, Baltimore, pp 245-321

HALL JE (1986a) Control of sodium excretion by angiotensin II: intrarenal mechanisms and blood pressure regulation. Am J Physiol 250: R960-R972

HALL JE (1986b) Regulation of glomerular filtration rate and sodium excretion by angiotensin II. Fed Proc 45: 1431-1437

HALL JE, GRANGER JP (1983) Renal hemodynamic actions of angiotensin II: interaction with tubuloglomerular feedback. Am J Physiol 245: R166-R173

HALL JE, GRANGER JP (1986a) Renal hemodynamics and arterial pressure during chronic intrarenal adenosine infusion in conscious dogs. Am J Physiol 250: F32-F39

HALL JE, GRANGER JP (1986b) Adenosine alters glomerular filtration control by angiotensin II. Am J Physiol 250: F917-F923

HALL JE, GUYTON AC, COWLEY AWJ (1977) Dissociation of renal blood flow and filtration rate autoregulation by renin depletion. Am J Physiol 232: F215-F221

HALL JE, GUYTON AC, TRIPPODO NC, LOHMEIER TE, MCCAA RE, COWLEY AWJ (1977a) Intrarenal control of electrolyte excretion by angiotensin II. Am J Physiol 232: F538-F544

HALL JE, GUYTON AC, JACKSON TE, COLEMAN TG, LOHMEIER TE, TRIPPODO NC (1977b) Control of glomerular filtration rate by renin-angiotensin system. Am J Physiol 233: F366-F372

HALL JE, COLEMAN TG, GUYTON AC, KASTNER PR, GRANGER JP (1981) Control of glomerular filtration rate by circulating angiotensin II. Am J Physiol 241: R190-R197

HALL JE, GRANGER JP, HESTER RL (1985) Interactions between adenosine and angiotensin II in controlling glomerular filtration. Am J Physiol 248: F340-F346

HALL JE, MIZELLE HL, WOODS LL (1986) The renin-angiotensin system and long-term regulation of arterial pressure. J Hypertens 4: 387-397

HAMADA M, BURMESTER HA, GRACI KA, FROHLICH ED, COLE FE (1987) Atrial natriuretic peptide binding properties of purified rat glomerular membranes. Life Sci 40: 1731-1737

HAMMERSEN G, KARSUNKY KP, FISCHINGER J, ROSENTHAL J, TAUGNER R (1971) Einfluß der Natriumkonzentration auf die Reninabgabe aus Nierenrindenschnitten und isolierten Glomerula. Pflügers Arch 328: 344-355

HANAHAN DJ (1986) Platelet activating factor: A biologically active phosphoglyceride. Ann Rev Biochem 55: 483-509

HAND AR, OLIVER C (1981) The Golgi apparatus: Protein transport and packaging in secretory cells. Methods Cell Biol 23: 137-153

HANDA RK, JOHNS EJ (1985) Interaction of the renin-angiotensin system and the renal nerves in the regulation of rat kidney function. J Physiol (Lond) 369: 311-321

HANDA N, FUKUNAGA R, YONEDA S, KIMURA K, KAMADA T, ICHIKAWA Y, TAKAHA M, SONODA T, TOKUNAGA K, KURODA C, ONISHI S (1986) State of systemic hemodynamics in a case of juxtaglomerular cell tumor. Clin Exp Hypertens [A] 8: 1-19

HANNA W, TEPPERMAN B, LOGAN AG, ROBINETTE MA, COLAPINTO R, PHILLIPS MJ (1979) Juxtaglomerular cell tumor (reninoma) with paroxysmal hypertension. CMA J 120: 957-959

HANNER RH, RYAN GB (1980) Ultrastructure of the renal juxtaglomerular complex and peripolar cells in the axolotl *(Ambystoma mexicanum)* and toad *(Bufo marinus)*. J Anat 130: 445-455

HARA M, MEYER D (1975) The size of the juxtaglomerular apparatus in glomerulonephritis with the nephrotic syndrome. A morphometrical study of renal biopsies. Virchows Arch [A] 367: 1-14

HARADA E, LESTER GE, RUBIN RP (1979) Stimulation of renin secretion from the intact kidney and from isolated glomeruli by the calcium ionophore A 23187. Biochim Biophys Acta 583: 20-27

HARDER A, PAKALAPATI G, DEBUCH H (1981) Influence of chloroquine treatment on enzymes and phospholipids from rat liver cell fractions. Biochem Biophys Res Commun 99: 9-15

HARDMAN JA, HORT YJ, CATANZARO DF, TELLAM JTl, BAXTER JD, MORRIS BJ, SHINE J (1984) Primary structure of the human renin gene. DNA 3: 457-468

HARGENS AR, TUCKER BJ, BLANTZ RC (1977) Renal lymph protein in the rat. Am J Physiol 233: F269-F273

HARRIS PJ, NAVAR LG (1985) Tubular transport responses to angiotensin. Am J Physiol 248: F621-F630

HARRIS PJ, YOUNG JA (1977) Dose-dependent stimulation and inhibition of proximal tubular sodium reabsorption by angiotensin II in the rat kidney. Pflügers Arch 367: 295-297

HARRIS PJ, NAVAR LG, PLOTH DW (1984) Evidence for angiotensin-stimulated proximal tubular fluid reabsorption in normotensive and hypertensive rats: effect of acute administration of captopril. Clin Sci 66: 541-544

HARRIS PJ, THOMAS D, MORGAN TO (1987) Atrial natriuretic peptide inhibits angiotensin-stimulated proximal tubular sodium and water reabsorption. Nature 326: 697-698

HARTMAN FA (1924) The general physiology and experimental pathology of the suprarenal glands. In: BARKER LF, HOSKINS RG, MOSENTHAL HO (eds) Endocrinology and metabolism. Appleton, New York, pp 101-125

HARTROFT PM (1963) Juxtaglomerular cells. Circ Res 12: 525-538

HARTROFT PM (1966) The juxtaglomerular complex. Annu Rev Med 17: 113-122

HARTROFT PM (1968) The juxtaglomerular complex as an endocrine gland. In: BLOODWORTH JMB (ed) Endocrine pathology. Williams and Wilkins, Baltimore, pp 641-677

HARTROFT PM, EDELMAN R (1960) Renal juxtaglomerular cells in sodium deficiency. Effect of hypophysectomy: localization of renin by the fluorescent antibody technique.

In: MOYER F (ed) Edema. Saunders, Philadelphia, pp 63-68

HARTROFT PM, HARTROFT WS (1953) Studies on renal juxtaglomerular cells I. Variations produced by sodium chloride and desoxycorticosterone acetate. J Exp Med 97: 415-427

HARTROFT P, NEWMARK LN (1961) Electron microscopy of renal juxtaglomerular cells. Anat Rec 139: 185-200

HARTROFT PM, SUTHERLAND LE, HARTROFT WS (1964) Juxtaglomerular cells as the source of renin: further studies with the fluorescent antibody technique and the effect of passive transfer of antirenin. Can Med Assoc J 90: 163-166

HASEGAWA H, TATEISHI H, MASSON GMC (1973) Evidence for an angiotensinogen-stimulating factor after nephrectomy. Can J Physiol Pharmacol 51: 563-566

HASEGAWA Y, KHOSLA MC, STEPHENS GA, BUMPUS FM (1986) Characterization of angiotensin I in turtle plasma. Hypertension 8 [Suppl 1]: 101-104

HASSID A, KONIECZKOWSKI M, DUNN MJ (1979) Prostaglandin synthesis in isolated rat kidney glomeruli. Proc Natl Acad Sci USA 76: 1155-1159

HATT PY (1961) Activité sécrétoire de la paroi des artérioles rénales. Démonstration cytologique au cours de l'ischémie rénale expérimentale. C R Acad Sci (Paris) 252: 1851-1853

HATT PY (1963) L'appareil juxtaglomérulaire dans l'hypertension artérielle expérimentale. Rev Fr Etudes Clin Biol 8: 358-365

HATT PY (1967) The juxtaglomerular apparatus. In: DALTON AJ, HAGUENAU F (eds) Ultrastructure of the kidney. Academic, New York, pp 101-141

HATT PY, DONTCHEFF A (1959) Contribution de la microscopie électronique à l'étude du mécanisme de l'hypertension artérielle expérimentale d'origine rénale chez le rat. Arch Mal Coeur 52: 490-503

HATT PY, DVOJAKOVIC M, CORNET P (1962) Contribution de la microscopie électronique à l'étude du mécanisme de l'hypertension artérielle experimentale d'origine rénale II. L'ischémie rénale chez le lapin. Pathol Biol (Paris) 10: 23-40

HAUGER-KLEVENE JH (1970) ACTH, cyclic AMP, dexamethasone and actinomycin D effect on renin release. Acta Physiol Pharmacol Latinoam 20: 373-381

HAUGER-KLEVENE JH (1970) High plasma renin activity in an oat cell carcinoma: a renin-secreting carcinoma? Cancer 26: 1112-1114

HAYARI Y, KUKULANSKY T, GLOBERSON A (1984) Effects of in vivo indomethacin treatment in aging mice. Prostaglandins Leukotrienes Med 15: 69-78

HAYASHI T, NAKAYAMA T, NAKAJIMA T, SOKABE H (1978) Comparative studies on angiotensins. V. Structure of angiotensin formed by the kidney of Japanese goosefish and its identification by dansyl method. Chem Pharm Bull (Tokyo) 26: 215-219

HEACOX R, HARVEY AM, VANDER AJ (1967) Hepatic inactivation of renin. Circ Res 21: 149-152

HEAGERTY AM, OLLERENSHAW JD (1987) The phosphoinositide signalling system and hypertension. J Hypertens 5: 515-524

HEALY DP, PRINTZ MP (1984) Distribution of immunoreactive angiotensin II, angiotensin I, angiotensinogen, and renin in the central nervous system of intact and nephrectomized rats. Hypertension 6 [Suppl 1]: 130-136

HEALY JK, DOUGLAS JB, ARNOLD JE (1969) The effect of angiotensin on isolated rabbit renal tubules. Clin Sci 37: 583-592

HECHTER O (1945) Concerning the hypersensitivity of adrenalectomized rats to vascular stress. Endocrinology 36: 77-87

HELBER A, MEYER D, SCHÜRHOLZ J, BOHLE A (1970) Zur Struktur und Funktion des juxtaglomulären Apparates. Quantitative lichtoptische Untersuchungen an juxtaglomerulärem Apparat und Macula densa der Nieren von Patienten mit primärem Hyperaldosteronismus, Nierenarterienstenose, Pseudo-Bartter-Syndrom und Morbus Addison. Dtsch Med Wochenschr 95: 2280-2288

HELENIUS A, MELLMAN I, WALL D, HUBBARD A (1983) Endosomes. Trends Biochem Sci 8: 245-250

HELLER J, HORACEK V (1986) Angiotensin II: Preferential efferent constriction? Renal Physiology 9: 357-365.

Helmchen U (1987) Pathologie der renalen Ischämie und des akuten Nierenversagens. Z Kardiol 76 [Suppl 4]: 99-105

Helmchen U, Kneissler U (1976) Role of the renin-angiotensin system in renal hypertension. An experimental approach. In: Current topics in pathology. Glomerulonephritis. GRUNDMANN E, KIRSTEN WH (Eds) Vol 61: 203-238 Springer Berlin, Heidelberg, New York

HELMCHEN U, KNEISSLER U, HELMCHEN U, BOHLE A (1977) Three-dimensional electron microscopy of the structure of the normal mesangium of the rat glomerulus. Kidney Int 11: 215-216 (Abstract)

HELMCHEN U, GRÖNE HJ, KIRCHERTZ EJ, BADER H, BOHLE RM, KNEISSLER U, KHOSLA MC (1982) Contrasting renal effects of different antihypertensive agents in hypertensive rats with bilaterally constricted renal arteries. Kidney Int 22 [Suppl 12] S-198-S-205

HENDERSON RM (1974) Types of cell contacts in arterial smooth muscle. Experientia 31: 103-105

HENDERSON RM (1975) Cell-to-cell contacts. In: DANIEL EE, PATON DM (eds) Methods in pharmacology. Smooth Muscle vol 3: Plenum Press, New York, London, pp 47-77

HENRICH WL (1981) Role of prostaglandins in renin secretion. Kidney Int 19: 822-830

HENRICH WL, CAMPBELL WB (1983) The systemic β-adrenergic pathway to renin secretion: relationships with the renal prostaglandin system. Endocrinology 113: 2247-2254

HENRICH WL, CAMPBELL WB (1984) Relationship between PG and β-adrenergic pathways to renin release in rat renal cortical slices. Am J Physiol 247: E343-E348

HENRICH WL, CAMPBELL WB (1986) Importance of calcium in renal renin release. Am J Physiol 251: E98-E103

HENRICH WL, MCALISTER BS, SMITH PB, LIPTON J, CAMPBELL WB (1987) Direct inhibitory effect of atriopeptin III on renin release in primate kidney. Life Sci 41: 259-264

HENRY LP, KEYL MJ, BELL RD (1969) Flow and protein concentration of capsular renal lymph in the conscious dog. Am J Physiol 217: 411-413

HENRY DP, AOI W, WEINBERGER MH (1977) The effects of dopamine on renin release in vitro. Endocrinology 101: 279-283

HERBERT SC, ANDREOLI TE (1984) Control of NaCl transport in the thick ascending limb. Am J Physiol 246: F745-F756

HERRMANN HC, MORRIS BJ, REID IA (1980) Effect of angiotensin II and sodium depletion on angiotensinogen production. Am J Physiol 238: E145-E149

HESS R (1965) Arylamidase activity related to angiotensinase. Biochim Biophys Acta 99: 316-324

HESS R, SCARPELLI DG, PEARSE AGE (1958) The cytochemical localization of oxidative enzymes. II. Pyridine nucleotide-linked dehydrogenases. J Biophys Biochem Cytol 4: 753-760

HESSE B, NIELSEN I (1977) Suppression of plasma renin activity by intravenous infusion of antidiuretic hormone in man. Clin Sci Mol Med 52: 357-360

HESSE IFA, JOHNS EJ (1985) The role of alpha adrenoceptors in the regulation of renal tubular sodium reabsorption and renin secretion in the rabbit. Br J Pharmacol 84: 715-724

HEYERAAS KJ, AUKLAND K (1987) Interlobular arterial resistance: Influence of renal arterial pressure and angiotensin II. Kidney Int 31: 1291-1298

HEYERAAS TØNDER KJ, AUKLAND K (1979/80) Interlobular arterial pressure in the rat kidney. Renal Physiol 2: 214-221

HIGGINS JT, DAVIS JO, URQUHART J (1964) Demonstration by pressor and steroidogenic assays of increased renin in lymph of dogs with secondary hyperaldosteronism. Circ Res 14: 218-227

HILGENFELDT U, HACKENTHAL E (1979) Purification and characterization of rat angiotensinogen. Biochim Biophys Acta 579: 375-385

HILL PA, COGHLAN JP, SCOGGINS BA, RYAN GB (1983) Ultrastructural changes in the sheep renal juxtaglomerular apparatus in response to sodium depletion or loading. Pathology 15: 463-473

HINKO A, FRANCO-SAENZ R, MULROW PJ (1984) Biphasic alteration of renin release by calcium. Proc Soc Exp Biol Med 175: 454-457

HINTZE TH, CURRIE MG, NEEDLEMAN P (1985) Atriopeptins: renal-specific vasodilators in conscious dogs. Am J Physiol 248: H587-F591

HIRATA Y, ISHII M, SUGIMOTO T, MATSUOKA H, ISHIMITSU T, ATARASHI K, SUGIMOTO T, MIYATA A, KANGAWA K, MATSUO H (1987) Relationship between the renin-aldosterone system and atrial natriuretic polypeptide in rats. Clin Sci 72: 165-170

HIROSE M, ARAKAWA K, KIKUCHI M, KAWASAKI I, OMOTO T, KATO H, NAGAYAMA T (1974) Primary reninism with renal hamartomatous alteration. JAMA 230: 1288-1292

HIROSE S, KIM SJ, MIYAZAKI H, PARK YS, MURAKAMI K (1985) In vitro biosynthesis of human renin and identification of plasma inactive renin as an activation intermediate. J Biol Chem 260: 16400-16405

HIRSCH JG, FEDORKO ME, COHN ZA (1968) Vesicle fusion and formation at the surface of pinocytic vacuoles in macrophages. J Cell Biol 38: 629-632

HIRST GDS, NEILD TO (1978) An analysis of excitatory junctional potentials recorded from arterioles. J Physiol (Lond) 280: 87-104

HIRUMA M, IKEMOTO F, YAMAMOTO K (1986) Rat atrial natriuretic factor stimulates renin release from renal cortical slices. Eur J Pharmacol 125: 151-153

HISA H, SATOH S (1983) Effects of a non-vasoconstrictor dose of phenylephrine on prostaglandin E2 and renin release in anaesthetized dogs. Clin Exp Pharmacol Physiol 10: 177-180

HISA H, TAKAHASHI K, SATOH S (1984) Evidence for prostaglandin-independent mechanisms in renin release mediated by alpha adrenoceptors during renal nerve stimulation in anesthetized dogs. J Pharmacol Exp Ther 229: 547-550

HO BYM, SHAM JSK, CHIU KW (1984) The vasopressor action of the renin-angiotensin system in the rat snake, *Ptyas korros*. Gen Comp Endocrinol 56: 313-320

HOBART PM, FOGLIANO M, O'CONNOR BA, SCHAEFER IM, CHIRGWIN JM (1984) Human renin gene: structure and sequence analysis. Proc Natl Acad Sci USA 81: 5026-5030

HOCK D, FORSSMANN WG (1984) Zur peptidergen Innervation der Niere. Verh Anat Ges 78: 463-464

HOFBAUER KG, ZSCHIEDRICH H, RAUH W, GROSS F (1973) Conversion of angiotensin I into angiotensin II in the isolated perfused rat kidney. Clin Sci 44: 447-456

HOFBAUER KG, ZSCHIEDRICH H, HACKENTHAL E, GROSS F (1974) Function of the renin-angiotensin system in the isolated perfused rat kidney. Circ Res 34/35 [Suppl 1]: 193-202

HOFBAUER KG, ZSCHIEDRICH H, GROSS F (1976) Regulation of renin release and intrarenal formation of angiotensin. Studies in the isolated perfused rat kidney. Clin Exp Pharmacol Physiol 3: 73-93

HOFBAUER KG, KONRADS A, SCHWARZ K, WERNER U (1982) Role of cyclic AMP in the regulation of renin release from the isolated perfused rat kidney. Klin Wochenschr 56 [Suppl 1]: 51-59

HÖKFELT T, SCHULTZBERG M, ELDE R, NILSSON G, TERENIUS L, SAID SI, GOLDSTEIN M (1978) Peptide neurons in peripheral tissues including the urinary tract: immunohistochemical studies. Acta Pharmacol Toxicol 43 [Suppl 2]: 79-89

HOLDAAS H, DIBONA GF, KIIL F (1981) Effect of low-level renal nerve stimulation on renin release from nonfiltering kidneys. Am J Physiol 241: F156-F161

HOLDAAS H, LANGARD O, EIDE I, KIIL F (1981) Mechanism of renin release during renal nerve stimulation in dogs. Scand J Clin Lab Invest 41: 617-625

HOLDAAS H, LANGARD O, EIDE I, KIIL F (1982) Conditions for enhancement of renin release by isoproterenol, dopamine, and glucagon. Am J Physiol 242: F267-F273

HOLLIFIELD JW, PAGE DL, SMITH C, MICHELAKIS AM, STAAB E, RHAMY R (1975) Renin-secreting clear cell carcinoma of the kidney. Arch Intern Med 135: 859-864

HOLST DV (1972) Renal failure as the cause of death in *Tupaia belangeri* exposed to persistent social stress. J Comp Physiol 78: 236-273

HOLTZMAN E (1976) Lysosomes: A survey. Springer, Wien, New York

HOLTZMAN E (1981) Membrane circulation: An overview. Methods Cell Biol 23: 379-397

HOPE A, TYSSEBOTN I (1983) The effect of water deprivation on local renal blood flow and filtration in the laboratory rat. Circ Shock 11: 175-186

HOPKINS CR (1969) The fine structural localization of acid phosphatase in the prolactin cell of the teleost pituitary following the stimulation and inhibition of secretory activity. Tissue Cell 1: 653-671

HORIUCHI K, TANAKA H, YAMAMOTO K, UEDA J (1971) Distribution of renin in the dog kidney. Life Sci 10: 727-734

HORKY K, ROJO-ORTEGA JM, RODRIGUEZ J, BOUCHER R, GENEST J (1971) Renin, renin substrate, and angiotensin I-converting enzyme in the lymph of rats. Am J Physiol 220: 307-311

HORNYCH H, BEAUFILS M, RICHET G (1972) The effect of exogenous angiotensin II on superficial and deep glomeruli in the rat kidney. Kidney Int 2: 336-343

HOSIE KF, BROWN JJ, HARPER AM, LEVER AF, MACADAM RF, MACGREGOR J, ROBERTSON JIS (1970) The release of renin into the renal circulation of the anaesthetized dog. Clin Sci 38: 157-174

HOYER JR, SEILER MW (1979) Pathophysiology of Tamm-Horsfall protein. Kidney Int 16: 279-289

HOYER JR, SISSON SP, VERNIER RL (1979) Tamm-Horsfall

glycoprotein: Ultrastructural immunoperoxidase localization in the rat kidney. Lab Invest 41: 168-173

HRUBAN Z, SLESERS A, HOPKINS E (1972) Drug-induced and naturally occurring myeloid bodies. Lab Invest 27: 62-70

HSUEH WA, CARLSON EJ, DZAU VJ (1983) Characterization of inactive renin from human kidney and plasma. Evidence of a renal source of circulating inactive renin. J Clin Invest 71: 506-517

HSUEH WA, GOLDSTONE R, CARLSON EJ, HORTON R (1985) Evidence that the β-adrenergic system and prostaglandins stimulate renin release through different mechanisms. J Clin Endocrinol Metab 61: 399-403

HUANG WC, PLOTH DW, NAVAR LG (1982) Angiotensin-mediated alterations in nephron function in Goldblatt hypertensive rats. Am J Physiol 243: F553-F560

HUANG CL, IVES HE, COGAN M (1985) Atrial natriuretic factor causes glomerular hyperfiltration, blunted tubuloglomerular feedback, and increased glomerular cGMP generation. Am Soc Nephrol 18: 235A (Abstract)

HUBBARD JW, COX RH, LAWLER JE, BLANK ML, SNYDER F (1983) Antihypertensive effects of I-hexadecyl-2-acetyl-sn-glycerol-3-phosphocholine on plasma renin activity and catecholamine responses in spontaneously hypertensive rats. Life Sci 32: 221-232

HUBER GC (1917) On the morphology of the renal tubules of vertebrates. Anat Rec 13: 305-339

HUGHES ML, ICHIKAWA I (1986) Interglomerular heterogeneity of filtration fraction among superficial nephrons: Fact or artifact. Kidney Int 29: 814-819

HUTTON JC (1986) Calcium-binding proteins and secretion. Cell Calcium 7: 339-352

HUXLEY VH, TUCKER VL, VERBURG KM, FREEMAN RH (1987) Increased capillary hydraulic conductivity induced by atrial natriuretic peptide. Circ Res 60: 304-307

ICHIKAWA I (1982) Hemodynamic influence of altered distal salt delivery on glomerular microcirculation. Kidney Int 22 [Suppl 12]: 109-113

IGIC RP, ROBINSON CJG, ERDÖS EG (1975) Angiotensin I converting enzyme in the choroid plexus and in the retina. Sixth Int Congr Pharmacology, Helsinki, Abstracts, p 176

IGIC RP, ROBINSON CJG, ERDÖS EG (1977) Angiotensin I converting enzyme activity in the choroid plexus and in the retina. In: BUCKLEY JP, FERRARIO CM (eds) Central actions of angiotensin and related hormones. Pergamon Oxford, New York, pp 23-27

IIMURA O, SHIMAMOTO K, HOTTA D, NAKATA T, MITO T, KUMAMOTO Y, DEMPO K, OGIHARA T, NARUSE K (1986) A case of adrenal tumor producing renin, aldosterone, and sex steroid hormones. Hypertension 8: 951-956

IKEMOTO F, KAWAMURA M, TAKAORI K, YAMAMOTO K (1980) Factors affecting molecular weight conversion of renin. Jpn Circ J 44: 371-374

IKEMOTO F, TAKAORI K, IWAO H, YAMAMOTO K (1982a) Sulphhydryl oxidation in the mechanism of molecular-weight conversion of renin in dog kidney. Clin Sci 62: 157-162

IKEMOTO F, TAKAORI K, IWAO H, YAMAMOTO K (1982b) Intrarenal localization of renin binding substance in rats. Life Sci 31: 1011-1016

IKEMOTO F, DZAU VJ, HABER E, TAKAORI K, YAMAMOTO K (1983) Immunoaffinity chromatography of canine high-molecular-weight renin: partial purification and characterization. Clin Sci 65: 117-120

IMAGAWA J, MIYAUCHI T, SATOH S (1985) Participation of prostaglandin and adrenergic nervous system in renin release induced by changes in renal arterial pressure in rats. Renal Physiol 8: 140-149

IMAGAWA J, KUROSAWA H, SATOH S (1986) Effects of nifedipine on renin release and renal function in anesthetized dogs. J Cardiovasc Pharmacol 8: 636-640

IMAI T, MIYAZAKI H, HIROSE S, HORI H, HAYASHI T, KAGEYAMA R, OHKUBO H, NAKANISHI S, MURAKAMI K (1983a) Cloning and sequence analysis of cDNA for human renin precursor. Proc Natl Acad Sci USA 80: 7405-7409

IMAI T, MIYAZAKI H, HIROSE S, MURAKAMI K (1983b) Cell-free translation of human renin mRNA. Clin Exp Hypertens [A] 5: 961-968

IMANISHI M, ABI Y, OKAHARA T, YUKIMURA T, YAMAMOTO K (1980) Effects of prostaglandin I2 und E2 on renal hemodynamics and function and renin release. Jpn Circ J 44: 875-882

IMBS JL, SCHMIDT M, SCHWARTZ J (1975) Effect of dopamine on renin secretion in the anesthetized dog. Eur J Pharmacol 33: 151-157

INAGAKI M, HIDAKA H (1984) Two types of calmodulin antagonists: a structurally related interaction. Pharmacolgy 29: 75-84

INAGAMI T (1982) Renin in the brain and neuroblastoma cells: An endogenous and intracellular system. Neuroendocrinology 35: 475-482

INAGAMI T, MURAKAMI K (1980) Prorenin Biomed Res 1: 456-475

INAGAMI T, HIROSE S, MURAKAMI K, MATOBA T (1977) Native form of renin in the kidney. J Biol Chem 252: 7733-7737

INAGAMI T, OKAMOTO H, OHTSUKI K, SHIMAMOTO K, CHAO J, MARGOLIUS HS (1982) Human plasma inactive renin: purification and activation by proteases. J Clin Endocrinol Metab 55: 619-627

INAGAMI T, KAWAMURA M, NARUSE K, OKAMURA T (1986) Localization of components of the renin-angiotensin system within the kidney. Fed PRoc 45: 1414-1419

INGELFINGER JR, PRATT RE, ELLISON K, DZAU VJ (1986) Sodium regulation of angiotensinogen mRNA expression in rat kidney cortex and medulla. J Clin Invest 78: 1311-1315

ITOH S, CARRETERO OA (1985a) Role of the macula densa in renin release. Hypertension 7 [Suppl 1]: 49-54

ITOH S, CARRETERO OA, MURRAY RD (1985a) Renin release from isolated afferent arterioles. Kidney Int 27: 762-767

ITOH S, CARRETERO OA, MURRAY RD (1985b) Possible role of adenosine in the macula densa mechanism of renin release in rabbits. J Clin Invest 76: 1412-1417

ITSKOVITZ HD, CAMPBELL WB (1976) Vasodilators, intrarenal distribution of blood flow, and renal function in isolated perfused canine kidneys. Proc Soc Exp Biol Med 153: 161-165

ITSKOVITZ HD, HILDRETH EA, SELLERS AM, BLAKEMORE WS (1963) The granularity of juxtaglomerular cells in human hypertension. Histologic and clinical correlations. Ann Intern Med 59: 8-23

ITSKOVITZ HD, HEBERT LA, MCGIFF JC (1973) Angiotensin as a possible intrarenal hormone in isolated dog kidneys. Circ Res 32: 550-555

IWAO H, ABE Y, YAMAMOTO K (1974) Effect of intrarenal arterial infusion of calcium on renin release in dogs. Jpn J Pharmacol 24: 482-484

IWAO H, MINAMI T, IKEMOTO F, TAKAORI K, NAKAMURA N, YAMAMOTO K (1982a) High molecular weight renin identified in cytosol fractions of mouse renal cortices. Biochem Biophys Res Commun 106: 933-939

IWAO H, NAKAMURA N, IKEMOTO F, YAMAMOTO K, MIZUHIRA V, ONO M, SUGIURA Y (1982b) Distribution of exogenously administered renin in mouse kidney. Clin Exp Hypertens [A] 4: 2449-2456

IWAO H, NAKAMURA N, IKEMOTO F, YAMAMOTO K (1983a) Subcellular localization of exogenously administered renin in mouse kidney. Jpn Circ J 47: 1198-1202

IWAO H, NAKAMURA N, IKEMOTO F, YAMAMOTO K (1983b) Whole body autoradiographic distribution of exogenously administered renin in mice. J Histochem Cytochem 31: 776-782

IWAYAMA T (1971) Nexuses between areas of the surface membrane of the same arterial smooth muscle cell. J Cell Biol 49: 521-525

JACKSON EK, HERZER WA, ZIMMERMAN JB, BRANCH RA, OATES JA, GERKENS JF (1981) 6-keto-prostaglandin E1 is more potent than prostaglandin I2 as a renal vasodilator and renin secretagogue. J Pharmacol Exp Ther 216: 24-27

JACKSON EK, HERZER WA, ZIMMERMAN JB, OATES JA, BRANCH RA, GERKENS JF (1982) Effects of indomethacin on beta adrenoceptor-stimulated renin release in the dog. J Pharmacol Exp Ther 222: 414-418

JACKSON EK, GERKENS JF, BRASH AR, BRANCH RA (1982) Acute renal artery constriction increases renal prostaglandin I2 biosynthesis and renin release in the conscious dog. J Pharmacol Exp Ther 222: 410-413

JACOBS K, SHOEMAKER C, RUDERSDORF R, NEILL SD, KAUFMAN RJ, MUFSON A, SEEHRA J, JONES SS, HEWICK R, FRITSCH EF, KAWAKITA M (1985) Isolation and characterization of genomic and cDNS clones of human erythropietin. Nature 313: 806-810

JACOBSEN NO, JØRGENSEN F, THOMSEN ÅC (1966) An electron microscopic study of small arteries and arterioles in the normal human kidney. Nephron 3: 17-39

JAKOBS KH, BAUER S, WATANABE Y (1985) Modulation of adenylate cyclase of human platelets by phorbol ester. Impairment of the hormone-sensitive inhibitory pathway. Eur J Biochem 151: 425-430

JELÍNEK J, HACKENTHAL R, HILGENFELDT U, SCHAECHTELIN G, HACKENTHAL E (1986) The renin-angiotensin system in the perinatal period in rats. J Dev Physiol 8: 33-41

JENSEN PK, STEVEN K (1977) Angiotensin II induced reduction of peritubular capillary diameter in the rat kidney. Pflügers Arch 371: 245-250

JINDRA AJ, KVETNANSKY R (1982) Stress-induced activation of inactive renin. J Biol Chem 257: 5997-5999

JOHNS EJ (1985) Role of the renal nerves in modulating renin release during pressure reduction at the feline kidney. Clin Sci 69: 185-195

JOHNS EJ, SINGER B (1973) Effect of propranolol and theophylline on renin release caused by furosemide in the cat. Eur J Pharmacol 23: 67-73

JOHNSON MD, FREESE JW, SCHMITT DE (1984) Effects of a β_1-adrenoceptor agonist, prenalterol, on renal function and renin secretion rate in anesthetized dogs. J Cardiovasc Pharmacol 6: 627-633

JOHNSON PC (1980) The myogenic response. In: The cardiovascular system, vol 2. Vascular smooth muscle. American Physiological Society, Bethesda, pp 409-442 (Handbook of physiology, sect 2)

JOHNSON PC (1986) Autoregulation of blood flow. Circ Res 59: 483-495

JOHNSTON CI, DAVIS JO, HARTROFT PM (1967) Renin-angiotensin system, adrenal steroids and sodium depletion in a primitive mammal, the american opossum. Endocrinology 81: 633-642

JOHNSTON PA, BERNARD DB, PERRIN NS, ARBEIT L, LIEBERTHAL W, LEVINSKY NG (1981a) Control of rat renal vascular resistance during alterations in sodium balance. Circ Res 48: 728-733

JONES SM, TORRETTI J, WILLIAMS JS, WEINBERGER SF (1979) Regional renin release by the cat kidney in vitro and in vivo. Am J Physiol 237: F188-F195

JONES WR, O'MORCHOE CCC (1983) Ultrastructural evidence for a reabsorptive role by intrarenal veins. Anat Rec 207: 253-262

JORDAN JM, GUNNELS JC (1985) Juxtaglomerular apparatus tumor: a rare but curable cause of secondary hypertension. South Med J 78: 1353-1356

JUAN H, SAMETZ W (1980) Histamine-induced release of arachidonic acid and of prostaglandins in the peripheral vascular bed. Mode of action. Naunyn-Schmiedebergs Arch Pharmacol 314: 183-190

KAGEYAMA R, OHKUBO H, NAKANISHI S (1984) Primary structure of human preangiotensinogen deduced from the cloned cDNA sequence. Biochemistry 23: 3603-3609

KAISSLING B (1980) Ultrastructural organization of the transition from the distal nephron to the collecting duct in the desert rodent *Psammomys obesus*. Cell Tissue Res 212: 475-495

KAISSLING B, KRIZ W (1982) Variability of intercellular spaces between macula densa cells: a transmission electron microscopic study in rabbits and rats. Kidney Int 22 [Suppl 12]: 9-17

KAISSLING B, KRIZ W (1985) Structure-function correlation in transporting epithelia. In: SELDIN DW, GIEBISCH G (eds) The kidney: physiology and pathophysiology vol 1. Raven, New York, pp 307-315

KALINYAK JE, PERLMAN AJ (1987) Tissue-specific regulation of angiotensinogen mRNA accumulation by dexamethasone. J Biol Chem 262: 460-464

KALLSKOG O, LINDBOM LO, ULFENDAHL HR, WOLGAST M (1976) Hydrostatic pressures within the vascular structures of the rat kidney. Pflügers Arch 363: 205-210

KALOYANIDES GJ, BASTRON RD, DIBONA GF (1974) Impaired autoregulation of blood flow and glomerular filtration rate in isolated dog kidney depleted of renin. Circ Res 35: 400-412

KAMITANI T, KATAMOTO M, TATSUMI M, KATSUTA K, ONO T, KIKACHI H, KAMADA S (1984) Mechanisms of the hypotensive effect of 1-O-octadecyl-2-acetylglycero-3-phosphorylcholine. Eur J Pharmacol 98: 357-366

KANETA M, ABE K, ITO T (1981) Appearance of dense granules in the rough endoplasmic reticulum of the juxtaglomerular cells in mice administered with captopril. Cell Tissue Res 220: 219-222

KANETA M, ABE K, ITO T (1983) Juxtaglomerular cells in mice after long-term treatment with captopril. An electron microscopic study. Jpn Circ J 47: 1071-1076

KAPLAN A, FRIEDMAN M (1942) Studies concerning the site of renin formation in the kidney. III. The apparent site of renin formation in the tubules of the mesonephros and metanephros of the hog fetus. J Exp Med 76: 307-316

KAPPELGAARD AM, GIESE J, IBSEN H, NIELSEN MD, RABOL A (1978) Different secretion patterns of active and inactive renin in man. Clin Sci Mol Med 55: 143-146

KASISKE BL, O'DONNELL MP, KEANE WF (1986) Renal effects of angiotensin II: modulation by calcium and cyclooxygenase products. Am J Physiol 251: F1043-F1048

KASSER A, WAEBER B, NUSSBERGER J, BURRIS J, BRUNNER HR (1985) Enhanced renin secretion in adrenalectomized rats with glucocorticoid-induced hypertension. Clin Exp Hypertens [A] 7: 1619-1628

KASTNER PR, HALL JE, GUYTON AC (1984) Control of glomerular filtration rate: role of intrarenally formed angiotensin II. Am J Physiol 246: F897-F906

KATADA T, GILMAN AG, WATANABE Y, BAUER S, JAKOBS KH (1985) Protein kinase C phosphorylates the inhibitory guanine-nucleotide-binding regulatory component and apparently suppresses it function in hormonal inhibition of adenylate cyclase. Eur J Biochem 151: 431-437

KATADA T, KUSAKABE K, OINUMA M, UI M (1987) A novel mechanism for the inhibition of adenylate cyclase via inhibitory GTP-binding proteins. J Biol Chem 262: 11897-11900

KATUNUMA N, KOMINAMI E (1983) Structures and functions of lysosomal thiol proteinases and their endogenous inhibitor. Curr Top Cell Regul 22: 71-101

KATUNUMA N, KOMINAMI E (1985) Lysosomal thiol cathepsins and their endogenous inhibitors. Distribution and localization. In: KHAIRALLAH EA, BOND JS, BIRD JWC (eds) Intracellular Protein catabolism. Liss, New York, pp 71-79

KATZ SA, MALVIN RL (1982) Secretion of newly synthesized renin. Endocrinology 111: 201-207

KATZ SA, MALVIN RL (1982a) Independence of β-adrenergic stimulation of renin release on renin synthesis. Am J Physiol 243: F434-F439

KAUKEL E, HILZ H (1972) Permeation of dibutyryl cAMP into Hela cells and its conversion to monobutyryl cAMP. Biochem Biophys Res Commun 46: 1011-1018

KAUL CL, GREWAL RS (1976) Effect of some antihypertensive drugs and catecholamine depletors on the plasma renin activity in the rat. Arch Int Pharmacodyn Ther 224: 91-101

KAWAMURA M, INAGAMI T (1983) Calmodulin antagonists stimulate renin release from isolated rat glomeruli. Endocrinology 112: 1857-1859

KAWAMURA M, IKEMOTO F, YAMAMOTO K (1980) Preliminary evidence for the conversion of dog renin into higher-molecular-weight form by cold storage. Clin Sci 58: 451-456

KAWAMURA M, AKABANE S, ITO K, OGINO K, IKEDA M (1983) The existence of inactive (trypsin-activated) renin in dog plasma and renin granules from the kidney. Clin Exp Hypertens [A] 5: 603-612

KAWAMURA M, NAKAMURA M, INAGAMI T (1985) Evidence for existence of angiotensins I and II in mature renin granules from rat kidney cortex. Biochem Biophys Res Commun 131: 628-633

KAWAMURA M, AKABANE S, MATSUSHIMA Y, ITO K, OMAE T, INAGAMI T (1986) Is renin secreted by exocytotic mechanism through mature renin granules from juxtaglomerular cells? Jpn Circ J 50: 1165-1169

KAYLOR CT, CARTER JM (1967) The juxtaglomerular apparatus in fetal and newborn mice. Anat Rec 159: 171-178

KAZIMIERCZAK J (1971) Development of the renal corpuscle and the juxtaglomerular apparatus. Acta Pathol Microbiol Immunol Scand [A] [Suppl] 218: 1-65

KEETON TK, CAMPBELL WB (1980) The pharmacologic alteration of renin release. Pharmacol Rev 32: 81-227

KEETON TK, CAMPBELL WB (1984) Control of renin release and its alteration by drugs. In: ANTONACCIO MJ (ed) Cardial pharmacology, 2nd edn. Raven, New York, pp 65-118

KEISER JA, VANDER AJ, GERMAIN CL (1983) Clearance of renin in unanesthetized rats: effects of chronic lead exposure. Toxicol Appl Pharmacol 69: 127-137

KENNEDY C, BURNSTOCK G (1984) Evidence for an inhibitory prejunctional P1-purinoceptor in the rat portal vein with characteristics of the A2 rather than of the A1 subtype. Eur J Pharmacol 100: 363-368

KEYL MJ, SCOTT JB, DABNEY JM, HADDY FJ, HARVEY RB, BELL RD, GINN HE (1965) Composition of canine renal hilar lymph. Am J Physiol 209: 1031-1033

KHAIRALLAH PA, HALL MM (1977) Angiotensinases. In: GENEST JG, KOIN E, KUCHEL O (eds) Hypertension. McGraw Hill, New York, pp 179-183

KHAIRALLAH PA, PAGE IH (1967) Plasma angiotensinases. Biochem Med 1: 1-8

KHOSLA MC (1985) Synthesis and pharmacology of nonmammalian angiotensins and their evolutionary development. Peptides 6 [Suppl 3]: 289-293

KHOSLA MC, NISHIMURA H, HASEGAWA Y, BUMPUS FM (1985) Identification and synthesis of [1-asparagine, 5-valine, 9-glycine] angiotensin I produced from plasma of American eel *Anguilla rostrata*. Gen Comp Endocrinol 57: 223-233

KHURI FJ, DAVIS RS, COCKETT ATK (1978) Lymphatic network of kidney. II. Effect of diuretics on intrarenal renin release. Urology 12: 621-625

KIFOR I, DZAU VJ (1987) Endothelial renin-angiotensin pathway: Evidence for intracellular synthesis and secretion of angiotensins. Circ Res 60: 422-428

KIHARA I, KITAMURA S, HOSHINO T, SEIDA H, WATANABE T (1968) A hitherto unreported vascular tumor of the kidney: a proposal of "juxtaglomerular cell tumor". Acta Pathol Jpn 18: 197-206

KIIL F (1975) Influence of autoregulation on renin release and sodium excretion. Kidney Int 8 [Suppl]: 208-218

KIM SJ, HIROSE S, MIYAZAKI H, UENO N, HIGASHIMORI K, MORINAGA S, KIMURA T, SAKAKIBARA S, MURAKAMI K (1985) Identification of plasma inactive renin as prorenin with a site-directed antibody. Biochem Biophys Res Commun 126: 641-645

KIM S, IWAO H, NAKAMURA N, IKEMOTO F, YAMAMOTO K (1987a) Whole-body autoradiographic distribution of exogenously administered renal renin in rats. J Histochem Cytochem 35: 549-557

KIM S, IWAO H, NAKAMURA N, IKEMOTO F, YAMAMOTO K (1987b) Fate of circulating renin in conscious rats. Am J Physiol 252: E136-E146

KIMBROUGH HMJ, VAUGHAN EDJ, CAREY RM, AYERS CR (1977) Effect of intrarenal angiotensin II blockade on renal function in conscious dogs. Circ Res 40: 174-178

KIMURA N, SASANO N (1986) Hyperplasia of renin-containing cells in a malignant pheochromocytoma: an immunohistochemical and semiquantitative study. Hum Pathol 17: 425-428

KIRCHHEIM H, GROSS R, HACKENBERG M, HACKENTHAL E (1980) Effect of carotid sinus baroreceptors on kidney blood-flow, renal sympathetic nerve discharge, and renin release in the conscious dog. In: SLEIGHT PE (ed) Arterial baroreceptors and hypertension. Oxford University Press, pp 168-173

KIRCHHEIM H, GROSS R, HACKENBERG HM, HACKENTHAL E, HUBER J (1981) Autoregulation of renin release and its modification by renal sympathetic nerves in conscious dogs. Kidney Int 20: 152 (Abstract)

KIRCHHEIM H, FINKE R, HACKENTHAL E, LOWE W, PERSSON P (1985) Baroreflex sympathetic activation increases threshold pressure for the pressure-dependent renin release in conscious dogs. Pflügers Arch 405: 127-135

KIRCHHEIM HR, EHMKE H, FISCHER S, HACKENTHAL E, LOWE W, PERSSON P (1986) Autoregulation of renal blood flow, glomerular filtration rate and renin release in conscious dogs. Renal Physiol 9: 84

KIRCHHEIM HR, EHMKE H, HACKENTHAL E, LOWE W, PERSSON P (1987) Autoregulation of renal blood flow, glomerular filtration rate and renin release in conscious dogs. Pflügers Arch 410: 441-449

KIRK KL, BELL PD, BARFUSS DW, RIBADENEIRA M (1985) Direct visualization of the isolated and perfused macula densa. Am J Physiol 248: F890-F894

KISCH ES, DLUHY RG, WILLIAMS GH (1976) Regulation of renin release by calcium and ammonium ions in normal man. J Clin Endocrinol Metab 43: 1343-1350

KITAMURA E, KIKKAWA R, FUJIWARA Y, IMAI T, SHIGETA Y (1986) Effect of angiotensin II infusion on glomerular angiotensin II receptor in rats. Biochim Biophys Acta 885: 309-316

KLETT C, KOMISCHKE K, HACKENTHAL E (1986) Is there an angiotensinogen-stimulating factor in plasma of nephrectomized rats? IRCS Med Sci 14: 827-828

KNAPP HR, OELZ O, ROBERTS LJ, SWEETMAN BJ, OATES JA, REED PW (1977) Ionophores stimulate prostaglandin and thromboxane biosynthesis. Proc Natl Acad Sci USA 74: 4251-4255

KNEPEL W, REIMANN W, NUTTO D (1982) On the mechanism of the vasopressin-induced inhibition of renin release. Horm Metab Res 14: 157-160

KOBAYASHI M, SUZUKI S, HASHIBA K (1983) Role of renal kallikrein in control of renin release in conscious rats. Am J Physiol 244: E262-E265

KOJIMA I, KOJIMA K, RASMUSSEN H (1985) Mechanism of inhibitory action of TMB-8 (8-(NN-diethylamino)octyl-3,4,5-trimethoxybenzoate) on aldosterone secretion in adrenal glomerulosa cells. Biochem J 232: 87-92

KOJIMA I, SHIBATA H, OGATA E (1986) Action of TMB-8 (8-(N,N-diethylamino)octyl-3,4,5-trimethoxybenzoate) on cytoplasmic free calcium in adrenal glomerulosa cell. Biochim Biophys Acta 888: 25-29

KOKUBU T, TAKADA Y, UNNO M, HIWADA K (1983) Biochemical and immunological studies of human angiotensin-converting enzyme. In: BEVAN JA et al. (eds) Vacsular neuroeffector mechanisms. Raven, New York, pp 403-406

KON V, ICHIKAWA I (1985) Hormonal regulation of glomerular filtration. Annu Rev Med 36: 315-331

KON Y, HASHIMOTO Y, KITAGAWA H, KUDO N (1984) Morphology and quantification of juxtaglomerular cells of the chicken kidney. Jpn J Vet Sci 46: 189-196

KON Y, HASHIMOTO Y, KITAGAWA H, KUDO N (1986a) Localization of acid phosphatase in chicken juxtaglomerular cells and its functional meaning. Jpn J Vet Sci 48: 925-931

KON Y, HASHIMOTO Y, KITAGAWA H, KUDO N, MURAKAMI K (1986b) Immunohistochemical demonstration of juxtaglomerular cells in the kidneys of domestic mammals and fowls. Jpn J Vet Res 34: 111-123

KONO T, FUMITAKE I, FUMIMASS O, HIROO I, JIVO E (1981) Suppression of captopril-induced increase in plasma renin activity by des-asp-l-,Ileu-8-angiotensin II in man. J Clin Endocrinol Metab 52: 354-358

KONRADS A, HOFBAUER KG, WERNER U, GROSS F (1978) Effects of vasopressin and its deamino-D-arginine analogue on renin release in the isolated perfused rat kidney. Pflügers Arch 377: 81-85

KONRADS A, HUMMERICH W, VLAHO M, WAMBACH G, BOTTCHER W, MEURER KA (1980) Extrarenal prorenin and renin in anephric patients. Klin Wochenschr 58: 233-236

KOPP U, DIBONA GF (1983) Interaction of renal adrenoceptors and prostaglandins in reflex renin release. Am J Physiol 244: F418-F424

KOPP U, DIBONA GF (1984) Interaction between neural and nonneural mechanisms controlling renin secretion rate. Am J Physiol 246: F620-F626

KOPP U, AURELL M, NILSSON IM, ABLAD B (1980) The role of beta-1-adrenoceptors in the renin release response to graded renal sympathetic nerve stimulation. Pflügers Arch 387: 107-113

KOPP U, AURELL M, SJOLANDER M, ABLAD B (1981) The role of prostaglandins in the alpha and beta-adrenoceptor mediated renin release reponse to graded renal nerve stimulation. Pflügers Arch 391: 1-8

KOPP U, BRADLEY T, HJEMDAHL P (1983) Renal venous outflow and urinary excretion of norepinephrine, epinephrine and dopamine during graded nerve stimulation. Am J Physiol 244: E52-E60

KORZETS A, NOURIEL H, STEINER Z, GRIFFEL B, KRAUS L, FREUND U, KLAJMAN A (1986) Resistant hypertension associated with a renin-producing ovarian Sertoli cell tumor. Am J Clin Pathol 85: 242-247

KOSEKI C, KANAI Y, HAYASHI Y, OHNUMA N, IMAI M (1986) Intrarenal localization of receptors for rat atrial natriuretic polypeptide: An autoradiographic study with (125J)-labeled ligand injected in vivo into the rat aorta. Jpn J Pharmacol 42: 27-33

KOTCHEN TA, STRICKLAND AL, RICE TW, WALTERS DR (1972) A study of the renin-angiotensin system in newborn infants. J Pediatr 80: 938-946

KOTCHEN TA, MAULL KI, LUKE R, REES D, FLAMENBAUM W (1974) Effect of acute and chronic calcium administration on plasma renin. J Clin Invest 54: 1279-1286

KOYAMA S, HOSOMI H (1986) Renal opiate receptor mediation of renin secretion to renal nerve stimulation in the dog. Am J Physiol 250: R973-R979

KOYAMA S, MANUGIUM V, AMMONS WS, SANTIESTEBAN HL, MANNING JW (1983) Effects of naloxone on baroreflex, sympathetic tone and blood pressure in the cat. Eur J Pharmacol 90: 367-376

KREISBERG JI (1983) Contractile properties of the glomerular mesangium. Fed Proc 42: 3053-3057

KREISBERG JI, KARNOWSKY MJ (1983) Glomerular cells in culture. Kidney Int 23: 439-447

KREISBERG JI, VENKATACHALAM M, TROYER D (1985) Contractile properties of cultured glomerular mesangial cells. Am J Physiol 249: F457-F463

KRISHNAMURTHY VG, BERN HA (1973) Juxtaglomerular cell changes in the euryhaline freshwater fish *Tilapia mossambica* during adaptation to sea water. Acta Zool 54: 9-14

KRIZ W (1969) Die Lymphbahnen der Säugerniere. Anat Anz 125 [Suppl]: 25-32

KRIZ W (1987) A periarterial pathway for intrarenal distribution of renin. Kidney Int 31 [Suppl 20]: 51-56

KRIZ W, BACHMANN S (1985) Pre- and postglomerular arterioles of the kidney. J Cardiovascul Pharmacol 7 [Suppl 3]: 24-30

KRIZ W, DIETERICH HJ (1970) Das Lymphgefäßystem der Niere bei einigen Säugetieren. Licht- und elektronenmikroskopische Untersuchungen. Z Anat Entwickl Gesch 131: 111-147

KRIZ W, KAISSLING B (1985) Structural organization of the

mammalian kidney. In: Seldin DW, Giebisch G (eds) The kidney: physiology and pathophysiology vol 1. Raven, New York, pp 265-306

Kriz W, Napiwotzky P (1979) Structural and functional aspects of the renal interstitium. Contrib Nephrol 16: 104-108

Kriz W, Sakai T (1988) Has the macula densa signal a mechanical component? In: Persson AEG, Boberg U (eds) The juxtaglomerular apparatus, 11th Fernström Symposium. Elsevier, Amsterdam, pp 27-37

Krompecher-Kiss E, Bucher O (1977) Comparison of the activities of some dehydrogenases in the juxtaglomerular complex of kidneys of Wistar rats and desert rats *(Meriones unguiculati)*. Histochemistry 53: 265-269

Kroon DB (1960) Origin of the PAS-positive granulated ε-cells of the juxtaglomerular apparatus. Acta Anat 41: 138-156

Kugler P (1981) Localization of aminopeptidase A (angiotensinase A) in the rat and mouse kidney. Histochemistry 72: 269-278

Kugler P (1982a) Ultracytochemistry of aminopeptidase A (angiotensinase A) in the kidney glomerulus and juxtaglomerular apparatus. Histochemistry 74: 199-212

Kugler P (1982b) Aminopeptidase A is angiotensinase A. II. Biochemical studies on aminopeptidase A and M in rat kidney homogenate. Histochemistry 74: 247-261

Kugler P (1982c) Histochemistry of angiotensinase A in the glomerulus and the juxtaglomerular apparatus. Kidney Int 22 [Suppl 12]: 44-48

Kugler P (1983) Topochemistry of aminopeptidase A (angiotensinase A) in the kidney cortex of the golden hamster. Anat Anz 153: 301-304

Kugler P, Schiebler TH (1984) Quantitataive histochemistry of the angiotensinase A (APA) in the renal glomeruli of rats after stimulation of the renin-angiotensin system. Cytometry 5: 392-395

Kurtz A (1986a) Intracellular control of renin release - an overview. Klin Wochenschr 64: 838-846

Kurtz A (1986b) Transmembrane signalling of atrial natriuretic peptide in rat renal juxtaglomerular cells. Klin Wochenschr 64 [Suppl 4]: 37-41

Kurtz A (1987) Adenosine stimulates guanylate cyclase activity in vascular smooth muscle cells. J Biol Chem 262: 6296-6300

Kurtz A, Pfeilschifter J, Bauer C (1984) Is renin secretion governed by the calcium permeability of the juxtaglomerular cell membrane? Biochem Biophys Res Commun 124: 359-366

Kurtz A, Pfeilschifter J, Hutter A, Bührle C, Nobiling R, Taugner R, Hackenthal E, Bauer C (1986a) Role of protein kinase C in inhibition of renin release caused by vasoconstrictors. Am J Physiol 250: C563-C571

Kurtz A, Della Bruna R, Pfeilschifter J, Taugner R, Bauer C (1986b) Atrial natriuretic peptide inhibits renin release from juxtaglomerular cells by a cGMP-mediated process. Proc Natl Acad Sci USA 83: 4769-4773

Kurtz A, Della Bruna R, Pfeilschifter J, Bauer C (1986c) Effect of synthetic atrial natriuretic peptide on rat renal juxtaglomerular cells. J Hypertens 4 [Suppl 2]: 57-60

Lacasse J, Ballak M, Mercure C, Gutkowska J, Chapeau C, Foote S, Menard J, Corvol P, Cantin M, Genest J (1985) Immunocytochemical localization of renin in juxtaglomerular cells. J Histochem Cytochem 33: 323-332

Lafont H, Chanussot F, Dupuy C, Lechene P, Lairon D, Carbonnier-Augeire M, Chabert C, Portugal H, Pauli AM, Hauton JC (1984) Influence of acute injection of chloroquine on the biliary secretion of lipids and lysosomal enzyme in rats. Lipids 19: 195-201

Lagios MD (1974) Granular epithelioid (juxtaglomerular) cell and renovascular morphology of the coelacanth *Latimeria chalumnae* Smith (Crossopterygii) compared with that of other fishes. Gen Comp Endocrinol 22: 296-337

Lai FM, Shepherd CA, Cervoni P, Wissner A (1983) Hypotensive and vasocilatory activity of (+/−) 1-O-octadecyl-2-acetylglyceryl-3-phosphorylcholine in the normotensive rat. Life Sci 32: 1159-1166

Lam ASC, Bédard YC, Buckspan MB, Logan AG, Steinhardt MI (1982) Surgically curable hypertension associated with reninoma. J Urol 128: 572-575

Lameire NH, Lifschitz MD, Stein JH (1977) Heterogeneity of nephron function. Annu Rev Physiol 39: 159-184

Lamers APM, Stadhouders AM (1985) Histochemistry of the juxtaglomerular apparatus in the toad *Bufo bufo*. Acid phosphatase activity of the epitheloid cells in the glomerular afferent arterioles. Basic Appl Histochem 29: 93-103

Lamers APM, van Dongen WJ, van Kemenade JAM (1973) The morphology of the juxtaglomerular apparatus in the toad, *Bufo bufo*. A light microscopic study. Z. Zellforsch. 138: 545-555

Lamers APM, van Dongen WJ, van Kemenade JAM (1974) An ultrastructural study of the juxtaglomerular apparatus in the toad, *Bufo bufo*. Cell Tissu Res 153: 449-464

Lamers APM, Speijers GJA, Stadhouders AM (1977) Histochemistry of the juxtaglomerular apparatus in the toad *Bufo bufo*. The glucose-6-phosphate dehydrogenase activity of the macula densa. Cell Tissue Res 184: 435-444

Lamers APM, Verhofstad AAJ, Stadhouders AM, Michelakis AM (1985) Immunohistochemical demonstration of renin in the juxtaglomerular apparatus of three *Bufo* species. Cell Tissue Res 239: 677-682

Lang RE, Unger T, Ganten D (1987) Atrial natriuretic peptide: a new factor in blood pressure control. J Hypertens 5: 255-271

Langard O, Holdaas H, Eide I, Kiil F (1981) Conditions for humoral alpha-adrenoceptor stimulated renin release in anaesthetized dogs. Scand J Clin Lab Invest 41: 527-534

Lansman JB, Hallam TJ, Rink TJ (1987) Single stretch-activated ion channels in vascular endothelial cells as mechanotransducers? Nature 325: 811-813

Lappe RW, Smits JFM, Todt JA, Debets JJM, Wendt RL (1985) Failure of atriopeptin II to cause arterial vasodilation in the conscious rat. Circ Res 56: 606-612

Laragh JH, Sealey JE (1973) The renin-angiotensin-aldosterone hormonal system and regulation of sodium, potassium, and blood pressure homeostasis. In: Orloff J, Berliner RW (eds) Handbook of physiology. Renal physiology. American Physiological Society Washington, pp 831-908

Laragh JH, Angers M, Kelly WG, Liebermann S (1960) Hypotensive agents and pressor substances. The effect of epinephrine, norepinephrine, angiotensin II and others on the secretory rate of aldosterone in man. JAMA 174: 234-240

Larsson C, Weber PC (1978) Renal prostaglandins and renin release. Acta Biol Med Germ 37: 857-862

Latta H (1973) Ultrastructure of the glomerular and juxtaglomerular apparatus. In: Orloff J, Berliner RW (eds) Handbook of physiology. Renal physiology. American Physiological Society Washington, pp 1-29

Latta H, Fligiel S (1985) Mesangial fenestrations, sieving, filtration, and flow. Lab Invest 52: 591-598

LATTA H, JOHNSTON WH (1978) Granular and agranular cell counts in the juxtaglomerular apparatuses of rats with unilateral renovascular hypertension. Lab Invest 39: 219-224

LATTA H, LEE TC (1983) Effects of excessive sodium chloride on the juxtaglomerular apparatus and blood pressure of uninephrectomized rats. Lab Invest 49: 99-106

LATTA H, MAUNSBACH AB (1962a) The juxtaglomerular apparatus as studied electron microscopically. J Ultrastruct Res 6: 547-561

LATTA H, MAUNSBACH AB (1962b) Relations of the centrolobular region of the glomerulus to the juxtaglomerular apparatus. J Ultrastruct Res 6: 562-578

LATTA H, MAUNSBACH AB, MADDEN SC (1960) The centrolobular region of the renal glomerulus studied by electron microscopy. J Ultrastructa Res 4: 455-472

LAVRAS AAC, FICHMAN M, HIRAICHI E, BOUCAULT MA, TOBO T (1977) Components of the renin-angiotensin system in the plasma of *Bothrops jararaca*. Agents Actions 8: 141-145

LAYCHOCK SG, PUTNEY JWJ (1982) Roles of phospholipid metabolism in secretory cells. In: CONN PM (ed) Cellular regulation of secretion and release. Academic, New York, pp 53-105

LEARY WP, LEDINGHAM JG (1969) Removal of angiotensin by isolated perfused organs of the rat. Nature 222: 959-960

LEBEL M, TALBOT J. GROSE J, MORIN J (1977) Adenocarcinoma of the kidney and hypertension: report of 2 cases with special emphasis on renin. J Urol 118: 923-927

LEBOFF MS, DLUHY RG, HOLLENBERG NK (1982) Abnormal renin short feedback loop in essential hypertension is reversible with converting enzyme inhibition. J Clin Invest 70: 335-341

LEBOFF MB, CHANG J, HENRY M, BEAUDOIN D, SWISTON L, BROWN EM (1986) Role of cytosolic calcium in the control of cAMP content by calcium in bovine parathyroid cells. Mol Cell Endocrinol 45: 127-135

LECKIE BJ (1978) An endogenous protease activating plasma inactive renin. Clin Sci Mol Med 55: 133s-134s

LECKIE BJ (1981) Inactive renin: an attempt at a perspective. Clin Sci 60: 119-130

LECKIE BJ, MCCONNELL A (1975) A renin inhibitor from rabbit kidney. Conversion of a large inactive renin to a smaller active enzyme. Circ Res 36: 513-519

LECKIE B, BROWN JJ, FRASER R, KYLE K, LEVER AF, MORTON JJ, ROBERTSON JIS (1978) A renal carcinoma secreting inactive renin. Clin Sci Mol Med 55: 159S-161S

LEDOUX S, GUTKOWSKA J, GARCIA R, THIBAULT G, CANTIN M, GENEST J (1982) Immunohistochemical localization of tonin in rat salivary glands and kidney. Histochemistry 76: 329-339

LEE JC, HURLEY S, HOPPER J (1965) JGA granular cells (mouse): Ultrastructural histochemistry and morphology of granules. Fed Proc 24: 434

LEE JC, HURLEY S, HOPPER J (1966) Secretory activity of the juxtaglomerular granular cells of the mouse. Morphologic and enzyme histochemical observations. Lab Invest 15: 1459-1476

LEE MR (1971) Renin-secreting kidney tumours. Lancet 2: 254-255

LEE MR (1982) Dopamine and the kidney. Clin Sci 62: 439-448

LEE RMKW, GARFIELD RE, FORREST JB, DANIEL EE (1984) Smooth muscle cell herniation in the contracted arterial wall of spontaneously hypertensive and normotensive rats. Acta Anat 119: 65-72

LEE S, VERNIER RL (1980) Immunoelectron microscopy of the glomerular mesangial uptake and transport of aggregated human albumin in the mouse. Lab Invest 42: 44-58

LEE-HUANG S (1984) Cloning and expression of human erythropoietin cDNA in Escherichia coli. Proc Natl Acad Sci USA 81: 2708-2712

LEIPER JM, THOMSON D, MACDONALD MK (1977) Uptake and transport of imposil by the glomerular mesangium in the mouse. Lab Invest 37: 526-533

LEITMAN DC, ANDRESEN JW, KUNO T, KAMISAKI Y, CHANG JK, MURAD F (1986) Identification of multiple binding sites for atrial natriuretic factor by affinity cross-linking in cultured endothelial cells. J Biol Chem 261: 11650-11655

LELKES PI, FRIEDMAN JE, ROSENHECK K, OPLATA A (1986) Destabilization of actin filaments as a requirement for the secretion of catecholamines from permeabilized chromaffine cells. FEBS Let 208: 357-363

LESTER GE, RUBIN RP (1977) The role of calcium in renin secretion from the isolated perfused cat kidney. J Physiol (Lond) 269: 93-108

LEVENS NR, PEACH MJ, CAREY RM (1981) Role of the intrarenal renin-angiotensin system in the control of renal function. Circ Res 48: 157-167

LEVENSON DJ, DZAU VJ (1987) Effects on angiotensin-converting enzyme inhibition on renal hemodynamics in renal artery stenosis. Kidney Int 31 [Suppl 20]: S-173-S-179

LEVER AF, PEART WS (1962) Renin and angiotensin-like activity in renal lymph. J Physiol (Lond) 160: 548-563

LEVINE M, LENTZ KE, KAHN JR, DORER FE, SKEGGS LT (1978) Studies on high molecular weight renin from hog kidney. Circ Res 42: 368-375

LEW PD, MONOD A, WALDVOGEL FA, DEWALD B, BAGGIOLINI M, POZZAN T (1986) Quantitative analysis of the cytosolic free calcium dependency of exocytosis from three subcellular compartments in intact human neutrophils. J Cell Biol 102: 2197-2204

LEW R, SUMMERS RJ (1985) Autoradiographic localization of beta-adrenoceptor subtypes in guinea-pig kidney. Br J Pharmacol 85: 341-348

LEWICKI JA, PRINTZ JM, PRINTZ MP (1983) Clearance of rabbit plasma angiotensinogen and relationship to CSF angiotensinogen. Am J Physiol 244: H577-H585

LEYSSAC PP (1978a) A micropuncture study of glomerular filtration and tubular reabsorption of endogenous renin in the rat. Renal Physiol 1: 181-188

LEYSSAC PP (1978b) Micropuncture study of renin release at the single nephron level. Evidence for some release directly into the circulating blood. Renal Physiol 1: 61-73

LEYSSAC PP (1984) Intrarenal feedback mechanisms controlling intratubular pressure and flow. Renal Physiol 7: 260-261 (Abstract)

LEYSSAC PP (1986) Changes in single nephron renin release are mediated by tubular fluid flow rate. Kidney Int 30: 332-339

LIFSCHITZ ME, PATEK RV, FADEM SZ, STEIN JH (1978) Urinary prostaglandin E excretion: effect of chronic alterations in sodium intake and inhibition of prostacyclin synthesis in the rabbit. Prostaglandins 16: 607-619

LIJNEN P, GROESENEKEN D, FAGARD R, STAESSEN J, AMERY A (1984) Effect of indomethacin on active and inactive renin in sodium replete man at rest and during exercise. In: GOLDBERG DM, WERNER M (eds) Selected topics in clinical enzymology. Walter de Gruyter, Berlin

LIJNEN P, STAESSEN J, FAGARD R, GROESENEKEN D, M'BU-

YAMBA-KABANGU JR, GRAUWELS R, AMERY A (1985) Active and acid-activable inactive renin during inhibition by indomethacin of prostaglandin synthesis in sodium-replete man. Eur J Clin Invest 15: 141-145

LINAS SL (1984) Role of prostaglandins in renin secretion in the isolated kidney. Am J Physiol 246: F811-F818

LINDNER A, TREMANN JA, PLANTIER J, CHAPMAN W, FORREY A, HAINES G, PALMIERI G (1978) Effects of parathyroid hormone on the renal circulation and renin secretion in unanesthetized dogs. Miner Electrolyte Metab 1: 155-165

LINDOP GBM (1987) Morphological aspects of renin synthesis, processing, storage and secretion. Kidney Int 31 [Suppl 20]: 18-24

LINDOP GBM, DOWNIE TT (1984) New morphological evidence for the synthesis and storage of renin in the human kidney: an ultrastructural immunocytochemical study. J Hypertens 2: 7-10

LINDOP GBM, FLEMING S (1984) Renin in renal cell carcinoma - an immunocytochemical study using an antibody to pure human renin. J Clin Pathol 37: 27-31

LINDOP GBM, GARDINER DS (1986) La cellule humaine sécrétrice de rénine. Ann Endocrinol (Paris) 47: 133-144

LINDOP GBM, LEVER AF (1986) Anatomy of the renin-angiotensin system in the normal and pathological kidney. Histopathology 10: 335-362

LINDOP GBM, MORE IAR, LECKIE B (1983a) An ultrastructural and immunocytochemical study of a renal carcinoma secreting inactive renin. J Clin Pathol 36: 639-645

LINDOP GBM, STEWART JA, DOWNIE TT (1983b) The immunocytochemical demonstration of renin in a juxtaglomerular cell tumour by light and electron microscopy. Histopathology 7: 421-431

LINDOP GBM, FLEMING S, GIBSON AAM (1984) Immunocytochemical localization of renin in nephroblastoma. J Clin Pathol 37: 738-742

LINDOP GBM, LECKIE B, WINEARLS CG (1986) Malignant hypertension due to a renin-secreting renal cell carcinoma - an ultrastructural and immunocytochemical study. Histopathology 10: 1077-1088

LITWIN JA (1980) Cell membrane features of rabbit arterial smooth muscle. Cell Tissue Res 212: 341-350

LJUNGQVIST A, WÅGERMARK J (1966) Renal juxtaglomerular granulation in the human foetus and infant. Acta Pathol Microbiol Immunol Scand [A] 67: 257-266

LLACH F, WEIDMANN P, REINHART R, MAXWELL MH, COBURN JW, MASSRY SG (1974) Effect of acute and longstanding hypocalcemia on blood pressure and plasma renin activity in man. J Clin Endocrinol Metab 38: 841-847

LÖNNERHOLM G, WISTRAND PJ (1984) Carbonic anhydrase in the human kidney: A histochemical and immunocytochemical study. Kidney Int 25: 886-898

LOGAN A, CHATZILIAS A (1980) The role of calcium in the control of renin release from the isolated rat kidney. Can J Physiol Pharmacol 58: 60-66

LOHMEIER TE, DAVIS JO, FREEMAN RH (1975) Des-aspl-angiotensin II: Possible role in mediating responses of the renin-angiotensin system. Proc Soc Exp Biol Med 149: 515-518

LOJDA Z, GOSSRAU R (1980) Study on aminopeptidase A. Histochemistry 67: 267-290

LOUDON M, BING RF, THURSTON H, SWALES JD (1983) Arterial wall uptake of renal renin and blood pressure control. Hypertension 5: 629-634

LOUTZENHISER R, EPSTEIN M (1985) Effects of calcium antagonists on renal hemodynamics. Am J Physiol 249: F619-F629

LOUTZENHISER R, EPSTEIN M, HORTON C, HAMBURGER R (1985) Nitrendipine-induced stimulation of renin release by the isolated perfused rat kidney. Proc Soc Exp Biol Med 180: 133-136

LOUTZENHISER R, EPSTEIN M, HORTON C (1987) Inhibition by diltiazem of pressure-induced afferent vasoconstriction in the isolated perfused rat kidney. Am J Cardiol 59: 72A-75A

LUDWIG G, SUZUKI F, MURAKAMI K, GANTEN D, HACKENTHAL E (1986) Kidney renin mRNA, renin secretion and plasma renin following adrenalectomy, salt depletion or enalapril treatment in the rat. J Hypertens 4 [Suppl 6]: 431-433

LUDWIG G, GANTEN D, MURAKAMI K, FASCHING U, HAKKENTHAL E (1987) Relationship between renin mRNA and renin secretion in adrenalectomized, salt-depleted, or converting enzyme inhibitor-treated rats. Mol Cell Endocrinol 50: 223-229

LUETSCHER JA, BIALEK JW, GRISLIS G (1982) Cathepsins B and H convert inactive renin to smaller, active renin. IRCS Med Sci 10: 171-172

LÜLLMANN-RAUCH R (1975) Lipidosislike renal changes in rats treated with chlorphentermine or with tricyclic antidepressants. Virchows Arch [B] 18: 51-60

LÜLLMANN-RAUCH R (1979) Drug-induced lysosomal storage disorders. In: DINGLE JT, JACQUES PJ, SHAW IH (eds) Lysosomes in applied biology and therapeutics. North-Holland, Amsterdam, pp 49-130

LUMBERS ER (1971) Activation of renin in human amniotic fluid by low pH. Enzyme 40: 329-336

LUNDBERG JM, TATEMOTO K (1982) Pancreatic polypeptide family (APP, BPP, NPY and PYY) in relation to sympathetic vasoconstriction resistant to alpha-adrenoceptor blockade. Acta Physiol Scand 116: 393-402

LUNDBERG JM, ANGGARD A, FAHRENKRUG J, HÖKFELT T, MUTT V (1980) Vasoactive intestinal polypeptide in cholinergic neurons of exocrine glands: functional significance of coexisting transmitters for vasocilation and secretion. Proc Natl Acad Sci USA 77: 1651-1655

LUNDBERG JM, TERENIUS L, HÖKFELT T, MARTLING CR, TATEMOTO K, MUTT V, POLAK J, BLOOM S, GOLDSTEIN M (1982) Neuropeptide Y (NPY)-like immunoreactivity in peripheral noradrenergic neurons and effects of NPY on sympathetic function. Acta Physiol Scand 116: 477-480

LUNDBERG JM, FRIED G, PERNOW J, THEODORSSON-NORHEIM E (1986) Co-release of neuropeptide Y and catecholamines upon adrenal activation in the cat. Acta Physiol Scand 126: 231-238

LUSH DJ, FRAY JCS (1984) Steady-state autoregulation of renal blood flow: A myogenic model. Am J Physiol 247: R89-R99

LYNCH DR, BRAAS KM, SNYDER SH (1986) Atrial natriuretic factor receptors in rat kidney, adrenal gland, and brain: autoradiographic localization and fluid balance dependent changes. Proc Natl Acad Sci USA 83: 3357-3361

LYONS HJ (1980) Studies on the mechanism of renin release from rat kidney slices: Calcium, sodium and metabolic inhibition. J Physiol 304: 99-108

LYONS HJ, CHURCHILL PC (1974) The influence of ouabain on in vitro renin secretion. Proc Soc Exp Biol Med 145: 1148-1150

LYONS HJ, CHURCHILL PC (1975a) Renin secretion from rat

renal cortical cell suspensions. Am J Physiol 228: 1835-1839

LYONS HJ, CHURCHILL PC (1975b) The influence of ouabain on in vitro renin secretion and intracellular sodium. Nephron 14: 442-450

MAACK T (1980) Physiological evaluation of the isolated perfused rat kidney. Am J Physiol 238: F71-F78

MAACK T (1985) Atrial natriuretic factor. Structure and functional properties. Kidney Int 27: 607-615

MAACK T, JOHNSON V, KAU ST, FIGUEIREDO J, SIGULEM D (1979) Renal filtration, transport, and metabolism of low-molecular-weight proteins: A review. Kidney Int 16: 251-270

MAACK T, MARION DN, CAMARGO MJF, KLEINERT HD, LARAGH JH, VAUGHAN EDJ, ATLAS SA (1984) Effects of auriculin (atrial natriuretic factor) on blood pressure, renal function, and the renin-aldosterone system in dogs. Am J Med 77: 1069-1075

MAACK T, ATLAS SA, CAMARGO MJF, COGAN MG (1986) Renal hemodynamic and natriuretic effects of atrial natriuretic factor. Fed Proc 45: 2128-2132

MACCALLUM DK, CONN JW, BAKER BL (1973) Ultrastructure of a renin-secreting juxtaglomerular cell tumor of the kidney. Invest Urol 11: 65-74

MACDONALD GJ, BLACKET RB (1967) Reduction of blood pressure by puromycin infused selectively into the ischaemic kidney in experimental renal hypertension. Cardiovasc Res 1: 215-218

MADEDDU P, GLORIOSO N, DESSI-FULGHERI P, OPPES M, TONOLO G, BANDIERA F, RAPPELI A (1984) Changes in circulating active and inactive renin after bilateral nephrectomy in the rat. IRCS Med Sci 12: 1143-1144

MAILLET M (1959) Innervation sympathique du rein: son rôle trophique. Acta Neuroveg (Wien) 20: 155-180

MAJERUS PW, CONNOLLY TM, DECKMYN H, ROSS TS, BROSS TE, ISHII H, BANSAL VS, WILSON DB (1986) The metabolism of phosphoinositide-derived messenger molecules. Science 234: 1519-1526

MALAYAN SA, REID IA (1982) Effects of a nonpressor analogue of vasopressin on plasma renin activity and salt and water excretion in water-loaded, anesthetized dogs. Life Sci 31: 2757-2763

MALAYAN SA, RAMSAY DJ, KEIL LC, REID IA (1980) Effects of increases in plasma vasopressin concentration on plasma renin activity, blood pressure, heart rate and plasma corticosteroid concentration in conscious dogs. Endocrinology 107: 1899-1904

MALLING C, POULSEN K (1977) A direct radioimmunoassay for plasma renin in mice and its evaluation. Biochim Biophys Acta 491: 532-541

MALTINTI G, ARZILLI F, CORTESE R, PEDRINELLI R, POLI L, SASSANO P, SALVETTI A (1977) Relationship between juxtaglomerular apparatus and plasma renin activity in human reno-vascular hypertension. Nephron 19: 220-227

MARGOLIUS HS (1984) The kallikrein-kinin system and the kidney. Annu Rev Physiol 46: 309-326

MARIN-GREZ M (1982) Multihormonal regulation of renal kallikrein. Biochem Pharmacol 31: 3941-3947

MARIN-GREZ M, BRIGGS JP, SCHUBERT G, SCHNERMANN J (1985) Dopamine receptor antagonists inhibit the natriuretic response to atrial natriuretic factor (ANF). Life Sci 36: 2171-2176

MARIN-GREZ M, FLEMING JT, STEINHAUSEN M (1986) Atrial natriuretic peptide causes pre-glomerular vasodilatation and post-glomerular vasoconstriction in rat kidney. Nature 324: 473-476

MARRE M, MISUMI J, RAEMSCH KD, CORVOL P, MENARD J (1982) Diuretic and natriuretic effects of nifedipine on isolated perfused rat kidneys. J Pharmacol Exp Ther 223: 263-270

MARUTA H, ARAKAWA K (1983) Confirmation of direct angiotensin formation by kallikrein. Biochem J 213: 193-200

MASSON P (1924) Le glomus neuro-artériel des régions tactiles de la peau et ses tumeurs. Lyon Chir 20: 257-282

MAST GJ, KONRAD G, NEISIUS D, TAUGNER R, ZIEGLER M (1983) Renale Hypertonie - eine drohende Komplikation stumpfer Nierenverletzungen. Unfallchirurgie 9: 280-287

MASUDA T, IMAI T, FUKUSHI T, SUDOH M, HIROSE M, MURAKAMI K (1982) Molecular cloning of DNA complementary to mouse submandibular gland renin mRNA. Biomed Res 3: 541-545

MATA LR (1976) Dynamics of HRPase absorption in the epithelial cells of the hamster seminal vesicles. J Microsc Biol Cell 25: 127-132

MATSUHASHI H (1979) Electron microscopic studies on the juxtaglomerular cells of the dehydrated mouse. Jpn J Vet Sci 41: 283-298

MATSUHASHI H, NISHIDA T, MOCHIZUKI K (1975) Juxtaglomerular cell granules in the developmental mesonephros and metanephros of swine embryos. Jpn J Vet 37: 261-269

MATSUHASHI H, NISHIDA T, MOCHIZUKI K (1977a) Comparative studies on granulation of juxtaglomerular cells of some mammalian kidneys and limitation of the specificity of Bowie staining. Jpn J Vet Sci 39: 379-388

MATSUHASHI H, NISHIDA T, MOCHIZUKI K (1977b) Enzyme activity of juxtaglomerular cell granules of the mouse. Jpn J Vet Sci 39: 657-659

MATSUMURA Y, MIYAWAKI N, MORIMOTO S (1984) Effects of W-7 and W-5 on renin release from rat kidney cortical slices. Jpn J Pharmacol 36: 268-271

MATSUMURA Y, MIYAWAKI N, SASAKI Y, MORIMOTO S (1985a) Inhibitory effects of norepinephrine, methoxamine and phenylephrine on renin release from rat kidney cortical slices. J Pharmacol Exp Ther 233: 782-787

MATSUMURA Y, SASAKI Y, SHINYAMA H, MORIMOTO S (1985b) The calcium channel agonist, Bay K 8644, inhibits renin release from rat kidney cortical slices. Eur J Pharmacol 117: 369-372

MATSUMURA Y, URIU T, SHINYAMA H, SASAKI Y, MORIMOTO S (1987) Inhibitory effects of calcium channel agonists on renin release from rat kidney cortical slices. J Pharmacol Exp Ther 241: 1000-1005

MAY CN, PEART WS (1986) Stimulation and suppression of renin release from incubations of rat renal cortex by factors affecting calcium flux. Br J Pharmacol 89: 173-182

MCCREDIE DA, POWELL HR, ROTENBERG E (1975) Effect of parathyroid extract on renin release in the dog. Clin Sci Mol Med 48: 461-463

MCDONALD KM, TAHER S, AISENBREY G, DE TORRENTE A, SCHRIER RW (1975) Effect of angiotensin II and an angiotensin II inhibitor on renin secretion in the dog. Am J Physiol 228: 1562-1567

MCGIFF JC (1981) Prostaglandins, prostacyclin, and thromboxanes. Annu Rev Pharmacol Toxicol 21: 479-509

MCGIFF JC, SPOKAS EG, WONG PYK (1982) Stimulation of renin release by 6-oxo-prostaglandin E1 and prostacyclin. Br J Pharmacol 75: 137-144

MCGRATH B, BODE K, LUXFORD A, HOWDEN B, JABLONSKI P

(1985) Effects of dopamine on renal function in the rat isolated perfused kidney. Clin Exp Pharmacol Physiol 12: 343-352

McIntyre GD, Pau B, Hallett A, Leckie BJ, Szelke M (1984) The purification of a high-molecular-weight, enzymatically inactive renin precursor from human kidney. J Hypertens 2: 305-310

McKelvey RW (1963) The presence of a juxtaglomerular apparatus in non-mammalian vertebrates. Anat Rec 145: 259-260

McKenna OC, Angelakos ET (1968) Adrenergic innervation of the canine kidney. Circ Res 22: 345-354

McKenzie IM, Heiman D, Winter JSD, McKenzie JK (1988) Inactive renin and aldosterone in Bartter's syndrome. Clin Invest Med 10: 303-308

McKenzie JC, Tanaka I, Misono KS, Inagami T (1985) Immunocytochemical localization of atrial natriuretic factor in the kidney, adrenal medulla, pituitary, and atrium of rat. J Histochem Cytochem 33: 828-832

McKenzie JK, Reisin E (1978) Acid-activated renin responses to hydrochlorothiazide, propranolol and indomethacin. Clin Sci Mol Med 55: 151s-153s

McLaren KM, MacDonald MK (1982) Histological and ultrastructural studies of the human juxtaglomerular apparatus in Bartter's syndrome and renal artery stenosis. J Pathol 136: 181-197

McLaren KM, MacDonald MK (1983) Histological and ultrastructural studies of the human juxtaglomerular apparatus in benign and malignant hypertension. J Pathol 139: 41-55

McManus JFA (1942) The juxtaglomerular complex. Lancet 2: 394-397

McManus JFA (1947) Further observations on the glomerular root of the vertebrate kidney. Q J Microsc Sci 88: 39-44

McManus JFA (1950) Medical diseases of the kidney. Lea and Febiger, Philadelphia

Means AR, Tash JS, Chafouleas JG (1982) Physiological implications of the presence, distribution, and regulation of calmodulin in eukaryotic cells. Physiol Rev 62: 1-39

Mellgren RL (1987) Calcium-dependent proteases: an enzyme system active at cellular membranes? Faseb J 1: 110-115

Mellman I, Fuchs R, Helenius A (1986) Acidification of the endocytic and exocytic pathways. Annu Rev Biochem 55: 663-700

Menard J, N'Goc PW, Bariety J, Guyenne PT, Corvol P (1979) Direct radioimmunoassay and immunocytochemical localization of renin in human kidneys. Clin Sci 57 [Suppl 5]: 105-108

Mendelsohn FAO (1976) A method for measurement of angiotensin II in tissues and its application to rat kidney. Clin Sci Mol Med 51: 111-125

Mendelsohn FAO (1979) Evidence for the local occurrence of angiotensin II in rat kidney and its modulation by dietary sodium intake and converting enzyme blockade. Clin Sci 57: 173-179

Mendelsohn FAO (1980) Failure of suppression of intrarenal angiotensin II in the contralateral kidney of one-clip two-kidney hypertensive rats. Clin Exp Pharmacol Physiol 7: 219-223

Mendelsohn FAO (1982) Angiotensin II: Evidence for its role as an intrarenal hormone. Kidney Int 22 [Suppl 12]: 78-81

Mendelsohn FAO (1985) Localization and properties of angiotensin receptors. J Hypertens 3: 307-316

Mendelsohn FAO, Aguilera G, Saavedra JM, Quirion R, Catt KJ (1983) Characteristics and regulation of angiotensin II receptors in pituitary, circumventricular organs and kidney. Clin Exp Hypertens [A] 5: 1081-1097

Mendelsohn FAO, Dunbar M, Allen A, Chou ST, Millan MA, Aguilera G, Catt KJ (1986) Angiotensin II receptors in the kidney. Fed Proc 45: 1420-1425

Menzie JW, Hoffman LH, Michelakis AM (1978) Immunfluorescent localization of renin in mouse submaxillary gland and kidney. Am J Physiol 234: E480-E483

Merker HJ (1965) Über das Vorkommen multivesikulärer Einschlusskörper im Vaginalepithel der Ratte. Z Zellforsch Mikrosk Anat 68: 618-630

Merrill DC, Skelton MM, Cowley AW (1986) Humoral control of water and electrolyte excretion during water restriction. Kidney Int 29: 1152-1161

Meyer D (1972) Morphometrische Untersuchungen am juxtaglomerulären Apparat menschlicher Nieren. Fischer, Stuttgart

Meyer D (1978) Quantitative morphology of the juxtaglomerular apparatus in various renal and extrarenal diseases. In: Gessler U, Seybold D (eds) Der juxtaglomeruläre Apparat. Morphologie, Physiologie, Pathophysiologie und Klinik. Dustri, München, pp 23-35

Meyer D, Reich H, Walvig F (1966) Vergleichende Untersuchungen zur Struktur des Gefäßpols der Nierenkörperchen bei Süß- und Seewasserfischen. Aktuelle Probleme der Nephrologie. Springer, Berlin Heidelberg New York, pp 566-579

Meyer D, Jerusalem C, Walvig F (1967) Untersuchungen zur Feinstruktur der granulierten epitheloiden Zellen präglomerulärer Arteriolen in den Nieren von Teleostiern. Z Zellforsch 83: 508-526

Michael AF, Keane WF, Raij L, Vernier RL, Mauer SM (1980) The glomerular mesangium. Kidney Int 17: 141-154

Michel JB, Dussaule JC, Choudat L, Auzan C, Nochy D, Corvol P, Menard J (1986) Effects of antihypertensive treatment in one-clip, two-kidney hypertension in rats. Kidney Int 29: 1011-1020

Michelakis AM (1971) The effect of sodium and calcium on renin release in vitro. Proc Soc Exp Biol Med 137: 833-836

Michelakis AM (1971) The effect of angiotensin on renin production and release in vitro. Proc Soc Exp Biol Med 138: 1106-1108

Michelakis AM, Caudle J, Liddle GW (1969) In vitro stimulation of renin production by epinephrine, norepinephrine, and cyclic AMP. Proc Soc Exp Biol Med 130: 748-753

Michelakis AM, Yoshida H, Menzie J, Murakami K, Inagami T (1974) A radioimmunoassay for the direct measurement of renin in mice and its application to submaxillary gland and kidney studies. Endocrinology 94: 1101-1105

Michielsen P (1962) Bijdrage tot de studie van de glomerulus. Arscia Brussels

Michielsen P, Creemers J (1967) The structure and function of the glomerular mesangium. In: Dalton AJ, Haguenau F (eds) Ultrastructure of the kidney, vol 2. Academic, New York, pp 57-72

Milavec-Krizman M, Evenou JP, Wagner H, Berthold R, Stoll AP (1985) Characterization of beta-adrenoceptor subtypes in rat kidney with new highly selective beta-1-blockers and their role in renin release. Biochem Pharmacol 34: 3951-3957

Millar JA, Leckie BJ, Semple PF, Morton JJ, Sonkodi S, Robertson JIS (1978) Active and inactive renin in human plasma. Renal arteriovenous differences and relationships with angiotensin and renin-substrate. Circ Res 43 [Suppl 1]: 120-127

Miller G, Hartroft PM (1961) Renal juxtaglomerular cells in acute dietary sodium deficiency. Fed Proc 20: 404

Miller MJS, Carrara MC, Westlin WF, McNeill H, McGiff JC (1986) Compartemental prostaglandin release by angiotensin II and arginine-vasopressin in rabbit isolated perfused kidney. Eur J Pharmacol 120: 43-50

Miller RA (1967) Regional responses of interrenal tissue and of chromaffin tissue to hypophysectomy and stress in pigeons. Acta Endocrinol 55: 108-118

Mimran A, Casellas D (1987) The renin-angiotensin system and nephron function heterogeneity. Kidney Int 31 (Suppl 20): 57-63

Mimran A, Leckie BJ, Fourcade JC, Baldet P, Navratil H, Barjon P (1978) Blood pressure, renin-angiotensin system and urinary kallikrein in a case of juxtaglomerular cell tumor. Am J Med 65: 527-536

Mink D, Schiller A, Kriz W, Taugner R (1984) Interendothelial junctions in kidney vessels. Cell Tissue Res 236: 567-576

Minuth M, Hackenthal E, Poulsen K, Rix E, Taugner R (1981) Renin Immunocytochemistry of the differentiating juxtaglomerular apparatus. Anat Embryol 162: 173-181

Misono KS, Chang JJ, Inagami T (1982) Amino acid sequence of mouse submaxillary gland renin. Proc Natl Acad Sci USA 79: 4858-4862

Mitchell GM, Stratford BF, Ryan GB (1982) Morphogenesis of the renal juxtaglomerular apparatus and peripolar cells in the sheep. Cell Tissue Res 222: 101-111

Mitchell JD, Baxter TJ, Blair-West JR, McCredie DA (1970) Renin levels in nephroblastoma (Wilms tumour). Arch Dis Child 45: 376-384

Mitchell KD, Navar LG (1987) Superficial nephron responses to peritubular capillary infusions of angiotensins I and II. Am J Physiol 252: F818-F824

Miyamori I, Ikeda M, Matsubara T, Okamoto S, Koshida H, Yasuhara S, Morise T, Takeda R (1987) The renal, cardiovascular and hormonal actions of human atrial natriuretic peptide in man; effects of indomethacin. Br J Clin Pharmacol 23: 425-431

Miyawaki N, Matsumura Y, Ohno Y, Morimoto S (1985) Renin release from kidney cortical slices in response to isoproterenol and glucagon is decreased in vitamin E-deficient rats. Life Sci 37: 923-930

Miyazaki H, Fukamizu A, Hirose S, Hayashi T, Hori H, Ohkubo H, Nakanishi S, Murakami K (1984) Structure of the human renin gene. Proc Natl Acad Sci USA 81: 5999-6003

Miyazaki M, Okunishi H, Nishimura K, Toda N (1984) Vascular angiotensin-converting enzyme activity in man and other species. Clin Sci 66: 39-45

Mizoguchi H, Dzau VJ, Siwek LG, Barger AC (1983) Effect of intrarenal administration of dopamine on renin release in conscious dogs. Am J Physiol 244: H39-H45

Mizuno K, Gotoh M, Hashimoto S, Fukuchi S (1985) Evidence for existence of inactive arterial renin-like enzyme in the rat. Clin Exp Hypertens [A] 7: 93-103

Mizuno K, Higashimori K, Imada T, Inagami T (1987) Direct relase of angiotensins I and II from isolated rat kidney perfused with angiotensinogen-free medium. Biochem Biophys Res Commun 149: 475-481

Mizuri S, Ozawa T, Hirata K, Takezawa K, Kawamura S (1987) Characteristic changes of the juxtaglomerular cells before and after treatment of pseudo-Bartter's syndrome due to furosemide abuse. Nephron 46: 23-27

Moffat DB, Creasey M (1971) The fine structure of the intra-arterial cushions at the origins of the juxtamedullary afferent arterioles in the rat kidney. J Anat 110: 409-419

Moffet B, McGowan RA, Gross KW (1986) Modulation of kidney renin messenger RNA levels during experimentally induced hypertension. Hypertension 8: 874-882

Molteni A, Rahill WJ, Koo JH (1974) Evidence for a vasopressor substance (renin) in human fetal kidneys. Lab Invest 30: 115-118

Molteni A, Mullis KB, Zakheim RM, Mattioli L (1976) The effect of changes in dietary sodium on lung and serum angiotensin-I-converting enzyme in the rat. Lab Invest 35: 569-573

Moore LC, Yarimizu S, Schubert G, Weber PC, Schnermann J (1980) Dynamics of tubuloglomerular feedback adaptation to acute and chronic changes in body fluid volume. Pflügers Arch 387: 39-45

Moore LC, Clausen C, Rich A (1988) Mathematical model of the transport-coupling hypothesis of tubuloglomerular feedback signal transmission. In: Persson AEG, Boberg U (eds) The juxtaglomerular apparatus, 11th Fernström Symposium. Elsevier, Amsterdam, pp 137-151

More IAR, Jackson AM, MacSween RNM (1974) Renin-secreting tumor associated with hypertension. Cancer 34: 2093-2102

Morgan T, Davis JM (1975) Renin secretion at the individual nephron level. Pflügers Arch 359: 23-31

Morgan T, Gillies A (1977) Factors controlling the release of renin. A micropuncture study in the cat. Pflügers Arch 368: 13-18

Morgan T, Davis J, Gillies A (1982) Release of renin into the circulation. Kidney Int 22 [Suppl 12]: 63-66

Morgunov N, Baines AD (1981) Renal nerves and catecholamine excretion. Am J Physiol 240: F75-F81

Morild I, Bohle A, Christensen JA (1985a) Structure of the avian kidney. Anat Rec 212: 33-40

Morild I, Mowinckel R, Bohle A, Christensen JA (1985b) The juxtaglomerular apparatus in the avian kidney. Cell Tissue Res 240: 209-214

Morild I, Christensen JA, Halvorsen OJ, Farstad M (1987) Effect of volume depletion on the afferent arterioles in the avian kidney. Virchows Arch [A] 411: 149-155

Morild I, Christensen JA, Mikeler E, Bohle A (1988) Peripolar cells in the avian kidney. Virchows Arch (A) 412: 471-477

Morimoto S, Yamamoto K, Horiuchi K, Tanaka H, Ueda J (1970) A release of renin from dog kidney cortex slices. Jpn J Pharmacol 20: 536-545

Morimoto S, Abe R, Fukuhara A, Tanaka K, Yamamoto K (1979) Effect of sodium restriction on plasma renin activity and renin granules in rat kidney. Am J Physiol 237: F367-F371

Morris BJ (1978) Activation of human inactive ("pro-")renin by cathepsin D and pepsin. J Clin Endocrinol Metab 46: 153-157

Morris BJ (1986) New possibilities for intracellular renin and inactive renin now that the structure of the human renin gene has been elucidated. Clin Sci 71: 345-355

Morris BJ, Catanzaro DF (1980) Activation of inactive renin by nerve growth factor. IRCS Med Sci 8: 433

Morris BJ, Catanzaro DF (1986) Evolution of renin. Clin Exp Pharmacol Physiol 13: 365-370

MORRIS BJ, JOHNSTON CI (1976a) Renin substrate in granules from rat kidney cortex. Biochem J 154: 625-637

MORRIS BJ, JOHNSTON CI (1976b) Isolation of renin granules from rat kidney cortex and evidence for an inactive form of renin (prorenin) in granules and plasma. Endocrinology 98: 1466-1474

MORRIS BJ, LUMBERS ER (1972) The activation of renin in human amniotic fluid by proteolytic enzymes. Biochim Biophys Acta 289: 385-391

MORRIS BJ, MCGIRR JG (1981) Direct evidence, using prophe-arg CH2Cl, that plasma kallikrein has a role in acid activation of inactive renin in plasma from normal subjects. Biomed Res 2: 552-559

MORRIS BJ, REID IA (1978) A "renin-like" enzymatic action of cathepsin D and the similarity in subcellular distributions of "renin-like" activity and cathepsin in midbrain of dogs. Endocrinology 103: 1289-1296

MORRIS BJ, NIXON RL, JOHNSTON CI (1976) Release of renin from glomeruli isolated from rat kidney. Clin Exp Pharmacol Physiol 3: 37-47

MORRIS BJ, IWAMOTO HS, REID IA (1979a) Localization of angiotensinogen in rat liver by immunocytochemistry. Endocrinology 105: 796-800

MORRIS BJ, REID IA, GANONG WF (1979b) Inhibition by α-adrenoceptor angonists of renin release in vitro. Eur J Pharmacol 59: 37-45

MORRIS BJ, DE ZWART RT, BROWN CR (1980) Activation of inactive renin by acrosin. IRCS Med Sci 8: 721

MORRIS BJ, CATANZARO DF, RICHARDS RI, MASON A, SHINE J (1981) Kallikrein and renin: molecular biology and biosynthesis. Clin Sci 61: 351-353

MORRIS BJ, CATANZARO DF, MULLINS JJ, HARDMAN J, SHINE J (1983) Synthesis of mouse renin as a 2-3-33-3 kilodalton pre-pro-two-chain molecule and use of its cDNA to identify the human gene. Clin Exp Pharmacol Physiol 10: 293-297

MORRIS BJ, CATANZARO DF, HARDMAN J, MESTEROVIC N, TELLAM J, HORT Y, SHINE J (1984a) Human renin gene sequence, gene regulation and prorenin processing. J Hypertens 2 [Suppl 3]: 231-233

MORRIS BJ, CATANZARO DF, HARDMAN J, MESTEROVIC N, TELLAM J, HORT Y, BENNETTS BH, SHINE J (1984a) Structure of human renin and expression of the renin gene. Clin Exp Pharmacol Physiol 11: 369-374

MOSS AH, PETERSON LJ, SCOTT CW, WINTER K, OLIN DB, GARBER RL (1982) Delayed diagnosis of juxtaglomerular cell tumor hypertension. NC Med J 43: 705-707

MUKAI M, IRI H, NAKAJIMA T, HIROSE S, TORIKATA C, KAGEYAMA K, UENO N, MURAKAMI K (1983) Alveolar soft part sarcoma. A review on its histogenesis and further studies based on electron microscopy, immunohistochemistry, and biochemistry. Am J Surg Pathol 7: 679-689

MULLANE KM, MONCADA S (1980) Prostacyclin - Release and the modulation of some vasoactive hormones. Prostaglandins 20: 25-49

MÜLLER J, BARAJAS L (1972) Electron microscopic and histochemical evidence for a tubular innervation in the renal cortex of the monkey. J Ultrastruct Res 41: 533-549

MÜLLER-SUUR R, GUTSCHE MU, SAMWER K, OELKERS W, HIERHOLZER K (1975) Tubuloglomerular feedback in rat kidneys of different renin contents. Pflügers Arch 359: 33-56

MULLINS JJ, BURT DW, WINDASS JD, MCTURK P, GEORGE H, BRAMMAR WJ (1982) Molecular cloning of two distinct renin genes from the DBA/2 mouse. EMBO J 1: 1461-1466

MUNDEL P, ELGER M, SAKAI T, KRIZ W (1988) Microfibrils are a major component of the mesangial matrix in the glomerulus of the rat kidney. Cell Tissue Res 254: 183-187

MUNKACSI I (1969) Distribution of the intrarenal monoaminergic nerves in the kidneys of the desert rat *(Dipodomys merriami)* and the white rat *(Rattus norvegicus)*. Acta Anat 73: 56-68

MUNTZ KH, GARCIA C, HAGLER HK (1985) Alpha 1-receptor localization in rat heart and kidney using autoradiography. Am J Physiol 249: H512-H519

MÜNZEL PA, HEALY DP, INSEL PA (1984) Autoradiographic localization of beta-adrenergic receptors in rat kidney slices using 125I-iodocyanopindolol. Am J Physiol 246: F240-F245

MURAD F (1986) Cyclic guanosine monophosphate as a mediator of vasodilation. J Clin Invest 78: 1-5

MURAKAMI E, HIWADA K, KOKUBU T (1980) Effects of insulin and glucagon on production of renin substrate by the isolated rat liver. J Endocrinol 85: 151-153

MURAKAMI K, TAKAHASHI S, SUZUKI F, HIROSE S, INAGAMI T (1980) Intermediate molecular weight renin and renin-binding protein(s) in the hog kidney. Biomed Res 1: 392-399

MURAKAMI K, HIROSE S, MIYAZAKI H, IMAI T, HORI H, HAYASHI T, KAGEYAMA R, OHKUBO H, NAKANISHI S (1984) Complementary DNA sequences of renin. State-of-the Art review. Hypertension 6 [Suppl 1]: 95-100

MURAKAMI T (1971) Application of the scanning electron microscope to the study of the fine distributions of the blood vessels. Arch Histol Jpn 32: 445-454

MURAKAMI T, KIKUTA A, OHTSUKA A, KANESHIGE T (1985a) Renal vasculature as observed by SEM of vascular casts. In: DIDIO LJA, MOTTA PM (eds) Basic, clinical, and surgical nephrology. Martinus Nijhoff, Boston, pp 83-98

MURAKAMI T, KIKUTA A, AKITA S, SANO T (1985b) Multiple efferent arterioles of the human kidney glomerulus as observed by scanning electron microscopy of vascular casts. Arch Histol Jpn 48: 443-447

MURRAY RD, CHURCHILL PC (1984) Effects of adenosine receptor agonists in the isolated, perfused rat kidney. Am J Physiol 247: H343-H348

MURRAY RD, CHURCHILL PC (1985) Concentration dependency of the renal vascular and renin secretory responses to adenosine receptor agonists. J Pharmacol Exp Ther 232: 189-193

MURRAY RD, MALVIN RL (1979) Intrarenal renin and autoregulation of renal plasma flow and glomerular filtration rate. Am J Physiol 236: F559-F566

MURTHY VV, GILBERT JC, GOLDBERG LI, KUO JF (1973) Dopamine-sensitive adenylate cyclase in canine renal artery. J Pharm Pharmacol 29: 567-571

NAFTILAN AJ, OPARIL S (1978) Inhibition of renin release from rat kidney slices by the angiotensins. Am J Physiol 235: F62-F68

NAFTILAN AJ, OPARIL S (1981) Effects of sodium intake and Goldblatt hypertension on renin release in rat kidney slices. Am J Physiol 240: F501-F507

NAFTILAN AJ, OPARIL S (1982) The role of calcium in the control of renin release. Hypertension 4: 670-675

NAGAI H, MATSUNAGA M, OGAWA K, KUWAHARA T, KANATSU K, PAK CH, HARA A, TAMURA T, KONO T, KAWAI C (1984) High level of plasma inactive renin in Bartter's syndrome. Jpn Circ J 48: 633-637

NAGATSU IT, GILLESPIE IL, GEORGE JM, FOLK JE, GLENNER GG (1965) Serum aminopeptidase "angiotensinase" and hypertension. II. Aminoacid B-naphthylamid hydrolysis by normal and hypertensive serum. Biochem Pharmacol 14: 853-861

NAGATSU IT, YAMAMOTO T, GLENNER GG, MEHL JW (1970) Purification of aminopeptidase A in human serum and degradation of angiotensin II by the purified enzyme. Biochim Biophys Acta 198: 255-270

NAIRN RC, FRASER KB, CHADWICK CS (1959) The histological localization of renin with fluorescent antibody. Br J Exp Pathol 40: 155-163

NAKADA T, SHIGEMATSU H, BARTTER FC, DELEA CS (1980) Nephropathologic characteristics of a woman with Bartter's syndrome after prolonged treatment with spironolactone. Nephron 26: 78-84

NAKAJIMA T, NAKAYAMA T, SOKABE H (1971) Examination of angiotensin-like substances from renal and extrarenal sources in mammalian and nonmammalian species. Gen Comp Endocrinol 17: 458-466

NAKAMURA M, JACKSON EK, INAGAMI T (1986) Role of vascular angiotensin II released by β-adrenergic stimulation in rats. J Cardiovasc Pharmacol 8 [Suppl 10]: 1-5

NAKAMURA N, SOUBRIER F, MENARD J, PANTHIER JJ, ROUGEON F, CORVOL P (1985) Nonproportional changes in plasma renin concentration, renal renin content, and rat renin messenger RNA. Hypertension 7: 855-859

NAKAMURA Y, NISHIMURA H, KHOSLA MC (1982) Vasodepressor action of angiotensin in conscious chickens. Am J Physiol 243: H456-H462

NAKANE H, NAKANE Y, CORVOL P, MENARD J (1980a) Sodium balance and renin regulation in rats: role of intrinsic renal mechanisms. Kidney Int 17: 607-614

NAKANE H, NAKANE Y, ROUX A, CORVOL P, MENARD J (1980b) Effects of selective and nonselective beta adrenergic agents on renin secretion in isolated perfused rat kidney. J Pharmacol Exp Ther 212: 34-38

NAKAYAMA T, NAKAJIMA T, SOKABE H (1977) Comparative studies on angiotensins. IV. Structure of snake *(Elaphe climocophora)* angiotensin. Chem Pharm Bull 25: 3255-3260

NARUSE K, INAGAMI T, CELIO MR, WORKMAN RJ, TAKII Y (1982) Immunohistochemical evidence that angiotensins I and II are formed by intracellular mechanism in juxtaglomerular cells. Hypertension 4 [Suppl 2]: 70-74

NARUSE K, NARUSE M, OBANA K, DEMURA R, DEMURA H, INAGAMI T, SHIZUME K (1986) Renin in the rat pituitary coexists with angiotensin II and depends on testosterone. Endocrinology 118: 2470-2476

NARUSE M, OBAMA K, NARUSE K, SUGINO N, DEMURA H, SHIZUME K, INAGAMI T (1985a) Antisera to atrial natriuretic factor reduces urinary sodium excretion and increases plasma renin activity in rats. Biochem Biophys Res Commun 132: 954-960

NARUSE M, SHIZUME K, INAGAMI T (1985b) Renin and angiotensins in cultured mouse adrenocortical tumour cells. Acta Endocrinologica 108: 545-549

NASJLETTI A, MALIK KU (1981) The renal kallikrein-kinin and prostaglandin systems interaction. Annu Rev Physiol 43: 597-609

NASJLETTI A, MASSON GMC (1971) Hepatic origin of renin substrate. Can J Physiol Pharmacol 49: 931-932

NASJLETTI A, MASSON GMC (1972) Studies on angiotensinogen formation in a liver perfusion system. Circ Res 30 [Suppl 2]: 187-202

NASJLETTI A, MCGIFF JC, COLINA-CHOURIO J (1978) Interrelations of the renal kallikrein-kinin system and renal prostaglandins in the conscious rat. Circ Res 43: 799-807

NAVAR LG (1986) Physiological role of the intrarenal renin-angiotensin system. Fed Proc 45: 1411-1413

NAVAR LG, ROSIVALL L (1984) Contribution of the renin-angiotensin system to the control of intrarenal hemodynamics. Kidney Int 25: 857-868

NAVAR LG, BELL PD, THOMAS CE, PLOTH DW (1978) Influence of perfusate osmolality on stop-flow pressure feedback responses in the dog. Am J Physiol 235: F352-F358

NAVAR LG, LAGRANGE RA, BELL PD, THOMAS CE, PLOTH DW (1979) Glomerular and renal hemodynamics during converting enzyme inhibition (SQ 20,881) in the dog. Hypertension 1: 371-377

NAVAR LG, BELL PD, PLOTH DW (1981) Role of feedback mechanism in renal autoregulation and sensing step in feedback pathway. Fed Proc 40: 93-98

NAVAR LG, JIRAKULSOMCHOK D, BELL PD, THOMAS CE, HUANG WC (1982a) Influence of converting enzyme inhibition on renal hemodynamics and glomerular dynamics in sodium-restricted dogs. Hypertension 4: 58-68

NAVAR LG, BELL PD, BURKE TJ (1982b) Role of macula densa feedback mechanism as a mediator of renal autoregulation. Kidney Int 22 [Suppl 12]: 157-164

NAVAR LG, EVAN AP, ROSIVALL L (1983) Microcirculation of the kidneys. In: MORTILLARO NA (ed) The Physiology and pharmacology of the microcirculation. Academie, New York, pp 397-488

NAVAR LG, CHAMPION WJ, THOMAS CE (1986a) Effects of calcium channel blockade on renal vascular resistance responses to changes in perfusion pressure and angiotensin-converting enzyme inhibition in dogs. Circ Res 58: 874-881

NAVAR LG, ROSIVALL L, CARMINES PK, OPARIL S (1986b) Effects of locally formed angiotensin II on renal hemodynamics. Fed Proc 45: 1448-1453

NAVAR LG, GILMORE JG, JOYNER WL, STEINHAUSEN M, EDWARDS RM, CASELLAS D, CARMINES PK, ZIMMERHACKL LB, YOKOTA SD (1986c) Direct assessment of renal microcirculatory dynamics. Fed Proc 45: 2856-2858

NAVAR LG, CARMINES PK, HUANG WC, MITCHELL KD (1987) The tubular effects of angiotensin II. Kidney Int 31 [Suppl 20]: 81-88

NEEDLEMAN P, ADAMS SP, COLE BR, CURRIE MG, GELLER DM, MICHENER ML, SAPER CB, SCHWARTZ D, STANDAERT DG (1985) Atriopeptins as cardiac hormones. Hypertension 7: 469-482

NEURATH H (1984) Evolution of proteolytic enzymes. Science 224: 350-357

NEWSTEAD J, MUNKACSI I (1969) Electron microscopic observations on the juxtamedullary efferent arterioles and arteriolae rectae in kidneys of rats. Z Zellforsch 97: 465-490

NIIRO GK, JAROSZ HM, O'MORCHOE PJ, O'MORCHOE CCC (1986) The renal cortical lymphatic system in the rat, hamster, and rabbit. Am J Anat 177: 21-34

NILSSON O (1965) The adrenergic innervation of the kidney. Lab Invest 14: 1392-1395

NISHIMURA H (1978) Physiological evolution of the renin-angiotensin system. Jpn Heart J 19: 806-822

NISHIMURA H (1980) Comparative endocrinology of renin and angiotensin In: JOHNSON JA, ANDERSON RR (eds) The renin-angiotensin system. Plenum, New York, pp 29-77

NISHIMURA H, BAILEY JR (1982) Intrarenal renin-angiotensin system in primitive vertebrates. Kidney Int 22 [Suppl 12]: 185-192

NISHIMURA H, OGURI M, OGAWA M, SOKABE H, IMAI M

(1970) Absence of renin in kidneys of elasmobranchs and cyclostomes. Am J Physiol 218: 911-915

NISHIMURA H, OGAWA M, SAWYER WH (1973) Renin-angiotensin system in primitive bony fishes and a holocephalian. Am J Physiol 224: 950-956

NISHIMURA H, LUNDE LG, ZUCKER A (1979) Renin response to hemorrhage in the aglomerular toadfish, *Opsanus tau*. Am J Physiol 237: H105-H111

NISHIMURA H, NAKAMURA Y, SUMNER RP, KHOSLA MC (1982) Vasopressor and depressor actions of angiotensin in the anesthetized fowl. Am J Physiol 242: H314-H324

NISHIZUKA Y (1983) Calcium, phospholipid turnover and transmembrane signalling. Philos Trans R Soc Lond [Biol] 302: 101-112

NOBILING R, BÜHRLE CP (1987) The mesangial cell culture: tool for the study of the electrophysiological and pharmacological properties of mesangial cells. Differentiation 36: 47-56

NOBILING R, BÜHRLE CP, HACKENTHAL E, HELMCHEN U, STEINHAUSEN M, WHALLEY A, TAUGNER R (1986) Ultrastructure, renin status, contractile and electrophysiological properties of the afferent glomerular arteriole in the rat hydronephrotic kidney. Virchows Arch [A] 410: 31-42

NOCHY D, BARRES D, CAMILLERI JP, BARIETY J, CORVOL P, MENARD J (1983) Abnormalities of renin-containing cells in human glomerular and vascular renal diseases. Kidney Int 23: 375-379

NOCHY D, BARIETY J, CAMILLERI JP, BARRES D, CORVOL P, MENARD J (1984) Diminished number of renin-containing cells in kidney biopsy samples from hypertensive women immediately postpartum: an immunomorphologic study. Kidney Int 26: 85-87

NOLLY HL, FASCIOLO JC (1972) The renin-angiotensin system through the phylogenetic scale. Comp Biochem Physiol [A] 41: 249-254

NOLLY HL, REID IA, GANONG WF (1974) Effect of theophylline and adrenergic blocking drugs on the renin response to norepinephrine in vitro. Circ Res 35: 575-579

NONOGUCHI H, KNEPPER MA, MANGANIELLO VC (1987) Effects of atrial natriuretic factor on cyclic guanosine monophosphate and cyclic adenosine monophosphate accumulation in microdissected nephron segments from rats. J Clin Invest 79: 500-507

NØRGAARD T (1976) Correlation of enzyme histochemical and structural segmentation in the proximal convoluted tubule of the rat kidney. Acta Pathol Microbiol Immunol Scand [A] 84: 172-182

NØRGAARD T (1979) Quantitative measurement of glucose-6-phosphate dehydrogenase in cortical fractions of the rabbit nephron. Histochemistry 63: 103-113

NOVIKOFF AB, ESSNER E (1962) Pathological changes in cytoplasmic organelles. Fed Proc 21: 1130-1142

NUSSEY SS, ANG VTY, JENKINS JS, CHOWDREY HS, BISSET GW (1984) Brattleboro rat adrenal contains vasopressin. Nature 310: 64-66

OATES HF, STOKES GS, GLOVER RG (1972) Plasma renin response to acute blockade of angiotensin II in the anaesthetized rat. Clin Exp Pharmacol Physiol 1: 155-160

OATES HF, FRETTEN JA, STOKES GS (1974) Disappearance rate of circulating renin after bilateral nephrectomy in the rat. Clin Exp Pharmacol Physiol 1: 547-549

OBANA K, NARUSE M, NARUSE K, SAKURAI H, DEMURA H, INAGAMI T, SHIZUME K (1985) Synthetic rat atrial natriuretic factor inhibits in vitro and in vivo renin secretion in rats. Endocrinology 117: 1282-1284

OBERLING C (1927) L'existence d'une housse neuro-musculaire au niveau des artères glomérulaires de l'homme. Compt Rend Soc Biol (Paris) 184: 1200-1202

OBERLING C (1944) Further studies on the preglomerular apparatus. Am J Pathol 20: 155-171

OBERLING C, HATT PY (1960a) Ultrastructure de l'appareil juxta-glomérulaire du rat. Compt Rend Soc Biol (Paris) 250: 229-230

OBERLING C, HATT PY (1960b) Etude de l'appareil juxtaglomérulaire du rat au microscope électronique. Ann Anat Pathol 5: 441-474

OCHI S, FUJIWARA Y, ORITA Y, TANAKA Y, SHIN SH, TAKAMA T, WADA A, UEDA N, KAMADA T (1987) Phosphoinositide turnover enhanced by angiotensin II in isolated rat glomeruli. Biochim Biophys Acta 927: 100-105

ODA M, ZIEGLER M (1965) Reningehalt der Nieren und des Blutes bei Kochsalzentzug und im Durst. Arch Exp Pathol Pharmacol 251: 174-176

O'DEA RFJ, HANSEN JA, MIRKIN BL (1984) Effect of calcium, sodium and isoproterenol on renin secretion from disaggregated rat renal cortical cells. Res Commun Chem Pathol Pharmacol 46: 187-205

ODY C, JUNOD AF (1977) Converting enzyme activity in endothelial cells isolated from pig pulmonary artery and aorta. Am J Physiol 232: C95-C98

OGAWA K, MATSUNAGA M, NAGAI H, HARA A, PAK CH, KAWAI C (1985) Effects of enalapril maleate on plasma level of inactive renin in renovascular hypertension. Clin Exp Hypertens [A] 7: 995-1005

OGAWA M, OGURI M (1978) Occurrence of the renin-angiotensin system in the vertebrates. Jpn Heart J 19: 791-798

OGAWA M, SOKABE H (1971) The macula densa site of avian kidneys. Z Zellforsch Mikrosk Anat 120: 29-36

OGAWA M, OGURI M, SOKABE H, NISHIMURA H (1972) Juxtaglomerular apparatus in the vertebrates. Gen Comp Endocrinol [Suppl] 3: 374-381

OGAWA N, ONO H (1985) No role for prostaglandins and bradykinin in the autoregulation of renal blood flow. Jpn J Pharmacol 39: 349-355

OGAWA N, ONO H (1986) Different effects of noradrenaline, angiotensin II and Bay K 8644 on the abolition of autoregulation of renal blood flow by verapamil. Naunyn-Schmiedebergs Arch Pharmacol 333: 445-449

OGAWA N, ONO H (1987) Role of Ca channel in the renal autoregulatory vascular response analysed by the use of BAY K 8644. Naunyn-Schmiedebergs Arch Pharmacol 335: 189-193

OGUNRO EA, LANMAN RB, SPENCER JR, FERGUSON AG, LESCH M (1979) Degradation of canine cardiac myosin and actin by cathepsin D isolated from homologous tissue. Cardiovasc Res 13: 621-629

OGURI M (1978) Presence of juxtaglomerular cells in the holocephalian kidney. Gen Comp Endocrinol 36: 170-173

OHASHI H, MATSUNAGA M, KUWAHARA T, PAK CH, KAWAI C (1985) Production and release of inactive renin by human vascular smooth muscle cells. Clin Exp Hypertens [A] 7: 1395-1407

OHKUBO H, KAGEYAMA R, UJIHARA M, HIROSE T, INAYAMA S, NAKANISHI S (1983) Cloning and sequence analysis of cDNA for rat angiotensinogen. Proc Natl Acad Sci USA 80: 2196-2200

OHKUBO H, NAKAYAMA K, TANAKA T, NAKANISHI S (1986) Tissue distribution of rat angiotensinogen mRNA and structural analysis of its heterogeneity. J Biol Chem 261: 319-323

OIEN AH, AUKLAND K (1983) A mathematical analysis of the myogenic hypothesis with special reference to autoregulation of renal blood flow. Circ Res 52: 241-252

OKAHARA T, ABE Y, YAMAMOTO K (1977) Effects of dibutyryl cyclic AMP and propranolol on renin secretion in dogs. Proc Soc Exp Biol Med 156: 213-218

OKAHARA T, ABE Y, IMANISHI M, MIURA K, YAMAMOTO K (1980) Effects of calcium ionophore, A23187, on prostaglandin E2 and renin release in dogs. Jpn Circ J 44: 394-399

OKAHARA T, MANCHANDIS MR, MICHELAKIS AM, YAMAMOTO K (1981) The renin-angiotensin system and renal prostaglandin E2 release in dogs. Proc Soc Exp Biol Med 166: 57-63

OKAMURA T, INAGAMI T (1984) Release of active and inactive renin from hog renal cortical slices in vitro. Am J Physiol 246: F765-F771

OKANO Y, ISHIZUKA Y, NAKSHIMA S, TOHMATSU T, TAKAGI H, NOZAWA Y (1985) Arachidonic acid release in rat peritoneal mast cells stimulated with antigen, ionophore A 23187, and compound 48/80. Biochem Biophys Res Commun 127: 726-732

OKEN DE, WOLFERT AI, LAVERI LA, CHOI SC (1985) Effects of intra-animal nephron heterogeneity on studies of glomerular dynamics. Kidney Int 27: 871-878

OKKELS MH (1929) Sur l'existence d'une spécialisation morphologique au niveau du pôle vasculaire du glomérule rénal chez la grenouille. C R Seances Acad Sci Paris 188: 193-195

OKKELS H, PÉTERFI T (1929) Beobachtungen über die Glomerulusgefäße der Froschniere. Z Zellforsch Mikrosk Anat 9: 327-331

OLIVER C (1983) Lysosomal heterogeneity in exocrine acinar cells. J Histochem Cytochem 31: 222-223

OLIVER C, HAND AR (1981) Membrane retrieval in exocrine acinar cells. Methods Cell Biol 23: 429-444

OLIVER JA, SCIACCA RR (1984) Local generation of angiotensin II as a mechanism of regulation of peripheral vascular tone in the rat. J Clin Invest 74: 1247-1251

OLIVER JA, SCIACCA RR, CANNON PJ (1983) Renal vasodilation by converting enzyme inhibition. Role of renal prostaglandins. Hypertension 5: 166-171

OLIVETTI G, GIOMELLI F, WIENER J (1985) Morphometry of superficial glomeruli in acute hypertension in the rat. Kidney Int 27: 31-38

OLSEN ME, HALL JE, MONTANI JP, GUYTON AC, LANGFORD HG, CORNELL JE (1985) Mechanisms of angiotensin II natriuresis and antinatriuresis. Am J Physiol 249: F299-F307

OLSEN ME, HALL JE, MONTANI JP, CORNELL JE (1987) Interaction between renal prostaglandins and angiotensin II in controlling glomerular filtration in the dog. Clin Sci 72: 429-436

OLSEN UB (1978) Kidney volume expansion and prostaglandin release by bradykinin. The effect of indomethacin pretreatment. Acta Physiol Scand 102: 129-136

OLSON RD, NIES AS, GERBER JG (1981) Alpha adrenergic-mediated renin release is prostaglandin-dependent. J Pharmacol Exp Ther 219: 321-325

OLSON RD, NIES AS, GERBER JG (1982) Beta adrenergically mediated release of renin in the dog is not confined to either beta-1 or beta-2 adrenoceptors. J Pharmacol Exp Ther 222: 606-611

OLSON RD, NIES AS, GERBER JG (1983) Catecholamine-induced renin release in the anesthetized mongrel dog is due to both alpha and beta adrenoceptor stimulation: evidence that only the alpha adrenoceptor component is prostaglandin mediated. J Pharmacol Exp Ther 224: 483-488

O'MORCHOE CCC, ALBERTINE KH, O'MORCHOE PJ (1978) Lymphatic organization in the canine renal cortex. Int Symp: The vascular and tubular organization of the kidney. June 25-28 (1978) Harvard Med School Boston. pp 30-31

O'MORCHOE CCC, O'MORCHOE PJ, ALBERTINE KH, JAROSZ HM (1981) Concentration of renin in the renal interstitium, as reflected in lymph. Renal Physiol 4: 199-206

ONDETTI MA, CUSHMAN DW (1984) Angiotensin-converting enzyme inhibitors: Biochemical properties and biological actions. CRC Crit Rev Biochem 16: 381-411

ONO H, KOKUBUN H, HASHIMOTO K (1974) Abolition by calcium antagonists of the autoregulation of renal blood flow. Naunyn-Schmiedebergs Arch Pharmacol 285: 201-207

OPARIL S, BAILIE MD (1973) Mechanism of renal handling of angiotensin II in the dog. Circ Res 33: 500-507

OPGENORTH TJ, ZEHR JE (1983) Role of calcium in the interaction of alpha and beta adrenoceptor-mediated renin release in isolated, constant pressure perfused rabbit kidneys. J Pharmacol Exp Ther 227: 144-149

OPGENORTH TJ, BURNETT JCJ, GRANGER JP, SCRIVEN TA (1986) Effects of atrial natriuretic peptide on renin secretion in nonfiltering kidney. Am J Physiol 250: F798-F801

ORCI L, STAUFFACHER W, RUFENER C, LAMBERT AE, ROUILLER C, RENOLD AE (1971) Acid phosphatase activity in secretory granules of pancreatic beta cells of normal rats. Diabetes 20: 385-388

ORCI L, GABBAY KH, MALAISSE WJ (1972) Pancreatic beta-cell web: Its possible role in insulin secretion. Science 175: 1128-1130

ORCI L, PERRELET A, GORDEN P (1978) Less-understood aspects of the morphology of insulin secretion and binding. Rec Prog Horm Res 34: 95-121

ORCI L, HALBAN P, AMHERDT M, RAVAZZOLA M, VASSALLI JD, PERRELET A (1984a) Nonconverted, amino acid analog-modified proinsulin stays in a Golgi-derived clathrin-coated membrane compartment. J Cell Biol 99: 2187-2192

ORCI L, HALBAN P, AMHERDT M, RAVAZZOLA M, VASSALLI JD, PERRELET A (1984b) A clathrin-coated, Golgi-related compartment of the insulin secreting cell accumulates proinsulin in the presence of monensin. Cell 39: 39-47

ORJAVIK OS, AAS M, FAUCHALD P, HOVIG T, OYSTESE B, BRODWALL EK, FLATMARK A (1975) Renin-secreting renal tumor with severe hypertension. Acta Med Scand 197: 329-335

OSATHANONDH V, POTTER EL (1966) Developing human kidney as shown by microdissection. V. Development of vascular pattern of glomerulus. Arch Pathol 82: 403-411

OSBORN JL, NOORDEWIER B, HOOK JB, BAILIE MD (1978) Mechanism of prostaglandin E2 stimulation of renin secretion. Proc Soc Exp Biol Med 159: 249-252

OSBORN JL, HOOK JB, BAILIE MD (1980) Regulation of plasma renin in developing piglets. Dev Pharmacol Ther 1: 217-228

OSBORN JL, DIBONA G, THAMES MD (1982a) Role of renal alpha-adrenoceptors mediating renin secretion. Am J Physiol 242: F620-F626

OSBORN JL, THAMES MD, DIBONA GF (1982b) Role of macula densa in renal nerve modulation of renin secretion. Am J Physiol 242: R367-R371

OSBORN JL, HOLDAAS H, THAMES MD, DIBONA GF (1983) Renal adrenoceptor mediation of antinatriuretic and renin secretion responses to low frequency renal nerve stimulation in the dog. Circ Res 53: 298-305

OSBORN JL, KOPP UC, THAMES MD, DIBONA GF (1984) Interactions among renal nerves, prostaglandins, and renal arterial pressure in regulation of renin release. Am J Physiol 247: F706-F713

OSBORNE MJ, DROZ B, MEYER P, MOREL F (1975) Angiotensin II: Renal localization in glomerular mesangial cells by autoradiography. Kidney Int 8: 245-254

OSMOND DH, ROSS LJ, SCAIFF KD (1973) Increased renin activity after cold storage of human plasma. Can J Physiol Pharmacol 51: 705-708

OSMOND DH, SCAIFF KO, COOPER RM, ROSS LJ (1974) Trypsin-induced increase in human plasma renin activity. Proc Can Fed Biol Soc 17: 27

OSSWALD H (1984) The role of adenosine in the regulation of glomerular filtration rate and renin secretion. Trends Pharmacol Sci 5: 94-97

OSSWALD H, SCHMITZ HJ, KEMPER R (1978) Renal action of adenosine: Effect on renin secretion in the rat. Naunyn-Schmiedebergs Arch Pharmacol 303: 95-99

OSSWALD H, NABAKOWSKI G, HERMES H (1980) Adenosine as a possible mediator of metabolic control of glomerular filtration rate. Int J Biochem 12: 263-267

OSSWALD H, HERMES HH, NABAKOWSKI G (1982) Role of adenosine in signal transmission of tubuloglomerular feedback. Kidney Int 22 [Suppl 12]: 136-142

OSTER P, HACKENTHAL E, HEPP R (1973) Radioimmunoassay of angiotensin II in rat plasma. Experientia 29: 353-354

PABST R, STERZEL RB (1983) Cell renewal of glomerular cell types in normal rats. An autoradiographic analysis. Kidney Int 24: 626-631

PAGE CP, ARCHER CB, PAUL W, MORLEY J (1984) Paf-acether: a mediator of inflammation and asthma. Trends Pharmacol Sci 5: 239-241

PAGE IH (1939) On the nature of the pressor action of renin. J Exp Med 70: 521-542

PAGE IH, HELMER OM (1940) Angiotonin activator, renin and angiotonin inhibitor and the mechanism of angiotensin tachyphylaxis in normal, hypertensive and nephrectomized animals. J Exp Med 71: 495-520

PANDEY KN, MISONO KS, INAGAMI T (1984) Evidence for intracellular formation of angiotensins: Coexistence of renin and angiotensin-converting enzyme in Leydig cells of rat testis. Biochem Biophys Res Commun 122: 1337-1343

PANTHIER JJ, HOLM I, ROUGEON F (1982) The mouse Rn locus: S allele of the renin regulator gene results from a single structural gene duplication. EMBO J 1: 1417-1421

PANTHIER JJ, DREYFUS M, TRONIK-LE ROUX D, ROUGEON F (1984) Mouse kidney and submaxillary gland renin genes differ in their 5′ putative regulatory sequences. Proc Natl Acad Sci USA 81: 5489-5493

PARK CS, MALVIN RL (1978) Calcium in the control of renin release. Am J Physiol 235: F22-F25

PARK CS, MALVIN RL, MURRAY RD, CHO KW (1978) Renin secretion as a function of renal renin content in dogs. Am J Physiol 234: F506-F509

PARK CS, HAN DS, FRAY JCS (1981) Calcium in the control of renin secretion: Ca2+ influx as an inhibitory signal. Am J Physiol 240: F70-F74

PARK CS, HONEYMAN TW, CHUNG ES, LEE JS, SIGMON DH, FRAY JCS (1986a) Involvement of calmodulin in mediating inhibitory actions of intracellular Ca2+ on renin secretion. Am J Physiol 251: F1055-1062

PARK CS, SIGMON DH, HAN DS, HONEYMAN TW, FRAY JCS (1986b) Control of renin secretion by Ca2+ and cyclic AMP through two parallel mechanisms. Am J Physiol 251: R531-R536

PARMENTIER M, INAGAMI T, POCHET R (1983) A 45000 molecular weight human renin precursor is synthesized in a cell-free translation system. Clin Sci 65: 475-477

PATRASSI GM, MANTERO F, FALLO F, DERKX FHM (1985) Captopril-induced changes on active and inactive renin in a patient with factor XII congenital deficiency. Res Exp Med 185: 217-220

PEACH MJ, SINGER HA, LOEB AL (1985) Mechanisms of endothelium-dependent vascular smooth muscle relaxation. Biochem Pharmacol 34: 1867-1874

PEART WS (1978) Intra-renal factors in renin release. Contrib Nephrol 12: 5-15

PEART WS, QUESADA T, TENYI I (1975) The effects on cyclic adenosine 3′,5′-monophosphate and guanosine 3′,5′-monophosphate and theophylline on renin secretion in the isolated perfused kidney of the rat. Br J Pharmacol 54: 55-60

PEART WS, QUESADA T, TENYI I (1977) The effects of EDTA and EGTA on renin secretion. Br J Pharmacol 59: 247-252

PEDERSEN JC, PERSSON AEG, MAUNSBACH AB (1980) Ultrastructure and quantitative characterization of the cortical interstitium in the rat kidney. In: MAUNSBACH AB, OLSEN TS, CHRISTENSEN EI (eds) Functional ultrastructure of the kidney. Academic, New York, pp 443-456

PEDRAZA-CHAVERRI J, ALATORRE-GONZALEZ MC, PENA JC, GARCIA-SAINZ JA (1986) Pertussis toxin enhances the beta-adrenergic and blocks the alpha2-adrenergic regulation of renin secretion in renal cortical slices. Life Sci 38: 1005-1011

PELAYO JC, FILDEY R, EISNER GM, JOSE PA (1983) The effect of dopamine blockade on renal sodium excretion. Am J Physiol 245: F247-F263

PELLETIER G (1973) Secretion and uptake of peroxidase by rat adrenohypophyseal cells. J Ultrastruct Res 43: 445-459

PERNOW B (1983) Substance P. Pharmacol Rev 35: 85-141

PERNOW J, LUNDBERG JM (1986) Neuropeptid Y constricts human skeletal muscle arteries via a nifedipine-sensitive mechanisms independent of extracellular calcium? Acta Physiol Scand 128: 655-656

PERRIN D, KEITH O, DOMINIQUE LA (1987) Anti a-fodrin inhibits secretion from permeabilized chromaffin cells. Nature 326: 498-501

PERSSON AEG, BOBERG U, HAHNE B, MÜLLER-SUUR R, NORLÉN BJ, SELÉN G (1982) Interstitial pressure as a modulator of tubuloglomerular feedback control. Kidney Int 22 [Suppl 12]: 122-128

PETER K (1907) Über die Nierenkanälchen des Menschen und einiger Säugetiere. Anat Anz 30 (Erg-H): 114-124

PETER K (1909) Untersuchungen über Bau und Entwicklung der Niere. G. Fischer, Jena

PETER S (1976) Ultrastructural studies on the secretory process in the epithelioid cells of the juxtaglomerular apparatus. Cell Tissue Res 168: 45-53

PETER S, MÖHRING J (1978) The juxtaglomerular apparatus of rats with hereditary hypothalamic diabetes insipidus. Cell Tissue Res 188: 335-339

PETER S, LAZAR J, GROSS F, FORSSMANN WG (1974a) Studies on the juxtaglomerular apparatus II. Quantitative morphology after adrenalectomy. Cell Tissue Res 151: 457-469

PETER S, LAZAR J, GROSS F, FORSSMANN WG (1974b) Studies on the juxtaglomerular apparatus III. Quantitative morphology after treatment with deoxycorticosterone (DOC). Cell Tissue Res 151: 471-480

Peters-Haefeli L (1971) Rate of inactivation of endogenous or exogenous renin in normal and in renin-depleted rats. Am J Physiol 221: 1339-1345

Peterson DR, Oparil S, Fluoret G, Carone FA (1977) Handling of angiotensin II and oxytocin by renal tubular segments perfused in vitro. Am J Physiol 232: F319-F324

Peterson DR, Chrabaszcz G, Peterson WR, Oparil S (1979) Mechanism for renal tubular handling of angiotensin. Am J Physiol 236: F365-F372

Peterson DR, Hjelle JT, Carone FA, Moore PA (1984) Renal handling of plasma high density lipoprotein. Kidney Int 26: 411-421

Pettinger WA, Marchelle M, Augusto L (1971) Renin suppression by DOC and NaCl in the rat. Am J Physiol 221: 1071-1074

Pettinger WA, Tanaka K, Keeton K, Campbell WB, Brooks SN (1975) Renin release, an artifact of anesthesia and its implications in rats. Proc Soc Exp Biol Med 148: 625-630

Pfaller W (1981) Structure function correlation on rat kidney. Adv Anat Embryol Cell Biol 70: 1-106

Pfeilschifter J, Kurtz A, Bauer C (1984) Activation of phospholipase C and prostaglandin synthesis by (arginine)vasopressin in cultures. Biochem J 223: 855-859

Pfeilschifter J, Kurtz A, Bauer C (1985) Inhibition of renin secretion by platelet activating factor (acetylglyceryl ether phosphorylcholine) in cultured rat renal juxtaglomerular cells. Biochem Biophys Res Commun 127: 903-910

Pfeilschifter J, Kurtz A, Bauer C (1986) Role of phospholipase C and protein kinase C in vasoconstrictor-induced prostaglandin synthesis in cultured rat renal mesangial cells. Biochem J 234: 125-130

Phat VN, Camilleri JP, Bariety J, Galtier M, Baviera E, Corvol P, Menard J (1981) Immunohistochemical characterization of renin-containing cells in the human juxtaglomerular apparatus during embryonal and fetal development. Lab Invest 45: 387-390

Phillips G, Mukherjee TM (1972) A juxtaglomerular cell tumor: light and electron microscopic studies of a renin-secreting kidney tumor containing both juxtaglomerular cells and mast cells. Pathology 4: 193-204

Phillips MI, Weyhenmeyer J, Felix D, Ganten D, Hoffman WE (1979) Evidence for an endogenous brain renin-angiotensin system. Fed Proc 38: 2260-2266

Pinet F, Corvol MT, Dench F, Bourguignon J, Feunteun J, Menard J, Corvol P (1985) Isolation of renin-producing human cells by transfection with three simian virus 40 mutants. Proc Natl Acad Sci USA 82: 8503-8507

Pinet F, Mizrahi J, Menard J, Corvol P (1986) Role of cyclic AMP in renin secretion by human transfected juxtaglomerular cells. J Hypertens 4 [Suppl 6]: 421-423

Pirotzky E, Bidault J, Burtin C, Gubler MC, Benveniste J (1984) Release of platelet-activating factor, slow-reacting substance, and vasoactive amines from isolated rat kidneys. Kidney Int 25: 404-410

Pirotzky E, Page C, Morley J, Bidault J, Benveniste J (1985) Vascular permeability induced by Paf-acether (platelet-activating factor) in the isolated perfused rat kidney. Agents Action 16: 17-18

Pisano JJ, Corthorn J, Yates K, Pierce JV (1978) The kallikrein-kinin system in the kidney. Contrib Nephrol 12: 116-125

Pitcock JA, Hartroft PM (1958) The juxtaglomerular cells in man and their relationship to the level of plasma sodium and to the zona glomerulosa of the adrenal cortex. Am J Pathol 34: 863-873

Pitcock JA, Hartroft PM, Newmark LN (1959) Increased renal pressor activity (renin) in sodium deficient rats and correlation with juxtaglomerular cell granulation. Proc Soc Exp Biol Med 100: 868-869

Ploth DW (1983) Angiotensin-dependent renal mechanisms in two-kidney, one-clip renal vascular hypertension. Am J Physiol 245: F131-F141

Ploth DW, Navar LG (1979) Intrarenal effects of the renin-angiotensin system. Fed Proc 38: 2280-2285

Ploth DW, Roy RN (1982a) Renal and tubuloglomerular feedback effects of (Sar 1, Ala 8) angiotensin II in the rat. Am J Physiol 242: F149-F157

Ploth DW, Roy NR (1982b) Renin-angiotensin influences on tubulo-glomerular feedback activity in the rat. Kidney Int 22 [Suppl 12]: 5114-5121

Ploth DW, Schnermann J, Dahlheim H, Hermle M, Schmidmeier E (1977) Autoregulation and tubuloglomerular feedback in normotensive and hypertensive rats. Kidney Int 12: 253-267

Ploth DW, Rudolph J, LaGrange R, Navar LG (1979) Tubuloglomerular feedback and single nephron function after converting enzyme inhibition in the rat. J Clin Invest 64: 1325-1335

Podjarny E, Shapira J, Rathaus M, Kariv N, Bernheim J (1986) Effect of angiotensin II on prostanoid synthesis in isolated rat glomeruli. Clin Sci 70: 527-530

Poe M, Liesch JM (1983) Mouse submaxillary gland renin contains a noncovalently attached fatty acid. J Biol Chem 258: 9856-9860

Pohlová I, Jelínek J (1974) Components of the renin-angiotensin system in the rat during development. Pflügers Arch 351: 259-270

Poisner AM, Wood GW, Poisner R (1982) Release of inactive renin from human fetal membranes and isolated trophoblasts. Clin Exp Hypertens [A] 4: 2007-2017

Pollock DM, Arendshorst WJ (1986) Effect of atrial natriuretic factor on renal hemodynamics in the rat. Am J Physiol 251: F795-F801

Porter JP, Ganong WF (1988) Vasoactive intestinal peptide and renin secretion. Am NY Acad Sci 527: 465-477

Porter JP, Reid IA, Said SI, Ganong WF (1982) Stimulation of renin secretion by vasoactive intestinal peptide. Am J Physiol 243: F306-F310

Porter JP, Said SI, Ganong WF (1983) Vasoactive intestinal peptide stimulates renin secretion in vitro: evidence for a direct action of the peptide on the renal juxtaglomerular cells. Neuroendocrinology 36: 404-408

Porter JP, Thrasher TN, Said SI, Ganong WF (1985) Vasoactive intestinal peptide in the regulation of renin secretion. Am J Physiol 249: F84-F89

Potkay SGJP (1973) Autoregulation of glomerular filtration in renin-depleted dogs. Proc Soc Exp Biol Med 143: 508-513

Potter DM, Dunn PM, McDonald WJ (1979) Evidence favoring the existence of two high molecular weight precursor forms of dog kidney renin. Endocrinology 105: 348-351

Poulsen K, Jacobsen J (1983) Renin precursors. J Hypertens 1: 3-5

Poulsen K, Jacobsen J (1986) Is angiotensinogen a renin inhibitor and not the substrate for renin? J Hypertens 4: 65-69

POULSEN K, NIELSEN AH (1981) Renin in the mouse kidney has a molecular weight of 40000. Clin Sci 60: 41-46

POULSEN K, VUUST J, LYKKEGARD S, NIELSEN AH, LUND T (1979) Renin is synthetized as a 50000 dalton single chain polypeptide in cell free translation system. FEBS Lett 98: 135-138

POULSEN K, VUUST J, LUND T (1980) Renin precursor from mouse kidney identified by cell-free translation of messenger RNA. Clin Sci 59: 297-299

POWELL HR, MCCREDIE DA, ROTENBERG E (1978) Renin release by parathyroid hormone in the dog. Endocrinology 103: 985-989

PRATT RE, OUELLETTE AJ, DZAU VJ (1983) Biosynthesis of renin: multiplicity of active and intermediate forms. Proc Natl Acad Sci USA 80: 6809-6813

PRATT RE, CARLETON JE, HEUSSER C, RICHIE JP, DZAU VJ (1986) Biosynthesis of multiple forms of renin in human kidney. J Hypertens 4 [Suppl 6]: 456-458

PREUSS KC, GROSS GJ, BROOKS HL, WARLTIER DC (1985) Slow channel calcium activators, a new group of pharmacological agents. Life Sci 37: 1271-1278

PREVOST M, GALLEZ D (1984) The role of repulsive hydration forces on the stability of aqueous black films. J Chem Soc Faraday Trans II 80: 517-533

PRICAM C, HUMBERT F, PERRELET A, ORCI L (1974) Gap junctions in mesangial and lacis cells. J Cell Biol 63: 349-354

PROUD D, NAKAMURA S, CARONE FA, HERRING PL, KAWAMURA M, INAGAMI T, PISANO JJ (1984) Kallikrein-kinin and renin-angiotensin systems in rat renal lymph. Kidney Int 25: 880-885

PROVOOST AP, DE JONG W (1978) Differential development of renal, DOCA-salt, and spontaneous hypertension in the rat after neonatal sympathectomy. Clin Exp Hypertens [A] 1: 177-189

PULLMAN TN, OPARIL S, CARONE FA (1975) Fate of labeled angiotensin II microinfused into individual nephrons in the rat. Am J Physiol 228: 747-751

QUESADA T, GARCIA TORRES L, ALBA F, GARCIA DEL RIO C (1979) The effects of dopamine on renin release in the isolated perfused rat kidney. Experientia 35: 1205

QUIRION R, GAUDREAU P, ST-PIERRE S, RIOUX F (1982) Localization of neurotensin binding sites in rat kidney. Peptides 3: 765-769

RABINOWE SL, TAYLOR T, DLUHY RG, WILLIAMS GH (1983) Beta-endorphin stimulates plasma renin release in normal human subjects (Abstract). Clin Res 31: 680

RADKE KJ, WILLIS LR, ZIMMERMAN GW, WEINBERGER MH, SELKURT EE (1986) Effects of histamine-receptor antagonists on histamine-stimulated renin secretion. Eur J Pharmacol 123: 421-426

RADZIWILL R, STUZMANN M, HILGENFELDT U, HACKENTHAL E (1986) Converting enzyme inhibitor-induced changes of plasma angiotensinogen concentration in the rat. Eur J Pharmacol 122: 59-64

RAIJ L, KEANE WF (1985) Glomerular mesangium: its function and relationship to angiotensin II. Am J Med 79 [Suppl 3C]: 24-30

RAINE AEG, ALLEN JM, LEDINGHAM JGG, BLOOM SR (1984) Renovascular distribution of NPY and its vasoconstrictor and natriuretic properties. Clin Exp Hypertens [A] 6: 1957-1960

RAMANATHAN K, GANTT C, GROSSMANN A (1973) Six year follow-up of a child with Bartter syndrome. Am J Dis Child 126: 230-235

RAPOPORT RM (1986) Cyclic guanosine monophosphate inhibition of contraction may be mediated through inhibition of phosphatidylinositol hydrolysis in rat aorta. Circ Rec 58: 407-410

RAPPELLI A, PEART WS (1968) Renal excretion of renin in the rat. Circ Res 23: 531-537

RASMUSSEN H, BARRETT PQ (1984) Calcium messenger system: An integrated view. Physiol Rev 64: 938-984

RASMUSSEN H, WAISMAN DM (1983) Modulation of cell function in the calcium messenger system. Rev Physiol Biochem Pharmacol 95: 111-148

RASMUSSEN H, KOJIMA K, KOJIMA W, ZAWALISCH W, APFELDORF W (1984) Calcium as intracellular messenger: sensitive modulation, C-kinase pathway, and sustained cellular response. Adv Cycl Nucleotide Res 18: 159-193

RASMUSSEN H, TAKUWA Y, PARK S (1987) Protein kinase C in the regulation of smooth muscle contraction. Faseb J 1: 177-185

RASSIER ME, LI T, ZIMMERMANN BG (1986) Analysis of influence of extra- and intrarenally formed angiotensin II on renal blood flow. J Cardiovasc Pharmacol 8 [Suppl 10]: 106-110

RE R (1987) The myocardial intracellular renin-angiotensin system. Am J Cardiol 59: 56-58

RE R, FALLON JT, DZAU V, OUAY SC, HABER E (1982) Renin synthesis by canine aortic smooth muscle cells in culture. Life Sci 30: 99-106

REALE E, MARINOZZI V, BUCHER O (1963) A propos de l'ultrastructure de l'appareil juxtaglomérulaire du rein. Acta Anat 52: 22-33

REGOLI D, BARABE J (1980) Pharmacology of bradykinin and related kinins. Pharmacol Rev 32: 1-46

REID IA (1984) Actions of angiotensin II on the brain: mechanisms and physiologic role. Am J Physiol 246: F533-F543

REID IA, STOCKIGT JR, GOLDFIEN A, GANONG WF (1972) Stimulation of renin secretion in dogs by theophylline. Eur J Pharmacol 17: 325-332

REID IA, MORRIS BJ, GANONG WF (1978) The renin-angiotensin system. Annu Rev Physiol 40: 377-410

REINECKE M, FORSSMANN WG (1988) Neuropeptide (neuropeptide Y, neurotensin, vasoactive intestinal polypeptide, substance P, calcitonin gene-related peptide, somatostatin) immunohistochemistry and ultrastructure of renal nerves. Histochemistry 89: 1-9

RHODIN JAG (1967) The ultrastructure of mammalian arterioles and precapillary sphincters. J Ultrastruct Res 18: 181-223

RIBSTEIN J, HUMPHREYS MH (1980) Endogenous opioids and electrolyte excretion after contralateral renal exclusion. Am J Physiol 244: F392-F398

RICHARDS HK, GRACE SA, NOBLE AR, MUNDAY KA (1979) Inactive renin in rabbit plasma: effect of haemorrhage. Clin Sci 56: 105-108

RICHARDS HK, LUSH DJ, NOBLE AR, MUNDAY KA (1981a) Inactive renin in rabbit plasma: effect of frusemide. Clin Sci 60: 393-398

RICHARDS HK, NOBLE AR, MUNDAY KA (1981b) Isoprenaline-induced secretion of active and inactive renin in anaesthetized rabbits and by kidney cortex slices. Clin Sci 61: 679-684

RICHOUX JP, CORDONNIER JL, BOUHNIK J, CLAUSER E, CORVOL P, MENARD J, GRIGNON G (1983) Immunocytochemical localization of angiotensinogen in rat liver and kidney. Cell Tissue Res 233: 439-451

Richoux JP, Amsaguine S, Grignon G, Bouhnik J, Menard J, Corvol P (1987) Earliest renin containing cell differentiation during ontogenesis in the rat. An immunocytochemical study. Histochemistry 88: 41-46

Riedel B, Bucher O (1967) Die Ultrastruktur des juxtaglomerulären Apparates des Meerschweinchens. Z Zellforsch 79: 244-258

Rightsel WA, Okamura T, Inagami T, Pitcock JA, Takii Y, Brooks B, Brown P, Muirhead EE (1982) Juxtaglomerular cells grown as monolayer cell culture contain renin, angiotensin I-converting enzyme, and angiotensins I and II/III. Circ Res 50: 822-829

Rioux F, Kérouac R, St-Pierre S (1982) Analysis of the biphasic depressor-pressor effect and tachycardia caused by neurotensin in ganglion-blocked rats. Neuropeptides 3: 113-127

Rix E, Ganten D, Schüll B, Unger T, Taugner R (1981) Converting-enzyme in the choroid prexus, brain, and kidney: Immunocytochemical and biochemical studies in rats. Neurosci Lett 22: 125-130

Robertson ALJ, Smeby RR, Bumpus FM, Page IH (1965) Renin production by organ cultures of renal cortex. Science 149: 650-651

Robertson PW, Klidjian A, Harding LK, Walters G, Lee MR, Robb-Smith AHT (1967) Hypertension due to a renin-secreting renal tumour. Am J Med 43: 963-976

Rocchini A, Barger P (1979) Renin release with carotid occlusion in the conscious dog: role of renal arterial pressure. Am J Physiol 236: H108-H111

Rodriguez-Puyol D, Arriba G, Blanchart A, Santos JC, Caramelo C, Fernandez-Cruz A, Hernando L, Lopez-Novoa JM (1986) Lack of a direct regulatory effect of atrial natriuretic factor on prostaglandins and renin release by isolated rat glomeruli. Biochem Biophys Res Commun 138: 496-501

Rojo-Ortega JM, Hatt PY, Genest J (1968) A propos de l'innervation des cellules juxtaglomérulaires. Étude au microscope électronique dans diverses conditions expérimentales chez le rat. Pathol Biol (Paris) 16: 497-504

Rojo-Ortega JM, Hayduk K, Genest J (1973a) Étude morphodynamique sur l'ontogénèse du système rénine-angiotensine. Union Med Can 102: 787-804

Rojo-Ortega JM, Yeghiayan E, Genest J (1973b) Lymphatic capillaries in the renal cortex of the rat. Lab Invest 29: 336-341

Rollhäuser H, Kriz W, Heinke W (1964) Das Gefäß-System der Rattenniere. Z Zellforsch 64: 381-403

Romen W, Thoenes W (1970) Histiocytäre und fibrocytäre Eigenschaften der interstitiellen Zellen der Nierenrinde. Virchows Arch [B] 5: 365-375

Romen W, Heine WD, Hollenz M (1978) The regeneration of the cells of the macula densa after subtotal nephrectomy in the rat. Virchows Arch [B] 27: 249-253

Rosen S, Tisher CC (1968) Observations on the rhesus monkey glomerulus and juxtaglomerular apparatus. Lab Invest 18: 240-248

Rosenbauer KA (1965) Die granulierten Zellen am Gefäßpol der Nierenkörperchen. Ergeb Allg Path Pathol Anat 46: 81-155

Rosenbauer KA, Krönig B (1967) Der Effekt von Aldosteron, Desoxycorticosteron und Hydrocortison auf die granulierten epitheloiden Zellen in der Wandung des Vas afferens der Mäuseniere. Z Gesamte Exp Med 144: 353-366

Rosenfeld MG, Mermod JJ, Amara SG, Swanson LW, Sawchenko PE, Rivier J, Vale WW, Evans RM (1983) Production of a novel neuropeptide encoded by the calcitonin gene via tissue-specific RNA processing. Nature 304: 129-135

Rosenlöf K (1986) Immunological comparison of renin substrate with erythropoietin. Scand J Clin Lab Invest 46: 497-504

Rosenlöf K, Fyhrquist F, Grönhagen-Riska C, Böhling T, Haltia M (1985) Erythropoietin and renin substrate in cerebellar haemangioblastoma. Acta Med Scand 218: 481-485

Rosivall L, Navar LG (1983) Effects on renal hemodynamics of intra-arterial infusions of angiotensins I and II. Am J Physiol 245: F181-F187

Rosivall L, Taugner R (1986) The morphological basis of fluid balance in the interstitium of the juxtaglomerular apparatus. Cell Tissue Res 243: 525-533

Rosivall L, Taugner R (1988) Fluid balance in the interstitium of the Goormaghtigh cell field. In: Persson AEG, Boberg U (eds) The juxtaglomerular apparatus, 11th Fernström Symposium. Elsevier, Amsterdam, pp 39-49

Rosivall L, Rinder DF, Champion J, Khosla MC, Navar LG, Oparil S (1983) Intrarenal angiotensin I conversion at normal and reduced renal blood flow in the dog. Am J Physiol 245: F408-F415

Rosivall L, Carmines PK, Navar LG (1984) Effects of renal arterial angiotensin I infusion on glomerular dynamics in sodium replete dogs. Kidney Int 26: 263-268

Rosivall L, Youngblood P, Navar LG (1986) Renal autoregulatory efficiency during angiotensin-converting enzyme inhibition in dogs on a low sodium diet. Renal Physiol 9: 18-28

Ross BD (1978) The isolated perfused rat kidney. Clin Sci Mol Med 55: 513-521

Rosset E, Veyrat R (1971) Effect of vasopressin (ADH), aldosterone, norepinephrine (NE), angiotensin I (AI) and II (AII) on renin release (RR) by human kidney (HK) slices in vitro. Acta Endocrinol [Suppl] 155: Abstract no 179

Rossi NF, Churchill PC, Churchill MC (1987a) Pertussis toxin reverses adenosine receptor-mediated inhibition of renin secretion in rat renal cortical slices. Life Sci 40: 481-487

Rossi NF, Churchill PC, Jacobson KE, Leahy AE (1987b) Further characterization of the renovascular effects of N6-cyclohexyladenosine in the isolated perfused rat kidney. J Pharmacol Exp Ther 240: 911-915

Rostand SG, Work J, Luke RG (1985) Effects of reduced chloride reabsorption on renin release in the isolated rat kidney. Pflügers Arch 405: 46-51

Roth J, Bendayan M, Orci L (1978) Ultrastructural localization of intracellular antigens by the use of protein A-gold complex. J Histochem Cytochem 26: 1074-1081

Rougeon F, Chambraud B, Foote S, Panthier JJ, Nageotte R, Corvol P (1981) Molecular cloning of a mouse submaxillary gland renin cDNA fragment. Proc Natl Acad Sci USA 78: 6367-6371

Rouiller C, Orci L (1971) The structure of the juxtaglomerular complex. In: Rouiller C, Muller AF (eds) The kidney. Morphology, biochemistry, physiology IV. Academic, New York, pp 1-80

Roy MW, Guthrie GPJ, Holladay FP, Kotchen TA (1983) Effects of verapamil on renin and aldosterone in the dog an rat. Am J Physiol 245: E410-E416

Rubanyi G, Vanhoutte PM (1985) Endothelium-removal

decreases relaxations of canine coronary arteries caused by adrenergic agonists and adenosine. J Cardiovasc Pharmacol 7: 139-144

RUBIN I, LAURITZEN E, LAURITZEN M (1980) Studies on the native forms of renin in the rat kidney. Biochim Biophys Acta 612: 126-136

RUDDY MC, ATLAS SA, SALERNO FG (1982) Hypertension associated with a renin-secreting adenocarcinoma of the pancreas. N Engl J Med 307: 993-997

RUDEHILL A, SOLLEVI A, FRANCO-CERECEDA A, LUNDBERG JM (1986) Neuropeptide Y (NPY) and the pig heart: release an coronary vasoconstrictor effects. Peptides 7: 821-826

RUMPF KW, SCHACHTERLE B, SCHMIDT S, BECKER K, SCHELER F (1978) Different responses of active and inactive plasma renin to various stimuli. Clin Sci Mol Med 55: 155-157

RUYTER JHC (1925) Über einen merkwürdigen Abschnitt der Vasa afferentia in der Mäuseniere. Z Zellforsch Mikrosk Anat 2: 242-248

RUYTER JHC (1964) Studies on an improved lead phosphate technique for the demonstration of non-specific acid phosphatase in non-deparaffinized organ and tissue sections. Histochemie 3: 521-537

RYAN GB, COGHLAN JP, SCOGGINS BA (1979) The granulated peripolar epithelial cell: a potential secretory component of the renal juxtaglomerular complex. Nature 277: 655-656

RYAN GB, ALCORN D, COGHLAN JP, HILL PA, JACOBS R (1982) Ultrastructural morphology of granule release from juxtaglomerular myoepithelioid and peripolar cells. Kidney Int 22 [Suppl 12]: 3-8

RYAN JW, RYAN US (1980) Biochemical and morphological aspects of the actions and inactivation of kinins and angiotensins. In: GROSS F, VOGEL G (eds) Enzymatic release of vasoactive peptides. Raven, New York, pp 259-274

RYAN JW, RYAN US, SCHULTS DR, WHITAKER C, CHUNG A (1975) Subcellular localization of pulmonary angiotensin-converting enzyme (kininase II). Biochem J 146: 497-499

RYAN MJ, BRODY MJ (1970) Distribution of histamine in the canine autonomic nervous system. J Pharmacol Exp Ther 174: 123-132

SAKAI T, KRIZ W (1987) The structural relationship between mesangial cells and basement membrane of the renal glomerulus. Anat Embryol 176: 373-386

SALAZAR FJ, FIKSEN-OLSEN MJ, OPGENORTH TJ, GRANGER JP, BURNETT JCJ, ROMERO JC (1986) Renal effects of ANP without changes in glomerular filtration rate and blood pressure. Am J Physiol 251: F532-F536

SALIDO EC, BARAJAS L, LECHAGO J, LABORDE NP, FISHER DA (1986a) Immunocytochemical localization of epidermal growth factor in mouse kidney. J Histochem Cytochem 34: 1155-1160

SALIDO EC, BARAJAS L, LECHAGO J, LABORDE NP, FISHER DA (1986b) Immunocytochemical localization of nerve growth factor in mouse kidney. J Neurosci Res 16: 457-465

SANFILIPPO F, PIZZO SV, CROKER BP (1982) Immunohistochemical studies of cell differentiation in a juxtaglomerular tumor. Arch Pathol Lab Med 106: 604-607

SARUTA T, MATSUKI S (1975) The effects of cyclic AMP, theophylline, angiotensin II and electrolytes upon renin release from rat kidney slices. Endocrinol Jpn 22: 137-140

SATO T, MCDOWELL EM, MCNEIL JS, FLAMENBAUM W, TRUMP BF (1977) Studies on the pathophysiology of acute renal failure III. A study of the juxtaglomerular apparatus of the rat nephron following administration of mercuric chloride. Virchows Arch [B] 24: 279-293

SATOH H, MATSUKURA K, YAMADA R, SATOH S (1981) Influence of angiotensins (I, II and III) on prostaglandin production by minced rat renal medulla. Prostaglandins 21: 973-984

SATOH H, TODA Y, SATOH S (1982) Stimulation of renin release from dog renal cortical slices with 1-isoproterenol and dibutyryl cyclic adenosine 3',5'-monophosphate. Jpn J Pharmacol 32: 945-949

SATOH H, TAKAHASHI K, TODA Y, SATOH S (1984) Prostacyclin-independence in -adrenoceptor-mediated renin release from dog renal cortical slices. Life Sci 35: 1519-1524

SAVIN VJ (1986) In vitro effects of angiotensin II on glomerular function. Am J Physiol 251: F627-F634

SCHAECHTELIN G, REGOLI D, GROSS F (1963) Bio-assay of circulating renin-like pressor material by isovolemic cross circulation. Am J Physiol 205: 303-306

SCHAECHTELIN G, REGOLI D, GROSS F (1964) Quantitative assay and disappearance rate of circulating renin. Am J Physiol 206: 1361-1364

SCHAFFENBURG CA, HAAS E, GOLDBLATT H (1960) Concentration of renin in kidney and angiotensinogen in serum of various species. Am J Physiol 199: 788-792

SCHAMBELAN M, HOWES EL Jr, STOCKIGT JR, NOAKES CA, BIGLIERI EG (1973) Role of renin and aldosterone in hypertension due to a renin-secreting tumour. Am J Med 55: 86-92

SCHARSCHMIDT LA, LIANOS E, DUNN MJ (1983) Arachidonate metabolites and the control of glomerular function. Fed Proc 42: 3058-3063

SCHARSCHMIDT LA, DOUGLAS JG, DUNN MJ (1986) Angiotensin II and eicosanids in the control of glomerular size in the rat and human. Am J Physiol 250: F348-F356

SCHAZ K, STOCK G, SIMMON W, SCHLOR KH, UNGER T, ROCKHOLD R, GANTEN D (1980) Enkephalin effects on blood pressure, heart rate, and baroreceptor reflex. Hypertension 2: 397-407

SCHENK DB, PHELPS MN, PORTER JG, SCARBOROUGH RM, MCENROE GA, LEWICKI JA (1985) Identification of the receptor for atrial natriuretic factor on cultured vascular cells. J Biol Chem 260: 14887-14890

SCHERBERICH JE, STUCKHARDT C, WOLF G, KUGLER P, SCHOEPPE W (1986) Angiotensinase A of human kidney: histo-/biochemical studies of an enzyme possibly involved in the regulation of intrarenal renin-angiotensin system. Nieren- und Hochdruckkrankheiten 15: 386 (Abstract)

SCHERF H, NIES AS, SCHWERTSCHLAG U, HUGHES M, GERBER JG (1986) Hemodynamic effects of platelet activating factor in the dog kidney in vivo. Hypertension 8: 737-741

SCHILLER A, TAUGNER R (1979) Are there specialized junctions in the pars maculata of the distal tubule? Cell Tissue Res 200: 337-344

SCHILLER A, FORSSMANN WG, GANTEN D, TAUGNER R (1978) Ultrastructure of the juxtaglomerular apparatus, renin and renal catecholamines in the rat and tree-shrew (Tupaia belangeri). Pflügers Arch [Suppl] 373: 33 (Abstract)

SCHILLER A, FORSSMANN WG, TAUGNER R (1980) The tight junctions of renal tubules in the cortex and outer medulla. A quantitative study of the kidneys of six species. Cell Tissue Res 212: 395-413

SCHLONDORFF D (1987) The glomerular mesangial cell: an expanding role for a specialized pericyte. FASEB J 1: 272-281

SCHLONDORFF D, NEUWIRTH R (1986) Platelet-activating factor and the kidney. Am J Physiol 251: F1-F11

SCHLONDORFF D, ROCZNIAK S, SATRIANO JA, FOLKERT VW (1980) Prostaglandin synthesis by isolated rat glomeruli: effect of angiotensin II. Am J Physiol 238: F486-F495

SCHLONDORFF D, GOLDWASSER P, SATRIANO JA (1985a) Synthesis of platelet-activating factor by isolated glomeruli and cultured mesangial cells of rat kidney. Clin Res 33: 588 (Abstract)

SCHLONDORFF D, PEREZ J, SATRIANO JA (1985b) Differential stimulation of PGE2 synthesis in mesangial cells by angiotensin and A23187. Am J Physiol 248: C119-C126

SCHLONDORFF D, GOLDWASSER P, NEUWIRTH R, SATRIANO JA, CLAY KL (1986) Production of platelet-activating factor in glomeruli and cultured glomerular mesangial cells. Am J Physiol 250: F1123-F1127

SCHLOSS G (1945/46) Der Regulationsapparat am Gefässpol des Nierenkörperchens in der normalen menschlichen Niere. Acta Anat 1: 365-410

SCHMID HE, GRAHAM LA (1962) Juxtaglomerular cell changes in dogs with antirenin titers. Circ Res 11: 853-856

SCHMIDT D, HAMMERSEN G, KARSUNKY KP, TAUGNER R (1971a) Reninfreisetzung aus isolierten Glomerula und Sekretgranula der Schweineniere. In: BOHLE A, SCHUBERT GE (eds) Fortschritte der Nephrologie. Schattauer, Stuttgart, New York, pp 429-433

SCHMIDT D, KARSUNKY KP, SCHNEIDER D, SOELL G, TAUGNER R (1971b) Structure and function of renin-containing granules isolated from pig kidney. Naunyn-Schmiedebergs Arch Pharmak 269: 487

SCHMIDT D, FORSSMANN WG, TAUGNER R (1972) Juxtaglomerular granules of the newborn rat kidney. Pflügers Arch 331: 226-232

SCHMIDT M, GIESEN-CROUSE EM, KRIEGER JP, WELSCH C, IMBS JL (1986) Effect of angiotensin converting enzyme inhibitors on the vasoconstrictor action of angiotensin I on isolated rat kidney. J Cardiovasc Pharmacol 8 [Suppl 10]: 100-105

SCHMIDT P, KOPSDA H, KRONENBERG KH, MEYER D, ZAZGORNIK J, KOTZAUREK R, NEUMANN E (1973) Bartter-Syndrom im Erwachsenenalter. Dtsch Med Wochenschr 98: 723-726

SCHNEIDER EG, DAVIS JO, BAUMBER JS, JOHNSON JA (1970) The hepatic metabolism of renin and aldosterone. Circ Res 26 [Suppl 1]: 175-183

SCHNEIDER W, THOENES W (1971) Macula densa und granulierte Zellen des juxtaglomerulären Apparates bei experimentellem Drosselungshochdruck. Morphometrische Untersuchungen. Virchows Arch [A] 353: 221-233

SCHNERMANN J, BRIGGS J (1985) Function of the juxtaglomerular apparatus: local control of glomerular hemodynamics. In: SELDIN DW, GIEBISCH G (eds) The kidney: physiology and pathophysiology. Raven, New York, pp 669-697

SCHNERMANN J, BRIGGS J (1986) Role of the renin-angiotensin system in tubuloglomerular feedback. Fed Proc 45: 1426-1430

SCHNERMANN J, WRIGHT FS, DAVIS JM, STACKELBERG WV, GRILL G (1970) Regulation of superficial nephron filtration rate by tubuloglomerular feedback. Pflugers Arch 318: 147-175

SCHNERMANN J, SCHUBERT G, HERMLE M, HERBST R, STOWE NT, YARIMIZU S, WEBER PC (1979) The effect of inhibition of prostaglandin synthesis on tubuloglomerular feedback in the rat kidney. Pflugers Arch 379: 269-286

SCHNERMANN J, BRIGGS JP, WEBER PC (1984a) Tubuloglomerular feedback, prostaglandins, and angiotensin in the autoregulation of glomerular filtration rate. Kidney Int 25: 53-64

SCHNERMANN J, BRIGGS JP, SCHUBERT C, MARIN-GREZ M (1984b) Opposing effects of captopril and aprotinin on tubuloglomerular feedback responses. Am J Physiol 247: F912-F918

SCHOR N, ICHIKAWA I, BRENNER BM (1981a) Mechanisms of action of various hormones and vasoactive substances on glomerular ultrafiltration in the rat. Kidney Int 20: 442-451

SCHOR N, ICHIKAWA I, RENNKE HG, TROY JL, BRENNER BM (1981b) Pathophysiology of altered glomerular function in aminoglycoside-treated rats. Kidney Int 19: 288-296

SCHRAMM M, TOWART R (1985) Modulation of calcium channel function by drugs. Life Sci 37: 1843-1860

SCHRYVER S, SANDERS E, BEIERWALTES WH, ROMERO CJ (1984) Cortical distribution of prostaglandin and renin in isolated dog glomeruli. Kidney Int 25: 512-518

SCHUREK HJ, ALT JM (1981) Effect of albumin on the function of perfused rat kidney. Am J Physiol 240: F569-F576

SCHÜRHOLZ J, SCHMID HJ, MEYER D, BOHLE A (1969) Light and electron microscopic investigations of the juxtaglomerular apparatus in a patient with a "pseudo-Bartter"-syndrome. In: PETERS G, ROCH-RAMEL F (eds) Progress in nephrology V. Springer, Berlin, Heidelberg, New York, pp 312-321

SCHUSTER VL (1986) Effects of angiotensin on proximal tubular reabsorption. Fed Proc 45: 1444-1447

SCHUSTER VL, KOKKO JP, JACOBSON HR (1984) Angiotensin II directly stimulates sodium transport in rabbit proximal convoluted tubuli. J Clin Invest 73: 507-515

SCHWARTZ J, REID IA (1986) Role of the vasoconstrictor and antidiuretic activities of vasopressin in inhibition of renin secretion in conscious dogs. Am J Physiol 250: F92-F96

SCHWARTZ WN, BIRD JWC (1977) Degradation of myofibrillar proteins by cathepsin B and D. Biochem J 167: 811-820

SCHWERTSCHLAG U, HACKENTHAL E (1982a) Histamine stimulates renin release from the isolated perfused rat kidney. Naunyn-Schmiedebergs Arch Pharmacol 319: 239-242

SCHWERTSCHLAG U, HACKENTHAL E (1982b) Forskolin stimulates renin release from the isolated perfused rat kidney. Eur J Pharmacol 84: 111-113

SCHWERTSCHLAG U, HACKENTHAL E (1983) Trifluoperazine antagonizes inhibition of renin release by angiotensin II. Clin Exp Pharmacol Physiol 10: 605-608

SCHWERTSCHLAG U, HACKENTHAL E, HACKENTHAL R, ROHS GH (1978) The effects of calcium and calcium-ionophores (X 537 A and A 23187) on renin release in the isolated perfused rat kidney. Clin Sci Mol Med 55: 163-166

SCHWERTSCHLAG U, SEYBERTH HW, MÜLLER H, GRUNEWALD R, ROHS HG, HACKENTHAL E (1980) Intrarenal conversion of $PGF_{2\alpha}$ into PGE and renin release in the isolated perfused rat kidney. Clin Sci 59: 117-119

SCHWERTSCHLAG U, SCHERF H, GERBER JG, MATHIAS M, NIES AS (1986) L-platelet activating factor (L-PAF) induced changes of renal vascular resistance (RVR), vascular reactivity and renin release (RR) in the isolated perfused rat kidney. Kidney Int 29: 388 (Abstract)

SCHWERTSCHLAG U, SCHERF H, GERBER JG, MATHIAS M, NIES AS (1987) L-Platelet activating factor induces changes of renal vascular resistance, vascular reactivity, and renin release in the isolated perfused rat kidney. Circ Res 60: 534-539

SCHWYZER R, ISELIN B, KAPPELER H, RINIKER B, RITTEL W, ZUBER A (1958) Synthese hochwirksamer Oktapeptide mit der vermutlichen Aminosäuresequenz des noch unbekannten Hypertensins II aus Rinderserum (val 5-Hypertensin II und val 5-Hypertensin II-asp-beta-amid). Helv Chim Acta 41: 1287-1295

SEALEY JE (1980) Prorenin activation by renal and plasma kallikreins. In: GROSS F, VOGEL G (eds) Enzymatic release of vasoactive peptides. Raven, New York, pp 117-136

SEALEY JE, LARAGH JH (1975) "Prorenin" in human plasma? Methodological and physiological implications. Circ Res 36/37 [Suppl 1]: 10-16

SEALEY JE, MOON C, LARAGH JH, ATLAS SA (1977) Plasma prorenin in normal, hypertensive, and anephric subjects and its effect on renin measurements. Circ Res 40 [Suppl 1]: 41-45

SEALEY JE, ATLAS SA, LARAGH JA, OZA NB, RYAN JW (1978) Human urinary kallikrein converts inactive to active renin and is a possible physiological activator of renin. Nature 275: 144-145

SEALEY JE, ATLAS SA, LARAGH JH (1980) Prorenin and other large molecular weight forms of renin. Endocr Rev 1: 365-391

SEALEY JE, ATLAS SA, LARAGH JH (1983) Prorenin in plasma and kidney. Fed Proc 42: 2681-2689

SEALEY JE, GLORIOSO N, ITSKOVITZ J, LARAGH JH (1986) Prorenin as a reproductive hormone. New form of the renin system. Am J Med 81: 1041-1046

SEAMON K, DALY JW (1981) Activation of adenylate cyclase by the diterpene forskolin does not require the guanine nucleotide regulatory protein. J Biol Chem 256: 9799-9801

SEGLEN PO, GORDON PB (1980) Effects of lysosomotropic monoamines, diamines, amino alcohols, and other amino compounds on protein degradation and protein synthesis in isolated rat hepatocytes. Mol Pharmacol 18: 468-475

SEKAR MC, HOKIN LE (1986) The role of phosphoinositides in signal transduction. J Membrane Biol 89: 193-210

SELKURT EE, HOCKEL GM, WEINBERGER MH (1979) Some evidence for interrelationship of histamine and prostaglandin on renal function. Proc Soc Exp Biol Med 160: 328-331

SESSLER FM, JAQUEZ JA, MALVIN RL (1986) Different production and decay rates of six renin forms isolated from rat plasma. Am J Physiol 250: E551-E557

SEYBERTH HW, RASCHER W, SCHWEER H, KÜHL PG, MEHLS O, SCHÄRER K (1985) Congenital hypokalemia with hypercalciuria in preterm infants: a hyperprostaglandinuric tubular syndrome different from Bartter's syndrome. J Pediatr 107: 694-701

SEYMOUR AA, ZEHR JE (1979) Influence of renal prostaglandin synthesis on renin control mechanisms in the dog. Circ Res 45: 13-25

SEYMOUR AA, BLAINE EH, MAZACK EK, SMITH SG, STABILITO II, HALEY AB, NAPIER MA, WHINNERY MA, NUTT RF (1985) Renal and systemic effects of synthetic atrial natriuretic factor. Life Sci 36: 33-44

SHADE RE, DAVIS JO, JOHNSON JA, GOTSHALL RW, SPIELMAN WS (1973) Mechanism of action of angiotensin II and antidiuretic hormone on renin secretion. Am J Physiol 224: 926-929

SHADE RE, DAVIS JO, JOHNSON JA, WITTY RJ (1972) Effects of aterial infusion of sodium and potassium on renin secretion in the dog. Circ Res 31: 719-727

SHETH KJ, TANG TT, BLAEDEL ME, GOOD TA (1978) Polydipsia, polyuria, and hypertension associated with renin-secreting Wilms tumour. J Pediatr 92: 921-924

SHINYAMA H, MATSUMURA Y, SASAKI Y, ICHIHARA T, MORIMOTO S (1987) Renin release in anesthetized rats is enhanced by the calmodulin antagonist W-7. Life Sci 40: 1687-1694

SHOEMAKER VH (1972) Osmoregulation and excretion in birds. In: FARNER DS, KING JR (eds) Avian biology, Vol. 2. Academic, New York, pp 527-574

SIEMENSEN HC, FRISCH E, SIEMENSEN JA (1972) Einfluss der Leber auf die Kinetik des Renin-Angiotensin-Systems der Ratte. Res Exp Med (Berl) 159: 23-34

SIKRI KL, FOSTER CL, BLOOMFIELD FJ, MARSHALL RD (1979) Localization by immunofluorescence and by light- and electron-microscopic immunoperoxidase techniques of Tamm-Horsfall glycoprotein in adult hamster kidney. Biochem J 181: 525-532

SIKRI KL, FOSTER CL, MACHUGH N, MARSHALL RD (1981) Localization of Tamm-Horsfall glycoprotein in the human kidney using immuno-fluorescence and immuno-electron microscopial techniques. J Anat 132: 597-605

SILVERMAN AJ, BARAJAS L (1974) Effect of reserpine on the juxtaglomerular granular cells and renal nerves. Lab Invest 30: 723-731

SIMIONESCU M, SIMIONESCU N, PALADE GE (1975) Segmental differentiation of cell junctions in the vascular endothelium. The microvasculature. J Cell Biol 67: 863-885

SIMIONESCU M, SIMIONESCU N, PALADE GE (1976) Segmental differentiations of cell junctions in the vascular endothelium. Arteries and veins. J Cell Biol 68: 705-723

SIMPSON FO (1970) Renal vasculature and hypertensive mechanism: Histological aspects. Circ Res 26-27 [Suppl 2]: 235-244

SIMPSON FO, DEVINE CE (1966) The fine structure of autonomic neuromuscular contacts in arterioles of sheep renal cortex. J Anat 100: 127-137

SIMPSON RV, GOODFRIEND TL (1984) Angiotensin and prostaglandin interactions in cultured kidney tubules. J Lab Clin Med 103: 255-271

SINAIKO AR (1983) Influence of adrenergic nervous and prostaglandin systems on hydralazine-induced renin release. Life Sci 33: 2269-2275

SINGHAL PC, SCHARSCHMIDT LA, GIBBONS N, HAYS RM (1986) Contraction and relaxation of cultured mesangial cells on a silicone rubber surface. Kidney Int 30: 862-873

SIRAGY HM, LAMB NE, ROSE EJ, PEACH MJ, CAREY RM (1986) Intrarenal angiotensin II influences renal function by direct tubular action. J Hypertens 4 [Suppl 6]: 60-62

SKAANE P, JAUSS E, MEYER D (1975) Morphometric studies of the juxtaglomerular apparatus in perimembranous glomerulonephritis. Acta Pathol Microbiol Immunol Scand [A] 83: 301-308

SKEGGS LT, KAHN JR, SHUMWAY NP (1956a) The preparation and function of the hypertensin-converting enzyme. J Exp Med 103: 295-299

SKEGGS LT, LENTZ KE, KAHN JR, SHUMWAY NP, WOODS KR (1956b) The amino acid sequence of hypertensin II. J Exp Med 104: 193-197

SKEGGS LT, LENTZ KE, HOCHSTRASSER H, KAHN JR (1964) The chemistry of renin substrate. Can Med Assoc J 90: 185-189

SKELTON FR, CHANDRA S, HUBBARD JC, BERNARDIS LL (1967) Studies on the genesis of the juxtaglomerular cell granule. Circ Res 20-21 [Suppl 2]: 29-46

SKINNER SL, MCCUBBIN JW, PAGE IH (1964) Control of renin secretion. Cir Res 15: 64-76

SKINNER SL, LUMBERS ER, SYMONDS EM (1972) Analysis of

changes in the renin-angiotensin system during pregnancy. Clin Sci 42: 479-488

SKINNER SL, CRAN EJ, GIBSON R, TAYLOR R, WALTERS WAW, CATT KJ (1975) Angiotensins I and II, active and inactive renin, renin substrate, renin activity, and angiotensinase in human liquor amnii and plasma. Am J Obstet Gynecol 121: 626-630

SKORECKI KL, BALLERMANN BJ, RENNKE HG, BRENNER BM (1983) Angiotensin II receptor regulation in isolated renal glomeruli. Fed Proc 42: 3064-3070

SKØTT O (1986a) Calcium and osmotic stimulation in renin release from isolated rat glomeruli. Pflügers Arch 406: 485-491

SKØTT O (1986b) Episodic release of renin from single isolated superfused rat afferent arterioles. Pflügers Arch 407: 41-45

SKØTT O (1987) Effects of amines, monensin and nigericin on the renin release from isolated superfused rat glomeruli. Pflügers Arch 409: 93-99

SKØTT O, BAUMBACH L (1985) Effects of adenosine on renin release from isolated rat glomeruli and kidney slices. Pflügers Arch 404: 232-237

SKØTT O, BRIGGS JP (1987) Direct demonstration of macula densa-mediated renin secretion. Science 237: 1618-1620

SKØTT O, TAUGNER R (1987) Effects of extracellular osmolality on renin release and on the structure of the epithelioid cell granules. Cell Tiss Res 249: 325-329

SLATER EE, HABER E (1978) A large form of renin from normal human kidney. J Clin Endocrinol Metab 47: 105-109

SLATER EE, HABER E (1979) Inactive renin - "through a glass darkly". N Engl J Med 301: 429-430

SMITH FG, LUPU AN, BARAJAS L, BAUER R, BASHORE RA (1974) The renin-angiotensin system in the fetal lamb. Pediatr Res 8: 611-620

SMITH JM, MOUW DR, VANDER AJ (1979) Effect of parathyroid hormone on plasma renin activity and sodium excretion. Am J Physiol 236: F311-F319

SMITH JM, MOUW DR, VANDER AJ (1983) Effect of parathyroid hormone on renin secretion. Proc Soc Exp Biol Med 172: 482-487

SMITH JP (1956) Anatomical features of the human renal glomerular efferent vessel. J Anat 90: 290-292

SMITH RE, FARQUHAR MG (1966) Lysosome function in the regulation of the secretory process in cells of the anterior pituitary gland. J Cell Biol 31: 319-347

SMITH WL, BELL TG (1978) Immunohistochemical localization of the prostaglandin-forming cyclooxygenase in renal cortex. Am J Physiol 235: F451-F457

SOFFER RL (1976) Angiotensin-converting enzyme and the regulation of vasoactive peptides. Annu Rev Biochem 45: 73-94

SOKABE H, OGAWA M (1974) Comparative studies of the juxtaglomerular apparatus. Int Rev Cytol 37: 271-327

SOKABE H, OGAWA M, OGURI M, NISHIMURA H (1969) Evolution of the juxtaglomerular apparatus in the vertebrate kidneys. Tex Rep Biol Med 27: 867-885

SOKABE H, OIDE H, OGAWA M, UTIDA S (1973) Plasma renin activity in japanese eels *(Aguilla japonica)* adapted to seawater or in dehydration. Gen Comp Endocrinol 21: 160-167

SOLTÉSZ BM, GOMBA SZ, SZOKOL M (1979) Lysosomal enzymes in the juxtaglomerular cell granules. Experientia 35: 533-534

SOSA RE, VOLPE M, MARION DN, GLORIOSO N, LARAGH JH, VAUGHAN EDJ, MAACK T, ATLAS S (1985) Effect of atrial natriuretic factor on renin secretion, plasma renin and aldosterone in dogs with acute unilateral renal artery constriction. J Hypertens 3 [Suppl 3]: 299-302

SOWERS JR (1984) Dopamine regulation of renin and aldosterone secretion: A review. J Hypertens 2 [Suppl 1]: 67-73

SOWERS JR, SOLLARS EG, TUCK ML, ASP ND (1980) Dopaminergic modulation of renin activity and aldosterone and prolactin secretion in the spontaneously hypertensive rat. Proc Soc Exp Biol Med 164: 598-603

SOWERS JR, BARRETT JD, SAMBHI MP (1981) Dopaminergic modulation of renin release. Clin Exp Hypertens [A] 3: 15-25

SPAHR J, DEMERS LM, SHOCHAT SJ (1981) Renin producing Wilms' tumour. J Pediatr Surg 16: 32-34

SPANIDIS A, WUNSCH H, KAISSLING B, KRIZ W (1982) Three-dimensional shape of a Goormaghtigh cell and its contact with a granular cell in the rabbit kidney. Anat Embryol 165: 239-252

SPARKS JC, SUSIC D (1977) The effects of propranolol on plasma renin activity and renal renin concentration in rats on normal and sodium deficient diets. Pharmacol Res Commun 9: 479-487

SPERBER J (1960) Excretion. In: MARSHALL AJ (ed) Biology and comparative physiology of birds, vol 1. Acadimic, London, pp 469-492

SPIELMAN WS (1984) Antagonistic effect of theophylline on the adenosine-induced decrease in renin release. Am J Physiol 247: F246-F251

SPIELMAN WS, OSSWALD H (1979) Blockade of postocclusive renal vasoconstriction by an angiotensin II antagonist: evidence for an angiotensin-adenosine interaction. Am J Physiol 237: F463-F467

SPIELMAN WS, THOMPSON CI (1982) A proposed role for adenosine in the regulation of renal hemodynamics and renin release. Am J Physiol 242: F423-F435

SPIELMAN WS, AREND LJ, FORREST JNJ (1987) The renal and epithelial actions of adenosine. In: GERLACH E, BECKER BF (eds) Topics and perspectives in adenosine research. Springer, Berlin Heidelberg New York Tokyo, pp 249-260

SPRINGATE JE, FILDES RD, HONG SK, Feld LG, ACARA M (1987) Renal effects of atrial natriuretic factor in domestic fowl. Life Sci 40: 915-920

SQUIRES JP, ULBRIGHT TM, DESCHRYVER-KECSKEMETI K, ENGLEMAN W (1984) Juxtaglomerular cell tumor of the kidney. Cancer 53: 516-523

SRAER JD, SRAER J, ARDAILLOU R, MIMOUNE O (1974) Evidence for renal glomerular receptors for angiotensin II. Kidney Int 6: 241-246

STANIER MW (1960) The function of the mammalian mesonephros. J Physiol (Lond) 151: 472-478

STANLEY EF, EHRENSTEIN G (1985) A model for exocytosis based on the opening of calcium-activated potassium channels in vesicles. Life Sci 37: 1985-1995

STANTON BA (1985) Role of adrenal hormones in regulating distal nephron structure and ion transport. Fed Proc 44: 2717-2722

STEIN JH (1985) The pathogenetic spectrum of Bartter's syndrome. Kidney Int 28: 85-93

STEINER DF (1985) Prohormonal processing mechanisms. In: HAKANSON R, THORELL J (eds) Biogenetics of neurohormonal peptides. Academic, London, pp 121-128

STEINER DF, DOCHERTY K, CHAN SJ, SAN SEGUNDO B, CARROLL B (1983) Intracellular proteolytic mechanisms in the biosynthesis of hormones and peptide neurotransmitters. In: KOCH G, RICHTER D (eds) Biochemical and clinical aspects of neuropeptides: synthesis, processing, and gene structure. Academic, London, pp 3-13

Steinhausen M, Tanner GA (1976) Microcirculation and tubular urine flow in the mammalian kidney cortex (in vivo microscopy). Springer, Berlin Heidelberg New York, pp 279-335

Steinhausen M, Snoei H, Parekh N, Baker R, Johnson PC (1983) Hydronephrosis: a new method to visualize vas afferens, efferens, and glomerular network. Kidney Int 23: 794-806

Steinhausen M, Kücherer H, Parekh N, Weis S, Wiegman DL, Wilhelm KR (1986a) Angiotensin II control of the renal microcirculation: Effect of blockade by saralasin. Kidney Int 30: 56-61

Steinhausen M, Weis S, Fleming J, Dussel R, Parekh N (1986b) Responses of in vivo renal microvessels. Kidney Int 30: 361-370

Steinhausen M, Sterzel RB, Fleming JT, Kuhn R, Weis S (1987) Acute and chronic effects of angiotensin II on the vessels of the split hydronephrotic kidney. Kidney Int 31 [Suppl 20]: 64-73

Stephenson R, Sole M, Baines AD (1982) Neural and extraneural catecholamine production by rat kidneys. Am J Physiol 242: F261-F266

Sternberger LA (1974) Immunocytochemistry. Englewood Cliffs: Prentice Hall

Sternberger LA (1986) Immunocytochemistry, 3rd edn. Wiley, New York

Steven K (1974) Effect of peritubular infusion of angiotensin II on rat proximal nephron function. Kidney Int 6: 73-80

Stiles GL (1986) Adenosine receptors: structure, function and regulation. Trends Pharmacol Sci 7: 486-490

Stoof JC, Kebabian JW (1984) Two dopamine receptors: biochemistry, physiology and pharmacology. Life Sci 35: 2281-2296

Stowe N, Schnermann J, Hermle M (1979) Feedback regulation of nephron filtration rate during pharmacologic interference with the renin-angiotensin and adrenergic systems in rats. Kidney Int 15: 473-486

Straus W (1964) Occurrence of phagosomes and phago-lysosomes in different segments of the nephron in relation to the reabsorption, transport, digestion, and extrusion of intravenously injected horseradish peroxidase. J Cell Biol 21: 295-308

Straus W (1967) Changes in the intracellular location of small phagosomes (micropinocytic vesicles) in kidney and liver cells in relation to time after injection and dose of horseradish peroxidase. J Histochem Cytochem 15: 381-393

Straus W (1971) Comparative analysis of the concentration of injected horseradish peroxidase in cytoplasmic granules of the kidney cortex, in the blood, urine, and liver. J Cell Biol 48: 620-632

Straus W (1979) Renal reabsorption and excretion of horseradish peroxidase. Kidney Int 16: 404-408

Struthers AD, Anderson JV, Payne N, Causon RC, Slater JDH, Bloom SR (1986) The effect of atrial natriuretic peptide on plasma renin activity, plasma aldosterone, and urinary dopamine in man. Eur J Clin Pharmacol 31: 223-226

Stutchfield J, Jones PM, Howell SL (1986) The effects of polymyxin B, a protein kinase C inhibitor, on insulin secretion from intact and permeabilised islets of Langerhans. Biochem Biophys Res Commun 136: 1001-1006

Stuzmann M, Radziwill R, Komischke K, Klett C, Hackenthal E (1986) Hormonal and pharmacological alteration of angiotensinogen secretion from rat hepatocytes. Biochim Biophys Acta 886: 48-56

Sudo JI (1981) Distributions of peptidases in the metabolization of peptide hormones, particularly angiotensin II, along the isolated single nephron of rat. Folia Pharmacol Jpn 78: 27-44

Sulimovici S, Pinkus LM, Susser FI, Roginsky MS (1984) Identification of calmodulin-sensitive and calmodulin-insensitive adenylate cyclase in rat kidney. Arch Biochem Biophys 234: 434-441

Summers RJ, Kuhar MJ (1983) Autoradiographic localization of beta-adrenoceptors in rat kidney. Eur J Pharmacol 91: 305-310

Summers RJ, Stephenson JA, Kuhar MJ (1985) Localization of beta adrenoceptor subtypes in rat kidney by light microscopic autoradiography. J Pharmacol Exp Ther 232: 561-569

Sutherland LE (1970) A fluorescent antibody study of juxtaglomerular cells using the freeze-substitution technique. Nephron 7: 512-523

Sutherland LE, Hartroft PM (1968) Comparative morphology of juxtaglomerular cells II. The presence of juxtaglomerular cells in embryos. Can J Zool 46: 257-263

Sutherland LE, Hartroft P, Balis JK, Baily JU, Lynch MJ (1970) Bartter's syndrome. Acta Paediatr Scand Suppl 201: 1-24

Suzuki M, Satoh S (1984) Suppression of bradykinin-induced renin release by indomethacin in anesthetized rats. Clin Exp Hypertens [A] 6: 1227-1235

Suzuki S, Franco-Saenz R, Tan SY, Mulrow PJ (1980) Direct action of rat urinary kallikrein on rat kidney to release renin. J Clin Invest 66: 757-762

Suzuki S, Franco-Saenz R, Tan SY, Mulrow PJ (1981a) Effect of indomethacin on plasma renin activity in the conscious rat. Am J Physiol 240: E286-E289

Suzuki S, Franco-Saenz R, Tan SY, Mulrow PJ (1981b) Direct action of kallikrein and other proteases on the renin-angiotensin system. Hypertension 3 [Suppl 1]: 13-17

Suzuki Y, Churg J, Grishman E, Mautner W, Dachs S (1963) The mesangium of the renal glomerulus. Electron microscopic studies of pathologic alterations. Am J Pathol 43: 555-578

Swain JA, Heyndrickx GR, Boettcher DH, Vatner SF (1975) Prostaglandin control of renal circulation in the unanesthetized dog and baboon. Am J Physiol 229: 826-830

Swales JD, Abramovici A, Beck F, Bing RF, Laudon M, Thurston H (1983) Arterial wall renin. J Hypertens 1 [Suppl 1]: 17-22

Swann HG, Norman RJ (1970) The periarterial spaces of the kidney. Tex Rep Biol Med 28: 317-334

Swerts JP, Perdrisot R, Malfroy B, Schwartz JC (1979) Is 'enkephalinase' identical with 'angiotensin-converting enzymes'? Eur J Pharmacol 53: 209-210

Szabó J, Dévényi I (1972) Ultrastructural data on different types of hyperplasia and hyperfunction of the juxtaglomerular apparatus. Acta Morphol Hung 20: 39-48

Szilagyi JE, Chelly J, Doursout MF (1986) Suppression of renin release by antagonism of endogenous opiates in the dog. Am J Physiol 250: R633-R637

Tagawa H, Vander AJ (1969) Effect of acetylcholine on renin secretion in salt-depleted dogs. Proc Soc Exp Biol Med 132: 1087-1090

Tagawa H, Vander AJ (1970) Effects of adenosine compounds on renal function and renin secretion in dogs. Circ Res 26: 327-338

TAGAWA H, VANDER AJ, BONJOUR JP, MALVIN RL (1971) Inhibition of renin secretion by vasopressin in unanesthetized sodium-deprived dogs. Am J Physiol 220: 949-951

TAKADA Y, HIWADA K, UNNO M, KOKUBU T (1982) Immunocytochemical localization of angiotensin converting enzyme at the ultrastructural level in the human lung and kidney. Biomed Res 3: 169-174

TAKAHASHI K, HISA H, SATOH S (1984) Effects of α-agonist on renin and prostaglandin E2 release in anesthetized dogs. Am J Physiol 247: E604-E608

TAKAHASHI S, MIURA R, MIYAKE Y (1985) A study on renin binding protein (RnBP) in the human kidney. J Biochem 97: 671-677

TAKAHASHI S, OHSAWA T, MURA R, MIYAKE Y (1983) Purification and characterization of renin binding protein (RnBP) from porcine kidney. J Biochem 93: 1583-1594

TAKAHASHI T, MIURA T, SUE A, SAITO K, SAKAUE M, YAMAGATA Y, FUKUCHI S, SATO Z, HIRAI T, TERASHIMA K, OKA K, IMAI Y (1976) A case of juxtaglomerular cell tumor diagnosed preoperatively. Nephron 17: 483-495

TAKAORI K, IKEMOTO F, YAMAMOTO K (1981) Biochemical properties of the renin binding substance of rat kidney. Clin Exp Hypertens [A] 3: 991-1000

TAKAORI K, IWAO H, IKEMOTO F, YAMAMOTO K (1982) Specific distribution of renin binding protein in rat and dog kidney. Clin Exp Hypertens [A] 4: 2097-2105

TAKEDA S, KUSANO E, MURAYAMA N, ASANO Y, HOSODA S, SOKABE S, KAWASHIMA H (1986) Atrial natriuretic peptide elevates cGMP contents in glomeruli and in distal tubules of rat kidney. Biochem Biophys Res Commun 136: 947-954

TAKEMOTO Y, NAKAJIMA T, HASEGAWA Y, WATANABE TX, SOKABE H, KUMAGOA S, SAKAKIBARA S (1983) Chemical structures of angiotensins formed by incubating plasma with the kidney and corpuscles of Stannius in the chum salmon, *Oncorhynchus keta*. Gen Comp Endocrinol 51: 219- 227

TAKII Y, INAGAMI T (1982) Purification of a completely inactive renin from hog kidney and identification as renin zymogen. Biochem Biophys Res Commun 104: 133-140

TAKII Y, TAKAHASHI N, INAGAMI T, YOKOSAWA N (1980) A new form of renin in normal human plasma: "big renin" is a mixture of inactive prorenin and the new active high molecular weight renin. Life Sci 26: 347-353

TANAKA M, IWAO H, IKEMOTO F, YAMAMOTO K (1985) High-molecular-weight renin converting enzyme from dog kidney: detection and fractionation. Life Sci 36: 1217-1224

TANAKA T, GRESIK EW, MICHELAKIS AM, BARKA T (1980) Immunocytochemical localization of renin in kidneys and submandibular glands of SWR/J and C57BL/6J mice. J Histochem Cytochem 28: 1113-1118

TANI E, YAMAGATA S, ITO Y (1977) Cell membrane structure of vascular smooth muscle of circle of Willis. Cell Tissue Res 179: 131-142

TARTAKOFF AM (1980) The Golgi complex: crossroads for vesicular traffic. Int Rev Exp Pathol 22: 227-251

TATEMOTO K (1982) Neuropeptide Y: complete amino sequence of the brain peptide. Proc Natl Acad Sci USA 79: 5485-5489

TATEMOTO K, CARLQUIST M, MUTT B (1982) Neuropeptide Y - a novel brain peptide with structural similarities to peptide YY and pancreatic polypeptide. Nature 296: 659-660

TAUGNER C, POULSEN K, HACKENTHAL E, TAUGNER R (1979) Immunohistochemical localization of renin in mouse kidney. Histochemistry 62: 19-27

TAUGNER R, GANTEN D (1982) The localization of converting enzyme in kidney vessels of the rat. Histochemistry 75: 191-201

TAUGNER R, HACKENTHAL E (1981) Angiotensin II in epithelioid (renin containing) cells of rat kidney. Histochemistry 72: 499-509

TAUGNER R, HACKENTHAL E (1988) On the character of the secretory granules in juxtaglomerular epithelioid cells. Int Rev Cytol 110: 93-131

TAUGNER R, METZ R (1986) Development and fate of the secretory granules of juxtaglomerular epithelioid cells. Cell Tissue Res 246: 595-606

TAUGNER R, BOLL U, ZAHN P, FORSSMANN WG (1976) Cell junctions in the epithelium of Bowman's capsule. Cell Tissue Res 172: 431-446

TAUGNER R, FORSSMANN WG, BILLICH H, BOLL U, GANTEN D, SELLER H (1978a) Innervation of the juxtaglomerular apparatus and the effect of renal nerve stimulation. In: COUPLAND RE, FORSSMANN WG (eds) Peripheral neuroendocrine interaction. Springer, Berlin Heidelberg New York, pp 153-163

TAUGNER R, SCHILLER A, KAISSLING B, KRIZ W (1978b) Gap junctional coupling between the JGA and the glomerular tuft. Cell Tissue Res 186: 279-285

TAUGNER R, FORSSMANN WG, GANTEN D, SCHILLER A (1980) Studies on the juxtaglomerular apparatus VI. Sympathetic innervation, catecholamines and the renin-angiotensin-system in rats and tree-shrews *(Tupaia belangeri)*. Cell Tissue Res 212: 375-382

TAUGNER R, HACKENTHAL E, NOBILING R, HARLACHER M, REB G (1981) The distribution of renin in the different segments of the renal arterial tree. Histochemistry 73: 75-88

TAUGNER R, HACKENTHAL E, HELMCHEN U, GANTEN D, KUGLER P, MARIN-GREZ M, NOBILING R, UNGER T, LOCKWALD I, KEILBACH R (1982a) The intrarenal renin-angiotensin-system. An immunocytochemical study on the localization of renin, angiotensinogen, converting enzyme and the angiotensins in the kidney of mouse and rat. Klin Wochenschr 60: 1218-1222

TAUGNER R, HACKENTHAL E, INAGAMI T, NOBILING R, POULSEN K (1982b) Vascular and tubular renin in the kidneys of mice. Histochemistry 75: 473-484

TAUGNER R, HACKENTHAL E, RIX E, NOBILING R, POULSEN K (1982c) Immunocytochemistry of the renin-angiotensin system: Renin, angiotensinogen, angiotensin I, angiotensin II, and converting enzyme in the kidneys of mice, rats, and tree shrews. Kidney Int 22 [Suppl 12]: 33-43

TAUGNER R, MARIN-GREZ M, KEILBACH R, HACKENTHAL E, NOBILING R (1982d) Immunoreactive renin and angiotensin II in the afferent glomerular arterioles of rats with hypertension due to unilateral renal artery constriction. Histochemistry 76: 61-69

TAUGNER R, BÜHRLE CP, GANTEN D, HACKENTHAL E, HARDEGG C, HARDEGG G, NOBILING R, UNGER T (1983a) Angiotensin-like activity in resistance vessels. Immunocytochemical study in Chinese hamsters. Histochemistry 78: 61-70

TAUGNER R, BÜHRLE CP, GANTEN D, HACKENTHAL E, HARDEGG C, HARDEGG G, NOBILING R (1983b) Immunohistochemistry of the renin-angiotensin-system in the kidney. Clin Exp Hypertens [A] 5: 1163-1177

TAUGNER R, BÜHRLE CP, HACKENTHAL E, MANNEK E, NOBILING R (1984a) Morphology of the juxtaglomerular apparatus and secretory mechanisms. Contrib Nephrol 43: 76-101

TAUGNER R, BÜHRLE CP, NOBILING R (1984b) Ultrastructural changes associated with renin secretion from the juxtaglomerular apparatus of mice. Cell Tissue Res 237: 459-472

TAUGNER R, KIRCHHEIM H, FORSSMANN WG (1984c) Myoendothelial contacts in glomerular arterioles and in renal interlobular arteries of rat, mouse and *Tupaia belangeri*. Cell Tissue Res 235: 319-325

TAUGNER R, MANNEK E, NOBILING R, BÜHRLE CP, HACKENTHAL E, GANTEN D, INAGAMI T, SCHRÖDER H (1984d) Coexistence of renin and angiotensin II in epithelioid cell secretory granules of rat kidney. Histochemistry 81: 39-45

TAUGNER R, BÜHRLE CP, NOBILING R, KIRSCHKE H (1985a) Coexistence of renin and cathepsin B in epithelioid cell secretory granules. Histochemistry 83: 103-108

TAUGNER R, WHALLEY A, ANGERMÜLLER S, BÜHRLE CP, HACKENTHAL E (1985b) Are the renin-containing granules of juxtaglomerular epithelioid cells modified lysosomes? Cell Tissue Res 239: 575-587

TAUGNER R, BÜHRLE CP, HACKENTHAL E, NOBILING R (1986a) Typical and atypical aspects of renin secretion from juxtaglomerular epithelioid cells. Klin Wochenschr 64: 829-837

TAUGNER R, MURAKAMI K, KIM SJ (1986b) Renin activation in juvenile secretory granules? Immunocytochemical experiments with an antiserum directed against the prosegment of human renin. Histochemistry 85: 107-109

TAUGNER R, YOKOTA S, BÜHRLE CP, HACKENTHAL E (1986c) Cathepsin D coexists with renin in the secretory granules of juxtaglomerular epithelioid cells. Histochemistry 84: 19-22

TAUGNER R, KIM SJ, MURAKAMI K, WALDHERR R (1987a) The fate of prorenin during granulopoiesis in epithelioid cells. Immunocytochemical experiments with antisera against renin and different portions of the renin prosegment. Histochemistry 86: 249-253

TAUGNER R, ROSIVALL L, BÜHRLE CP, GRÖSCHEL-STEWART U (1987b) Myosin content and vasoconstrictive ability of the proximal and distal (renin-positive) segments of the preglomerular arteriole. Cell Tissue Res 248: 579-588

TAUGNER R, METZ R, ROSIVALL L (1988a) Macroautophagic phenomena in renin granules. Cell Tissue Res 251: 229-231

TAUGNER R, NOBILING R, METZ R, TAUGNER F, BÜHRLE C, HACKENTHAL E (1988b) Hypothetical interpretation of the calcium paradox in renin secretion. Cell Tissue Res 252: 687-690

TAUGNER R, WALDHERR R, SEYBERTH HW, ERDÖS EG, MENARD J, SCHNEIDER D (1988c) The juxtaglomerular apparatus in Bartter's syndrome and related tubulopathies. An immunocytochemical and electron microscopic study. Virchows Arch [A] 412: 459-470

TAYLOR AA, DAVIS JO, BREITENBACH RP, HARTROFT PM (1970) Adrenal steroid secretion and a renal-pressor system in the chicken (Gallus domesticus). Gen Comp Endocrinol 14: 321

TAYLOR GM, CARMICHAEL DJS, PEART WS (1986) Active and inactive renin in anephric man: a comparison of molecular weight studies with normal human plasma and the effect of a specific monoclonal anti-renin antibody. J Hypertens 4: 703-712

TAYLOR G, PEART WS, PORTER KA, ZONDEK LH, ZONDEK T (1986) Concentration and molecular forms of active and inactive renin in human fetal kidney, amniotic fluid and adrenal gland: Evidence for renin-angiotensin system hyperactivity in 2nd trimester of pregnancy. J Hypertens 4: 121-129

TERRAGNO NA, TERRAGNO DA, MCGIFF JC (1977) Contribution of prostaglandins to the renal circulation in conscious, anesthetized, and laparotomized dogs. Circ Res 40: 590-595

TETU B, TOTOVIC V, BECHTELSHEIMER H, SMEND J (1984) Tumeur rénale à sécrétion de rénine. Ann Pathol 4: 55-59

TEWKSBURY DA, FROME WL, DUMAS ML (1978) Characterization of human angiotensinogen. J Biol Chem 253: 3817-3820

THAMES MD, DIBONA GF (1979) Renal nerves modulate the secretion of renin mediated by nonneuronal mechanisms. Circ Res 44: 645-652

THOENES W (1961) Zur Feinstruktur der Macula densa im Nephron der Maus. Z Zellforsch 55: 486-499

THURAU K (1963) Fundamentals of renal circulation. Proceedings 2nd international congress of nephrology, Prague. Excerpta Medica Foundation, Amsterdam

THURAU K (1964) Renal hemodynamics. Am J Med 36: 698-719

THURAU K (1981) Tubulo-glomerular feedback. Adv Physiol Sci 11: 75-82

THURAU K, MASON J (1974) The intrarenal function of the juxtaglomerular apparatus. In: THURAU K (ed) MTP International review of science. Butterworth, London, pp 357-389

THURAU K, SCHNERMANN J (1965) Die Natriumkonzentration an den Macula densa-Zellen als regulierender Faktor für das Glomerulumfiltrat. Klin Wochenschr 43: 410-413

THURAU K, SCHNERMANN J, NAGEL W, HORSTER M, WAHL M (1967) Composition of tubular fluid in the macula densa segment as a factor regulating the function of the juxtaglomerular apparatus. Circ Res 20/21 [Suppl 2]: 79-89

THURAU K, DAHLHEIM H, GRANGER P (1970) On the local formation of angiotensin at the site of the juxtaglomerular apparatus. Proceedings 4th international congress of nephrology, Stockholm. Karger, Basel, pp 24-30

THURAU K, DAHLHEIM H, GRÜNER A, MASON J, GRANGER P (1972) Activation of renin in the single juxtaglomerular apparatus by sodium chloride in the tubular fluid at the macula densa. Circ Res 30 [Suppl 2] 182-186

THURAU K, GRÜNER A, MASON J, DAHLHEIM H (1982) Tubular signal for the renin activity in the juxtaglomerular apparatus. Kidney Int 22 [Suppl 12]: 55-62

TIEDEMANN K, EGERER G (1984) Vascularization and glomerular ultrastructure in the pig mesonephros. Cell Tissue Res 238: 165-175

TIGERSTEDT R, BERGMAN PG (1898) Niere und Kreislauf. Skand Arch Physiol 8: 223-271

TISHER CC, BULGER RE, TRUMP BF (1968) Human renal ultrastructure III. The distal tubule in healthy individuals. Lab Invest 18: 655-668

TOBIAN L (1960a) Physiology of the juxtaglomerular cells. Ann Intern Med 52: 395-410

TOBIAN L (1960b) Interrelationship of electrolytes, juxtaglomerular cells, and hypertension. Physiol Rev 40: 280-312

TOBIAN L (1962) Relationship of juxtaglomerular apparatus to renin and angiotensin. Circulation 25: 189-192

TOBIAN L, THOMPSON J, TWEDT R, JANECEK J (1958) The granulation of juxtaglomerular cells in renal hypertension, desoxycorticosterone and postdesoxycorticosterone hypertension, adrenal regeneration hypertension, and adrenal insufficiency. J Clin Invest 37: 660-671

TOBIAN L, JANECEK J, TOMBOULIAN A (1959) Correlation between granulation of juxtaglomerular cells and extractable renin in rats with experimental hypertension. Proc Soc Exp Biol Med 100: 94-96

TOKUMORI Y, KURAHASHI A, MURAKAMI J, MOKUDA GO, IKEDA T, TAKEDA A, TOMINAGA M, MASHIB H (1983) Biphasic renin release from perfused rat kidney. Horm Metab Res 15: 310-311

TOMITA T, POISNER A, INAGAMI T (1987) Immunohistochemical localization of renin in renal tumors. Am J Pathol 126: 73-80

TOMLINSON S, MCNEIL S, WALKER SW, OLLIS CA, MERRITT JE, BROWN BL (1984) Calmodulin and cell function. Clin Sci 66: 497-508

TONNESEN MG, KLEMPNER MS, AUSTEN KF, WINTROUB BU (1982) Identification of a human neutrophil angiotensin II-generating protease as cathepsin G. J Clin Invest 69: 25-30

TORRES VE, NORTHRUP TE, EDWARDS RM, SHAH SV, DOUSA TP (1978) Modulation of cyclic nucleotides in isolated rat glomeruli. Role of histamine, carbamylcholine, parathyroid hormone and angiotensin II. J Clin Invest 62: 1334-1343

TORRETTI J (1982) Sympathetic control of renin release. Annu Rev Pharmacol Toxicol 22: 167-192

TOUGARD C, LOUVARD D, PICART R, TIXIER-VIDAL A (1985) Antibodies against a lysosomal membrane antigen recognize a prelysosomal compartment involved in the endocytic pathway in cultured prolactin cells. J Cell Biol 100: 786-793

TSUDA N (1969) Ultrastructural study of secretory granules in the juxtaglomerular cells - particularly on formation and extrusion -. Acta Med Nagasaki 13: 140-155

TSUDA N, NICKERSON PA, MOLTENI A (1971) Ultrastructural study of developing juxtaglomerular cells in the rat. Lab Invest 25: 644-652

TUCKER BJ, MUNDY CA, BLANTZ RC (1987) Adrenergic and angiotensin II influences on renal vascular tone in chronic sodium depletion. Am J Physiol 252: F811-F817

TURGEON C, SOMMERS SC (1961) Juxtaglomerular cell counts and human hypertension. Am J Pathol 38: 227-241

UEDA H, TAGAWA H, ISHII M, KANEKO Y (1967) Effect of renal denervation on release and content of renin in anesthetized dogs. Jpn Heart J 8: 156-167

UEDA J, NAKANISHI H, ABE Y (1978) Effect of glucagon on renin secretion in the dog. Eur J Pharmacol 52: 85-92

ULFENDAHL HR, ERICSON AC, KÄLLSKOG Ö, LI BN, LINDBOM LO, SJÖQUIST M, WOLGAST M (1980) Flow resistance within the renal vasculature, with special reference to glomerular filtration dynamics at various cortical depths. In: LEAF A, GIEBISCH G, BOLIS L, GORINI S (eds) Renal pathophysiology - recent advances. Raven, New York, pp 183-187

UNSICKER K, AXELSSON S, OWMAN C, SVENSSON KG (1975) Innervation of the male genital tract and kidney in the amphibia, *Xenopus laevis* Daudin, *Rana temporaria* L, and *Bufo bufo* L. Cell Tissue Res 160: 453-484

UVA B, VALLARINO M (1982) Renin-angiotensin system and osmoregulation in the terrestrial chelonian *Testudo hermanni* Gmelin. Comp Biochem Physiol [A] 71: 449-451

VALDÉS G, LOPEZ JM, MARTINEZ P, ROSENBERG H, BARRIGA P, RODRIGUEZ JA, OTIPKA N (1980) Renin-secreting tumour. Hypertension 2: 714-718

VAN BELLE H (1981) R 24571: a potent inhibitor of calmodulin-activated enzymes. Cell Calcium 2: 483-494

VANDER AJ (1967) Control of renin release. Physiol Rev 47: 359-382

VANDER AJ (1968a) Inhibition of renin release in the dog by vasopressin and vasotocin. Circ Res 23: 605-609

VANDER AJ (1968b) Direct effects of prostaglandin on renal function and renin release in anesthetized dog. Am J Physiol 214: 218-221

VANDER AJ, GEELHOED GW (1965) Inhibition of renin secretion by angiotensin II. Proc Soc Exp Biol Med 120: 399-403

VANDER AJ, MILLER R (1964) Control of renin secretion in the anesthetized dog. Am J Physiol 207: 537-546

VAN DE VOORDE J, LEUSEN I (1983) Role of the endothelium in the vasodilator response of rat thoracic aorta to histamine. Eur J Pharmacol 87: 113-120

VANDONGEN R (1975) Inhibition of renin secretion in the isolated rat kidney by antidiuretic hormone. Clin Sci Mol Med 49: 73-76

VANDONGEN R (1976) Suppression of renin secretion in the isolated rat kidney by cycloheximide. Eur J Pharmacol 40: 179-181

VANDONGEN R, GREENWOOD DM (1975a) The inhibition of renin secretion in the isolated rat kidney by clonidine hydrochloride (catapres). Clin Exp Pharmacol Physiol 2: 583-588

VANDONGEN R, GREENWOOD DM (1975b) The stimulation of renin secretion by diazoxide in the isolated rat kidney. Eur J Pharmacol 33: 197-200

VANDONGEN R, PEART WS (1974) Calcium dependence of the inhibitory effect of angiotensin on renin secretion in the isolated perfused kidney of the rat. Br J Pharmacol 50: 125-129

VANDONGEN R, PEART WS, BOYD GW (1973) Adrenergic stimulation of renin secretion in the isolated perfused rat kidney. Circ Res 32: 290-296

VANDONGEN R, PEART WS, BOYD GW (1974) Effect of angiotensin II and its nonpressor derivatives on renin secretion. Am J Physiol 226: 277-282

VANDONGEN R, STRANG KD, POESSE MH, BIRKENHAGER WH (1979) Suppression of renin secretion in the rat kidney by a nonvascular -adrenergic mechanism. Circ Res 45: 435-439

VAN DONGEN WJ, VAN DER HEIJDEN CA (1969) Demonstration of renal juxtaglomerular granules and evaluation of the index of granulation in the toad, *Bufo bufo.* Z Zellforsch 94: 40-45

VELLETRI P, BEAN BL (1982) The effects of captopril on rat aortic angiotensin-converting enzyme. J Cardiovasc Pharmacol 4: 315-325

VENKATACHALAM MA, KREISBERG JI (1985) Agonist-induced isotonic contraction of cultured mesangial cells after multiple passage. Am J Physiol 249: C48-C55

VERBERCKMOES R, VAN DAMME B, CLEMENT J, AMERY A, MICHIELSEN P (1976) Bartter's syndrome with hyperplasia of renomedullary cells: Successful treatment with indomethacin. Kidney Int 9: 302-307

VIKHERT AM, SEREBROVSKAYA YA (1964) Localization of renin in kidneys of healthy people and animals. Fed Proc 23 (Translation supplement): 178-182

VIKSE A, KIIL F (1985) Enhancement of renal prostaglandin E2 and renin release by autoregulatory dilation of preglomerular vessels in dogs. Renal Physiol 8: 169-178

VIKSE A, HOLDAAS H, HARTMANN A, KIIL F (1983) Segmental

distribution of vascular resistances during ureteral occlusion. Acta Physiol Scand 119: 147-158

VIKSE A, HOLDAAS H, SEJERSTED OM, KIIL F (1984) Relationship between PGE2 and renin release in dog kidneys. Effects of afferent arteriolar dilation and adrenergic stimulation. Acta Physiol Scand 121: 261-268

VIKSE A, BUGGE J, DAHL E, KIIL F (1985a) Dissociation between renal prostaglandin E2 and renin release. Effects of glucagon, dopamine and cyclic AMP in dogs. Acta Physiol Scand 125: 619-626

VIKSE A, HOLDAAS H, SEJERSTED OM, KIIL F (1985b) Hemodynamic conditions for renal PGE2 and renin release during α- and β-adrenergic stimulation in dogs. Acta Physiol Scand 124: 163-172

VILLARREAL D, DAVIS JO, FREEMAN RH, SWEET WD, DIETZ JR (1984) Effects of meclofenamate on the renin response to aortic constriction in the rat. Am J Physiol 247: R546-R551

VILLARREAL D, FREEMAN RH, DAVIS JO, VERBURG KM, VARI RC (1986) Renal mechanisms for suppression of renin secretion by atrial natriuretic factor. Hypertension 8 [Suppl 2]: 28-35

VINCENZI FF (1981) Calmodulin pharmacology. Cell Calcium 2: 387-409

VISKOPER RJ, ROSENFELD S, MAXWELL MH, DE LIMA J, LUPU AN, ROSENFELD JB (1976) Effect of Ca2+ binding by EGTA on renin release in the isolated perfused rabbit kidney. Proc Soc Exp Biol Med 152: 415-418

VOLPE M, ODELL G, KLEINERT HD, MÜLLER F, CAMARGO MJ, LARAGH JH, MAACK T, VAUGHAN ED, ATLAS SA (1985) Effect of atrial natriuretic factor on blood pressure, renin, and aldosterone in Goldblatt hypertension. Hypertension 7 [Suppl 1]: 43-48

WÅGERMARK J, UNGERSTEDT U, LJUNGQVIST A (1968) Sympathetic innervation of the juxtaglomerular cells of the kidney. Circ Res 22: 149-153

WAGNER PK, PIPPIG L, THOENES W (1979) Zur Histopathologie der Niere beim Pseudo-Bartter-Syndrom durch chronischen Diuretikaabusus. Klin Wochenschr 57: 135-142

WALDMAN SA, MURAD F (1987) Cyclic GMP synthesis and function. Pharmacol Rev 39: 163-198

WALLACE KB, HOOK JB, BAILIE MD (1980) Postnatal development of the renin-angiotensin system in rats. Am J Physiol 238: R432-R437

WARD PE, GEDNEY CD, DOWBEN RM, ERDÖS EG (1975) Isolation of membrane-bound renal kallikrein and kininase. Biochem J 151: 755-758

WARD PE, ERDÖS EG, GEDNEY CD, DOWBEN RM, REYNOLDS RC (1976a) Isolation of membrane-bound renal enzymes that metabolize kinins and angiotensins. Biochem J 157: 643-650

WARD PE, ERDÖS EG, GEDNEY CD, DOWBEN RM, REYNOLDS RC (1976b) Isolation of renal membranes that contain kallikrein, angiotensin I-converting enzyme (kininase II) and angiotensinase in the rat. Clin Sci Mol Med 51: 267s-270s

WARD PE, SCHULTZ W, REYNOLDS RC, ERDÖS EG (1977) Metabolism of kinins and angiotensins in the isolated glomerulus and brush border of rat kidney. Int Acad Pathol 36: 599-606

WARD PE, SHERIDAN MA, HAMMON KJ, ERDÖS EG (1980) Angiotensin I converting enzyme (kininase II) of the brush border of human and swine intestine. Biochem Pharmacol 29: 1525-1529

WARSHAW BL, ANAND SK, OLSON DL, GRUSHKIN CM, HEUSER ET, LIEBERMAN E (1979) Hypertension secondary due to a renin-producing juxtaglomerular cell tumour. J Pediatr 94: 247-250

WATKINS BE, DAVIS JO, LOHMEIER TE, FREEMAN RH (1976) Intrarenal site of action of calcium on renin secretion in dogs. Circ Res 39: 847-853

WEARE JA, GAFFORD JT, LU HS, ERDÖS EG (1982) Purification of human kidney angiotensin I converting enzyme using reverse-immunadsorption chromatography. Anal Biochem 123: 310-319

WEBB RL, DELLA PUCA R, MANNIELLO J, ROBSON RD, ZIMMERMAN MB, GHAI RD (1986) Dopaminergic mediation of the diuretic and natriuretic effects of ANF in the rat. Life Sci 38: 2319-2327

WEEKS JR, COMPTON LD (1979) The cardiovascular pharmacology of prostacyclin (PGI2) in the rat. Prostaglandins 17: 501-513

WEGMANN W (1970) Der juxtaglomeruläre Apparat der Niere bei primärem und sekundärem Hyperaldosteronismus. Virchows Arch [A] 349: 21-47

WEICHERT G (1965) Über die Wirkung fraktionierter Nierenextrakte von Vertebraten verschiedener Evolutionsstadien auf den Blutdruck der nephrektomierten Ratte. Pflügers Arch 284: 147-159

WEIGAND K, WERNZE H, FALGE C (1977) Synthesis of angiotensinogen by isolated rat liver cells and its regulation in comparison to serum albumin. Biochem Biophys Res Commun 75: 102-110

WEINBERGER HD, MARTIN BA, GATTONE II VH, CONNORS BA, LUFT FC, WILLIS LR (1986) The effect of angiotensin II on glomerular function and morphology in the rat. J Submicrosc Cytol 18: 29-34

WEISS B, PROZIALECK WC, WALLACE TL (1982) Interaction of drugs with calmodulin. Biochemical, pharmacological and clinical implications. Biochem Pharmacol 31: 2217-2226

WELCH WJ, OTT CE, GUTHRIE GP, KOTCHEN TA (1983a) Effect of adrenalectomy on renin release in the rat. Endocrinology 113: 2086-2091

WELCH WJ, OTT CE, GUTHRIE GP, KOTCHEN TA (1983b) Mechanism of increased renin release in the adrenalectomized rat. Adrenal insufficiency and renin. Hypertension 5 [Suppl 1]: 47-52

WELCH WJ, OTT CE, GUTHRIE GP, KOTCHEN TA (1985) Renin secretion and loop of Henle chloride reabsorption in the adrenalectomized rat. Am J Physiol 249: F596-F602

WELCH WJ, OTT CE, LORENZ JN, KOTCHEN TA (1986) Effects of chlorpropamide on loop of Henle function and plasma renin. Kidney Int 30: 712-716

WELLS JN, KRAMER GL (1981) Phosphodiesterase inhibitors as tools in cyclic nucleotide research: a precautionary note. Mol Cell Endocrinol 23: 1-9

WENTING GJ, DERKX FHM, TAN-TJIONG LH, VAN SEYEN AJ, MAN IN'T VELD AJ, SCHALEKAMP MADH (1987) Risks of angiotensin converting enzyme inhibition in renal artery stenosis. Kidney Int 31 [Suppl 20]: 180-183

WHORTON AR, MISONO K, HOLLIFIELD J, FROLICH JC (1977) Prostaglandins and renin release: I. Stimulation of renin release from rabbit renal cortical slices by PGI2. Prostaglandins 14: 1095-1104

WHORTON AR, LAZAR JD, SMIGEL MD, OATES JA (1981) Prostaglandins and renin release: III. effects of PGE1, E2, F2 and D2 on renin release from rabbit renal cortical slices. Prostaglandins 22: 455-468

WIDEMAN RF, BRAUN EJ, ANDERSON GL (1981) Microanato-

my of the renal cortex in the domestic fowl. J Morphol 168: 249-267
WIESENFELD-HALLIN Z, HÖKFELT T, LUNDBERG JM, FORSSMANN WG, REINICKE M, TSCHOPP FA, FISCHER J (1984) Immunreactive calcitonin gene-related peptide and substance P coexist in sensory neurones and interact in spinal behavioral responses. Neurosci Lett 52: 199-204
WIGGER HJ, STALCUP SA (1978) Distribution and development of angiotensin converting enzyme in the fetal and newborn rabbit. Lab Invest 38: 581-585
WILCOX CS, PEART WS (1987) Release of renin and angiotensin II into plasma and lymph during hyperchloremia. Am J Physiol 253: F734-F741
WILCOX CS, AMINOFF MJ, KURTZ AB, SLATER JDH (1974) Comparison of the renin response to dopamine and noradrenaline in normal subjects and patients with autonomic insufficiency. Clin Sci Mol Med 46: 481-488
WILCZYNSKI EA, OSMOND DH (1983) Plasma prorenin in humans and dogs. Hypertension 5: 277-285
WILKES BM (1987) Reduced glomerular angiotensin II receptor density in diabetes mellitus in rat: time course and mechanism. Endocrinology 120: 1291-1298
WILLIAMS BC, DUNCAN FM, DRURY PL, TRAIN LMC, EDWARDS CRW (1983) Dopamine stimulates renin release in isolated rat renal cortical cells by activation of specific dopaminergic receptors. J Hypertens 1 [Suppl 2]: 177-179
WILLIAMS GH, HOLLENBERG NK, MOORE TJ, DLUHY RG, BAVLI SZ, SOLOMON HS, MERSEY JH (1978) Failure of renin suppression by angiotensin II in hypertension. Circ Res 42: 46-52
WILLIAMSON JR (1986) Role of inositol lipid breakdown in the generation of intracellular signals. Hypertension 8 [Suppl 2]: 140-156
WILLINGHAM MC, PASTAN I (1984) Endocytosis and exocytosis: Current concepts of vesicle traffic in animal cells. Int Rev Cytol 92: 51-91
WILSON JX (1984a) The renin-angiotensin system in nonmammalian vertebrates. Endocr Rev 5: 45-61
WILSON JX (1984b) Coevolution of the renin-angiotensin system and the nervous control of blood circulation. Can J Zool 62: 137-147
WILSON SK (1986) The effects of angiotensin II and norepinephrine on afferent arterioles in the rat. Kidney Int 30: 895-905
WINER N, CHOKSHI DS (1971) Effects of cyclic AMP, sympathomimetic amines, and adrenergic receptor antagonists on renin secretion. Circ Res 29: 239-248
WINTROUB BU, KLICKSTEIN LB, DZAU VJ, WATT KWK (1984) Granulocyte-angiotensin system. Identification of angiotensinogen as the plasma protein substrate of leukocyte cathepsin G. Biochemistry 23: 227-232
WITTENBERG C, CHOSHNIAK I, SHKOLNIK A, THURAU K, ROSENFELD J (1986) Effect of dehydration and rapid rehydration on renal function and on plasma renin and aldosterone levels in the black Bedouin goat. Pflügers Arch 406: 405-408
WITTY RT, DAVIS JO, JOHNSON JA, PREWITT RL (1971) Effects of papaverin and hemorrhage on renin secretion in the non-filtering kidney. Am J Physiol 221: 1666-1671
WOLFF HP, VECSEI P, KRÜCK F, ROSCHER S, BROWN JJ, DÜSTERDIECK GO, LEVER AF, ROBERTSON JIS (1968) Psychiatric disturbance leading to potassium depletion, sodium depletion, raised plasma-renin concentration, and secondary hyperaldosteronism. Lancet 2: 257-261
WOLFF J, LONDOS C, COOPER DMF (1981) Adenosine receptors and the regulation of adenylate cyclase. Adv Cyclic Nucl Res 14: 199-214
WRIGHT FS (1984) Intrarenal regulation of glomerular filtration rate. J Hypertens 2 [Suppl 1]: 105-113
WÜLFROTH P, PETZELT C (1985) The so-called anticalmodulins fluphenazine, calmidazolium, and compound 48/80 inhibit the Ca2+-transport system of the endoplasmic reticulum. Cell Calcium 6: 295-310
WURFER K, HACKENTHAL E, METZ R, NOBILING R, SIMON T, TAUGNER R (1988) Interzonal and intrazonal heterogeneities in the renin status of the preglomerular arterioles in five species. Histochemistry 89: 283-287
WYMAN LC, TUM SUDEN C (1930) Studies on suprarenal insufficiency. VIII. The blood volume of the rat in suprarenal insufficiency, anaphylactic shock and histamine shock. Am J Physiol 94: 579-585
YAMADA K, ERDÖS EG (1982) Kallikrein and prekallikrein of the isolated basolateral membrane of rat kidney. Kidney Int 22: 331-337
YAMAGUCHI T, KURIHARA S, IKEKITA M, KIZUKI K, MORIYA H (1986) Angiotensin I converting enzyme activity in the kidney of bullfrog *(Rana catesbeiana)*. J Pharmacobiodyn 9: 585-592
YAMAMOTO H, HWANG O, VANBREEMEN C (1984) Bay K 8644 differentiates between potential and receptor operated Ca2+ channels. Eur J Pharmacol 102: 555-557
YAMAMOTO K, IKEMOTO F (1983) High molecular-weight renin and renin-binding protein. Trends Pharmacol Sci 4: 381-383
YANG GCH, MORRISON AB (1980) Three large dissectable rat glomerular models reconstructed from wide-field electron micrographs. Anat Rec 196: 431-440
YANG VV, O'MORCHOE PJ, O'MORCHOE CCC (1981) Transport of protein across lymphatic endothelium in the rat kidney. Microvasc Res 21: 75-91
YOKOSAWA N, TAKAHASHI N, INAGAMI T, PAGE DL (1979) Isolation of completely inactive plasma prorenin and activation by kallikreins, a possible link between renin and kallikrein. Biochim Biophys Acta 569: 211-219
YOKOTA I, DUNHAM EW (1983) Vasodilator effect of acetyl glyceryl ether phosphorylcholine (AGEPC) in the rat kidney. Pharmacologist 25: 239 (Abstract)
YOKOTA S, TSUJI H, KATO K (1985) Immuncytochemical localization of cathepsin D in lysosomes of cortical collecting tubule cells of the rat kidney. J Histochem Cytochem 33: 191-200
YOKOYAMA H, YAMANE Y, TAKAHARA J (1979) A case of ectopic renin-secreting orbital hemangiopericytoma associated with juvenile hypertension and hypokalemia. Acta Med Okoyama 33: 315-322
YOSHIDA H, MENZIE J, MICHELAKIS AM (1975) Distribution and disappearance rate of submaxillary renin. Proc Soc Exp Biol Med 150: 451-456
YUKIMURA T, MIURA K, MATSUSHIMA Y, IKEMOTO F, YAMAMOTO K (1984) Urinary excretion of renin and its biochemical properties in dogs. Hypertension 6: 837-842
YUM M, GANGULY A, DONOHUE JP (1984) Juxtaglomerular cells in renal angiomyolipoma. Ultrastructural observation. Urology 24: 283-286
YUN JCH, GILL JRJ, BARTTER FC, KELLY GD, KEISER HR (1982) Effect of bradykinin on renal function in dogs treated with indomethacin or propranolol. Renal Physiol 5: 31-43
ZAKI FG, KEIM GR, TAKII Y, INAGAMI T (1982) Hyperplasia of juxtaglomerular cells and renin localization in kidneys

of normotensive animals given captopril. Electron microscopic and immunohistochemical studies. Ann Clin Lab Sci 12: 200-215

Zamboni L, De Martino C (1968) A re-evaluation of the mesangial cells of the renal glomerulus. Z Zellforsch 86: 364-383

Zambraski EJ, Prosnitz EH, DiBona GF (1978) Lack of evidence for neurogenic renal vasodilatation in anaesthetized dogs. Proc Soc Exp Biol Med 158: 462-465

Zambraski EJ, Tucker MS, Lakas CS, Grassl SM, Scanes CG (1984) Mechanism of renin release in exercising dog. Am J Physiol 246: E71-E76

Zanchetti A, Stella A, Leonetti G, Morganti A, Terzoli L (1976) Control of renin release: a review of experimental evidence and clinical implications. Am J Cardiol 37: 675-691

Zavagli G, Aleotti A, Farinelli A (1983) Human renin granules: Ultrastructural aspects. Nephron 33: 29-33

Zeidel ML, Silva P, Brenner BM, Seifter JL (1987) cGMP mediates effect of atrial peptides on medullary collecting duct cells. Am J Physiol 252: F551-F559

Zhang C (1979) A case of juxtaglomerular cell tumor. J Chin Surg 17: 184-186

Zhuo J, Harris PJ, Skinner SL (1986) Modulation of proximal tubular reabsorption by angiotensin II. Clin Exp Pharmacol Physiol 13: 277-281

Zimmerberg J, Whitaker M (1985) Irreversible swelling of secretory granules during excytosis caused by calcium. Nature 315: 581-584

Zimmerberg J, Curran M, Cohen FS, Brodwick M (1987) Simultaneous electrical and optical measurements show that membrane fusion precedes secretory granule swelling during exocytosis of beige mouse mast cells. Proc Natl Acad Sci USA 84: 1585-1589

Zimmerhackl B, Parekh N, Kücherer H, Steinhausen M (1985) Influence of systemically applied angiotensin II on the microcirculation of glomerular capillaries in the rat. Kidney Int 27: 17-24

Zimmerman BG (1978) Effect of meclofenamate on renal vascular resistance in early Goldblatt hypertension in conscious and anesthetized dogs. Prostaglandins 15: 1027-1034

Zimmermann HD (1971) Development and differentiation of the glomerulus and juxtaglomerular apparatus in the fetal metanephros of man. Light and electron microscopic studies. Verh Dtsch Ges Pathol 55: 491-497

Zimmermann KW (1929) Über den Bau des Glomerulus der menschlichen Niere. Z Mikrosk Anat Forsch 18: 520-552

Zimmermann KW (1933) Über den Bau des Glomerulus der Säugerniere. Z Mikrosk Anat Forsch 32: 176-278

Zschiedrich H, Hofbauer KG, Baron GD, Hackenthal E, Gross F (1975) Relationship between perfusion pressure and renin release in the isolated rat kidney. Pflügers Arch 360: 255-266

Zschiedrich H, Hofbauer KG, Hackenthal E, Baron GD, Gross F (1975) Intrarenal formation of angiotensin II: Interference of saralasin and SQ 20881. Clin Sci Mol Med 48: 37s-40s

Subject Index

L

M

N

O

P

Q

R

S

T

V

W